International Scientific Forum (ISF 2019)

IOP Conference Series: Materials Science and Engineering
Volume 1127

Malacca, Malaysia
16 - 17 December 2019

ISBN: 978-1-7138-3408-3
ISSN: 1757-8981

Printed from e-media with permission by:

Curran Associates, Inc.
57 Morehouse Lane
Red Hook, NY 12571

Some format issues inherent in the e-media version may also appear in this print version.

Institute of Physics
Dirac House, Temple Back
Bristol BS1 6BE UK

Phone: 44 1 17 929 7481
Fax: 44 1 17 920 0979

techtracking@iop.org

Additional copies of this publication are available from:

Curran Associates, Inc.
57 Morehouse Lane
Red Hook, NY 12571 USA
Phone: 845-758-0400
Fax: 845-758-2633
Email: curran@proceedings.com
Web: www.proceedings.com

TABLE OF CONTENTS

IOP Conf. Series: Materials Science and Engineering 1127 (2021) 011001 doi:10.1088/1757-899X/1127/1/011001

Preface: International Scientific Forum 2019 (ISF2019)

The research articles in this proceeding have been presented at the International Scientific Forum 2019 (ISF2019), which took place at the Mudzaffar Hotel in Malacca, Malaysia, on 16-17 December 2019. The ISF2019 was organized by UM Power Energy Dedicated Advanced Center (UMPEDAC), University of Malaya in conjunction with the UMPEDAC's 10th anniversary establishment as the Higher Institution Centre of Excellence (HICoE) in Malaysia. This event was also co-organized together with other local universities, and supported by The Institution of Engineering and Technology (IET) Malaysia Section.

This conference had gathered more than 100 participants from countries such as India, Japan, Brunei Darussalam, Myanmar, Vietnam, and Thailand. With the theme 'Renewable Energy and Sustainability,' local and international researchers were provided the platform to exchange and share their knowledge, and to build research network for long-term collaboration opportunities related to renewable energy. Two keynote talks have been delivered. The first keynote was given by Professor Hideaki Ohgaki from Kyoto University, Japan, with the title 'Effect of Different Electrification Schemes on Quality of Life in ASEAN Rural Villages, and the second keynote was delivered by Professor Ir. Dr. Ab. Halim Abu Bakar from University of Malaya, Malaysia, with the topic 'Variable Inverter – Based Generation Integration.'

Oral presentations of the accepted papers were delivered in parallel sessions. Nearly seventy articles had been submitted to this conference, and this proceeding contains over fifty accepted articles after the peer-review process. This proceeding covers several major areas, namely: materials science, renewable energy, thermal engineering, and electrical engineering. We thank the ISF2019 Organizing Committee, reviewers and participants for their support and contribution throughout the entire process.

Dr. Siti Rohani Sheikh Raihan
Organizing Secretary (ISF2019)

IOP Conf. Series: Materials Science and Engineering 1127 (2021) 011001 doi:10.1088/1757-899X/1127/1/011001

Committee List

International Advisory Committee
Barry Williams (Strathclyde University, United Kingdom)
Boon Teck Ooi (McGill University, Canada)
Emil Levi (Liverpool John Moores University, United Kingdom)
Faz Rahman (University of New South Wales, Australia)
Hideaki Ohgaki (Kyoto University, Japan)
Pradeep Chandra Pant (Ministry of New and Renewable Energy, India)
S. K. Tyagi (Sardar Swaran Singh National Institute of Renewable Energy, India)
V. V. Tyagi (Shri Mata Vaishno Devi University, India)

Technical Committee
Jeyraj Selvaraj (University of Malaya, Malaysia)
Mohd Faris Abdullah (Universiti Teknologi Petronas, Malaysia)
Jasronita Jasni (Universiti Putra Malaysia, Malaysia)
Mohd Amran Mohd Radzi (Universiti Putra Malaysia, Malaysia)
Erwan Sulaiman (Universiti Tun Hussein Onn, Malaysia)
Raja Nor Firdaus (Universiti Teknikal Malaysia Melaka)
Gan Chin Kim (Universiti Teknikal Malaysia Melaka)
Muhamad Mansor (Universiti Tenaga Nasional, Malaysia)
Mohammad Faridun Naim Tajuddin (Universiti Malaysia Perlis, Malaysia)
Azimah Omar (University of Malaya, Malaysia)
Md Hasanuzzaman (University of Malaya, Malaysia)
Che Hang Seng (University of Malaya, Malaysia)
Tan Chia Kwang (University of Malaya, Malaysia)
Muhamad Zahim Sujod (Universiti Malaysia Pahang, Malaysia)
Muhamad Saipul Fakir (University of Malaya, Malaysia)
Syahrul Ashikin Azmi (Universiti Malaysia Perlis, Malaysia)
Khairus Syifa Hamdan (University of Malaya, Malaysia)
Shahid Mehmood (University of Malaya, Malaysia)
Md Shouquat Hossain (University of Malaya, Malaysia)
Adarsh Kumar Pandey (Sunway University, Malaysia)
Maaspaliza Azri (Universiti Teknikal Malaysia Melaka, Malaysia)

Organizing Committee
Nasrudin Abd. Rahim (Chairman)
Siti Rohani Sheikh Raihan
Md Shouquat Hossain
Attisa Sazali
Md. Hasanuzzaman
Norridah Amin
Tan Chia Kwang
Muhamad Saipul Fakir
Khairus Syifa Hamdan
Chuah Joon Huang

IOP Conf. Series: Materials Science and Engineering 1127 (2021) 011002 doi:10.1088/1757-899X/1127/1/011002

Peer review declaration

All papers published in this volume of *IOP Conference Series: Materials Science and Engineering* have been peer reviewed through processes administered by the Editors. Reviews were conducted by expert referees to the professional and scientific standards expected of a proceedings journal published by IOP Publishing.

- **Type of peer review: Double-blind**
- **Conference submission management system: UMConference Management System**
- **Number of submissions received: 67**
- **Number of submissions sent for review: 63**
- **Number of submissions accepted: 53**
- **Acceptance Rate (Number of Submissions Accepted / Number of Submissions Received X 100): 80%**
- **Average number of reviews per paper: 2**
- **Total number of reviewers involved: 60**
- **Any additional info on review process:**

Oral presentation is required for the accepted paper to be included in content accepted for publication. After screening with Turnitin plagiarism software, papers score < 30% will be sent for double blind review process.

The average number of papers per reviewer is 3. Reviewers are required to give feedback score based on the following acceptance criteria:
a) Suitability of title to the contents
b) Abstract sufficiently informative
c) Clarity of research objectives
d) Appropriateness of methodology (data, methodology procedures)
e) Data analysis
f) Discussion of results
g) Theoretical soundness
h) Contribution of study
i) Presentation quality (language, sentence, structure, etc.)
j) Bibliography adequate and relevant
At least two reviewers will be appointed for each paper. In the event of conflicting opinions, additional reviewer(s) will be appointed.

- **Contact person for queries:**

1) Dr. Siti Rohani Sheikh Raihan
UM Power Energy Dedicated Advanced Centre (UMPEDAC)

IOP Conf. Series: Materials Science and Engineering 1127 (2021) 011002 doi:10.1088/1757-899X/1127/1/011002

University of Malaya, Malaysia
E-mail: srohani_sr@um.edu.my

2) **Dr. Tan Chia Kwang**
UM Power Energy Dedicated Advanced Centre (UMPEDAC)
University of Malaya, Malaysia
E-mail: cktan@um.edu.my
3) **Dr. Md. Hasanuzzaman**
UM Power Energy Dedicated Advanced Centre (UMPEDAC)
University of Malaya, Malaysia
E-mail: hasan@um.edu.my

IOP Conf. Series: Materials Science and Engineering 1127 (2021) 012001 doi:10.1088/1757-899X/1127/1/012001

Investigation of engine characteristics of a variable compression ratio diesel engine using biodiesel and alcohol blends

Sumit Taneja*, Princepreet Singh †

*Assistant Professor, Mechanical Engineering Department, Manipal University Jaipur

†Mechanical Engineering Department, PEC, Chandigarh

E-mail: sumit4088@mail.com

Abstract. Increasing environmental pollution and reducing fossil fuels are major concerns of today's age. To address both the above said issues, the present research focuses on fossil fuel compensation with Biofuels additives. The objective is to analyze combined effect of biodiesel, alcohols and other additives on the diesel engine performance and emission characteristics. The prime target of present research is to analyze the emission characteristics of the diesel engine fueled with Biodiesel (20%)-diesel, Biodiesel (20%)-ethanol (10%)-diesel and Biodiesel (20%)-butanol (10%)-diesel blends. For this purpose, various loads and compression ratios have been applied on each of the blends, to get the optimum result for better performance and reduction in emissions. Engine tests were conducted at variable loads, starting from no load (0 Kg) to full load (15 Kg) at rpm of 1500 on a single cylinder diesel engine. Two compression ratios CR16 and CR18 were used for the research. Engine performance parameters namely BSFC, BTE and engine emissions namely NO_X, CO, HC and CO_2 were measured. Performance and emission variations of the modified blends were compared with the reference fuel (pure diesel) for the quantitative assessment of the improvements.

1. Introduction

The reserves of the fossil fuels are diminishing in an incredible pace. So, designers and researchers are finding distinctive strategies to optimize the fuel efficiency and to diminish the problem of carbon emissions. As per an estimate, the fossil fuel reserves will last for 60 years for natural gas, 40 years for oil and 218 years for coal [1, 2].

Also, concerns for the environment have significantly increased throughout the world over the last few decades. Due to excessive use of fossil fuels, various environmental changes are taking place globally such as ozone layer depletion, greenhouse effect, climate changes, acid rain etc. So, scientists throughout the world realized that, something constructive need to be done soon to reduce the Greenhouse Gas and toxic emissions such as carbon monoxide, carbon dioxide, oxides of nitrogen, hydrocarbons and particulate matter. Various types of alternative fuels like bio ethanol, biodiesel, hydrogen, LPG have been proposed by researchers and engineers. These fuels are substitute to the crude oil, so they are considered to be very important fuels.

Among all the alternative fuels available to the customers in India, only two most importantly used are ethanol, in the form of ethanol-gasoline blends and biodiesel, in the form of diesel-biodiesel blends. In compression ignition engines, biodiesel became popular due to its various advantages and benefits in terms of curbing emissions from the engines. But neat biodiesel and biodiesel blends with high percentage can cause a number of engine performance issues, including injector choking, fuel filter plugging, elastomer seal hardening and cracking, piston ring sticking and breaking, and engine lubricant dilution and degradation. Hence, blending with diesel is the best option to conserve fossil fuels as well as for reducing emissions from C.I engine [3-5].

Low energy content, low cetane number, decreased flash point, worse miscibility, higher volatility and worsened lubricity prevent ethanol from being used as a fuel in diesel engines. One of the best alternatives to ethanol is butanol. Butanol releases more energy per unit mass, is less evaporative and has a lower auto-ignition temperature. It has higher calorific value, higher cetane number, and improved lubricating ability compared to methanol and ethanol. It is highly miscible with diesel, FAME and vegetable oils and it is less corrosive. When mixed with diesel or FAME slightly increases specific fuel consumption. It was found that Diesel- butanol-FAME blend lowers the cetane number and viscosity of the blend with increase in the butanol content. Also, the specific fuel consumption increases with the use of these blends and the harmful emissions (CO, Particulate matter, soot) are reduced. Combustion temperature gets lowered by FAME-Butanol blends and hence both soot and NO_X emissions also reduce [6, 7].

2. Experiment

The experiment comprises of selection and collection of plant's seed for the oil after treatment with quenching agents with determination of extraction method.

2.1 Materials

Ethanol, 1-Butanol, were purchased from the local chemical supplier in the city beautiful, Chandigarh. Soyabean biodiesel (FAME) was purchased from Agroprocessor ltd. (Nagpur, India). Petrodiesel (Indian Oil Corporation) was obtained from a local petrol pump.

2.2 Preparation of Blends

Four fuels of 1.5 litres each were prepared in the chemical laboratory as engine needs minimum of 1.5 litres fuel for performing tests. For Blend 1 (BD), 300 ml (20%) of biodiesel and 1200 ml (80%) of diesel were mixed. For Blend 2 (BbuD), 300 ml (20%) of biodiesel, 150 ml (10%) of ethanol and 1050 ml (80%) of diesel were mixed. For Blend 3 (BED), 300 ml (20%) of biodiesel, 150 ml (10%) of butanol and 1050 ml (80%) of diesel were mixed. Fuels are shown in the Figure-1.

Diesel BD BbuD BED

Figure 1. Diesel and its blends with biodiesel and alcohol

After making blends, fuel properties such as density and calorific values were measured using CCD 13 – Density Meter and bomb calorimeter respectively. These values are required to be input in the performance prediction software used in the engine. Fuel properties of the various test fuels used in this study are given in Table-1.

Table 1. Fuel properties of the blends

Fuel	HHV (KJ/kg)	Density @ 15 °C
Diesel (100%)	45500	832
Biodiesel (20%)	44296	843
Biodiesel (20%) Ethanol (10%)	42730	838
Biodiesel (20%) Butanol (10%)	43480	840

2.3 Experimental set-up

Engine used for the experimentation was single cylinder, four stroke, and variable compression ratio direct injection diesel engine. Tilting block arrangement was used to change the compression ratio. Engine test rig was connected to eddy current dynamometer and load cell was used for changing load with the help of dynamometer. Cooling water was supplied with the help of two rotameters for calorimeter and cylinder head water flow measurement. The technical specifications of the engine are given in Table-2.

Table 2. Specifications of the engine

Model	TV1
Make	Kirloskar Engine
Type	Water cooled , Four stroke, diesel
No. of cylinder	single
Bore	87.5 mm
Stroke	110 mm
Combustion principle	C.I
Capacity	.661 litres
Peak pressure	77.5 kg/cm^2
Fuel timing for std. engine	23^0 BTDC

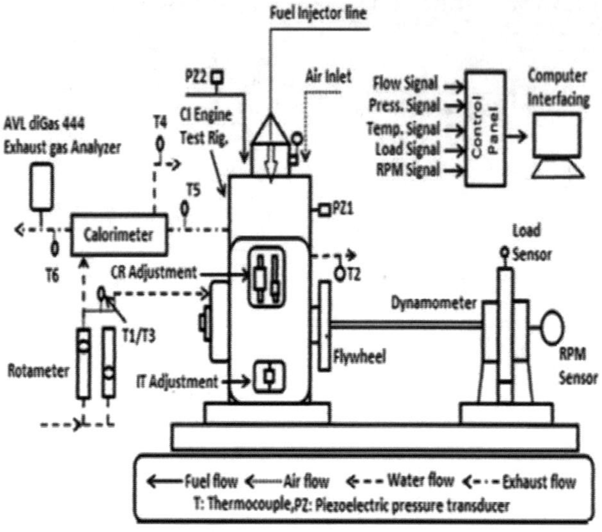

Figure 2. Engine Layout

Fuel line pressure and Combustion pressure were measured by two piezometric sensors that were fitted on fuel injector and cylinder head respectively. The setup consists of an air box, manometer, fuel tank, and central processing unit. The Data acquisition device made by national instrument was connected to the computer and all the signals were transmitted to the software 'Enginesoft LV Version 9.0' (Apex Innovation Pvt. Ltd). AVL DI Gas 444 Analyzer (India) was used for performing the exhaust gas analysis. The exhaust gases namely CO, CO_2, HC, and NO_x were analyzed. Figure-2 shows the Engine Layout.

2.4 Experimental procedure

To validate the fact that CR has impact upon Emissions, CR16 and CR18 were used as compression ratios for various Biodiesel-alcohol blends. Then, comparative analysis with Variable Compression Ratio Diesel engine was conducted at standard specifications.

The initial test was done with pure diesel followed by BD, BED and BbuD at both compression ratios. During the measurement engine was tested with no load followed by 20%, 40%, 60%, 80%, and 100% load conditions. Output readings of in-cylinder pressure, temperatures, engine speed, air flow rate, fuel flow rate, , and fuel line pressure were automatically recorded by the engine software (Enginesoft LV Version 9.0) and engine software used these values for the generation of the performance parameters.

Among the emission characteristics NO_x, CO, HC and CO_2 were measured and among the performance characteristics namely brake thermal efficiency and brake specific fuel consumption were measured.

3. Results and Discussions – Emission Characteristics

3.1 Carbon monoxide (CO)

CO emission is a result of incomplete combustion and slow burning of the soot during final combustion phase. CO emissions reduced as the load increased and this trend was due to the increase in combustion temperature which leads to increased complete combustion at higher loads [8].

From the Figures-3 (a)/ (b) it has been shown that, with the presence of biodiesel in the fuel, CO emissions of blend decrease due to increase in the oxygen content and lower carbon to hydrogen ratio as compared to pure diesel [15]. Biodiesel has a higher cetane number, which results in the lower possibility of formation of rich fuel zone and thus reduces CO emissions [8].

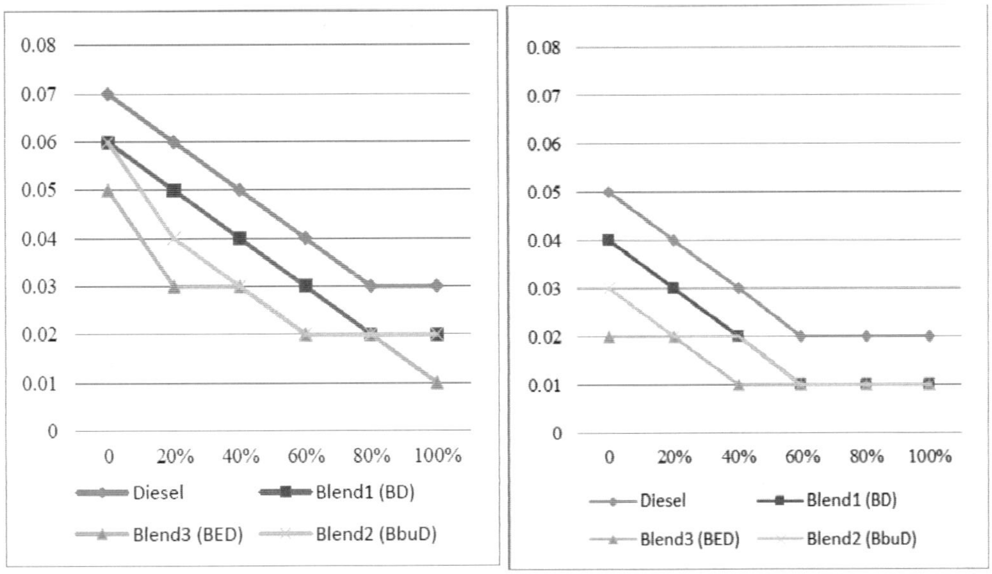

Figure 3. (a) CO vs Load at CR16; (b) CO vs Load at CR18

3.2 Carbon Dioxide (CO2)

The CO_2 emissions variation with engine load has been depicted in Figure- 4 (a)/ (b). The steady increase in CO_2 emissions trend with constant increase in the engine load for all the test fuels has shown that the combustion was more complete and effective at higher loads. CO_2, being the byproduct of complete combustion of CO and HC, always increases with the more oxidation of CO and HC. This increase of CO_2 concentration in the exhaust indicates that the proper combustion of fuel has occurred in the combustion chamber of the engine [20].

At CR16, the CO_2 emission for biodiesel and alcohol blended fuels was higher than the diesel as they are oxygenated fuels i.e. they carry high oxygen content thus more amount of CO_2 was produced as combustion byproduct as shown in the Figure- 4(a). However, the higher carbon dioxide emissions is a matter of less concern because it is recovered by the nature cycle for biodiesel crops, as previously suggested by many authors [8]. But at full load, Blend 1 resulted in lower CO_2 emissions than diesel. Biodiesel is a low carbon fuel as compared to pure diesel. During complete combustion due to the lower carbon to hydrogen ratio dominates the effect of increased oxygen content thus reducing the CO_2 emissions at full load [8].

The CO_2 emission behaviour at CR-18 for all loads is quite similar to that at CR-16 except for biodiesel blended fuel where CO_2 emissions are lower than diesel at low loads of upto 40%. Another exception was for Blend 3, where in spite of ethanol carrying additional oxygen, Blend 3 is having lower CO_2 emissions than Blend 2 at lower loads of up to 40%.

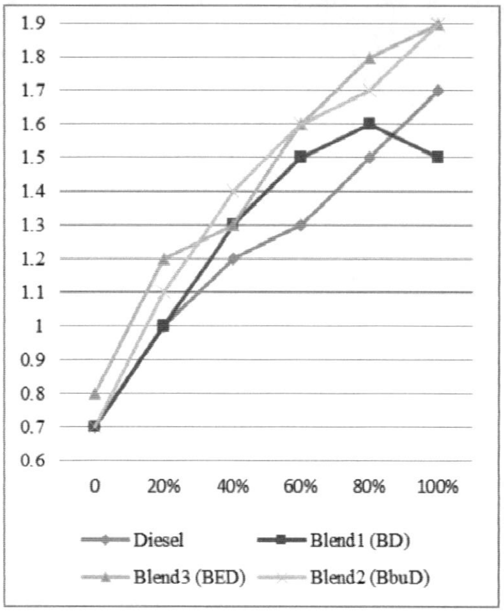

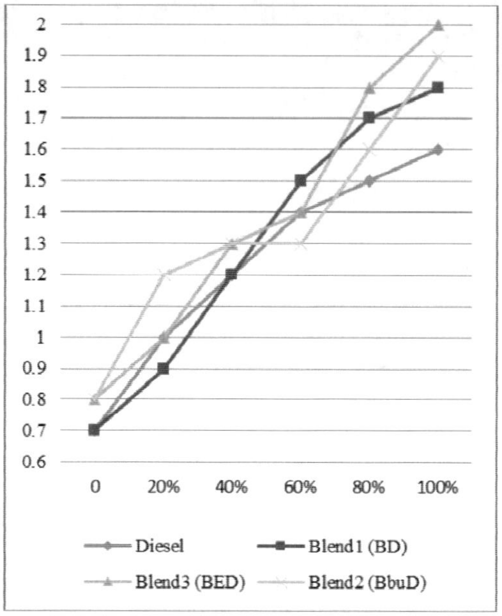

Figure 4. (a) CO_2 vs Load at CR16; (b) CO_2 vs Load at CR18

The effect increase or decrease in CO_2 emission is dependent upon the presence of oxygen content and carbon to hydrogen ratio. Biodiesel, ethanol and butanol are oxygenated blends and have lower carbon-hydrogen ratio as compared to diesel. If effect of oxygenated nature of blend dominates the lower carbon-hydrogen ratio effect, then CO_2 emission tends to increase. But if effect of lower carbon to hydrogen ratio dominates the increased oxygen effect, then CO_2 emission for the blend tends to decrease [8], [21].

3.3 Oxides of Nitrogen

With the constant increase in the load the overall air-fuel ratio decreased which resulted in the higher adiabatic flame temperature of the gases inside the cylinder and this led to increase in NO_X formation, as also stated by Xue et al [8]. Same reason for this trend was explained by Sehmus et al [12], that NO_X emissions increased with the increase of the engine load for all fuels. This was due to the higher combustion temperature inside the cylinder, as greater amount of fuel is burnt at higher loads. Also, NO_X levels obtained at CR16 are lower as compared to CR18. This reduction had been mainly caused by the incomplete combustion at low cylinder pressure at CR16. At CR16 all the oxygenated blends showed more NO_X emissions as compared to diesel fuel as shown in Figure 5(a). It is due to the fuel borne oxygen present in the blended fuels [8].

Also, NO_X levels obtained at CR16 are lower as compared to CR18. This reduction had been mainly caused by the incomplete combustion at low cylinder pressure at CR16. It is due to the fuel borne oxygen present in the blended fuels.

A possible reasoning for lower NO_X emissions Blend 2 (Biodiesel-Butanol-Diesel) as compared to Blend 1 (Biodiesel-Diesel) can be attributed to lower calorific value of butanol leading to lower combustion temperature as stated by G.Tuccar [16]. S. Imtenan et al [13] also stated the same reasoning. At lower loads for CR18, NO_X emissions increased for Blend 2 as compared to biodiesel showing the dominating effect of lower cetane number and thus longer ignition delay, leading to higher combustion temperature as stated by G.Tuccar [16].

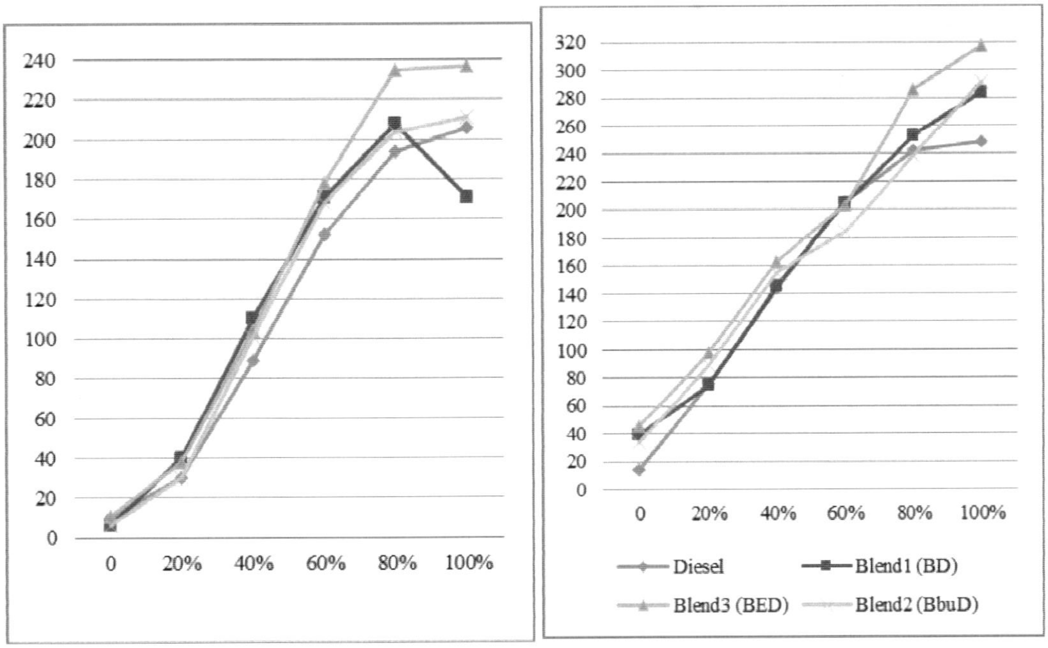

Figure 5. (a) NOX vs Load at CR16; (b) NOX vs Load at CR18

Possible explanation for the highest NO_X emissions by Blend 3 i.e. Biodiesel-Ethanol-Diesel blend as shown in Figures- 5 (a)/(b) can be attributed to the highest percentage of oxygen present in ethanol, leading to fuel dilution, hence leaner fuel leading to higher combustion temperature thus maximizing the NO_X levels. The reason for this can be attributed to the highest percentage of oxygen present in ethanol, leading to fuel dilution, hence leaner fuel leading to higher combustion temperature thus maximizing the NO_X levels. The same logical reasoning was also stated by S. Imtenan [13].

4. Conclusions

Following conclusions can be drawn from this research work:

- In general, CO emissions reduced for all the fuels with the increase in load. With respect to pure diesel, all blends exhibited lower CO emissions. Blend 1 (BD), Blend 2 (BbuD) and Blend 3 (BED) shows maximum CO reduction of 33.3%, 50% and 66.67% respectively with respect to pure diesel at Compression Ratio 16 and 50%, 50% and 66.67% respectively at Compression Ratio 18.
  Also, Blend 2 (BbuD) and Blend 3 (BED) showed even lower CO emissions as compared to Blend 1 (BD). Blend 2 (BbuD) and Blend 3 (BED) shows maximum reduction of 33.3% and 50% respectively at both compression ratios, in comparison with the Blend 1 (BD).
- In general, CO_2 emissions increased for all fuels with increase in load. With respect to pure diesel, all Blends exhibited higher CO_2 emissions. Blend 1 (BD), Blend 2 (BbuD) and Blend 3 (BED) showed maximum CO_2 increase of 15.4%, 23% and 23% respectively at Compression Ratio 16 and 13.3%, 18.7% and 25% respectively at CR18 with respect to pure diesel.
  Though Blend 2 (BbuD) and Blend 3 (BED) have showed higher CO_2 emissions as compared to Blend 1 (BD). But among the Blend 2 (BbuD) and Blend 3 (BED), Blend 2 has shown a maximum of 12.5% and 11% reduction in CO_2 emissions at Compression Ratio 16 and Compression Ratio 18 respectively with respect to Blend 3.
- In general, NO_X emissions increased for all fuels with increase in load. Blend 1 (BD), Blend 2 (BbuD) and Blend 3 (BED) showed maximum increase of 33.3%, 13.5 and 26.7% respectively

at Compression Ratio 16 and 14.5%, 17.3% and 28.9% respectively at CR 18 with respect to pure diesel.

Among the blended fuels, Blend 3 (BED) showed higher NO_X emissions (Maximum increase- 38%), whereas Blend 2 (BbuD) showed lower NO_X emissions (Maximum reduction- 25%) with respect to Blend 1(BD). Blend 2 (BbuD) shows average reduction of 16.52% and 12.08% with respect to Blend 3 (BED) at Compression Ratio 16 and Compression Ratio 18 respectively.

References

[1] J. Goldemberg , T. B. Johnsson, A. K. N. Reddy, R. H. Williams. "Energy for the new millennium", R Swedish Science. 2001; 30(6), pp. 330-337.

[2] R. Gilbert , A. Perl. "A report prepared for national round table on the environment and the economy, University of Calgary", Energy and transport futures. 2005; pp. 91-96.

[3] Directive 2009/28/EC, on the promotion of the use of energy from renewable sources and amending and subsequently repealing directives 2001/77/EC and 2003/30/EC. 2009.04.23. Off. J. European Union. 2009; 140, pp. 16–62.

[4] FIE Manufacturers, Fatty acid methyl ester fuels as a replacement or extender for diesel fuels diesel fuel injection. Diesel Fuel Injection Equipment (FIE) Manufacturers Common Position Statement, 2012.

[5] Engine Manufacturers Association (EMA). "Technical statement on use of biodiesel fuel", Engine Manufacturer Association (EMA). EMADOCS, 27902.14, 2009.

[6] N. Nylund & P. Aakko. "Alcohols/ethers as oxygenates in diesel fuel: Properties of blended fuels and evaluation of practical experiences", IEA advanced motor fuels Annex xxvi final report, report TEC 3, 2005.

[7] L. R. McCormick, R. Parish. "Advanced Petroleum Based Fuels Program and Renewable Diesel Program. Milestone Report: Technical Barriers to the Use of Ethanol in Diesel Fuel", NREL/MP-540-32674, 2001.

[8] J. Xuea, E. T. Grift, C. A.. Hansena. "Effect of biodiesel on engine performances and emissions", Renewable and Sustainable Energy Reviews. 2011;15, pp. 1098–1116.

[9] S. B. Chavan, R. R. Kumbhar, A. Kumar, Y. Sharma. "Study of Biodiesel Blends on Emission and Performance Characterization of a Variable Compression Ratio Engine", ACS Energy fuels, 5b00742, 2015.

[10] Y. Di, C. S. Cheung & Z. Huang. "Comparison of the Effect of Biodiesel-Diesel and Ethanol-Diesel on the Particulate Emissions of a Direct Injection Diesel Engine", Aerosol Science and Technology. 2009; 43, pp. 455– 465.

[11] D. C. Rakopoulos, C. D. Rakopoulos, E. G. Giakoumis, A. M. Dimaratos, D. C. Kyritsis. "Effects of butanol–diesel fuel blends on the performance and emissions of a high-speed DI diesel engine", Energy Conversion and Management. 2010; 51, pp. 1989–1997

[12] S. Altun, C. O. Yasar, H. Adin. "Effect of n-Butanol Blending with a Blend of Diesel and Biodiesel on Performance and Exhaust Emissions of a Diesel Engine", Article in Industrial & Engineering Chemistry Research. 2011;50, pp. 9425-9430.

[13] S. Imtenan, H. H. Masjuki, M. Varman, I. M. Rizwanul Fattah, H. Sajjad, M. I. Arbab. "Effect of *n*-butanol and diethyl ether as oxygenated additives on combustion– emission-performance characteristics of a multiple cylinder diesel engine fuelled with diesel–jatropha biodiesel blend", 10th International Conference on Mechanical Engineering, Procedia Engineering. 2014; 90: pp. 472- 477.

[14] M. Al-Hassan, H. Mujafet, M. Al-Shannag. "An Experimental Study on the Solubility of a Diesel-Ethanol Blend and on the Performance of a Diesel Engine Fueled with Diesel-Biodiesel - Ethanol Blends", Jordan Journal of Mechanical and Ind Engineering. 2012;6(2): pp. 147-62.

[15] E. Uludamar, C. Ozgur, T. Ozgur, K. Aydin. "Performance and emissions characteristics of a diesel engine fuelled with ethanol additive in diesel-soybean biodiesel fuel blend", The

Journal of Macro Trends in Energy and Sustainability. 2015; 3(1), pp. 212-218.

[16] G. Tuccar, T. Ozgur, K. Aydin. "Effect of diesel– microalgae biodiesel–butanol blends on performance and emissions of diesel engine", Fuel. 2014;132, pp. 47–52.

[17] J. Cedik, M. Pexa, J. Marik, V. Honig, S. Hornickova, K. Kubin. "Influence of butanol and FAME blends on operational characteristics of compression ignition engine", Agronomy Research. 2015;13(2), pp. 541–549.

[18] D. C. Rakopoulos, C. D. Rakopoulos, E. C. Kakaras, E. G. Giakoumis. "Effects of ethanol–diesel fuel blends on the performance and exhaust emissions of heavy duty DI diesel engine", Energy Conversion and Management. 2008;49, pp. 3155–3162.

[19] V. Hariram, R. V. Shangar. "Influence of compression ratio on combustion and performance characteristics of direct injection compression ignition engine", Alexandria Engineering Journal. 2015; 54, pp. 807–814.

[20] H. Kumar, L. J. Konwar, M. Aslam and A. Sharma. "Performance, combustion and emission characteristics of a direct injection VCR CI engine using a Jatropha curcas oil microemulsion: A comparative assessment with JCO B100, JCO B20 and petrodiesel". RSC Advances. 2016; 6, pp. 3764-66.

[21] B. Tesfa, F. Gu, R. Mishra and A. Ball. "Emission Characteristics of a CI Engine Running with Range of Biodiesel Feedstocks", Energies. 2018;7, pp. 334-350.

[22] A. Shirneshan, M. Almassi, B. Ghobadian, B. Ali, G. Najafi. "Effects of Biodiesel and Engine Load on Some Emission Characteristics of a Direct Injection Diesel Engine", Current World Environment. 2012;7(2), pp. 207-212.

[23] A. Keskin, A. Yasar, İ. Resitoglu, M. A. Akar. "The Influence of Diesel Fuel biodiesel-ethanol-butanol Blends on the Performance and Emission Characteristics of a Diesel Engine", Energy Sources, Part A: Recovery, Utilization, and Environmental Effects. 2013; 35(19), pp. 313-318.

The Waste Management of Polyethylene Terephthalate (PET) Plastic Waste: A Review

N Thachnatharen*, S Shahabuddin †, N Sridewi*

*Faculty of Defence Science and Technology, National Defence University of Malaysia, Kuala Lumpur 57000, Malaysia

†Department of Science, School of Technology, Pandit Deendayal Petroleum University, Knowledge Corridor, Raisan Village, Gandhinagar, 382007 Gujarat, India

E-mail: thachna.tharen1366@gmail.com

Abstract. This paper reviews on the current techniques used for the management of Polyethylene Terephthalate (PET) plastic waste. The increasing amount of PET wastes and growing environmental concern had placed emphasis on various recycling methods such as chemical and mechanical recycling of PET waste. The focus of this review is on the various mechanical and chemical recycling methods for PET wastes. The conclusion with suggestion to the involved authority on the proper PET waste management to avoid pollution to the environment have been presented in last section.

1. Introduction

An unimaginable life without plastic usage in our daily life is created by our society. It is due to its ample advantages such as good chemical, mechanical and thermal properties with light weight and low production cost. The invention of the Bakelite is the first synthetic plastic in 1907, plays an important role in the beginning of the plastic industry. It carries a great usage prospect in various of field such as packaging, medical, manufacturing, automotive and the list goes on. It is been the most generally used item in our daily life.

The Polyethylene Terephthalate (PET) plastic is the most generally used thermoplastic polymer in the world. According to IUPAC polymer nomenclature, PET is known as poly (oxyethylene oxyterephthaloyl) based on its systematic structure. It is well known as Polyester in the textile industries. It is also added with glass fibers and carbon to increase its material strength. Even without the addition of those materials in PET, it is still very strong for its lightweight nature.[1] These criteria of PET make it to be effective item in packaging as less materials needed for the process. It only adds very minimal weight to the packaged item using it, and less fuel is needed for shipping those items when using PET packaging. It is transparent, virtually shatterproof and semi- crystalline in nature.[2]

PET is manufactured by the polymerization of ethylene glycol (EG) and terephthalic acid (TPA). The ethylene glycol is extracted from ethylene and it is in colorless liquid form. The figure 1 shows the complete reaction involved in the production of PET. The terephthalic acid is obtained from xylene and it is in crystalline solid form.

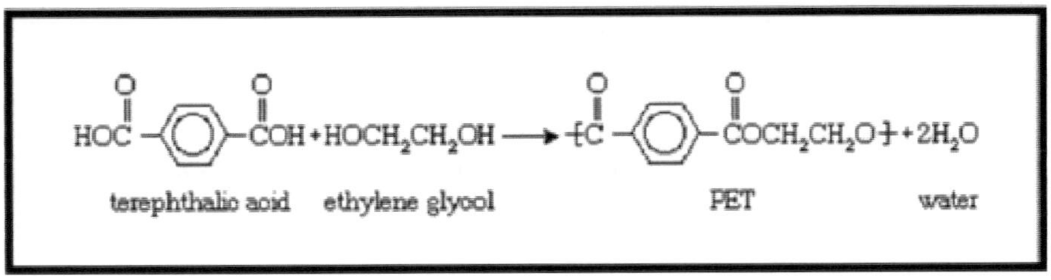

Figure 1. shows the overall reaction in production of PET

2. The Waste Management of PET Waste

2.1. Chemical Recycling

Chemical recycling of PET is a process where the polymers of waste PET is broken down into useful chemicals in the form of oligomers and monomer, which involves various types of depolymerization techniques. The advantage of chemical recycling is that the quality of virgin PET can be achieved.[3].The chemical recycling of PET waste consist of the depolymerization using several technologies such as methanolysis, hydrolysis, glycolysis, aminolysis and ammonolysis.[4] Figure 2 shows the chemical recycling method of PET.

2.1.1. Methanolysis Methanolysis of the PET waste is a process where PET is depolymerized with methanol to yield DMT (dimethyl terephthalate) and ethylene glycol (EG), in the presence of catalysts under a pressure of 2–4 MPa and a temperature of 180–280 °C. Then the reaction mixture is cooled. The DMT is obtained from the mixture by precipitation, centrifugation and crystallization. The reaction is catalyzed by zinc acetate, cobalt acetate, magnesium acetate, and lead oxide. However, zinc acetate is the most commonly used catalyst. The other method is supercritical methanolysis where it uses methanol vapor instead of methanol. A batch reactor was used at temperatures of 300-350 °C under an estimated pressure of 20 MPa for a reaction time of 2–120 min.[5]

2.1.2. Hydrolysis Hydrolysis is another chemical recycling method of PET waste, where it yields terephthalic acid (TPA) and ethylene glycol (EG). This reaction can be done in acidic, basic and neutral condition. In alkaline hydrolysis, PET waste is reacted using NaOH or KOH to obtain terephthalate salt which by further acidification gives monomer TPA. The reaction generally runs for 3-5 hours at temperature of 210-250 °C, under pressure of 1.4-2 MPa.[6]

In neutral hydrolysis, hot water or hot steam is used. This process needs high pressure autoclaves at temperature of 200–300 °C and pressures of 1–4 MPa. The main products formed are EG and TPA. TPA is separated from the post reaction mixture by filtration, while a substantial volume of diluted EG is recovered through distillation or by extraction.[7] Then, in acidic hydrolysis concentrated sulphuric acid is used at temperature range between 130-170 °C. This process can take place in a pressure-less apparatus.[8]

2.1.3. Glycolysis Glycolysis is one of the commonly used method in chemical recycling of PET waste. The result of deep glycolysis by EG is primarily bis-2- (hydroxyethyl) terephthalate (BHET), which can be directly used in PET synthesis. The process is usually done in a temperature range between 180 and 250 °C with excess EG and in the presence of transesterification catalyst, usually zinc or lithium acetate.[9]

2.1.4. Aminolysis In aminolysis of PET waste, different aqueous amine solutions to yield the corresponding diamides of TPA and EG. This reaction usually uses primary amine aqueous solution such as methylamine, ethylamine and ethanolamine in the temperature range of 20-100 °C. The catalyst used in this process is glacial acetic acid, sodium acetate and potassium sulphate. Pure bis (2-

hydroxy ethylene) terephthalamide (BHETA) is obtained with sufficiently high yields with the presence of catalysts.[10]

2.1.5. Ammonolysis In ammonolysis, ammonia is reacted with PET waste to produce a terephthaldiamide. This main product can be converted to terephthalonitrile or another corresponding chemical. The reaction is carried out under pressure of about 2 MPa in a temperature range of 120-180 °C for 1-7 h. The produced amide is filtered, rinsed with distilled water and dried in temperature of 80 °C. The product have very high purity level with is not less than 99%.[11]

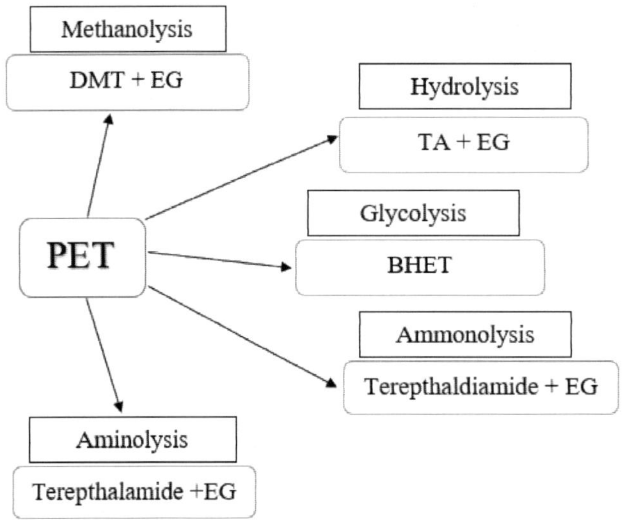

Figure 2. shows the various process involved in PET waste management by chemical recycling

2.2. Mechanical recycling

Mechanical recycling is one of the effective methods of PET waste management due to its non-biodegradable characteristic. There are several main steps involved in the mechanical recycling of PET waste namely plastic collection, manual sorting, chipping, washing and pelleting. Figure 3 show the process involved in the mechanical recycling.

2.2.1. Plastic collection The PET waste can be obtained from 2 main sources. It is postindustrial PET waste and postconsumer PET waste. The postindustrial PET waste is the rejected waste due to damages and industrial cut off. Mostly this kind of PET waste is collected directly from the industries. Meanwhile, the postconsumer PET waste is the waste that had been used by people and this waste can be collected from recycling bin and domestic garbage bin.[3]

2.2.2. Manual sorting In this step, the plastic waste is manually sorted into different kind of plastic classification. Other types of plastics, rocks and metals are removed from the PET plastic in order to reduce the impurities from the PET waste.[12, 13] It also helps to reduce problems caused by the impurities for the further steps of PET waste recycling.

2.2.3. Chipping The sorted PET waste is sent to a cylindrical vessel fitted with sharp blades. Here the plastic is cut in smaller sizes for further process.

2.2.4. Washing The small chips of PET plastic are washed in large washing vessel with the temperature 90 °C for the time range of 12-30 minutes to remove the glue, labels and dirt contained.[14] The washing detergent is an alkaline based detergent used to remove dirt, grease and degrades any organic remnants.

2.2.5. Pelleting The cleaned PET plastic chips are then melted and extrude into a porous metal outlet which makes the melted plastic into long strands. It is the cooled with sprayed water and cut into small pellets. Then the PET pallets are melted to produce any other items such as fibers, automobile parts and plastic pillars

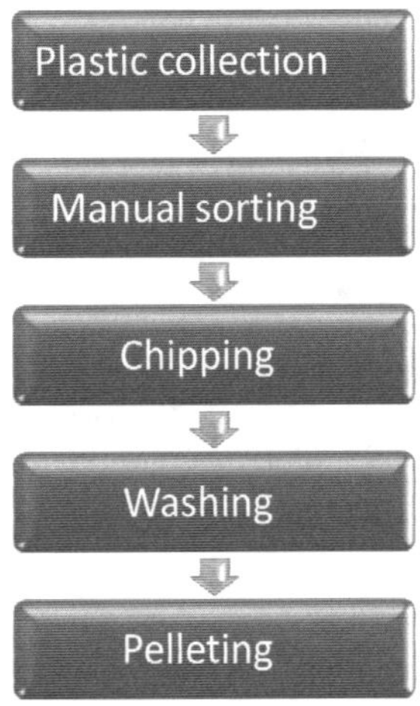

Figure 3. shows the steps involved in mechanical recycling

2.3. Degradation of PET waste

The degradation process of PET results in reduced of molecular weight due to chain scission and discoloration. This process can occur in a natural condition due to excessive exposure to temperature, moisture, UV, bacteria and visible light but in a prolonged condition. To speed up the degradation process of PET waste, the degrading element can be induced manually.

2.3.1. Biodegradation Biodegradation is effective alternative to PET waste management as it is cheap, more efficient and does not produce secondary pollutants, such as associated with incineration and landfilling of PET waste.[15] The metabolic diversity of polymer- degrading microorganism make them a useful solution for many environmental pollution.[16]. In biodegradation, the PET waste is exposed to the polymer-degrading microorganism which makes them to adherence an colonize on the PET waste surface.

The microbial degradation by hydrolysis involves 2 steps. First, the enzyme binds to the polymer substrate then subsequently catalyzes a hydrolytic cleavage. Then, the PET waste polymer is degraded into low molecular weight oligomers, dimers and monomers and finally converted into CO2 and H2O.[17] The biodegradation of plastic waste technique is widely used and lots of research is still ongoing worldwide.

2.3.2. Photocatalytic degradation Photodegradation of PET waste occurs naturally in environment due to the exposure to ultraviolet (UV) radiation from the sun. The UV radiation provide activation energy required to initiate the incorporation of oxygen atom in PET waste polymer chain[18, 19].This process make the PET waste to become brittle and break down into smaller pieces. This process continues until the polymer chains reach sufficiently low molecular weight which soon can be converted into carbon dioxide by the aid of microorganism.[20]However, the entire process is very slow and could take up to 50 or more years to fully degrade[21].

In photocatalytic degradation, photocatalyst is used in order to speed up the natural photodegradation of PET waste. Titanium dioxide (TiO2) is the most commonly used photocatalyst. The process of photocatalytic degradation of PET waste occurs when the photocatalyst which mixed into the PET waste create electron-hole pairs when subjected to UV radiation. These pairs can be transferred to the surface of TiO2 where they react with absorbed O2 and H2O and subsequently produce O2 •–, HOO• and OH•. These radicals reacts with PET waste and results in oxidation and decomposition.[22] Figure 3 show the mechanism involves in the photocatalytic degradation of PET. However, this technique of PET waste management is not widely used and still under research.

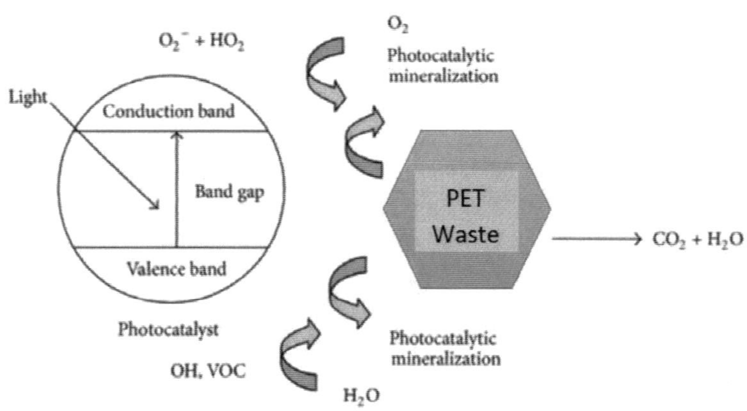

Figure 4. shows the photocatalytic degradation of PET waste mechanism

3. Conclusion

The PET provides of vast amount of benefits due to its property of lightweight, cheap production cost and have good thermal stability. This PET materials are easily available and affordable for most of the people worldwide because of its low cost. This results in mass production of this PET materials due to its high demand around the world. This creates a negative impact to the environment because large volume of PET waste are being disposed into the garbage, drains and rivers. This pollutes the environment as the PET materials is not biodegradable in the environment.

This PET pollution affects the humans, animals and the non-livings such as soil, air and water. But PET waste can be recycled after it had been used. This recyclability property of PET waste must be seen as valuable resources for the country. The integrative effort from public, government and private sectors can solve the problem regarding the waste management of PET waste.

Acknowledgements

We would like acknowledge the Center for Research Management and Innovation of UPNM for the financial support through the Graduate Research Assistantship.

References

[1] Reddish, W., *The dielectric properties of polyethylene terephthalate (terylene).* Transactions of the Faraday Society, 1950. **46**: p. 459-475.

[2] Ravindranath, K. and R. Mashelkar, *Polyethylene terephthalate—I. Chemistry, thermodynamics and transport properties.* Chemical Engineering Science, 1986. **41**(9): p. 2197-2214.

[3] Mudgal, S., L. Lyons, and M. Kong, *Study on an increased mechanical recycling target for plastics.* Final Report Prepared for Plastic Recyclers Europe: Bio-Intelligence Service, 2013.

[4] Koo, B.-M., et al., *Material and structural performance evaluations of Hwangtoh admixtures and recycled PET fiber-added eco-friendly concrete for CO2 emission reduction.* Materials, 2014. **7**(8): p. 5959- 5981.

[5] Foti, D., *Use of recycled waste pet bottles fibers for the reinforcement of concrete.* Composite Structures, 2013. **96**: p. 396-404.

[6] Spaseska, D. and M. Civkaroska, *Alkaline Hydrolysis of Poly (EthyleneTerephthalate) Recycled from the Post consumer Soft-Drink Bottles.* Journal of the university of chemical technology and metallurgy, 2010. **45**(4): p. 379-384.

[7] Güçlü, G., et al., *Hydrolysis of waste polyethylene terephthalate and characterization of products by differential scanning calorimetry.* Thermochimica Acta, 2003. **404**(1-2): p. 193-205.

[8] Ochi, T., S. Okubo, and K. Fukui, *Development of recycled PET fiber and its application as concrete-reinforcing fiber.* Cement and Concrete Composites, 2007. **29**(6): p. 448-455.

[9] Shukla, S. and A.M. Harad, *Glycolysis of polyethylene terephthalate waste fibers.* Journal of applied polymer science, 2005. **97**(2): p. 513-517.

[10] Hameed, R.A., *Aminolysis of polyethylene terephthalate waste as corrosion inhibitor for carbon steel in HCl corrosive medium.* Advances in Applied Science Research, 2011. **2**(3): p. 483-499.

[11] Mittal, A., et al., *Scanning electron microscopic study of hazardous waste flakes of polyethylene terephthalate (PET) by aminolysis and ammonolysis.* Journal of hazardous materials, 2010. **178**(1-3): p. 390-396.

[12] Gabriel, D.S. *How to increase plastic waste acceptance for mechanical recycling: An introduction to material value conservation and its phenomenon.* in *Key Engineering Materials*. 2016. Trans Tech Publ.

[13] Nkwachukwu, O.I., et al., *Focus on potential environmental issues on plastic world towards a sustainable plastic recycling in developing countries.* International Journal of Industrial Chemistry, 2013. **4**(1): p. 34.

[14] Suyama, T., et al., *Phylogenetic affiliation of soil bacteria that degrade aliphatic polyesters available commercially as biodegradable plastics.* Appl. Environ. Microbiol., 1998. **64**(12): p. 5008-5011.

[15] Michaud, L., et al., *Biodegradative potential and characterization of psychrotolerant polychlorinated biphenyl-degrading marine bacteria isolated from a coastal station in the Terra Nova Bay (Ross Sea, Antarctica).* Marine pollution bulletin, 2007. **54**(11): p. 1754-1761.

[16] Iranzo, M., et al., *2078601. The use of microorganisms in environmental remediation.* Annals of microbiology, 2001. **51**(2): p. 135-143.

[17] Zhang, J., et al., *A study on the biodegradability of polyethylene terephthalate fiber and diethylene glycol terephthalate.* Journal of Applied Polymer Science, 2004. **93**(3): p. 1089-1096.

[18] Andrady, A.L., *Microplastics in the marine environment.* Marine pollution bulletin, 2011. **62**(8): p. 1596-1605.

[19] Raquez, J.M., et al., *Oxidative degradations of oxodegradable LDPE enhanced with thermoplastic pea starch: Thermo-mechanical properties, morphology, and UV-ageing studies.* Journal of Applied Polymer Science, 2011. **122**(1): p. 489-496.

[20] Yamada-Onodera, K., et al., *Degradation of polyethylene by a fungus, Penicillium simplicissimum YK.* Polymer degradation and stability, 2001. **72**(2): p. 323-327.

[21] Müller, R.-J., I. Kleeberg, and W.-D. Deckwer, *Biodegradation of polyesters containing aromatic constituents.* Journal of biotechnology, 2001. **86**(2): p. 87-95.

[22] Kamrannejad, M.M., et al., *Photocatalytic degradation of polypropylene/TiO2 nano-composites.* Materials Research, 2014. **17**(4): p. 1039-1046.

Optimization of Extraction Process of Jatropha Oil by Using Quenching Agent

Richa Kothari[*], Shamshad Ahmad[†], M Samykano[#], V.V.Tyagi[‡], A K Pandey[¥], R.Saidur[¥]

*Department of Environmental Sciences, Central University of Jammu, Rahya-Suchani, Samba, Jammu, J&K

[†]Bioenergy and Wastewater Treatment Laboratory, Department of Environmental Sciences, Babasaheb Bhimrao Ambedkar University, Lucknow (U.P.), India.

[#]Faculty of Mechanical & Manufacturing Engineering, Universiti Malaysia Pahang, 26600 Pekan, Pahang, Malaysia

[‡]School of Energy Management, Shri Mata Vaishno Devi University, Kakryal, Katra, Jammu. J&K, India

[¥]Research Centre for Nano-Materials and Energy Technology(RCNMET), School of Science and Technology, Sunway University, No.5, Jalan University Bandar Sunway, Petaling Jaya, 47500 Selangor Darul Ehsan, Malaysia

E-mail: richakothari786@gmail.com

Abstract. Among the various energy plant species, Jatropha curcus is found to be best oil-bearing plant species with a wide range of applications and having enormous economic potentials for its seed, which can be converted into biodiesel via transesterification which is an alternative to petrodiesel. In this present study, the effect of three factors, Time (1-4 hour), temperature (60-120°C) and quenching agents (NaCl, KCl, and ZrCl2) and dose of quenching agent(0.5-2gm) use for the extraction of oil from 10grm of jatropha seed. Using the Box-Benken Design (BBD) approach of Response Surface Methodology (RSM), 17 experimental runs were generated. n-Hexane was used as solvent for each experiment carried out in the 100 ml soxhlet extraction apparatus. Results obtained from the experiments were modeled and analyzed by choosing a quadratic model. From the analysis of modeled found temperature extraction time and dose of quenching agent revealed the good agreement of value (R2=0.98) between observed and predicted value of the experiment and were to be $p>0.05$ that was based on 95% confidence interval. By the use of RSM-BBD an optimizing experiment 60°C temperature 3h time and 2gm $ZrCl_2$ quenching agent give best extraction of oil yield 41.41%.

1. Introduction

Due to decrease in reserves of fossil fuels and the growing environmental concern have made renewable energy as an attractive alternative energy source for the future in the commercial energy sector. Among the various bioenergy resources from list of renewable energy options, biodiesel is one of the promising alternative sources for diesel engine. It is defined as the mono-alkyl ester of long chain fatty acids derived from vegetable oils or animal fats and alcohol with or without a catalyst [1]. It is renewable, environmental friendly, non-toxic and readily available source. Physico-chemical properties of biodiesel are very close to those of diesel as cited by scientists and researchers [2], and consequently biodiesel or its blends can be used in diesel engines with a few or no modifications. As per the literature review, most commonly used blend is B20 comprises of 1:5 mixture of biodiesel and petroleum diesel, suggested for commercial use. The oil obtained from the substrate cannot be used as fuel directly because of it posses damage to the engine machinery at storage and combustion due to content of free fatty acid, high viscosity, acidity, etc. To convert this oil into fuel at commercial scale chemical modifications are cited by various researchers to improve the qualitative part like direct blending, transesterification, pyrolysis, and emulsification. Among these, transestrification is the key and foremost important step to reduce the cleaner and environmentally safe fuel from vegetable oil [3].

Presently, vegetable oils, animal fats, and microbial oils are dominant sources for biodiesel production. Most of vegetable oils are from edible oils (Soybean, Rapeseed). The prices of these oils are relatively high for fuel-grade production. Whereas, animal fats and microbial oils (yeast, fungi, algal) also have been converted to biodiesel [4] in search of less-expensive fuel production. Use of waste-cooking oil is also cited for biodiesel production by recent researchers in their work [5]. Therefore it is necessary to find a new feedstock for biodiesel production, which is now focused on Jatropha curcus, a non-edible plant of family Euphorbiaceous found to be growing in many parts of the country. It is rugged in nature and can be grown without large monitory requirement. The traditional consists in extracting the oil from the paste using boiling water as a solvent. According to current ratio, the price of Jatropha seen is much lower than the oil price, but as the demand of jatropha oil as biofuel increases, the cost will increases too with consideration of jatropha plantation and the product price [6]. Besides the cost also depends on the quality of the oil extracted from the seed. The increasing demand of jatropha oil also has opened up wide opportunity for global marketing which leads to the requirement of competitive products which comes in advantage in term of quality, cost and production time, hence it is best to identify a best extraction technique, as to extract higher yield of oil with higher quality at lower cost. Other than that, this small investigation with research is also conducted to improve existing extraction process towards more environmental friendly method. Existing extraction method can be improved by optimizing extraction parameters. Response surface methodology (RSM) has been used by Ahmad et al., (2018) [7] and Kirrolia et al. (2014) [8] for optimization of process. RSM is an necessary instrument for designing, formulating, developing, and analyzing the processes and production of any study. This model can be used to analyze combinations of given factor and their responses within the experimental range. In present study, jatropha seed was used extracted oil using hexane as solvent and the objective of this work was to determine the effect of temperature and time on the yield and quality of jatropha oil using RSM. Hence, focus on optimization of oil extraction from jatropha seed can enhance economic status of the seed.

2. Materials and methods

The experiment comprises of selection and collection of plant's seed for the oil after treatment with quenching agents with determination of extraction method.

2.1. Collection of jatropha seed

Jatropha seed collected from the research field station of Department of Environmental sciences, Babasaheb Bhimrao Ambedkar University, Lucknow, U.P. (India). Seed was dry at 50°C for 24 hours using hot air oven then finally jatropha seed were grinded and sieved with 0.5 and 0.1mm sieve plate.

2.2. Experimental design for optimization of culture conditions and data analysis

Response surface methodology (RSM) is a methodical statistical design tool to discover the collective or multivariate effect of two or more variables within experiment. In the present study, jatropha seed used for extraction of oil using three level, full-factorial Box Benken design (BBD) Design using Expert software (DESv11) (Student version) with three independent variables, the dose of quenching agent (gm), extraction time (h) and extraction temperature (°C) was applied in this study (Table 1), requiring

17 sets of experiment in triplicate runs comprising 6 replicates of center points, 6 axial (star points) and 5 factorial (cubic points).

Different analysis followed several statistical tests like Fischer's test (F-test), lack of fit, analysis of variance (ANOVA) and other tests were performed identifying the best-fitted model for optimization. In addition to that, the determination coefficient value (R2) of the model articulates the goodness of fit of the suggested model. The effect of the independent factor on dependent variable was analyzed by a Quadratic polynomial equation (Eq.1) expressed as:

$$Y = \beta_0 + \beta_A x_A + \beta_B x_B + \beta_C x_C + \beta_{AB} x_A x_B + \beta_{AC} x x_C + \beta_{AB} x_B x_C + \beta_{AA} x_A^2 + \beta_{BB} x_B^2 + \beta_{CC} x_C^2 \quad (1)$$

Where, Y is the predicted response and$x_A, x_{B,..........,} x_C$ are the variables, which affects the predicted response; $x_A^2, x_B^2 \ldots\ldots\ldots, x_C^2$are the square or quadratic effect of variables; $x_A x_B, x_A x_C \ldots\ldots, x_B x_C$ are the interaction effect of variables;,β_0 is the intercept;$\beta_A, \beta_B, \beta_C$ is the linear effect $\beta_{AA}, \beta_{BB}, \beta_{CC}$ is the effect of squared; $\beta_{AB}, \beta_{AC}, \beta_{BC}$ is the effect of interaction.

After the experimental a point prediction analysis (PPA) done for the optimization according to previous experimental analysis.

Table 1. CODED LIMIT FOR FACTORS USED IN THE EXPERIMENTAL DESIGN

Factor	**Unit**		**Coded level**		
			-1	**0**	**+1**
Dose of quenching agent	**gm**	X_1	0.5	1.5	2.5
Time	**Hour**	X_2	1	3.5	6
Temperature	**°C**	X_3	30	45	60

2.3. *Quenching agent*

In present study, Zirconium chloride (ZrCl2) (Analytical grade) used as a quenching agent for extraction of oil from jatropha seed.

2.4. *Extraction of oil from jatropha seed*

40 gm of cursed jatropha seed was taken in the cellulose thimble with 150 ml of solvent (Hexane) for extraction by the soxhlet extraction unit fig. (1). The reaction was processed for different time interval and different temperature according to BBD-RSM design. After completing the extraction process the percent yield of jatropha oil calculated by using Eq.(1).

$$\%\, yield\ of\ oil = \frac{(Yi - Yf)x100}{Yi}$$

Yi; Initial weight of seed, Yf; Final weight of oil

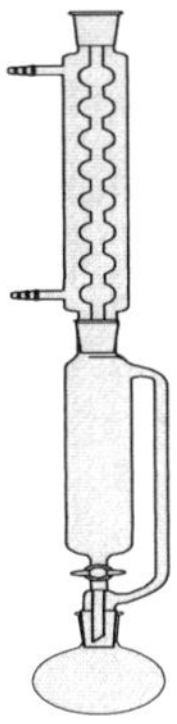

Figure 1. Soxhlet extraction unit

3. Results and Discussions

In order to achieve high yield of extraction from jatropha seed, the factor (dose of quenching agent, time and temperature) taken for extraction and optimizing the maximum % yield. The extraction yield varies between 14.54-46.56% respectively for different combination as shown in Table.2

Statistical and mathematical analysis of Yex% observed by the comparative variation between using predicted and actual observation as shown in Fig.2 and found that predicted and actual value of extraction yield is distributed around the central line and noticed the minimal error. The given regression analysis supporting the results and not any distortion or variation is observed as shown in Table.3 and assured with high degree of accuracy in the experimental data. The best fitted BBD model response for Extraction yield are composed of linear coefficient (X_1, X_2, and X_3) quadratic coefficient (X_1^2, X_2^2 and X_2^3) and Interactive coefficient (X_1X_2, X_1X_3 and X_2X_3) used for the development of predicted responses for the extraction yield by using a quadratic Eq.2.

$$\text{Extraction Yield} = -13.31+28.84*X_1-0.86*X_2+0.64*X_3+0.12*X_1*X_2+0.25*X_1*X_3+0.088*X_2*X_3-0.57*X_1^2-0.34*X_2^2-9.9x10^{-3}*X_3^2 \quad (2)$$

Table 2. MATRIX OF BOX BENKEN DESIGN (BBD) FOR CONDITION WITH OPTIMIZATION FOR EXTRACTION OF JATROPHA OIL PE

Factors				Responses	
S.N	**X_1**	**X_2**	**X_3**	**Y_{Exp}**	**Y_{obs}**
1	0.5	1	45	15.76	15.76
2	1.5	3.5	45	39.76	39.17
3	2.5	3.5	30	23.56	24.89
4	1.5	3.5	45	38.54	39.17
5	1.5	3.5	45	38.54	39.17
6	1.5	3.5	45	39.12	39.17
7	2.5	3.5	60	45.67	45.39
8	1.5	6	30	28.54	27.21
9	2.5	6	45	37.76	37.76
10	0.5	3.5	60	21.67	20.34
11	0.5	6	45	18.54	19.59
12	1.5	1	60	34.45	35.78
13	1.5	3.5	45	39.87	39.17
14	2.5	1	45	33.76	32.71
15	1.5	6	60	46.56	46.84
16	1.5	1	30	29.65	29.37
17	0.5	3.5	30	14.54	14.82

X_1: Quenching agent dose,X_2: Extraction time,X_3: Extraction temperature

Table 3. ANALYSIS OF VARIANCE (ANOVA) FOR EXTRACTION OF JATROPHA OIL

Source	df	Sum^2	Mean Square	F-value	p-value
Model	9	1633.	181.47	113.2	< 0.0001
X_1	1	616.7	616.71	384.8	< 0.0001
X_2	1	39.52	39.52	24.66	0.0016
X_3	1	338.7	338.78	211.4	< 0.0001
X_1X_2	1	0.372	0.3721	0.232	0.6446
X_1X_3	1	56.10	56.10	35.0	0.0006
X_2X_3	1	43.69	43.69	27.2	0.0012
X_1^2	1	470.9	470.91	293.8	< 0.0001
X_2^2	1	19.20	19.20	11.98	0.0105
X_2^3	1	20.95	20.95	13.0	0.0086
Lack fit	3	9.58	3.19	7.8	0.0378
Error	4	1.63	0.4086		
Cor Total	16	1644			
Std. Dev.		1.27			
Mean		32.13			
C.V. %		3.94			
R^2_{Adj}		0.99			
R^2_{Pred}		0.98			
R^2		0.90			

According to statistical and mathematical analysis the generated response show good agreement with predictive response shown in Table.3 and best-fitted model responses Linear responses (X_1, X_2, and X_3), interactive response (X_1X_3 and X_2X_3) and quadratic responses (X_1^2,X_2^2 and X_2^3). The appropriateness of all the parameter was verified by the analysis of variance (ANOVA) and F test was performed to each model for confirming the significance of the model due to high F value [9].

If the F value less than the 4, the model is not significant but in present study, the value of F value 113.24 which determine model is highly significant.

The Fitting efficiency of the Quadratic model regression equation was used to determine the coefficient (R^2) and an adjustable coefficient (R^2_{adj}). The goodness of fitting for the correctness of quadratic model which is predicted by predicted responses (R_{2Pred}; 0.98) and adjustable responses (R_{2adj}:0.99) indicating the high degree of agreement between experimental and predicted. The model could predict the response to explain 95% of the variability which describe the pure error (Prob>F) to check the significance of regression coefficient. The contribution of individual operating variable toward extraction of oil from jatropha seed maximum 62% contribution by quenching agent and 34% by temperature and 4 % extraction time. The deviation of contribution variable (Quenching agent >temperature>time) for extraction of oil. For the interactive behavior of variables shows the (Quenching Agent: Time> time: Temperature >Quenching Agent: Temperature) responsible for the extraction of oil from jatropha seed. Maximum interactive contribution by Quenching Agent: Time. Kartika et al. [10] only work on the temperature reaction time and methanol to seed ratio (2:1–6:1), was examined to extraction of jatropha oil not involve any type of quenching agent.

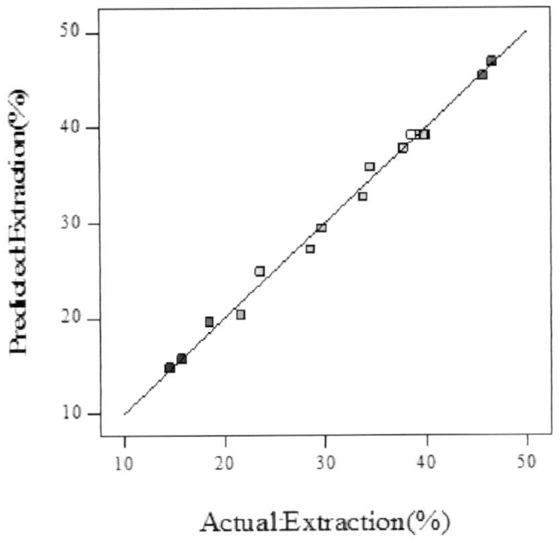

Figure 2. Relation between actual and predicted value of extraction of oil from jatropha seed

The 3D surface plot was drawn to show the interactive effect of variables toward the responses, it also shows the effect of independent variable on the dependent. It was observed from, the graph that interaction between every two variables with maximum prediction of extraction as shown in fig 3a-c.

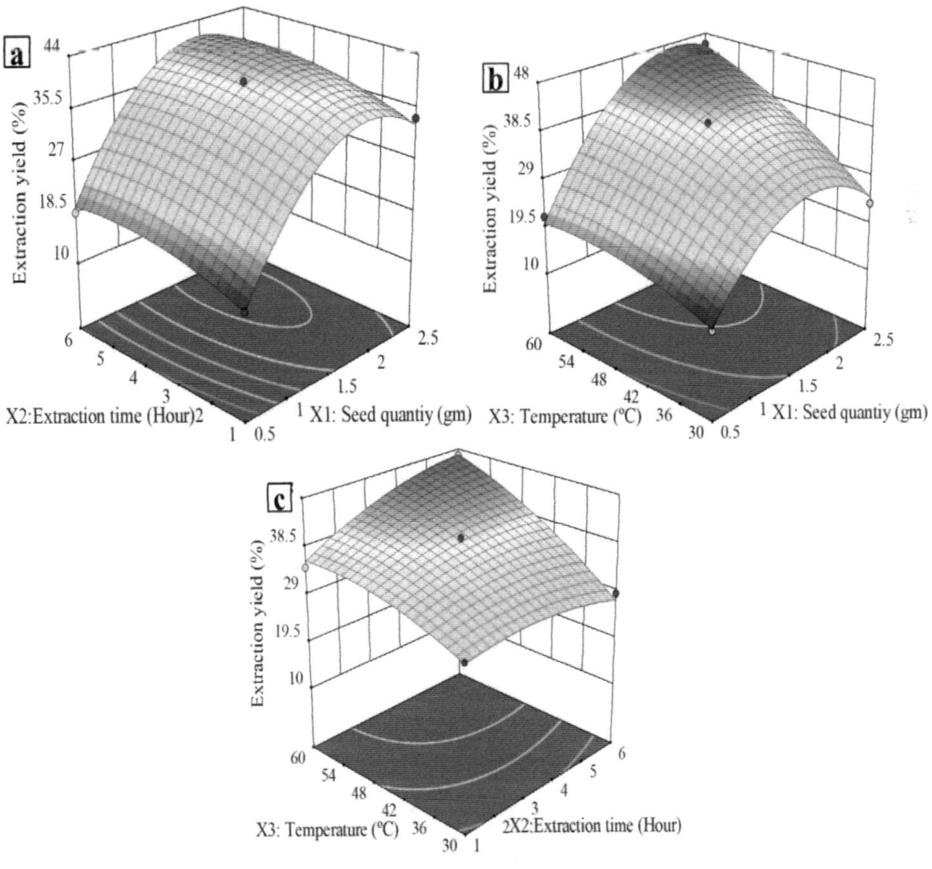

Figure 3. 3D Representation of Biodiesel yield Optimization of variables for extraction jatropha oil

Extraction of oil factor determined by the dose of quenching agent, extraction time and temperature were analyzed for the optimization condition using the point prediction analysis and found different result if variables are changing from lower to higher but for high extraction of oil with using minimal requirement of factor (Quenching agent, Time and Temperature) as shown in fig.3(a-c) provide different results. If minimum quantity of quenching agent (0.5g) used with 1 hour extraction time and 30ºC temperatures give only 14.67% extraction yield but the factor condition is reverse give maximum yield of extraction oil *i.e.* 49%.

4. Conclusions

In this present experimental study, second-order mathematical and statistical model used to predict the variable such as dose of quenching agent , extraction time and temperature for the extraction of oil from jatropha seeds based on box Benken design on response surface methodology (BBD-RSM). Result obtained from the mathematical analysis compared with the experimental study with good agreement of fitness because its estimated value was 0.99. The model developed by BBD-RSM was useful for predicting the optimal extraction condition to achieve maximum extraction. The achievement from the optimizing experiments and found best extraction of oil yield (41.41%) at 60ºC temperature with 3h time and 2gm $ZrCl_2$ quenching agent.

Acknowledgement

The Author and Co-author of this article are very thankful to UGC India for providing financial support and very thankful to head of Department of Environmental Science Babasaheb Bhimrao Ambedkar university Lucknow for providing consistent support for access of laboratory.

References

[1] Kamani MH, Eş I, Lorenzo JM, Remize F, Roselló-Soto E, Barba FJ, et al. Advances in plant materials, food by-products, and algae conversion into biofuels: use of environmentally friendly technologies. *Green Chemistry*. 2019;**21**(12):3213-31.

[2] Anguebes-Franseschi F, Bassam A, Abatal M, May Tzuc O, Aguilar-Ucán C, Wakida-Kusunoki AT, et al. Physical and Chemical Properties of Biodiesel Obtained from Amazon Sailfin Catfish (Pterygoplichthys pardalis) Biomass Oil. *Journal of Chemistry*. 2019;**2019**:7829630.

[3] Kesieme U, Pazouki K, Murphy A, Chrysanthou A. Biofuel as an alternative shipping fuel: technological, environmental and economic assessment. *Sustainable Energy & Fuels*. 2019;**3**(4):899-909.

[4] Li Q, Du W, Liu D. Perspectives of microbial oils for biodiesel production. *Applied Microbiology and Biotechnology*. 2008;**80**(5):749-56.

[5] Ahmad S, Kothari R, Pathak VV, Pandey MK. Fuel Quality Index: A Novel Experimental Evaluation Tool for Biodiesel Prepared from Waste Cooking Oil. *Waste and Biomass Valorization*. 2019;**10**(8):2237-47.

[6] Ramírez V, Martí-Herrero J, Romero M, Rivadeneira D. Energy use of Jatropha oil extraction wastes: Pellets from biochar and Jatropha shell blends. *Journal of Cleaner Production*. 2019;**215**:1095-102.

[7] Ahmad S, Pathak VV, Kothari R, Kumar A, Naidu Krishna SB. Optimization of nutrient stress using C. pyrenoidosa for lipid and biodiesel production in integration with remediation in dairy industry wastewater using response surface methodology. *3 Biotech*. 2018;**8**(8):326.

[8] Kirrolia A, Bishnoi NR, Singh R. Response surface methodology as a decision-making tool for optimization of culture conditions of green microalgae Chlorella spp. for biodiesel production. *Annals of microbiology*. 2014;**64**(3):1133-47.

[9] Ahmad S, Kothari R, Pathania D, Tyagi VV. Optimization of nutrients from wastewater using RSMfor augmentation of Chlorella pyrenoidosa with enhanced lipid productivity, FAME content, and its quality assessment using fuel quality index. *Biomass Conversion and*

Biorefinery. 2020;**10**(2):495-512.
[10] Amalia Kartika I, Yani M, Ariono D, Evon P, Rigal L. Biodiesel production from jatropha seeds: Solvent extraction and in situ transesterification in a single step. *Fuel*. 2013;**106**:111-7.

Effect of Doping Rare-Earth Element on the Structural, Morphological, Optical and Photocatalytic Properties of ZnO Nanoparticles in the Degradation of Methylene Blue Dye

I.Y. Habib[1], Nuraina Mohamad Zain[2], Chee Ming Lim[1], A. Usman[2], N T R N Kumara[1], Abdul Hanif Mahadi[1*]

[1] Centre for Advanced Material and Energy Sciences, Universiti Brunei Darussalam, Tungku Link, Gadong BE1410, Negara Brunei Darussalam.

[2] Chemical Sciences, Faculty of Science, Universiti Brunei Darussalam, Jalan Tungku Link, Gadong, BE 1410, Brunei Darussalam.

E-mail: hanif.mahadi@ubd.edu.bn

Abstract. ZnO has been widely explored as a remarkable photocatalyst for the degradation of wide ranges of organic pollutants. However, its photocatalytic activity could be enhanced through metal doping. In this study, a rare-earth element, Cerium was doped into the microstructure of ZnO nanoparticles (NPs) using deposition precipitation method to enhance its photocatalytic activity towards methylene blue degradation via UV light irradiation. The resulting degradation efficiencies (DEs) recorded in this study are ~ 64 %, 85 % and 55 % respectively. Similarly, the recorded rates are also found to be 9 x 10 $^{-3}$, 21 x10 $^{-3}$ and 6 x 10 $^{-3}$ respectively, which are well correlated with the DE values. This superior photocatalytic performance achieved by 1 % Ce doping could be attributed to the reduction in band gap of the pure ZnO NPs from 2.69 eV – 2.60 eV. Therefore, 1 % Ce was the optimum doping of Ce^{4+} ion into ZnO microstructure, which demonstrates that Ce^{4+} ion could be used as an effective tool for stabilizing the generation of the reactive oxygen species, slowing down the recombination of the photo-generated charge carriers as well as enhancing their separation. Therefore, the optimum Ce doping into the ZnO NPs could play a remarkable role in facilitating the structural, microstructural, optical and electronic properties of the ZnO semiconductor, , and eventually enhanced its photocatalytic activity.

1. Introduction

Metal oxide nanoparticles (NPs) such as TiO_2, CeO_2 and ZnO have been widely explored as efficient photocatalysts for the degradation of organic pollutants as well as treatment of water and waste water due to their remarkable optical, structural, morphological and electronic properties [1, 2]. ZnO in particular has been widely explored as suitable photocatalyst for the degradation of organic pollutants after TiO_2 due to its simplicity of fabrication and the susceptibility to be moulded into different morphologies while utilizing different methods of synthesis such as hydrothermal, precipitation, sol-gel and vapor deposition among others [3]. Such tendency for the ZnO crystals to be transformed into different sizes and morphologies has been found to be influenced by some reaction conditions such as pH and annealing temperature [4, 5]. Its applications irrespective of the methods of synthesis covers a number of disciplines including gas sensor, dye sensitized solar cells and photocatalysis [6-8]. These applications were influenced by the crystallinity, surface area, presence of defects such as oxygen vacancy and zinc interstitials [7, 9]. Among the aforementioned applications, the utilization of ZnO NPs in photocatalysis was found to be remarkable due to its high tendency to absorb visible and ultraviolet radiation, which could subsequently, promotes its optical and electronic properties. The photocatalytic property of the ZnO material towards a number of organic pollutants such as; Methylene blue, rhodamin B dye and direct red-23 dye have been successfully documented in literatures [8, 10, 11]. However, the photocatalytic activity of the ZnO material has been found to be relatively low due to its possession of wide band gap (3.37 eV), low surface area as well as fast electron-hole pair recombination [12, 13]. Several literatures reported the feasibility of enhancing the photocatalytic activity of ZnO material through rare earth metal doping. Rare-earth elements such as Dy, Er, La, Tb, Eu and Nd has been found to enhance the photocatalytic activity of ZnO NPs due to their remarkable light absorption properties, high electrical and thermal conductivity and high lewis acidity [14-19]. Doping a rare earth elements into the ZnO microstructure could virtually results to either displacement of some of the lattice Zn atoms in the ZnO crystal lattice or probably be dispersed at its surface thereby promoting efficient light absorption and eventually changes the optical, electronic and photocatalytic properties of the ZnO material [18, 20]. In this study, a rare-earth element (Ce) was chosen as the potential dopant into the microstructure of ZnO due to its possession of a stable (+4) oxidation state in addition to the possession of oxidation state of (+3), which is very much common among the rare-earth elements. Its oxide was also found to demonstrate remarkable physical and chemical properties such as high electrical and thermal conductivity and stability, wide surface area and energy band gap, high oxygen mobility and ability to form surface defects such as Ce^{3+} ion and oxygen vacancies [21, 22]. The pure and the Ce doped ZnO NPs were synthesized using precipitation and deposition precipitation methods respectively. The as-calcined samples were subsequently utilized in the photocatalytic degradation of MB dye. MB Dye was used as the model organic pollutant due to its ubiquitous nature and staining properties.

2. Methodology

2.1. Synthesis of undoped ZnO NPs by precipitation and Ce doped ZnO NPs by deposition precipitation

Materials used were analytically graded and they include; Zinc chloride [$ZnCl_2$, MW= 136.286 g/mol, $Ce(NO_3)_3.6H_2O$, MW = 434.218 g/mol; Merck Chemicals Ltd], Ammonia solution [NH_4OH, MW= 35.04 g/mol, HMBG] and de-ionized water [H_2O, MW= 18.02g/mol].

The synthesis method utilized in the present study was similar to that employed by Zhang et *al.* [23]. The ZnO NPs was synthesized using conventional precipitation method. Precisely 1.50g of $ZnCl_2$ was dissolved in 20 ml of distilled water in a 100 ml beaker to form 0.55 M clear homogeneous solution. Upon magnetic stirring at RT, 0.5 ml of NH_4OH was added drop-wise to form a cloudy precipitate and the reaction mixture was allowed to proceed for 24 h at 70°C. The resulting cloudy suspension was vacuum filtered, dried at room temperature (RT) and subsequently calcined in air furnace at 400°C for 2 h. The recorded pH of the suspension was 6.3. The as-calcined powder is designated undoped ZnO NPs. The Ce doped ZnO catalysts were synthesized using deposition precipitation method. In this experiment, precisely 0.50 g each of ZnO support catalyst was introduced into a specific amount of Ce $(NO_3)_3.6H_2O$ containing 20 ml of deionized H_2O and 0.5 ml of NH_4OH was added drop-wise to form cloudy precipitates. The reactions proceed with a pH of about 11 at all dopant concentrations. The corresponding suspensions were also filtered, dried at RT and subsequently calcined in air furnace at 400°C for 2 h. The calcined powders for the supported catalysts were designated 1 Ce/ZnO and 3 Ce/ZnO corresponding to; 1 wt % Ce doped ZnO NPs and 3 wt % Ce doped ZnO NPs respectively.

2.2. Experimental procedure for the photocatalytic degradation of methylene blue

The photocatalytic reactions were carried out in a transparent spherical glass reactor with 100 ml capacity. Precisely 20 mg each of either ZnO or Ce doped ZnO NPs was mixed with 100 ml of 10-ppm MB solution. The recorded pH of the solution is about 6. Exactly 100 ml of the above mixture was poured into a 100 ml transparent glass spherical reactor and subsequently irradiated with 365 nm UV light in a dark environment. During the period of irradiation, the reaction mixture was continuously stirred at RT and the mixture was irradiated at different irradiation times between 5 - 100 minutes. About 5 ml of the photo-irradiation mixture was taken at regular intervals (5, 10, 15, 20, 25, 30, 35, 40, 45, 50, 55, 60, 65, 70, 75, 80, 85, 90, 95 and 100 min), until the final irradiation time of 100 min was achieved. After the irradiation, the mixtures were centrifuged at 3000 rpm for 15 min to obtain the clean samples. The supernatants were collected by means of decantation process and subjected to absorption spectrum measurements using UV-visible spectrophotometer (Shimadzu UV-1601PC, Japan). The schematic of the MB degradation experiment is illustrated in Fig. 1. In order to evaluate the photocatalytic activity as well as the degree of the MB mineralization of the NPs, the following calculation based on degradation efficiency (DE) has been employed [24].

$$\mathbf{DE}\ (\%) = \frac{A_0 - A}{A_0}\ \mathbf{X\ 100} \quad (1)$$

Where DE stands for degradation efficiency in percent, A_0 stands for the MB absorption before UV light irradiation and A is the MB absorption after UV light irradiation of the reaction mixture as a function of irradiation time.

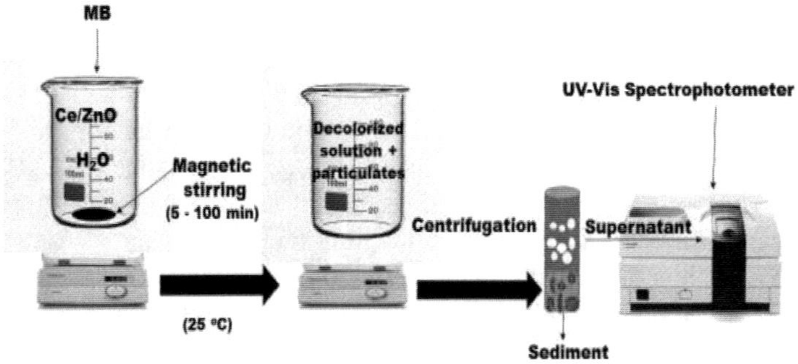

Figure 1. Schematic illustration of the MB photocatalytic degradation experiment for the Cr/ZnO NPs

2.3. Nanoparticles characterizations

The X-ray diffractions were carried out with a Shimadzu X-ray diffractometer XRD - 7000 using Cu Kα radiation (λ = 1.54060 Å) to investigate the structural properties of the undoped and Ce doped ZnO NPs. The ramping rate was set at 5°/min and the X-ray irradiation occurred at 2θ range between 20 ° to 80 °. The field emission scanning electron microscope (FE-SEM) coupled with the energy dispersive X-ray spectroscope (EDX) (Model; JSM-7610F) were employed to determine the morphology, the average particle size as well as the chemical compositions of the as-calcined samples. The ultraviolet-visible light absorption spectroscopy was used to measure the absorption of the NPs using Cary 500 UV-VIS-NIR Spectrophotometer at RT across 200 nm to 800 nm wavelength range.

3. Results and discussion

3.1. XRD analysis

XRD measurements were carried out on the as-calcined samples to investigate their structural properties. Fig 2. Presents the XRD patterns of the undoped ZnO, 1 Ce/ZnO, and 3 Ce/ZnO NPs. The XRD patterns show strong peaks at various reflections corresponding to the following 2θ values (100, 002, 101, 102, 110, 103, and 112). These reflections are assigned to the crystalline hexagonal wurtzite structure of ZnO material and are also indexed to the JCPDS card no. 89-0510 [25]. The formation of strong peaks confirm the formation of highly crystalline particles. It could be observed that peak intensity increases progressively with Ce doping, indicating that the ZnO crystallinity has improved. The complete absence of any secondary phase in the XRD patterns implies that the hexagonal wurtzite structure of ZnO material is maintained. The complete absence of peaks due to CeO_2 may be attributed to the presence of low amount of doped Ce precursor in the ZnO lattice crystal [5]. Lattice shift was not observed, indicating that the Ce atom was dispersed across the surface of the ZnO host material rather than substituting the Zn atoms present in the ZnO lattice [26]. This could be attributed to the presence of higher ionic radius of Ce ion, Ce^{4+}/Ce^{3+} (0.97 Å/1.03Å) compared to that of Zn^{2+} ion (0.74 Å) [27]. Similar result was obtained when Eu was doped into the microstructure of ZnO support material [26]. It was also highlighted that the Eu was located at the amorphous or low symmetry site of the ZnO host lattice rather than substituting the Zn atoms of the ZnO crystal lattice [26]. It is also interesting to note that both the undoped and the Ce doped ZnO structural parameters evaluated in this study were estimated by utilizing the XRD data based on the first 3 reflections (100, 002 and 101). The average crystallite sizes for the undoped and the Ce doped ZnO samples were calculated using the known Scherra equation [28].

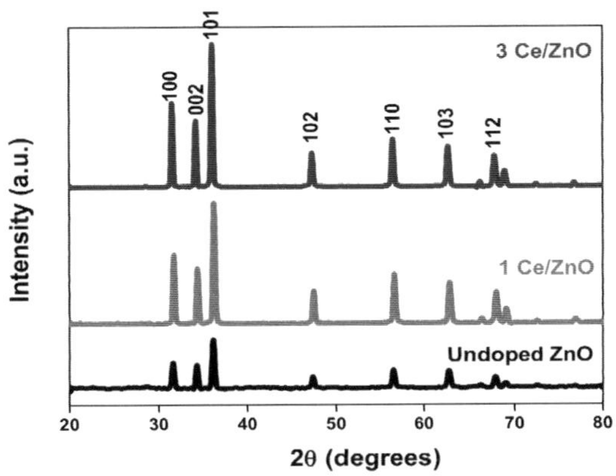

Figure 2. XRD patterns of the undoped ZnO, 1 Ce/ZnO and 3 Ce/ZnO.

$$D = \frac{\kappa\,\lambda}{\mathrm{FWHM\ COS\theta}} \tag{2}$$

Where D is the average crystallite size, κ is the dimensionless constant value (0.9), λ is the wavelength of the irradiation and FWHM is the full width at half maximum measured in radian.

The calculated average crystallite sizes recorded for the undoped, 1 Ce/ZnO and 3 Ce/ZnO NPs were 31, 51 and 59 nm respectively. These values were presented in Table 1. It is obvious that Ce doping results in the significant increase in the average crystallite size. Similar result was also obtained from literature through doping Tb into the lattice structure of ZnO NPs, with the net effect leading to the increase in the crystallite size and strain of the ZnO NPs.

The lattice parameters (a and c) were calculated using the following equations [29].

$$\frac{1}{d^2} = \frac{4}{3}\frac{(h^2+k^2+hk)}{a^2} + \frac{l^2}{c^2} \tag{3}$$

Where d stands for the inter planar spacing, and hkl stands for the miller indices. The values of the lattice parameters were also presented in Table 1. The volume, V of the unit cell of the hexagonal wurtzite structure of the ZnO was calculated using the following equation [29];

$$V = \frac{\sqrt{3}}{2}a^2c \tag{4}$$

The calculated V values for the undoped and Ce doped ZnO samples are also presented in Table 1. The position parameter 'u' and the Zn-O bond length (L) for the undoped and Ce doped ZnO samples were calculated using the following the equations;

$$u = \frac{a^2}{3c^2} \tag{5}$$

$$L = [\frac{a^2}{3} + (\frac{1}{2} - u)^2c^2]^{\frac{1}{2}} \tag{6}$$

The calculated values for both u and L are also presented in Table 1. The values of the lattice parameters, u, V and L were found to be relatively similar upon Ce doping as shown in Table 1, thus, suggesting the absence of Zn atoms substitution by the Ce atoms in the ZnO host lattice.
The specific surface area ($\mathbf{S_s}$) of the samples were also calculated using the known BET equation [22].

$$\mathbf{S_s} = \frac{6}{D_t * D_{XRD}} \tag{7}$$

The resulting specific surface area values for the undoped ZnO, 1 Ce/ZnO and 3 Ce/ZnO NPs were found to be 35, 21, and 18 m^2/g respectively as shown in Table 1. Such decrease in the specific surface area of the ZnO NPs upon Ce doping is attributed to the increase in the average crystallite size.

Table 1. The average crystallite size (D), lattice parameters, LPs (a and c), u parameter, unit cell volume (V), Zn-O bond length, and specific surface area Ss) for the undoped and Ce doped ZnO NPs.

Material	D	LPs			u	L (Å)	S_s
	(nm)	a(Å)	c(Å)	V (Å³)			
ZnO	31	3.255	5.196	47.675	0.131	3.673	35
1 %Ce	51	3.247	5.206	47.532	0.130	3.680	21
3 %Ce	59	3.244	5.196	47.355	0.130	3.674	18

3.2. FE-SEM analysis for the undoped, 1Ce/ZnO and 3 Ce/ZnO NPs

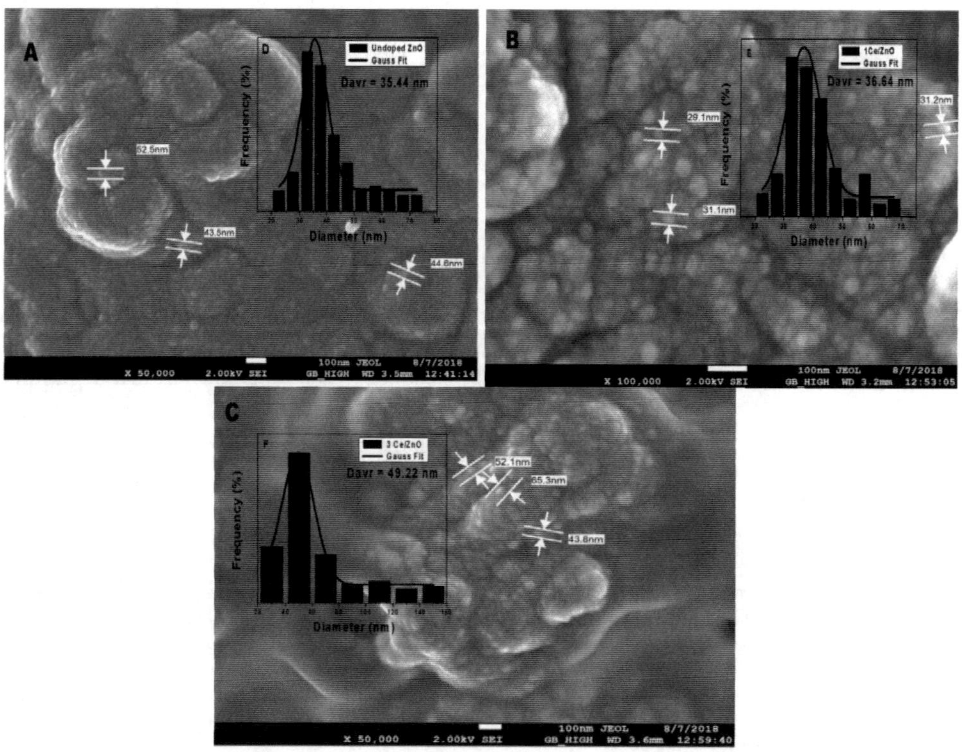

Figure 3. FE-SEM images for the (A) undoped ZnO (B) 1 Ce/ZnO (C) 3 Ce/ZnO and the diameter distribution of (D) undoped ZnO (E) 1 Ce/ZnO (F) 3 Ce/ZnO.

FE-SEM analysis was carried out on the prepared films to determine the particles morphology of the undoped and the Ce doped ZnO NPs. The FE-SEM images for all the analyzed samples are depicted in Fig. 3 (A-C). The same scales (100 nm) and different magnifications (50,000 – 100, 000) were used during the analysis until optimum imaging resolutions were achieved for all samples. It could be clearly seen that all the FE-SEM images show a spherical morphology and wide distributions of different sizes as shown in Fig. 3 (A-C). The recorded ranges of the particle sizes captured are 43.5 – 52.5, 29.1 – 31.2, and 43.8 – 65.3 corresponding to the undoped ZnO, 1 Ce/ZnO, and 3 Ce/ZnO NPs respectively as shown in Fig. 3 (A-C). The Gaussian fitted curves for the undoped and Ce doped ZnO samples were depicted in Fig. 3 (D-F). The estimated average particle sizes were found to be 35.44, 36.64 and 49.22 nm recorded for the undoped ZnO, 1 Ce/ZnO and 3 Ce/ZnO NPs respectively. Results of the average particle size show that the estimated average particle size recorded for the undoped ZnO NPs increases with Ce doping. This trend agreed with the average crystallite sizes calculated using the XRD data as presented in Table 1. The decrease in both the average crystallite and particle sizes of the NPs could be attributed to the decrease in the particles uniformity and increase in the surface roughness as shown in Fig. 3 (D-F). It is important to note that Ce doping did not result to change in morphology, thus, morphological property of ZnO NPs is unchanged in this study. Based on this analysis, both the FE-SEM micrographs and the statistical distributions confirm the formation of small average particle sizes that increases upon Ce doping while exhibiting spherical morphology.

3.3. EDX analysis for the undoped ZnO, 1Ce/ZnO and 3 Ce/ZnO NPs

The energy dispersive X-ray spectroscopy (EDX) was utilized to analyse the chemical compositions of the undoped ZnO, 1 Ce/ZnO and 3 Ce/ZnO NPs. The corresponding percent compositions for each individual constituent are listed in Table 2. The percent chemical compositions of Zn and O atoms, which are the main building blocks of the ZnO material, are presented in Table 2. The EDX data also shows the presence of Ce atom as shown in Table 2. The formation of relatively 1 to 1 ratio of Zn to O (Zn:O; 1:1) atoms in the whole samples indicates that the ZnO lattice distortion is minimized.

Table 2. EDX analysis revealing the percentage compositions for the undoped, 1 Ce/ZnO and 3 Ce/ZnO

S/NO	Element/ Symbol	Undoped ZnO Wt (%)	1Ce/ZnO Wt (%)	3Ce/ZnO Wt (%)
1	Oxygen/O	23.62	19.29	18.97
2	Zinc /Zn	76.38	79.01	76.14
3	Cerium/Ce	-	1.70	4.89

4. UV Vis analysis

The absorbance measurements were carried out within 200 - 800 nm range to investigate the absorption properties of the prepared films for the undoped and Ce doped ZnO NPs. The absorption spectra for the undoped and Ce doped ZnO films show a very sharp absorption peaks as depicted in Fig. 4 (A). All the wavelength of the absorptions occurred below 400 nm with the undoped ZnO film occurring at 350 nm as shown in Fig. 4 (A). Doping 1 % Ce results to a slight shift to the higher wavelength (355 nm) indicating an increase in the photo-generated electron-hole pairs as well as change in band gap. Further increase in the Ce content to 3 % did not result to change in wavelength in the Ce/ZnO composite material. The measured absorbance values were used to estimate the direct energy band gap using the known Tauc equation [29].

$$\alpha h\nu = A\,(h\nu - Eg)^{\frac{1}{2}} \quad (8)$$

Where α, A, $h\upsilon$ and E_g denote the absorption coefficient, constant, photon energy (eV), and energy band gap respectively. Extrapolation was done to obtain the E_g values on the photon energy axis (x-axis). The energy band gap of the undoped ZnO NP was found to be 2.69 eV as shown in Fig. 4 (B). This band gap was found to decrease upon 1 % Ce doping to 2.60 eV indicating a remarkable increase in the optical and electronic property of the undoped ZnO NPs. Koao et *al.* also reported similar result during their photocatalytic study using Ce doped ZnO nanoflower [30]. However, the energy band gap of the undoped ZnO film did not change upon further doping with 3 % Ce impurity as shown in Fig. 4 (B). It is obvious that the recorded energy band gap values are strongly correlated with the respective UV-Vis absorptions of the individual photocatalyst as displayed in Fig. 4 (A). Based on the band gap values recorded in this study, 1 Ce/ZnO film emerge as the optimum Ce doping due to its tendency to undergo a red shift and susceptibility to serve as a better photocatalyst among its analogues. This observed red shift exhibited by the 1 Ce/ZnO film may be ascribed to the presence of high absorption capacity, less amount of particles inter-growth as well as agglomeration as shown in Fig. 2 (B). The occurrence of such narrow energy band gaps exhibited by the undoped and Ce doped ZnO films indicate the presence of excellent optical and electronic properties. In comparison with literature, doping ZnO material with rare earth elements such as La, Eu and Dy were found to increase the energy band gap of ZnO material [16-18]. Possible reason could be attributed to the presence of Burstein Moss Effect, a phenomenon that virtually results to a shift in the fermi level as a result of increase in a dopant concentration [16]. However, band gap reduction upon doping ZnO microstructure with Tb was also evident from literature [31].

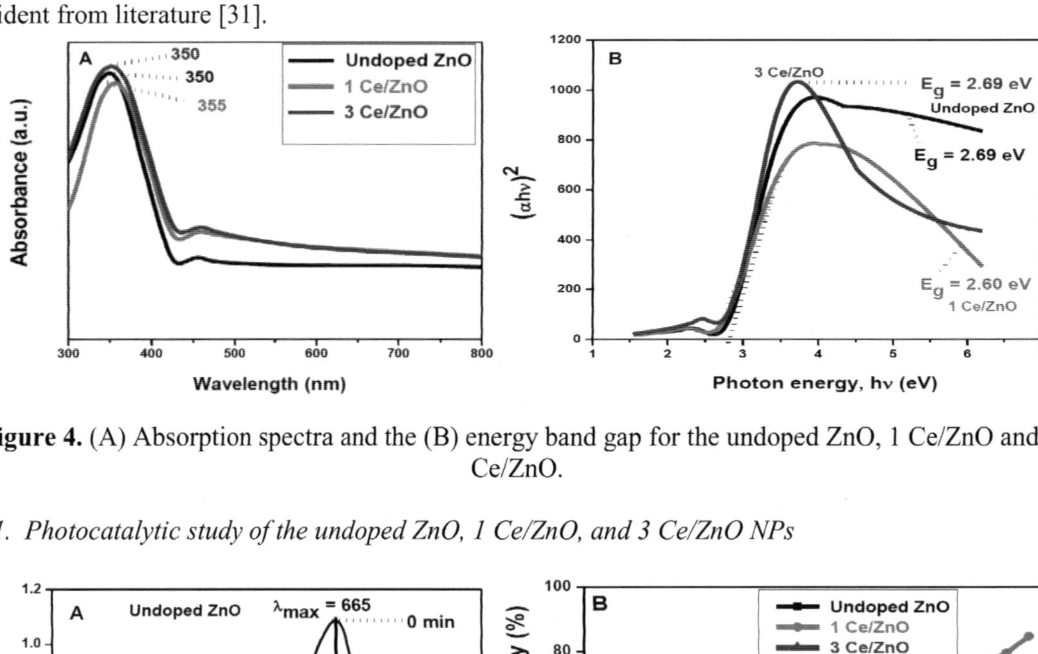

Figure 4. (A) Absorption spectra and the (B) energy band gap for the undoped ZnO, 1 Ce/ZnO and 3 Ce/ZnO.

4.1. Photocatalytic study of the undoped ZnO, 1 Ce/ZnO, and 3 Ce/ZnO NPs

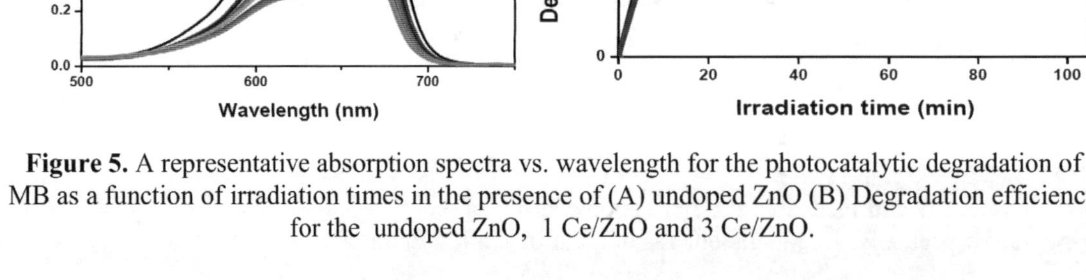

Figure 5. A representative absorption spectra vs. wavelength for the photocatalytic degradation of MB as a function of irradiation times in the presence of (A) undoped ZnO (B) Degradation efficiency for the undoped ZnO, 1 Ce/ZnO and 3 Ce/ZnO.

The photocatalytic degradation of 10 ppm MB dye took place under UV light irradiation between 5 - 100 min. The plots of the MB absorption spectra at the wavelength range between 200 - 800 nm for the undoped ZnO NPs has been presented in Fig. 5 (A). From the given photo-degradation absorption spectra, the most intense absorption bands occurred at 665 nm, which was within the range of the absorption λ_{max} commonly reported from literatures for MB degradation [8]. It could be observed clearly from Fig. 5 (A) that there is a very rapid photo-degradation achieved with respect to the undoped ZnO NPs. This could be seen from the decrease in the absorption band intensities as shown in Fig. 5 (D) presents the degradation efficiencies (DEs) of the undoped and Ce doped ZnO NPs, which compares the photocatalytic activities of the NPs during the photo-degradation period (5 – 100 min). The maximum estimated DE values achieved after 100 min of irradiation were found to be 63.58, 84.62, and 54.63 % corresponding to the undoped ZnO, 1Ce/ZnO, and 3 Ce/ZnO NPs, respectively. The trend in the increasing photocatalytic activity follows the order; 3 Ce/ZnO < undoped ZnO < 1 Ce/ZnO. This shows that, the 1 Ce/ZnO NPs demonstrates the highest photocatalytic activity. It could be noticed that 1 Ce/ZnO NP exhibited the most decreased and well separated MB absorption bands upon increasing irradiation time than the undoped and 3 Ce/ZnO NPs. This suggests the occurrence of a well efficient electron-hole pairs separation with the minimum possibility of electron-hole pairs recombination. The addition of Ce dopant into the lattice crystal of ZnO material could possibly enhance the structural and optical properties of the ZnO photocatalyst and will affect greatly the electronic properties of ZnO semiconductor **(thirty two)** [32]. Therefore, the presence of optimum level of Ce^{4+}/Ce^{3+} in the ZnO lattice could enhanced electron-hole pairs separation via trapping free electrons from the conduction band and subsequently increases photocatalytic activity [12]. From the chemical point of view of Ce metal having a variable oxidation states of +3 and +4, the presence of Ce^{4+} in the lattice structure of ZnO, may acts as an electron trapping agent by trapping temporarily the photo-generated electrons at the conduction band of ZnO, which could eventually enhances charge carriers separation [5]. The main rationale that led to the fastest and efficient photo-degradation achieved with 1 Ce/ZnO NP could be attributed to the reduced energy band gap and less agglomerated particles leading to a structurally enhanced surface that could efficiently slow down the recombination of electron-hole pairs and stabilizes the generation of the reactive radical species. Thus, the sudden reduction of energy band gap could eventually reduce the activation energy of the reaction and consequently enhance the photocatalytic activity [33]. Further increase in the amount of Ce (up to 3 %) in this study was found to decrease the undoped ZnO photocatalytic activity. Chang et *al.* also reported similar photocatalytic activity during their photocatalytic studies [34]. In comparison with literature, 1 % doping of other rare earth elements such as Nd into ZnO NPs demonstrated the highest degradation efficiency of 92 % under 300 min of UV light irradiation in the photo-degradation of MB dye [19]. Also doping 1 % of Eu^{3+} into ZnO NPs was found to exhibit the most efficient activity for tissue imaging in medicine [26]. Therefore, doping 1 % of a rare-earth element into the microstructure of ZnO NPs could open up a new avenue to synthesis of more efficient NPs that could be utilized in a number of applications.

4.2. Kinetics of the MB photocatalytic degradation in the presence of the undoped ZnO, 1 Ce/ZnO, and 3 Ce/ZnO NPs

The plots of the linear fittings of the ln $[MB]_t$ vs. irradiation and the rate of reactions for the undoped and Ce doped ZnO NPs is depicted in Fig. 6 (A and B). The values of the rate constant (k) for the undoped and Ce doped ZnO samples were obtained from the slopes of the linear fittings of the plot of ln $[MB]_t$ vs. irradiation time as shown in Fig. 6 (A). These rate constant values were found to be 9.09 x 10^{-3}, 21.10 x 10^{-3}, and 5.93 x $10^{-3}$$min^{-1}$ corresponding to the undoped, 1, and 3 Ce/ZnO NPs respectively. As shown in Fig. 6 (B), it could be seen that, 1 Ce/ZnO NPs gave the highest rate constant value of about 21.10 x 10^{-3} min^{-1}, hence its regarded as the most reactive and fastest photocatalyst for the MB photo-degradation achieved in this study.

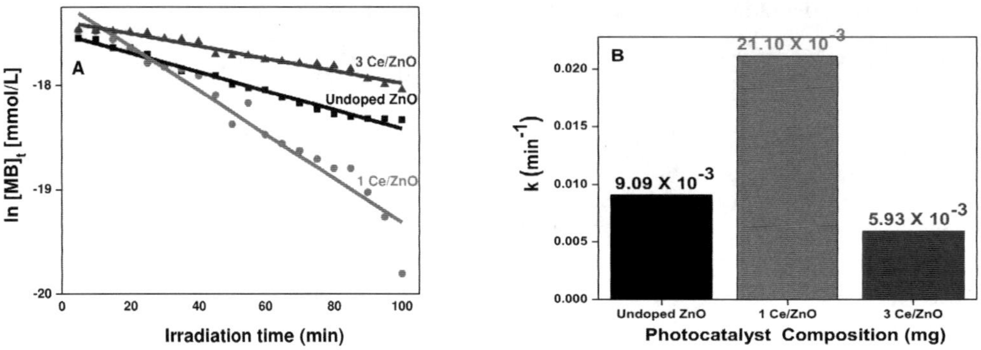

Figure 6. (A) Linear fittings of the ln [MB]t vs. irradiation time and the (B) Plot of k vs. photocatalyst composition for the undoped ZnO, 1 Ce/ZnO and 3 Ce/ZnO.

4.3. Photo-degradation mechanism of methylene blue

In order to effectively evaluate the reactivity and capability of a given photo-catalysts, different catalyst compositions and reaction conditions were involved. In this study, MB dye was used as the model organic pollutant, whereas the undoped and Ce doped ZnO NPs were utilized as the photo-catalysts under UV light irradiation as a function of time. Herein, we discuss the mechanism of MB photo-degradation. ZnO is an n-type semiconductor that has a very rich defect chemistry due to the presence of dominant electrons contributed by the oxygen vacancies and Zn interstitials [29]. In the ZnO hexagonal wurtzite structure, Zn atom occupies half of the tetrahedral sites leaving the octahedral site empty. This shows that, ZnO is highly capable of accommodating both intrinsic defect and extrinsic dopant [29]. The overall photo-degradation mechanism depends on the degree of the MB dye adsorption on the surface of the photo-catalyst. The ability of the photo-generated electrons to migrate to the conduction band of ZnO semiconductor upon UV light irradiation facilitates the adsorption process as well as the formation of reactive species. The photocatalytic process begins by migration of electrons (e^-) from the valence band to the conduction band of the ZnO semiconductor material, leaving behind the photo-generated holes (h^+) at the valence band upon UV light irradiation [35]. The molecular oxygen trapped the photo-generated electrons in the conduction band of ZnO, where it is then converted into the reactive oxygen species [5]. However, in the presence of Ce^{4+}, the photo-generated electron is firstly temporarily trapped and thereafter transferred to the molecular oxygen to form the superoxide radical ($O_2^{\cdot-}$), and the process is accompanied by reducing Ce^{4+} to Ce^{3+}. This is because the lewis acidity of Ce^{4+} is theoretically superior to that of O_2 [16].. The $O_2^{\cdot-}$ then reacts with water molecules (H_2O) to produce per-hydroxyl radical ($HO_2^{\cdot}$) and hydroxide ion (OH^-) [20]. The $HO_2^{\cdot}$, rearranges to form molecular oxygen (O_2) and hydrogen peroxide (H_2O_2) [20] and Finally the hydroxyl radical ($OH^{\cdot}$) was formed, which was subsequently utilized for the complete mineralization of the MB dye and eventually leading to the formation of CO_2 and H_2O and other possible mineralization products (MP). These could be expressed using the following chemical equations;

$$\mathbf{ZnO + h\upsilon \Rightarrow e^- + h^+} \quad (9)$$

$$\mathbf{Ce^{4+} + e^- \Rightarrow Ce^{3+}} \quad (10)$$

$$\mathbf{Ce^{3+} + O_2 \Rightarrow Ce^{4+} + O_2^{\cdot-}} \quad (11)$$

$$\mathbf{O_2^{\cdot-} + H_2O \Rightarrow OH^- + HO_2^{\cdot}} \quad (12)$$

$$\mathbf{2HO_2^{\cdot} \Rightarrow O_2 + H_2O_2} \quad (13)$$

$$\mathbf{H_2O_2 + O_2^{\cdot-} \Rightarrow O_2 + OH^- + OH^{\cdot}} \quad (14)$$

$$\mathbf{OH^{\cdot}} \quad + \mathbf{MB} \quad \Rightarrow \quad \mathbf{CO_2} \quad + \quad \mathbf{H_2O} \quad + \quad \mathbf{MP} \qquad (15)$$

The results of the photo-degradation experiments revealed that the 1 Ce/ZnO NP is the optimum Ce doping utilized in this study for better MB photo-degradation. The utilization of such photocatalyst and its analogues may go along way solving the predicament associated with water pollutants and contaminants such as MB dye from any given aqueous solution or industrial effluent.

5. Conclusion

The synthesis and characterizations of the undoped, 1, and 3 Ce doped ZnO NPs using precipitation and deposition precipitation methods have been successful. NPs have been characterized using; XRD, SEM, EDX and UV-Vis. The photocatalytic degradation studies have been successfully carried out upon UV light irradiation as a function of irradiation time between 5 -100 min in the presence of the undoped and Ce doped ZnO NPs using MB dye as the model pollutant at RT. The results of the photocatalysis were found to be superb and promising. The most effective MB photocatalytic degradation occurred at 100 min photo-irradiation. The corresponding DE values were calculated to be 63.58, 84.62, and 54.63 % corresponding to the undoped, 1% and 3% Ce/ZnO NPs respectively. Possible reasons for the high photocatalytic activity associated with the 1 Ce/ZnO NPs, could be attributed to the presence of well-defined and less agglomerated particles and reduced energy gap. This study shows that rare-earth element such as Ce could enhance the photocatalytic properties of the ZnO NPs when doped with an optimum amount. Therefore, doping 1 % of a rare-earth element into the microstructure of ZnO NPs could open up a new avenue to synthesis of more efficient NPs that could be utilized in a number of applications.

Acknowledgement
This work was fully supported by Universiti Brunei Darussalam grants UBD/RSCH/1.9/FICBF(b)/2018/002 and UBD/RSCH/1.9/FICBF(b)/2018/003. The authors would also like to thank the support by AUN-SEED/Net Japan ASEAN Collaborative Education Programme (JACEP) under the CEP 2019.

References

[1] Rajendran S, Khan MM, Gracia F, Qin J, Gupta VK, Arumainathan S. Ce 3+-ion-induced visible-light photocatalytic degradation and electrochemical activity of ZnO/CeO 2 nanocomposite. *Sci Rep*. 2016;**6**(1):1-11.

[2] Xu C, Rangaiah GP, Zhao XS. Photocatalytic Degradation of Methylene Blue by Titanium Dioxide: Experimental and Modeling Study. *Ind Eng Chem Res*. 2014;**53**(38):14641-9.

[3] Kołodziejczak-Radzimska A, Jesionowski T. Zinc oxide—from synthesis to application: a review. *Materials*. 2014;**7**(4):2833-81.

[4] Huang S, He B, Yan X, Khan I, Wang J, Gao M, et al. Hierarchical ZnO/Si nanowire arrays as an effective substrate for surface-enhanced Raman scattering application. *Sensors Actuators B: Chem*. 2018;**273**:48-55.

[5] Faisal M, Ismail AA, Ibrahim AA, Bouzid H, Al-Sayari SA. Highly efficient photocatalyst based on Ce doped ZnO nanorods: Controllable synthesis and enhanced photocatalytic activity. *Chem Eng J*. 2013;**229**:225-33.

[6] Marimuthu T, Anandhan N, Thangamuthu R. Electrochemical synthesis of one-dimensional ZnO nanostructures on ZnO seed layer for DSSC applications. *Appl Surf Sci*. 2018;**428**:385-94.

[7] Habib I, Tajuddin AA, Noor HA, Lim CM, Mahadi AH, Kumara N. Enhanced Carbon monoxide-sensing properties of Chromium-doped ZnO nanostructures. *Sci Rep*. 2019;**9**(1):1-12.

[8] Zulmajdi S, Ajak S, Hobley J, Duraman N, Harunsani M, Yasin H, et al. Kinetics of photocatalytic degradation of methylene blue in aqueous dispersions of TiO2 nanoparticles under UV-LED irradiation. *Am J Nanomater*. 2017;**5**:1-6.

[9] Khokhra R, Bharti B, Lee H-N, Kumar R. Visible and UV photo-detection in ZnO nanostructured thin films via simple tuning of solution method. *Sci Rep*. 2017;**7**(1):1-14.

[10] Rahman QI, Ahmad M, Misra SK, Lohani M. Effective photocatalytic degradation of rhodamine B dye by ZnO nanoparticles. *Mater Lett*. 2013;**91**:170-4.
[11] Sobana N, Thirumalai K, Swaminathan M. Kinetics of solar light assisted degradation of direct red 23 on activated carbon-loaded zinc oxide and influence of operational parameters. *Trans*. 2016;**4**(1):77-89.
[12] Ahmad M, Ahmed E, Zafar F, Khalid N, Niaz N, Hafeez A, et al. Enhanced photocatalytic activity of Ce-doped ZnO nanopowders synthesized by combustion method. *Journal of Rare Earths*. 2015;**33**(3):255-62.
[13] Zhu L, Zeng W. Room-temperature gas sensing of ZnO-based gas sensor: A review. *Sensors and Actuators A: Physical*. 2017;**267**:242-61.
[14] Hastir A, Kohli N, Singh RC. Comparative study on gas sensing properties of rare earth (Tb, Dy and Er) doped ZnO sensor. *J Phys Chem Solids*. 2017;**105**:23-34.
[15] Sun Y, Zhao Z, Li P, Li G, Chen Y, Zhang W, et al. Er-doped ZnO nanofibers for high sensibility detection of ethanol. *Appl Surf Sci*. 2015;**356**:73-80.
[16] Bhatia S, Verma N, Kumar R. Morphologically-dependent photocatalytic and gas sensing application of Dy-doped ZnO nanoparticles. *J Alloys Compd*. 2017;**726**:1274-85.
[17] Anandan S, Vinu A, Mori T, Gokulakrishnan N, Srinivasu P, Murugesan V, et al. Photocatalytic degradation of 2, 4, 6-trichlorophenol using lanthanum doped ZnO in aqueous suspension. *Catal Commun*. 2007;**8**(9):1377-82.
[18] Zong Y, Li Z, Wang X, Ma J, Men Y. Synthesis and high photocatalytic activity of Eu-doped ZnO nanoparticles. *Ceram Int*. 2014;**40**(7):10375-82.
[19] Yayapao O, Thongtem T, Phuruangrat A, Thongtem S. Ultrasonic-assisted synthesis of Nd-doped ZnO for photocatalysis. *Mater Lett*. 2013;**90**:83-6.
[20] Kumar R, Umar A, Kumar G, Akhtar M, Wang Y, Kim S. Ce-doped ZnO nanoparticles for efficient photocatalytic degradation of direct red-23 dye. *Ceram Int*. 2015;**41**(6):7773-82.
[21] Shajahan I, Ahn J, Nair P, Medisetti S, Patil S, Niveditha V, et al. Praseodymium doped ceria as electrolyte material for IT-SOFC applications. *Mater Chem Phys*. 2018;**216**:136-42.
[22] Murugan R, Kashinath L, Subash R, Sakthivel P, Byrappa K, Rajendran S, et al. Pure and alkaline metal ion (Mg, Ca, Sr, Ba) doped cerium oxide nanostructures for photo degradation of methylene blue. *Mater Res Bull*. 2018;**97**:319-25.
[23] Zhang R, Lu K, Zong L, Tong S, Wang X, Zhou J, et al. Control synthesis of CeO2 nanomaterials supported gold for catalytic oxidation of carbon monoxide. *Molecular Catalysis*. 2017;**442**:173-80.
[24] Khaksar M, Amini M, Boghaei DM, Chae KH, Gautam S. Mn-doped ZrO2 nanoparticles as an efficient catalyst for green oxidative degradation of methylene blue. *Catal Commun*. 2015;**72**:1-5.
[25] Suehiro S, Horita K, Pramata AD, Yuasa M, Ishiwata Y, Shimanoe K, et al. Efficient solution route to transparent ZnO semiconductor films using colloidal nanocrystals. *Journal of Asian Ceramic Societies*. 2016;**4**(3):319-23.
[26] Kaszewski J, Kiełbik P, Wolska E, Witkowski B, Wachnicki Ł, Gajewski Z, et al. Tuning the luminescence of ZnO: Eu nanoparticles for applications in biology and medicine. *Opt Mater*. 2018;**80**:77-86.
[27] Kannadasan N, Shanmugam N, Cholan S, Sathishkumar K, Viruthagiri G, Poonguzhali R. The effect of Ce4+ incorporation on structural, morphological and photocatalytic characters of ZnO nanoparticles. *Mater Charact*. 2014;**97**:37-46.
[28] Habib IY, Kumara N, Lim CM, Mahadi AH, editors. Dynamic light scattering and zeta potential studies of ceria nanoparticles. Solid State Phenomena; 2018: Trans Tech Publ.
[29] Basyooni MA, Shaban M, El Sayed AM. Enhanced gas sensing properties of spin-coated Na-doped ZnO nanostructured films. *Sci Rep*. 2017;**7**(1):1-12.
[30] Koao L, Dejene F, Swart H, Botha J. The effect of Ce3+ on structure, morphology and optical properties of flower-like ZnO synthesized using the chemical bath method. *J Lumin*. 2013;**143**:463-8.

[31] Hastir A, Kohli N, Singh RC. Temperature dependent selective and sensitive terbium doped ZnO nanostructures. *Sensors Actuators B: Chem*. 2016;**231**:110-9.
[32] Lin C-C, Chiang Y-J. Preparation of coupled ZnO/SnO2 photocatalysts using a rotating packed bed. *Chem Eng J*. 2012;**181**:196-205.
[33] Aslanzadeh S. Transition metal doped ZnO nanoclusters for carbon monoxide detection: DFT studies. *J Mol Model*. 2016;**22**(7):1-6.
[34] Chang C-J, Lin C-Y, Hsu M-H. Enhanced photocatalytic activity of Ce-doped ZnO nanorods under UV and visible light. *Journal of the Taiwan Institute of Chemical Engineers*. 2014;**45**(4):1954-63.
[35] Thepnurat M, Chairuangsri T, Hongsith N, Ruankham P, Choopun S. Realization of interlinked ZnO tetrapod networks for UV sensor and room-temperature gas sensor. *ACS applied materials & interfaces*. 2015;**7**(43):24177-84.

2-D Study of Near-field Generated by Surface Plasmon Resonance of Short Axis of Gold Nano-rods on SiO_2 and CeO_2/SiO_2 Substrates for Optimisation of Water Splitting Reaction

M N S M Idris[1], H P Chiang[2,3], Y F Chou Chau1, A H Mahadi,[1,*], C M Lim[1]

[1]Centre for Advanced Material and Energy Sciences, Universiti Brunei Darussalam, Jalan Tungku Link, Gadong, BE 1410, Brunei Darussalam,

[2]Institute of Optoelectronic Sciences, National Taiwan Ocean University, No. 2 Pei-Ning Rd., 202, Keelung, Taiwan,

[3]Institute of Physics, Academia Silica, Taipei 11529, Taiwan

E-mail: hanif.mahadi@ubd.edu.bn

Abstract. Distinguishing near-field intensity distribution and plasmon resonance peak wavelength of surface plasmon resonance of gold nano-rods (AuNRs) can provide information for the optimisation of localised surface plasmon and gap plasmon resonances of gold nano-rods. This work, shows the influence of how refractive indexes of the surrounding medium and adjacent surface, as well as, AuNRs pair inter-particle gap contributes to the generation of surface plasmon resonances. The simulation model presented consists of AuNRs pair situated on silicon dioxide (SiO_2) and cerium dioxide/silicon dioxide (CeO_2/SiO_2) substrates with air and water as the surrounding media. The results show high near-field intensities at AuNRs/CeO_2 interfaces with CeO_2 refractive index ($n = 2.38$), and the near-field intensities contributed by the gap resonance is minimal between the AuNRs of inter- particle gap of 10 nm, however, the near-field intensities become significant near to the AuNRs/CeO_2 interfaces because of reduced near-field interference. The simulation set-up provides the conditions for water splitting in thermochemical redox reaction of CeO_2/CeO_{2-x} resulting in the production of hydrogen. AuNRs pair with inter-particle gap of 5 nm situated on 10 nm thick CeO_2 of CeO_2/SiO_2 substrate shows the most favorable conditions for water splitting.

1. Introduction

Plasmonic gold nano-particles (AuNPs) have been widely studied in the field of catalysis [1, 2], electronics [3, 4], sensors [5, 6], optics [7] and biomedicine [8]. These wide research fields of AuNPs are by virtue of its extensive optical properties related to surface plasmons (SP) [9], permissive of surface functionalisation for organic and biological molecules [10], and high physical and chemical stability [11, 12]. Amongst these aspects of AuNPs, the surface plasmons are the major attractions. Besides surface plasmon polariton (SPP), the optical properties of AuNPs are linked to non-propagating surface plasmons called localised surface plasmon resonance (LSPR) for AuNPs, where the AuNPs size is considerably smaller than the wavelength of the incident electromagnetic

(EM) wave [13]. LSPR can be tuned by changing the size [13], shape [14, 15], arrangement [16] of the AuNPs, and the dielectric constant (refractive index) [13, 17] of the surrounding medium.

One of the most interesting AuNPs is gold nano-rods (AuNRs) because of their structure which constitutes of short and long axes. This structure allows two excitations of SP modes which are (i) between 520 nm and 540 nm for the short axis excitation; and (ii) in the range of 600 – 1600 nm for the long axis excitation, in which both depend on the AuNRs aspect ratio [18]. The SP excitation via incident electrical field (E-field) of electromagnetic (EM) wave along the short axis and long axis of AuNRs generate transverse LSPR and longitudinal LSPR, respectively [19]. Furthermore, when a metallic nano-rod (MNR) is brought closer to another MNR (MNR pair) with a few nano-meters separation (plasmonic cavity) [20], SP excitation along the short axis of the MNR pair can induce another SP resonance phenomenon which is known as gap plasmon resonance (GPR). It arises from electrical field (E-field) localisation within the gap of the MNRs pair associated with the polarisation of SP oscillations on the short axis of MNRs pair [21-24]. The GPR disappears when the polarisation of SP oscillations are along the long axis of the MNRs pair. Thus, it is beneficial to incorporate both LSPR and GPR of short axis of AuNRs pair in developing its plasmonic behavior for specific applications, example for water splitting.

The LSPR of AuNPs relax into radiative and non-radiative decays [25]. The non-radiative decay is related to thermal generation of AuNPs [9]. This also applies to AuNRs as the size of AuNRs falls within the size of AuNPs. The heat transfer follows the Joule heating mechanisms [26], whereby the square of the electric field (E-field) is directly proportional to the Joule heating. Therefore, E-field enhancement (near-field) caused by LSPR of AuNRs results in the increase in the thermal generation and heat transfer. In addition, the heat is distributed from AuNRs to the surrounding and adjacent surface. This heat can be utilised to drive two-step thermochemical redox reaction of metal oxides for the hydrogen generation by water splitting [27]. Moreover, the LSPR can facilitate charge injection which further assists the redox reaction process [28].

This work, therefore, focuses on the study of near-field intensities and distributions, plasmon resonance peak wavelengths, and thermal generation corresponding to LSPR and GPR of the short axis of AuNRs pair. This investigation looks into the different models of AuNRs pairs that are supported by silicon dioxide (SiO_2) and cerium dioxide/silicon dioxide (CeO_2/SiO_2) substrates in search of optimum conditions for water spitting and other promising applications. The inter-particle gap between AuNRs, surrounding medium, and the thickness of CeO_2 are varied in order to study the effect of GPR and refractive index of adjacent surface to the LSPR of AuNRs pairs.

2. Simulation section

2.1. Simulation method

Two-dimensional finite-element method (2D-FEM) is used to perform numerical simulation of near-field intensities and distributions, and plasmon resonance peaks of the short axis of AuNRs pairs on different substrates as shown in Figure 1. The short axis of AuNR is clarified in Figure 2(D). The input power of light source used is 1 mW. To study the response of the GPR between AuNRs pair, and the refractive index of the adjacent surface to the LSPR of AuNRs pairs, the AuNRs pairs are excited by E-field oscillating in the x-direction (i.e. short axis). The induced LSPR and GPR are in the x-direction of the AuNRs pairs as illustrated in Figure 1. The permittivity data of gold is acquired from the literature [29] and amended with size effect using the Drude model [30]. Triangular high order edge element formulation is used in the simulation with direct implementation of anisotropic perfectly matched layers (PMLs) [31] at the outer boundaries (in Figure 2) to compromise with anisotropic material in terms of magnetic permeability and dielectric permittivity, respectively. The scattering boundary condition (SBC) and matched boundary (MB) are the simulation boundaries as shown in Figure 1.

In order to investigate the near-field intensities and distributions, and plasmon resonance peaks of the short axis AuNRs pair on different adjacent surface, SiO_2 and CeO_2/SiO_2 are used as substrates as shown in Figure 2. The refractive index of CeO_2 films at different thickness is obtained from the literature [32]. Approximately, at incident wavelength of 530 nm, the refractive index, n, of CeO_2 films of 10 nm and 20 nm thick are 2.38 and 1.68, respectively. At 530 nm wavelength, the imaginary part, k of the refractive index is zero for both CeO_2 films thickness. The dispersion equation for the refractive index of SiO_2 is retrieved from Refs [33]. The wavelength range of the applied incident photon is between 300 nm to 1200 nm. Moreover, the nearfield intensities and plasmon resonance peaks are measured at the gap (point A), 0.5 nm from one of the AuNR (point B), and near Au/CeO_2 interfaces (points C & D) as depicted in Figure 1. The two AuNRs are identical and symmetrical pair, therefore, these four points represent similar near-field intensities and plasmon resonance peaks for both the AuNRs.

In the case of thermal simulation on heat generation and transfer of AuNRs pairs, 2D-FEM is employed by using the Joule heating with general heat transfer module. The simulation used the heat balance equation [26].

The input power of the light source is reduced from 1.0 mW to 0.017 μW and 8.0 μW for AuNRs pairs in air and water, respectively. In order to incorporate the contribution of LSPR of AuNRs pair to the heat generation, the average near-field intensity of AuNRs pair is taken as electric field strength which is then used to calculate the heat source of the AuNRs pair. The simulation temperature boundary conditions are held at 25 °C. The parameters such as thermal conductivity, density, heat capacity, relative permittivity of AuNRs, air and water that are used for the simulation, are retrieved from Refs [26, 29, 34] and tabulated in table 1.

Table 1. The thermal conductivity, density, specific heat, relative permittivity of AuNRs, air and water.

Material	Thermal Conductivity (W/mK)	Density (kg/m^3)	Heat Capacity (J/kgK)	Relative Permittivity
AuNRs	317	19300	129.0	(-4.5461) + i2.4577
Air	0.0262	2.7900	1000	1.00
Water	0.600	998.30	4182	1.78

2.2. *Simulation model*

The simulation model composed of AuNRs on SiO_2 substrate (AuNRs/SiO_2), AuNRs on CeO_2/SiO_2 substrate with ceria thickness of 10 nm (AuNRs/CeO_2-T10/SiO_2), and AuNRs on CeO_2/SiO_2 substrate with ceria thickness of 20 nm (AuNRs/CeO_2-T20/SiO_2) are shown in Figure 2(A), 2(B) and 2(C), respectively. The thickness of SiO_2 for all model is equivalent to 1 μm. The diameter and length of AuNRs pairs are kept constant at 10 nm and 100 nm. In this study, air and water are used as the surrounding media with refractive index of 1.00 and 1.33, respectively. The gap between AuNRs pairs is set to 5 nm and 10 nm. In the case of thermal simulation on heat generation and transfer, AuNRs pairs are simulated without the substrates in order to avoid complexity.

3. Results and discussion

3.1. *Near-field intensities and distribution of AuNRs pair*

The simulation results showed a variation of the near-field intensities and distributions, and the plasmon resonance peak wavelength in the vicinity of the AuNRs pairs with inter- particle gap of 5 nm and 10 nm in air and water media; adjacent to SiO_2 substrates; and CeO_2/SiO_2 substrates with ceria thickness of 10 nm and 20 nm, are presented in Figure 3 to 6. The plasmon resonance wavelength of AuNRs pairs is discussed in the latter section.

In the case of the AuNRs pair with inter-particle gap of 5 nm (Figure 3), the near-field intensities at the gap (point A) and surface (point B) of AuNRs pair for AuNRs/SiO_2 model in water (Figure 3(D)) are higher than AuNRs/SiO_2 model in air (Figure 3(A)). These effects are also observed for AuNRs/CeO_2-T10/SiO_2 model (Figure 3(D)) and AuNRs/CeO_2-T20/SiO_2 model (Figure 3(F)) in water medium. The refractive index of the surrounding medium affects the LSPR and GPR of AuNRs pair showing distinct near-field intensities. On the other hand, in Figure 3, the nearfield intensities that are close to the AuNRs pair interfaces (point C and D) of AuNRs/SiO_2 model (Figure 3(A)&(D)) and AuNRs/CeO_2-T20/SiO_2 model (Figure 3(C)&(F)) are lower than the AuNRs/CeO_2-T10/SiO_2 model (Figure 3 (B)&(E)) for both in air and water media. Moreover, the nearfield intensity at point D is higher than at point C for the AuNRs/CeO_2-T10/SiO_2 model. This is because of destructive interference of the near-field produced by both AuNRs at point C [35]. It is demonstrated that the high near-field intensities occur at AuNRs/CeO_2 interfaces as a consequence of high CeO_2 refractive index (n = 2.38). In addition, the nearfield intensities distributions of AuNRs pair with interparticle gap of 5 nm on SiO_2 and CeO_2/SiO_2 substrates are presented in Figure 4.

The GPR effects are minor for the AuNRs pairs with interparticle gap of 10 nm (Figure 5) and for all the models, where the simulation results show poor near-field intensities at the gap (point A) of the AuNRs pair. In contrast, the near-field intensities that are close to AuNRs pair interfaces (at point C and D) show significant results in particular for the AuNRs/CeO_2-T10/SiO_2 model (Figure 5(B) & (E)). It exhibits high near-field intensities at the interfaces when compared to the AuNR/SiO_2 model (Figure 5(A) & (D)), and the AuNRs/CeO_2-T20/SiO_2 model (Figure 5(C) & (F)) for both air and water. Furthermore, the near-field intensities at the interfaces of AuNRs/CeO_2-T10/SiO_2 model are almost equivalent and this is due to LSPR of the AuNRs pair, and showing less near-field interference. In addition, the influence of GPR of AuNRs pair becomes less as the inter-particle gap is increased to 10 nm. Therefore, only LSPR of AuNRs pair contributes to the generation of the near-field. Moreover, the refractive index, n, of CeO_2 affects the occurrence of high near-field intensities of the AuNRs pair at the AuNRs/CeO_2 interfaces for both the AuNRs pair inter-particle gap of 5 nm and 10 nm, where CeO_2 with n = 2.38 (10 nm thick) has been demonstrated to have high near-field intensities when compared to CeO_2 with n = 1.68 (20 nm thick). The near-field intensities distributions of AuNRs pair with inter-particle gap of 10 nm on SiO_2 and CeO_2/SiO_2 substrates are displayed in Figure 6.

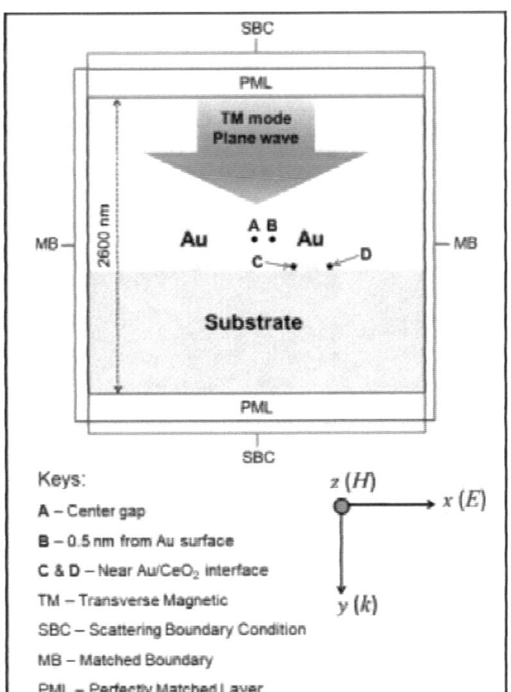

Figure 1. Schematic diagram of AuNRs pair on a substrate with the parameters for the simulation method. The incident TM mode plane wave has its magnetic field (*H*) in the z- direction, and electric field (*E*) in the x-direction propagating in the direction of wave vector (*k*) in y-direction. The length between PMLs is 2600 nm.

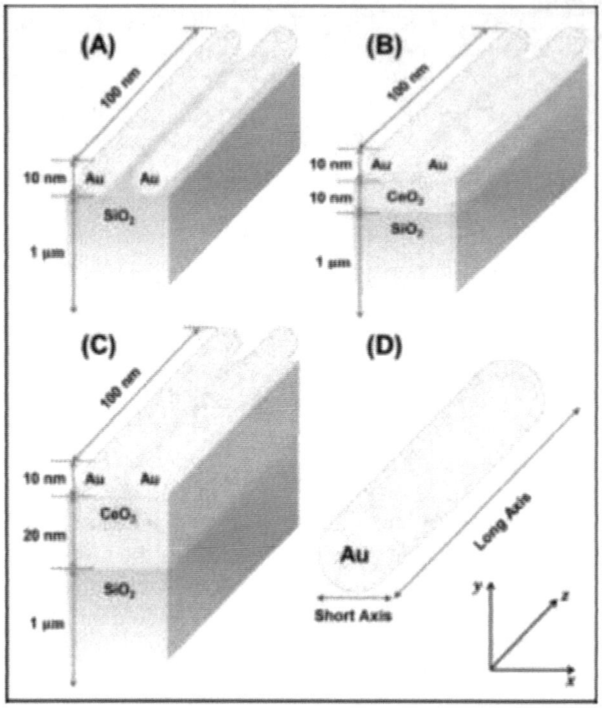

Figure 2. Schematic diagram of simulation models of AuNRs pairs on SiO_2 (A); and CeO_2/SiO_2 substrates with CeO_2 film thickness of 10 nm (B); and 20 nm (C), respectively. (D) showing the short axis and long axis of an AuNR.

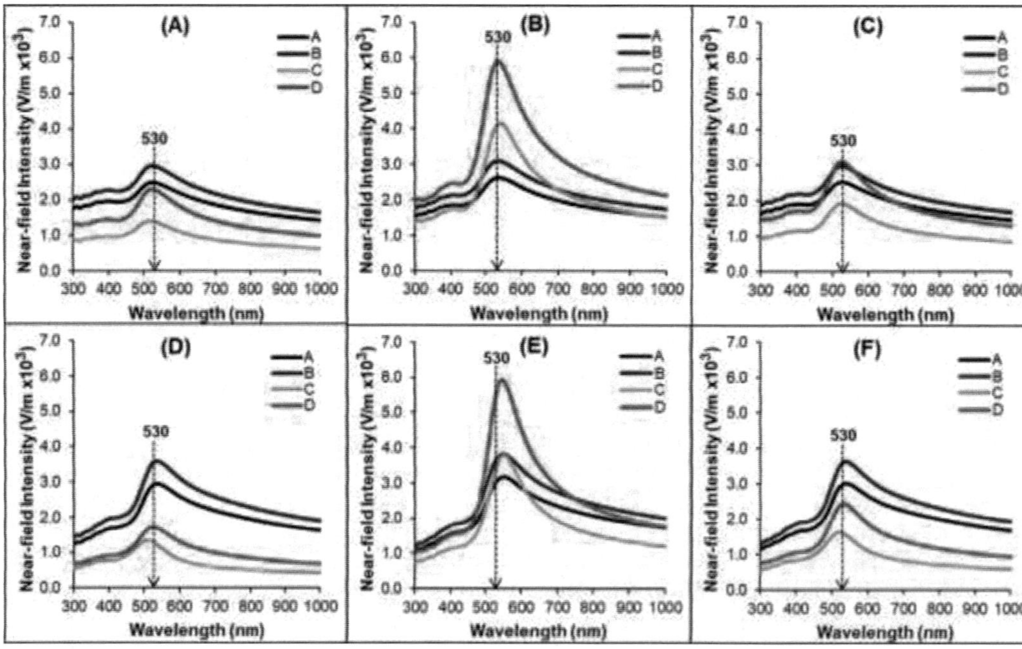

Figure 3. Simulation results of near-field intensities versus wavelength of AuNRs pair with inter-particle gap of 5 nm in air for AuNRs/SiO_2 (A), AuNRs/CeO_2-T10/SiO_2 (B), and AuNRs/CeO_2-T20/SiO_2 (C); and in water for AuNRs/SiO_2 (D), AuNRs/CeO_2-T10/SiO_2 (E), and AuNRs/CeO_2-

T20/SiO_2 (F). The measurement points for near-intensities are A (at the gap:black); B (0.5 nm from Au surface:blue); and, C and D (near AuNRs/CeO_2 interfaces:green and red, respectively).

3.2. Plasmon resonance peaks of AuNRs pair

The plasmon resonance peak of spherical gold nano-particles (AuNPs) has been reported in the visible range of the electromagnetic spectrum, typically at 532 nm [9]. The crosssection of the short axis of AuNRs is analogous to AuNPs. Hence, the plasmon resonance peak of AuNRs is close to that of AuNPs. The plasmon resonance peak wavelength of AuNPs is indicated as dashed arrows, and labelled as 530 nm instead of 532 nm, as shown in Figure 3 and Figure 5. This is because the smallest resolution used for the wavelength in the simulation is 10 nm.

The simulation results showed that most of the plasmon resonance peaks wavelength of AuNRs pairs are shifted away from 530 nm wavelength. This is associated with the change in the refractive index of the medium and the adjacent surface (Figure 3). In the case of AuNRs/SiO2, AuNRs/CeO2- T10/SiO2, and AuNRs/CeO2-T20/SiO2 models in air, the plasmon resonance peaks wavelength of AuNRs pairs at point A, B, C and D are consistent. However, in water medium, the plasmon resonance peaks wavelength varied except for AuNRs/CeO2-T10/SiO2. Moreover, the plasmon resonance peak wavelengths of AuNRs pair in AuNRs/CeO2-T20/SiO2 model remained at 530 nm at all measured points whereas AuNRs/SiO2 and AuNRs/CeO2-T20/SiO2 models in water medium showed the 530 nm plasmon resonance peak wavelength to be located at point D.

The majority of the measured points of the plasmon resonance peak wavelengths of AuNRs pair in AuNRs/SiO_2, AuNRs/CeO_2-T10/SiO_2, and AuNRs/CeO_2-T20/SiO_2 models are deviated from the 530 nm peak wavelength. These are caused by the change in the refractive index of the surrounding medium and the adjacent surface (substrate), as mentioned previously. The plasmon resonance peak wavelength of AuNRs pair in AuNRs/SiO_2 model showed similar peaks at 520 nm for all measured points. In the case of AuNRs/CeO_2-T10/SiO_2, and AuNRs/CeO_2-T20/SiO_2 models, the plasmon resonance peak wavelengths are inconsistent.

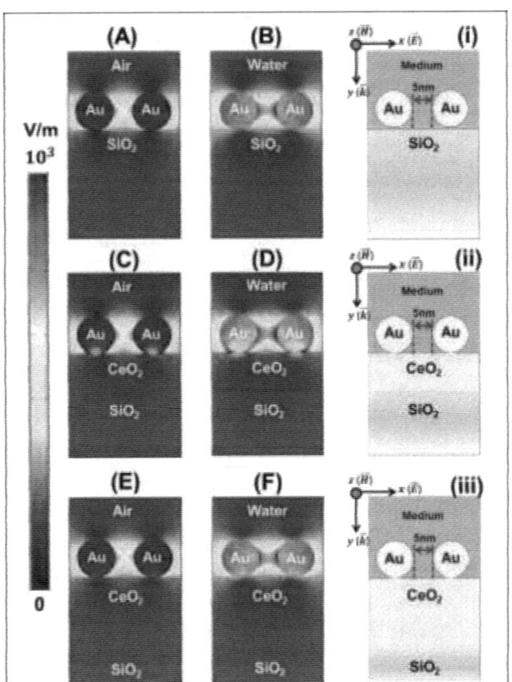

Figure 4. The Near-field intensities and distributions of AuNRs pair with inter-particle gap of 5 nm for AuNRs/SiO_2 in air (A), and water (B); AuNRs/CeO_2-T10/SiO_2 in air (C), and water (D); and

AuNRs/CeO_2-T20/SiO_2 in air (E), and water (F). The schematic design of AuNRs/SiO_2 (i), AuNRs/CeO_2-T10/SiO_2 (ii), and AuNRs/CeO_2-T20/SiO_2 (iii).

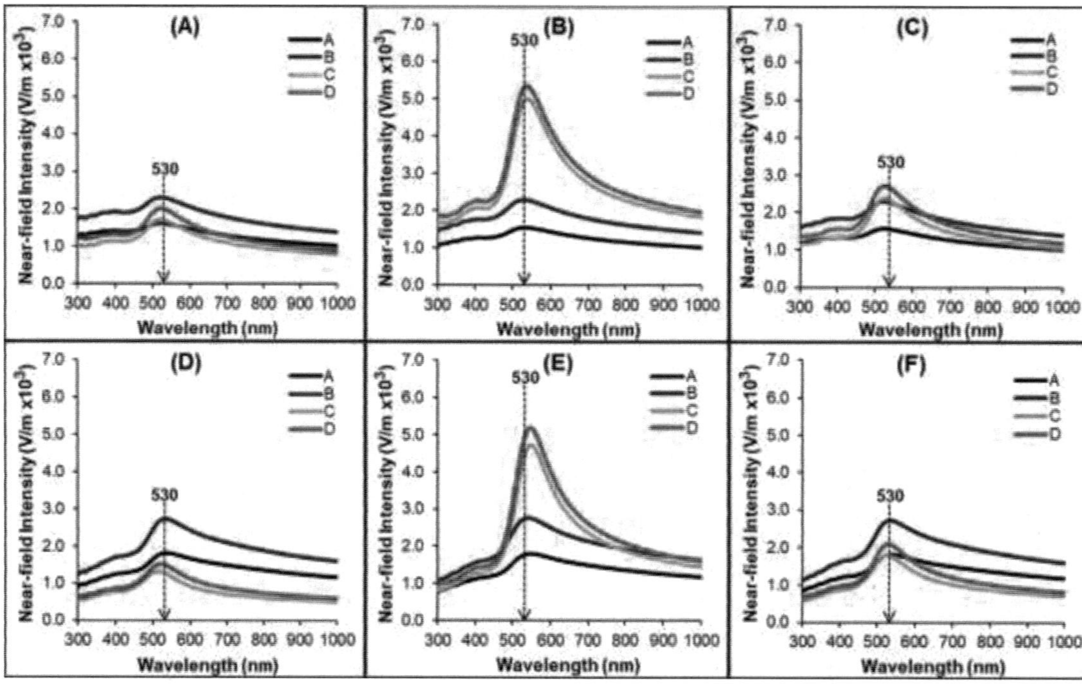

Figure 5. Simulation results of near-field intensities versus wavelength of AuNRs pair with inter-particle gap of 10 nm in air for AuNRs/SiO_2 (A), AuNRs/CeO_2-T10/SiO_2 (B), and AuNRs/CeO_2-T20/SiO_2 (C); and in water for AuNRs/SiO_2 (D), AuNRs/CeO_2-T10/SiO_2 (E), and AuNRs/CeO_2-T20/SiO_2 (F). The measurement points for near-intensities are A (at the gap:black); B (0.5 nm from Au surface:blue); and, C and D (near AuNRs/CeO_2 interfaces:green and red, respectively).

Furthermore, the 530 nm plasmon peak wavelength of AuNRs pairs are observed at point A and B of the AuNRs/SiO_2 model in water; and AuNRs/CeO_2-T10/SiO_2 model in air, as well as, at point C and D of AuNRs/CeO_2- T20/SiO_2 model in air and water media. It is also detected at point B of AuNRs/CeO_2-T20/SiO_2 model in water.

3.3. Local E-field enhancement of AuNRs pair

The local E-field enhancements of AuNRs pair are observed as near-field of high intensities and distributions whereby it is found within the gap of the AuNRs pair and the AuNRs/substrate interfaces correspond to the GPR and LSPR of the AuNRs pair, respectively. Based on the previous nearfield intensities and distribution analyses, the local E-field enhancements are allocated within the gap of the AuNRs pair in AuNRs/SiO_2, AuNRs/CeO_2-T10/SiO_2, and AuNRs/CeO_2- T20/SiO_2 models having inter-particle gap of 5 nm in water media. These models can be used in visible light Raman spectroscopy to detect low concentration of foreign molecules in water solution by amplifying the Raman signals. Furthermore, in the case of AuNRs/CeO_2-T10/SiO_2 model, the local E-field enhancement is also observed at the AuNRs/CeO2 interfaces. A charge (electron) injection has been reported at metal/metal oxide interfaces under visible light excitation [28, 36]. Therefore, it is possible for the charge to be transferred from the AuNRs to the CeO2 surface. In order to determine whether redox reaction at CeO2 surface has taken place or not, visible light Raman spectroscopy can be performed in which the generation of molecules caused by the redox reaction at AuNRs/CeO2 interfaces can be identified in this model.

3.4. Thermal generation and heat transfer of AuNRs pair

The heat source of AuNRs pair is directly proportional to the square of E-field strength [26]. The LSPR of AuNRs pair enhanced the local E-field strength in the form of near-field on the surface of AuNRs pair at plasmon resonance peak wavelength of 530 nm. The near-field is averaged and then used to determine the heat source of AuNRs pair for the thermal simulation. As a result, the AuNRs pairs produced heat with temperature higher than 500 °C as shown in Figure 7 and Figure 8. The E-field strength, and relative permittivity of surrounding media affect the heat source of AuNRs pair. An increase in the light input power resulted in an increase in the average near-field (i.e. local E-field strength). This caused a large increase in the heat source of AuNRs pair due to the proportionality which remarkably increases the temperature of AuNRs pair over its melting point. In addition, the variation in thermal conductivity of media allows different heat conduction which then influence the rate of heat transfer from the AuNRs pair to the surrounding. Therefore, in order for AuNRs pair to produce heat with temperature below its melting point, various simulations are performed in search of the right input power of light for AuNRs pair in different media. It is found that the suitable input power of light for AuNRs pair in air and water are 0.017 μW and 8.0 μW, respectively. Despite the scaled down of light input power from 1 mW to 0.017 μW and 8.0 μW, the near-field distributions and plasmon resonance peaks of the AuNRs pairs on SiO_2 and CeO_2/SiO_2 substrates in air and water are similar whereas only the near-field intensities (local E-field strength) are decremented. Thus, only the near-field intensities are taken into account (for the thermal simulation) without presenting the simulation results of near-field distributions and plasmon resonance peaks.

The temperature profile for AuNRs pairs with inter-particle gap of 5 nm is shown in Figure 7. The temperature of AuNRs pair (grey regions) in air and water medium reached 1056 °C and 976 °C, respectively. The temperature decreased to 57 °C as the distance from the AuNRs increased as shown in the green regions. The temperature at the inter-particle gap (white region) showed curves profile which centered at 978 °C and 904 °C for AuNRs pair in air and water, respectively. In the case of AuNRs pairs with inter-particle gap of 10 nm, the temperature profile is presented in Figure 8. The temperature of AuNRs pair (grey regions) reached 584 °C in air whereas 529 °C in water. The temperature decreased to 100 °C as the distance from the AuNRs increased as shown in the green regions. The temperature at the inter-particle gap (white region) showed curves profile which centered at 408 °C and 307 °C for AuNRs pair in air and water, respectively. The temperature curves are larger than AuNRs pair with inter-particle gap of 5 nm.

3.5. Water splitting in CeO2 based thermochemical reaction

Water splitting in CeO_2 (ceria) based thermochemical reaction consists of two processes; (1) the reduction of CeO_2 to CeO_{2-x}, and (2) the oxidation of CeO_{2-x} to CeO_2 [27]. In the first process, oxygen vacancies sites are formed, and this is followed by re-oxidation of CeO_{2-x} which can be induced by exposing the reduced ceria to water, and resulting in hydrogen production [37, 38]. Heat is essential in these processes. The heat generated from AuNRs pair can be utilised to drive the redox reactions. Based on the simulation results on the nearfield intensities and distributions, as well as, the thermal modelling of AuNRs pair, the optimum combination for water splitting in CeO_2 based thermochemical reaction is by coupling the AuNRs pair with CeO_2/SiO_2 substrate and with AuNRs inter-particle gap of 5 nm, and CeO_2 thickness of 10 nm (i.e. AuNRs/CeO_2-T10/SiO_2 model). In addition, a charge injection at AuNRs/CeO_2 interfaces coalesced with heat generation by the AuNRs pair at high temperature enhances the redox reaction of CeO_2. The water splitting process in CeO_2 based thermochemical reaction can be described by the two reactions below:

Thermal reduction of CeO_2 in air,

$$\frac{1}{x}CeO_2 \rightarrow \frac{1}{x}CeO_{2-x} + \frac{1}{2}O_2(g)$$

(2) Thermal oxidation of reduced CeO_{2-x} by water (vapor),

$$H_2O(g) + \frac{1}{x}CeO_{2-x} \rightarrow \frac{1}{x}CeO_2 + H_2(g)$$

According to temperature programmed reduction (TPR) of CeO_2, the thermal reduction of CeO_2 can occur at temperature of 497 °C (surface) and 827 °C (bulk) [39]. The AuNRs pair is capable of producing temperature of 1056 °C with light input power of 0.017 μW in air (Figure 7). Hence, surface and bulk reduction of CeO_2 to CeO_{2-x} (reaction 1) can be induced. The thermal oxidation of CeO_{2-x} by water (reaction 2) can occur at temperature of about 800 °C (bulk) [37, 40]. The heat produced by the AuNRs pair can reach a temperature of 976 °C in water with a given light input power of 8.0 μW (Figure 7). Therefore, the AuNRs pair in AuNRs/CeO_2-T10/SiO_2 model with AuNRs interparticle gap of 5 nm provides the most favorable conditions for splitting of water in CeO_2 based thermochemical reaction.

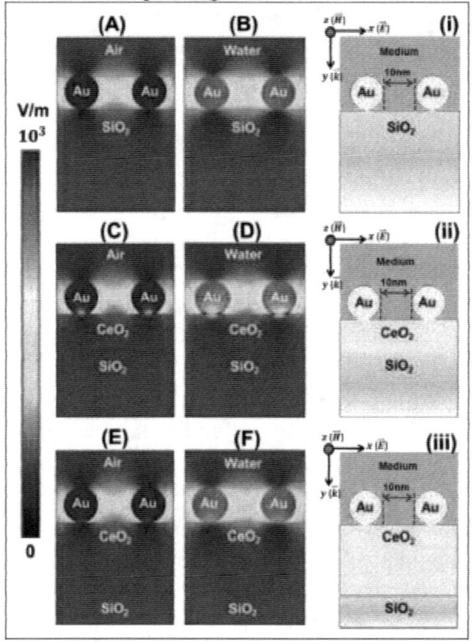

Figure 6. Near-field intensities and distributions of AuNRs pair with inter-particle gap of 10 nm for AuNRs/SiO_2 in air (A), and water (B); AuNRs/CeO_2-T10/SiO_2 in air (C), and water (D); and AuNRs/CeO_2-T20/SiO_2 in air (E), and water (F). The schematic design of AuNRs/SiO_2 (i), AuNRs/CeO_2- T10/SiO_2 (ii), and AuNRs/CeO_2-T20/SiO_2 (iii).

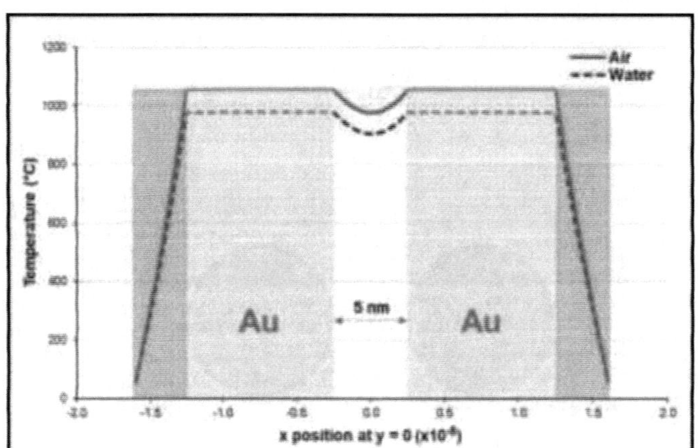

Figure 7. Temperature profiles for AuNRs pairs with inter- particle gap of 5 nm in air (0.017 μW) and water (8.0 μW) at light wavelength of 530 nm. The x-axis indicated the position and size of AuNRs pair by using coordinates (x, y) system.

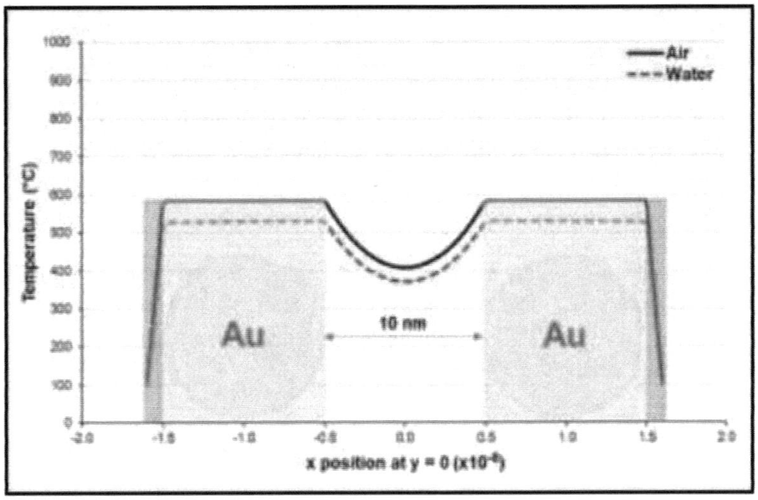

Figure 8. Temperature profiles for AuNRs pairs with inter- particle gap of 10 nm in air (0.017 μW) and water (8.0 μW) at light wavelength of 530 nm. The x-axis indicated the position and size of AuNRs pair by using coordinates (x, y) system.

4. Conclusion

In conclusion, under incident EM wave excitation, the nearfield intensities distribution and plasmon resonance peak wavelength related to LSPR and GPR of the short axis of AuNRs pair changed with the refractive index of the surrounding medium and adjacent surface. A tunable wavelength light source is required to excite SP resonance at a precise wavelength for optimal occurrence of LSPR and GPR, as well as, the local E-field enhancement along the short axis of the AuNRs pair. In addition, the local E-field enhancements of the AuNRs pair in AuNRs/SiO_2, AuNRs/CeO_2-T10/SiO_2, and AuNRs/CeO_2-T20/SiO_2 models with inter-particle gap of 5 nm in water media can be employed to visible Raman spectroscopy as molecules detector in water solution. In water splitting using CeO_2 based thermochemical reaction, light input power control is crucial for AuNRs pair to produce heat for redox reaction in CeO_2/CeO_2 system without melting the AuNRs pair. The optimum conditions for the water splitting are achieved in AuNRs/CeO_2-T10/SiO_2 model with AuNRs pair inter-particle gap of 5 nm. Thoroughly, the AuNRs pair inter-particle gap, support, medium and light input power all play important roles in resolving specific applications for the local E-field enhancement and thermal generation from LSPR and GPR of AuNRs pair.

Acknowledgement

This work was fully supported by Universiti Brunei Darussalam grants UBD/RSCH/1.9/FICBF(b)/2018/005 and UBD/RSCH/1.9/FICBF(b)/2018/002. The authors would also like to thank the support by AUN-SEED/Net Japan ASEAN Collaborative Education Programme (JACEP) under the CEP 2019.

References

[1] Albero J, García H. 2.05 - Catalysis by Supported Gold Nanoparticles. In: Andrews DL, Lipson RH, Nann T, editors. Comprehensive Nanoscience and Nanotechnology (Second Edition). Oxford: Academic Press; 2019. p. 91-108.

[2] Chen X, Zhu H, Groarke RJ. Catalysis by Supported Gold Nanoparticles. Reference Module in Materials Science and Materials Engineering: Elsevier; 2016.

[3] Wu W, Han S-T, Venkatesh S, Sun Q, Peng H, Zhou Y, et al. Biodegradable skin-inspired nonvolatile resistive switching memory based on gold nanoparticles embedded alkali lignin. *Org Electron.* 2018;**59**:382-8.

[4] Cho ES, Kim J, Tejerina B, Hermans TM, Jiang H, Nakanishi H, et al. Ultrasensitive detection of toxic cations through changes in the tunnelling current across films of striped nanoparticles. *Nature Materials*. 2012;**11**(11):978-85.
[5] Branagan D, Breslin CB. Electrochemical detection of glucose at physiological pH using gold nanoparticles deposited on carbon nanotubes. *Sensors Actuators B: Chem*. 2019;**282**:490-9.
[6] Ly TN, Park S. Highly sensitive gas sensor using hierarchically self-assembled thin films of graphene oxide and gold nanoparticles. *Journal of Industrial and Engineering Chemistry*. 2018;**67**:417-28.
[7] Churchyard S, Fang X, Vrcelj R. Laser ignitibility of energetic crystals doped with gold nanoparticles. *Optics & Laser Technology*. 2019;**113**:281-8.
[8] Elahi N, Kamali M, Baghersad MH. Recent biomedical applications of gold nanoparticles: A review. *Talanta*. 2018;**184**:537-56.
[9] Amendola V, Pilot R, Frasconi M, Maragò OM, Iatì MA. Surface plasmon resonance in gold nanoparticles: a review. *J Phys: Condens Matter*. 2017;**29**(20):203002.
[10] Dreaden EC, Alkilany AM, Huang X, Murphy CJ, El-Sayed MA. The golden age: gold nanoparticles for biomedicine. *Chem Soc Rev*. 2012;**41**(7):2740-79.
[11] Daniel M-C, Astruc D. Gold nanoparticles: assembly, supramolecular chemistry, quantum-size-related properties, and applications toward biology, catalysis, and nanotechnology. *Chem Rev*. 2004;**104**(1):293-346.
[12] Amendola V, Meneghetti M. Laser ablation synthesis in solution and size manipulation of noble metal nanoparticles. *PCCP*. 2009;**11**(20):3805-21.
[13] Maier SA. Plasmonics: fundamentals and applications: Springer Science & Business Media; 2007.
[14] Xia Y, Halas NJ. Shape-controlled synthesis and surface plasmonic properties of metallic nanostructures. *MRS Bull*. 2005;**30**(5):338-48.
[15] Wi J-S, Park J, Kang H, Jung D, Lee S-W, Lee TG. Stacked gold nanodisks for bimodal photoacoustic and optical coherence imaging. *ACS nano*. 2017;**11**(6):6225-32.
[16] Zohar N, Chuntonov L, Haran G. The simplest plasmonic molecules: Metal nanoparticle dimers and trimers. *Journal of Photochemistry and Photobiology C: Photochemistry Reviews*. 2014;**21**:26-39.
[17] Mayer KM, Hafner JH. Localized surface plasmon resonance sensors. *Chem Rev*. 2011;**111**(6):3828-57.
[18] Link S, Mohamed M, El-Sayed M. Simulation of the optical absorption spectra of gold nanorods as a function of their aspect ratio and the effect of the medium dielectric constant. *The Journal of Physical Chemistry B*. 1999;**103**(16):3073-7.
[19] Cao J, Sun T, Grattan KT. Gold nanorod-based localized surface plasmon resonance biosensors: A review. *Sensors Actuators B: Chem*. 2014;**195**:332-51.
[20] Okuno M, Tokimoto T, Eguchi M, Kano H, Ishibashi T-a. Intensity enhancement of vibrational sum frequency generation by gap-mode plasmon resonance. *Chem Phys Lett*. 2015;**639**:83-7.
[21] Chau Y-F, Jiang Z-H, Li H-Y, Lin G-M, Wu F-L, Lin W-H. Localized resonance of composite core-shell nanospheres, nanobars and nanospherical chains. *Progress In Electromagnetics Research*. 2011;**28**:183-99.
[22] Chau Y-F, Yeh H-H, Tsai DP. Near-field optical properties and surface plasmon effects generated by a dielectric hole in a silver-shell nanocylinder pair. *Appl Opt*. 2008;**47**(30):5557-61.
[23] Chau Y-F, Jiang Z-H. Plasmonics effects of nanometal embedded in a dielectric substrate. *Plasmonics*. 2011;**6**(3):581-9.
[24] Chau Y-F, Tsai DP. Three-dimensional analysis of silver nano-particles doping effects on super resolution near-field structure. *Opt Commun*. 2007;**269**(2):389-94.
[25] Kats MA, Yu N, Genevet P, Gaburro Z, Capasso F. Effect of radiation damping on the spectral response of plasmonic components. *Opt Express*. 2011;**19**(22):21748-53.
[26] Govorov AO, Richardson HH. Generating heat with metal nanoparticles. *Nano today*.

2007;**2**(1):30-8.

[27] Bhosale RR, Takalkar G, Sutar P, Kumar A, AlMomani F, Khraisheh M. A decade of ceria based solar thermochemical H2O/CO2 splitting cycle. *Int J Hydrogen Energy*. 2019;**44**(1):34-60.

[28] Tian Y, Tatsuma T. Plasmon-induced photoelectrochemistry at metal nanoparticles supported on nanoporous TiO 2. *Chem Commun*. 2004(16):1810-1.

[29] Johnson PB, Christy R-W. Optical constants of the noble metals. *Physical review B*. 1972;**6**(12):4370.

[30] Okamoto T. Near-field spectral analysis of metallic beads. *Near-field optics and surface plasmon polaritons*. 2001:97-123.

[31] Gresho PM, Sani RL. Incompressible flow and the finite element method. Volume 1: Advection-diffusion and isothermal laminar flow. 1998.

[32] Janicki V, Zorc H. Refractive index profiling of CeO2 thin films using reverse engineering methods. *Thin Solid Films*. 2002;**413**(1-2):198-202.

[33] Malitson IH. Interspecimen comparison of the refractive index of fused silica. *Josa*. 1965;**55**(10):1205-9.

[34] Engineering Toolbox. Air - Density at varying pressure and constant temperatures 2004 [Available from: https://www.engineeringtoolbox.com/air-temperaturepressure-density-d_771.html.

[35] Idris MNSiM, Chiang H-P, Muslim N, Chau Y-FC, Mahadi AH, Voo NY, et al. Raman Spectrometry of Scattering of Nano-Gold Ceria Films. *Advanced Science Letters*. 2018;**24**(11):8940-3.

[36] Tian Y, Tatsuma T. Mechanisms and applications of plasmon-induced charge separation at TiO2 films loaded with gold nanoparticles. *J Am Chem Soc*. 2005;**127**(20):7632-7.

[37] Zhao Z, Uddi M, Tsvetkov N, Yildiz B, Ghoniem AF. Redox kinetics study of fuel reduced ceria for chemical-looping water splitting. *The Journal of Physical Chemistry C*. 2016;**120**(30):16271-89.

[38] Scheffe JR, Welte M, Steinfeld A. Thermal reduction of ceria within an aerosol reactor for H2O and CO2 splitting. *Ind Eng Chem Res*. 2014;**53**(6):2175-82.

[39] Trovarelli A. Catalytic properties of ceria and CeO2-containing materials. *Catalysis Reviews*. 1996;**38**(4):439-520.

[40] Chueh WC, Falter C, Abbott M, Scipio D, Furler P, Haile SM, et al. High-flux solar-driven thermochemical dissociation of CO2 and H2O using nonstoichiometric ceria. *Science*. 2010;**330**(6012):1797-801.

IOP Conf. Series: Materials Science and Engineering 1127 (2021) 012006 doi:10.1088/1757-899X/1127/1/012006

Thermal Energy Storage in Phase Change Material Integrated Solar Collectors for Air Heating Application

V. V. Tyagi[1*], Dheeraj Kumar Nagilla[1], Jeyraj Selvaraj[2], Kapil Chopra[1,3], Richa Kothari[4], A. K. Pandey[5]

[1]School of Energy Management, Shri Mata Vaishno Devi University, Katra, Jammu & Kashmir, India-182320

[2]UM Power Energy Dedicated Advanced Center (UMPEDAC), Level 4, Wisma R&D, University of Malaya, 59990 Kuala Lumpur, Malaysia

[3]School of Mechanical Engineering, Shri Mata Vaishno Devi University, Katra, Jammu & Kashmir, India-182320

[4] Department of Environmental Sciences, Central University of Jammu, Jammu Bagla (Rahya Suchani) Samba (J&K), India-180001

[5]Research Centre for Nano-Materials and Energy Technology (RCNMET), School of Science and Technology, Sunway University, No. 5, JalanUniversiti, Bandar Sunway, Petaling Jaya, 47500 Selangor Darul Ehsan, Malaysia

E-mail: v.tyagi@smvdu.ac.in, vtyagi16@gmail.com

Abstract. Solar energy is an abundant source of renewable energy which can able to support the expansion of energy demand. This review paper represents a complete literature review on recent developments in the phase change material (PCM) based solar collectors for air heating applications such as building heating, greenhouse heating, etc. In this paper thermal efficiency of various types of storage based collector and their performance for air heating (SAH) applications have been discussed. Based on the literature, it has been found that compared to basic solar air heating (SAH) systems, the collector-storage based air heating systems are able to provide higher efficiency with their capacity to recover and utilize the stored heat at night or cold weather conditions. These storage based collectors can be efficiently utilized for applications requiring medium operating temperature range during daytime as well as night times.

1. Introduction

Among all the renewable energy sources available on the earth, solar energy is mostly available. Everyday earth receives a large quantity of solar energy from the sun. Of the total solar energy received by the earth, only a part of it is used by the world in the form of photosynthesis and day-lighting and one-third gets reflected into space and the remaining quantity is received by oceans, land, and clouds. Therefore, it is appropriate to collect solar energy and utilize it for the production of electricity, heat and cooling applications. This way of using solar energy for energy production has a nominal impact on the environment without releasing of harmful pollutants into the atmosphere. Also, apart from the environmental concern, solar energy is also termed as an appropriate source of energy to fill the gap between energy demand and supply due to the depletion of fossil fuels. Many researchers have explored and developed various technologies for the collection and usage of solar energy and still working on various technologies in order to increase the performance of the system [1,2]. Although, there are distinct confronts in the process of collection and storage of solar energy even though it is free. As we know that the availability of solar energy is only during the daytime, it must be collected and stored efficiently to recover and to use at times when solar energy is unavailable. Solar collector converts the absorbed solar energy into thermal energy which is transferred to working fluid. Hence, solar collectors are the principal and very crucial units of any solar thermal energy system. Essentially collectors are of two types, stationary and tracking [2] (Figure 1).

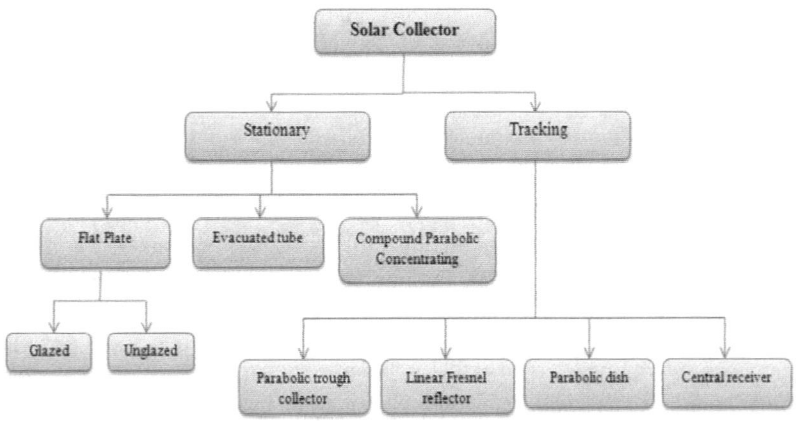

Figure 1. Types of solar collectors [1]

Substantial temperature ranges can be acquired by various collectors such as flat plate solar collector (FPSC) operates in the temperature range of 20-80 °C and evacuated tube solar collector (ETSC) works in the temperature range of 50-200 °C [3,4]. Of all the solar collectors, FPSCs are simple in design, require low maintenance cost, more productive and largely used but delivers less efficiency and outlet temperatures. On the other hand, ETSCs delivers a higher range of outlet temperatures and higher efficiencies compared to FPSCs. Solar collectors are being used for various domestic and industrial applications such as solar water heating [5], solar air heating, solar air conditioning, solar cooking, solar drying, solar desalination [6], steam generation and other applications [7,8]. This paper primarily focused on documentation of recent explorations on the integration of phase change material in the solar collector for air heating applications. Hence, this review primarily explores various storage based collector designs, performance for air heating applications in domestic and industrial sectors and also discussed the aspects that control the efficiency of both collector and LHTES unit, challenges of collector-storage usage along with economic considerations concerning the utilization of collector-storage system for air heating application. Till now, there is no review on collector-storage based air heating systems and according to the author's estimation, this is the first review paper on recent developments in the phase change material based solar collectors for air heating applications.

2. Employment of phase change materials in solar collectors for air heating applications

Air heating is one of the major energy-consuming processes in various applications such as building heating, greenhouse heating, food, textile, paper, pharmaceutical industries, etc. There are various types of collectors available in the market depending on the requirement of the operating temperature range. Since air heating application requires only a medium operating temperature range, that can be fulfilled by FPSCs and ETSCs [9]. Since solar energy is available only during the daytime, so there is a need for storage of solar energy to utilize during the night time or at a time when solar radiation is unavailable. Solar thermal energy storage systems store solar energy in the form of thermal energy as sensible heat, latent heat or the combination of both. Sensible heat thermal energy storage (SHTES) materials such as water, oils, molten salts, rocks, liquid metals store the thermal energy by raising their temperature. They make use of the heat capacity and change in temperature during the charging and discharging process. The specific heat of the material, temperature change and quantity of storage material determine the quantity of heat stored [10]. SHTES materials are cheap, but its energy density is low and it discharges at varied temperature ranges. Latent heat thermal energy storage (LHTES) materials also called as phase change materials such as mineral salts, paraffin, fatty acids, esters, etc., store/release thermal energy as latent heat based on the absorption or release of heat when it undergoes a phase change from solid-liquid or vice versa [11]. However, compared to the SHTES material, PCM possesses nearly constant operating temperature, high energy storage density and better heat transfer performance which are desirable properties for efficient thermal energy storage. Due to the better thermo-physical properties, phase change materials are utilized in various types of solar collectors to store thermal energy for later use [12]. But one of the major disadvantages with the phase change material is lower thermal conductivity. The thermal conductivity of phase change material can be enhanced by adding highly conductive materials in it [13,14]. Nanoparticles that possess a high thermal conductivity can be immersed in the phase change material which finally forms a novel nanocomposite phase change material with enhanced thermo-physical properties [15,16,17]. By taking advantage of the above-mentioned characteristics of PCM's, various researchers developed and studied various solar air heating systems by combining solar collectors with the phase change material. Benli et al., 2009; [18] replaced the conventional heating with a PCM ($CaCl_2 6H_2O$) storage unit to provide space heating during the night time or extremely cold days in Turkey. They investigated PCM's performance by considering the charging and discharging processes and also analyzed 5 different types of 10 solar collector units along with a PCM unit. Solar collector heats the air and then the hot air is passed into the PCM unit where it charged and then enters into the ambient air of the greenhouse. The results disclose that there is acquirement of about 6-9 °C of the temperature difference between inside and outside of the greenhouse and storage based system maintain this temperature difference up to 3-4 hours more than with the usual heating system. This reported that a storage based collector air heating system is very efficient during nights or extremely cold days in Turkey.

Figure 2. Experimental setup of greenhouse heating system using solar air collectors [18]

Arkar et al., 2005; [19] developed and investigated the performance of air heating system consisting of evacuated tube collectors and Rubitherm PX-21 as phase change material. They experimentally and numerically investigated the process of heat transfer in both collector and LHTES system and conducted the parametric analysis of the whole system for heating of an energy-efficient building which has a large share of passive heating. The evacuated tube solar collector arranged in series-parallel connection with $0.71 m^2$ of aperture area was placed on the building's roof as shown in the fig.3 and the Rubitherm PX-21 PCM with the specific mass of 5.17 kg/m is poured between the tubes of an aluminum concentric-tube whose dimensions are based on the quantity of PCM required.

The whole system showed in fig.4 is arranged in a way that the building's indoor air is heated up in the collectors and the hot air is made to pass the LHTES system made of concentric tubes where PCM is placed which absorb the heat from the hot air and stored heat as latent heat. Then the cooled air again enters into the collector and the cycle repeats. The latent heat, which is stored in the LHTES system can be utilized during night time for building heating purpose by allowing indoor air directly into the LHTES system by passing the solar collectors which can be repeated in short cycles. In this study quasi-dynamic method was used instead of steady-state method. By validating the experimental results with numerical results, it stated that the mathematical models appropriately specify the conduct of heat transfer in the solar collector. As conduction mode of heat transfer is the governing process in the LHTES concentric-tube as shown in fig.5, they examined heat transfer process mathematically by creating a conduction-dominated model along the unidimensional airflow direction by neglecting conduction along the axial direction due to lower thermal conductivity of PCM and by adopting the explicit method and finite-difference. Due to the microencapsulation of PCM, the convective mode of heat transfer in the liquid phase of PCM is not considered in this study. Using heat capacity method, by considering the PCM's melting temperature, latent heat and other thermo-physical properties of PCM they plotted the melting and solidification process of PCM. They found the negligible difference between experimentally measured and numerically calculated results. The maximum experimentally measures values of the air inlet, air outlet, PCM inlet, PCM outlet temperature were 73°C, 49°C, 46°C and 33 °C whereas maximum numerically calculated values are 73°C, 47°C, 45°C and 36°C, respectively. This study reveals that of the total produced heat during the daytime was about 54-67% of the recovered heat at the night time for building heating.

Figure 3. Evacuated tube solar air collector installed on a building's roof [19]

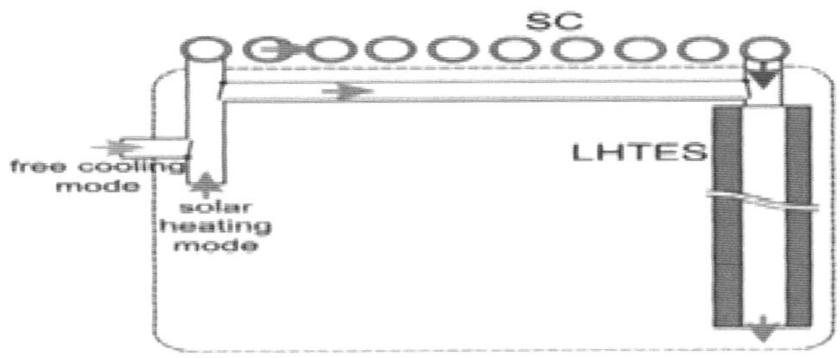

Figure 4. Solar Collector-LHTES storage system [19]

Figure 5. Aluminum Concentric tube LHTES system [19]

Belmonte et al., 2016; [20] numerically studied the performance of storage based solar air heating system which is comprised of a solar collector integrated with paraffin wax as PCM (as shown in the fig.6). In comparison to the packed bed of rocks/pebbles which are SHTES materials, phase change material based fluidized bed latent heat storage system charges /discharges faster with high energy density. From the collector, the hot air passes through the LHTES fluidized bed which is sufficient to fluidize the LHTES unit. They analyzed the effect of PCM's melting temperature, capacity of storage, collector's area, envelope of building on the contribution of solar energy. The results of this study admitted that this type of system is sufficient enough to supply the energy required for the house of a single-family in mild winter. They also found that for the collectors having the areal range of $10 - 20\text{m}^2$, the optimal airflow rate of about $100\,\text{m}^3/\text{hm}^2$ is appropriate to fluidize 2000 kg PCM fluidized bed. The TRNSYS simulation results unfolded that a single-family house in regions such as Barcelona and Madrid can utilized this collector-storage system with 2000 kg of PCM is sufficient to satisfy their 50% daily energy requirement in the mild winter. This type of PCM fluidized bed storage integrated with solar air collector is appropriate to reduce the building heating requirements using solar energy.

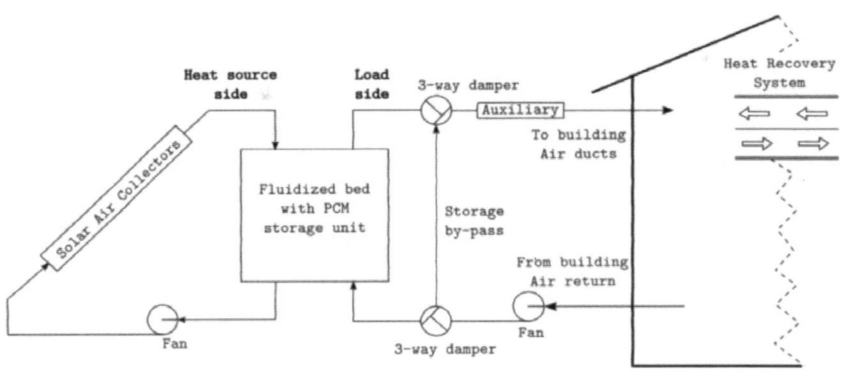

Figure 6. Solar collector-storage air heating system [20]

In order to increase the share of using solar energy for heating buildings Arkar et al., 2016; [21] developed an air heating system comprising of solar collectors and Rubitherm PX-21 PCM unit which is shown in fig.7. By developing a thermal response model for a building, they numerically studied the system's design and thermal performance by considering parameters such as solar heating share and solar heat usability. They mentioned that it is mandatory to integrate energy storage system with the solar air heating system of the building. They found that using solar collector which is 1/6th of the building floor area integrated with Rubitherm PX-21 as PCM can able to supply 34 kWh/m2 of energy to the building.

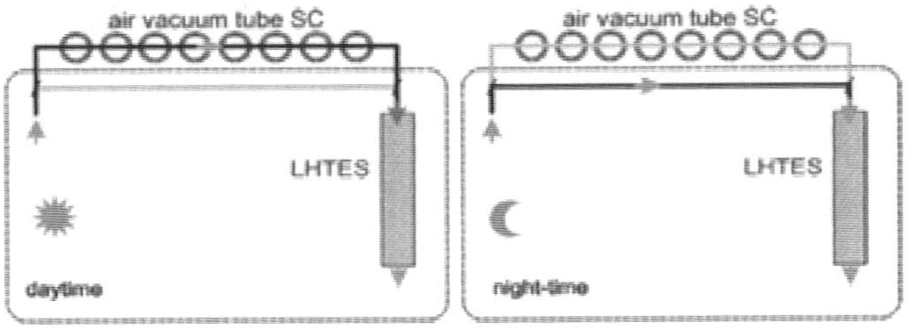

Figure 7. Design of evacuated tube collector and PCM storage unit [21]

Wang et al., 2017; [22] experimentally investigated the performance of a solar air heating system comprised of solar collector with flat micro-heat pipe arrays (FMHPA) and PCM unit which is shown in fig.8. In this study solar collector area of 0.93 m^2 and Paraffin wax as PCM is considered. During PCM's charging and discharging process, they tracked the distribution of temperature to observe the melting and solidification front. They calculated the efficiency and power during charging and discharging of PCM. This study comes out with a conclusion that of the charging and discharging power level of 393 and 344W, the obtained efficiencies were 59% and 91.6%, respectively. It was found that when the heat transfer fluid's flow rate is increased from 100m^3/h and 200m^3/h, the heat power of extraction of heat will rise by 10% and 26% respectively and the time of heat extraction will be lowered by 8% and 20%, respectively. There was loss of heat taking place form the glass cover and absorber through convection during charging and during discharging from the insulation through conduction.

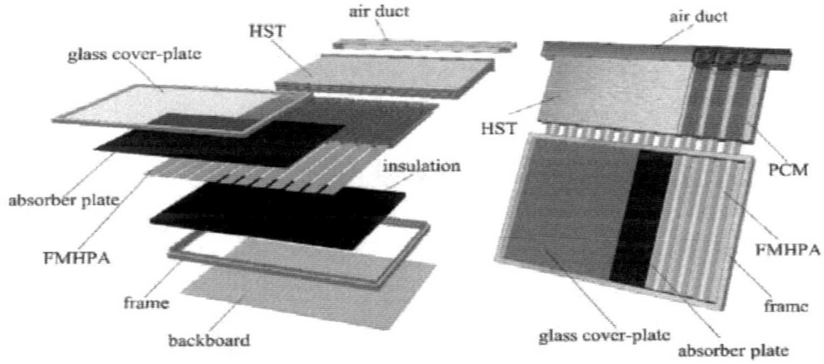

Figure 8. SAH system with FMHPA based solar collector-LHTES storage [22]

Agathokleous et al., 2019; [23] investigated the design and performance of flat plate solar air collector as shown in fig.9. Compared to the current collectors, the uniqueness of this system was embodied by the lesser initial cost and implemented the simple design. This system was well designed to assimilate with the building envelope. Along with energy performance, they also evaluated the economic performance of the system by creating a dynamic simulation model. The results revealed that considered collector having the shortest payback period of 6.2 years.

Figure 9. Exterior design and interior configuration of SAH system [23]

Khadraoui et al., 2016; [24] analyzed the performance of solar air heating system (SAH) comprising of flat plate solar air collector and paraffin wax PCM unit as shown in fig.10. The SAH system with PCM unit is compared with another SAH without the PCM unit to check the influence of PCM on the performance of the system. The results indicated that due to the presence of PCM unit, there is an increment in the outlet temperature of the SAH system. The daily energy efficiency of the SAH system without PCM and with PCM was 17% and 33%, respectively. They also found that during night time, in the SAH-PCM system there is an increase in outlet temperature of air by 3°C to 7°C compared to the simple SAH system. Also during the night time, due to the elevation of an outlet air temperature of SAH system with PCM, dehumidification of the ambient space taken place by increasing the evaporative capacity of air. This type of system is more favorable for drying purposes at night.

Figure 10. Experimental setup of SAH system with FPC collector-PCM unit [24]

Wang et al., 2017; [25] experimentally studied the performance of SAH system containing vacuum tubes, lauric acid PCM and flat micro-heat pipe arrays (FMHPA) as shown in fig.11. They utilized FMHPA to increase the rate of heat transfer in the system. In this test collector's, efficiency, PCM's charging and discharging time were calculated for different airflow rates. The results of their tests concluded that the enhancement in the collector's efficiency, heat transfer rate and decline of PCM's charging and discharging time took place at a higher airflow rate. They found that at high airflow rate, active air distribution increases the heat transfer rate, decreases the heat loss and decreases the charging and discharging time by increasing the rate of charging and discharging. At 240 m^3/h airflow rate, there is an achievement of the least charging time of 134 min at 682 W instantaneous heat transfer rate and least discharging time of 153 min. During discharging, they found that there is a provision of steady outlet temperature at 60 m^3/h airflow rate.

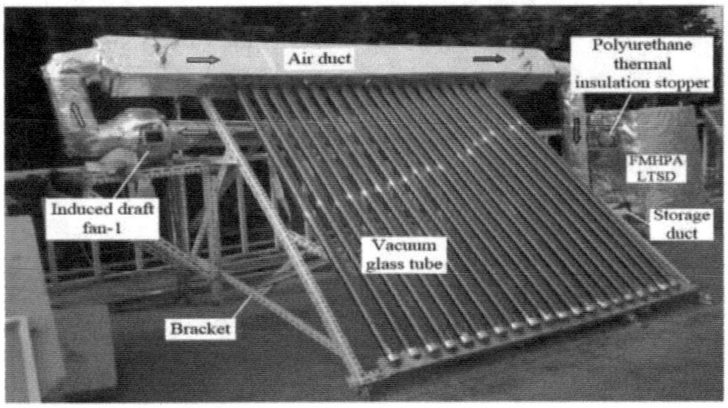

Figure 11. Solar collector-storage air heating system [25]

Arfaoui et al., 2017; [26] experimentally studied the performance of solar air heating (SAH) system as shown in fig.12 containing solar air collector and 2-packed bed of ($CaCl_26H_2O$) PCM spherical capsules aiming to raise the efficiency of the system. During experiments, they took the measurements of solar radiation, wind speed, relative humidity, the temperature of the inlet and outlet air and distribution of PCM packed bed temperature. To evaluate the PCM's performance they used the measured values and calculated the energy collected, stored and used during the process of charging and discharging of PCM. The SAH- 2 packed bed system acquired a daily energy efficiency of 47%. They also found that as the absorbed solar energy increases, there is an increment in the stored heat in PCM during the charging process. During the discharging process for night time, the system discharged 0.3 kW useful heat at a constant outlet air temperature of 27°C. They also found that the system is more productive for night hours during the discharging process by recovering the stored energy.

Figure 12. Experimental of SAH-2 packed bed PCM system [26].

Abuska et al., 2019; [27] developed a solar air collector combining flat plate solar collector with Rubitherm RT54HC PCM and aluminum honeycomb as shown in fig. 13. They studied the effect of the aluminum honeycomb (PCM's thermal conductivity enhancer) on the performance of the solar collector-PCM storage system under conditions of natural convection. They compared three different types of collectors which are 26 kg PCM integrated honeycomb collector (type I), a 26 kg PCM integrated collector (type II) and a simple flat plate air collector (type-III). They found that under daylight, the thermal efficiency of type-I, type-II and type-III of solar collectors were 13.6%, 10.9%, and 10.1%, respectively. Therefore, the aluminum honeycomb as a heat transfer enhancer is very effective, particularly during the discharging process. They also found that compared to type III collector, the energy production in type I and type II continued for a further 469 min and 539 min, respectively.

Figure 13. Experimental setup of solar air collector [27]

3. Conclusion

This paper offers an outline of the latest investigations on the performance of SAH system by integrating solar collector with the PCM unit and disclosed that this innovative system has vast possibilities in residential, industrial and agricultural sectors. Solar air heating systems have been used for heating energy-efficient buildings, greenhouses, food, paper, timber, and pharmaceutical industries and for many other uses. A collector-storage air heating system is extremely endorsed for medium temperature applications as they can absorb, store solar energy as latent heat during daytime and can recover heat at night or at cold weather conditions. The performance of the SAH system using various PCM based LHTES systems is also presented and found that the storage based collector air heating system responds greatly with the inclusion of PCMs.

Acknowledgements

The authors would like to thank the financial and technical assistance provided by AUN-SEED/Net Japan ASEAN Collaborative Education Programme (JACEP). One of the author (V.V. Tyagi) gratefully acknowledges the PDA grant of Shri Mata Vaishno Devi University Katra.

References

[1] Sabiha MA, Saidur R, Mekhilef S, Mahian O. Progress and latest developments of evacuated tube solar collectors. 2015 Nov 1. Renewable and Sustainable Energy Reviews, 51, 1038-1054.

[2] Kalogirou SA. Solar thermal collectors and applications. 2004 Jan 1. Progress in energy and combustion science, 30(3), 231-295.

[3] Kim YS, Balkoski K, Jiang L, Winston R. Efficient stationary solar thermal collector systems operating at a medium-temperature range. 2013 Nov 1. Applied energy, 111, 1071-1079.

[4] Kalogirou, Soteris A. Solar energy engineering: processes and systems. 2013 Oct 25. Academic Press.

[5] Hossain MS, Saidur R, Fayaz H, Rahim NA, Islam MR, Ahamed JU, Rahman MM. Review on solar water heater collector and thermal energy performance of circulating pipe. 2011 Oct 1. Renewable and sustainable energy reviews, 15(8), 3801-3812.

[6] Mosleh HJ, Mamouri SJ, Shafii MB, Sima AH. A new desalination system using a combination of heat pipe, evacuated tube and parabolic trough collector. 2015 July 15. Energy conversion and management, 99, 141-150.

[7] Chopra K, Tyagi VV, Pandey AK, Sari A. Global advancement on experimental and thermal analysis of evacuated tube collector with and without heat pipe systems and possible applications. 2018 Oct 15. Applied energy, 228, 351-389.
[8] Kalogirou, Soteris. The potential of solar industrial process heat applications. 2003 Dec 1. Applied Energy, 76(4), 337-361.
[9] Sharma A, Tyagi VV, Chen CR, Buddhi D. Review on thermal energy storage with phase change materials and applications. 2009 Feb 1. Renewable and Sustainable energy reviews, 13(2), 318-345.
[10] Zalba B, Marın JM, Cabeza LF, Mehling H. Review on thermal energy storage with phase change: materials, heat transfer analysis and applications. 2003 Feb 1. Applied thermal engineering, 23(3), 251-283.
[11] Sharma RK, Ganesan P, Tyagi VV, Metselaar HS, Sandaran SC. Developments in organic solid–liquid phase change materials and their applications in thermal energy storage. 2015 May 1. Energy Conversion and Management, 95, 193-228.
[12] Ferrer G, Solé A, Barreneche C, Martorell I, Cabeza LF. Review on the methodology used in thermal stability characterization of phase change materials. 2015 Oct 1. Renewable and Sustainable Energy Reviews, 50, 665-685.
[13] Kenisarin M, Mahkamov K. Solar energy storage using phase change materials. 2007 Dec 1. Renewable and sustainable energy reviews, 11(9), 1913-1965.
[14] Rathod MK, Banerjee J. Thermal stability of phase change materials used in latent heat energy storage systems: a review. 2013 Feb 1. Renewable and sustainable energy reviews, 18, 246-258.
[15] Khodadadi JM, Hosseinizadeh SF. Nanoparticle-enhanced phase change materials (NEPCM) with great potential for improved thermal energy storage. 2007 May 1. International communications in heat and mass transfer, 34(5), 534-543.
[16] Kibria MA, Anisur MR, Mahfuz MH, Saidur R, Metselaar IH. A review on thermophysical properties of nanoparticle dispersed phase change materials. 2015 May 1. Energy Conversion and Management, 95, 69-89.
[17] Sharma RK, Ganesan P, Tyagi VV, Metselaar HS, Sandaran SC. Thermal properties and heat storage analysis of palmitic acid-TiO2 composite as nano-enhanced organic phase change material (NEOPCM). 2016 Apr 25. Applied Thermal Engineering, 99, 1254-1262.
[18] Benli H, Durmuş A. Performance analysis of a latent heat storage system with phase change material for new designed solar collectors in greenhouse heating. 2009 Dec 1. Solar Energy, 83(12), 2109-2119.
[19] Arkar C, Medved S. Optimization of latent heat storage in solar air heating system with vacuum tube air solar collector. 2015 Jan 1. Solar energy, 111, 10-20.
[20] Belmonte JF, Izquierdo-Barrientos MA, Molina AE, Almendros-Ibáñez JA. Air-based solar systems for building heating with PCM fluidized bed energy storage. 2016 Oct 15. Energy and Buildings, 130, 150-165.
[21] Arkar C, Šuklje T, Vidrih B, Medved S. Performance analysis of a solar air heating system with latent heat storage in a lightweight building. 2016 Feb 25. Applied Thermal Engineering, 95, 281-287.
[22] Wang ZY, Diao YH, Liang L, Zhao YH, Zhu TT, Bai FW. Experimental study on an integrated collector storage solar air heater based on flat micro-heat pipe arrays. 2017 Oct 1. Energy and Buildings, 152, 615-628.
[23] Agathokleous R, Barone G, Buonomano A, Forzano C, Kalogirou SA, Palombo A. Building façade integrated solar thermal collectors for air heating: experimentation, modelling and applications. 2019 Apr 1. Applied energy, 239, 658-679.
[24] El Khadraoui A, Bouadila S, Kooli S, Guizani A, Farhat A. Solar air heater with phase change material: An energy analysis and a comparative study. 2016 Aug 25. Applied Thermal Engineering, 107, 1057-1064.

[25] Wang TY, Diao YH, Zhu TT, Zhao YH, Liu J, Wei XQ. Thermal performance of solar air collection-storage system with phase change material based on flat micro-heat pipe arrays. 2017 Jun 15. Energy Conversion and Management, 142, 230-243.
[26] Arfaoui N, Bouadila S, Guizani A. A highly efficient solution of off-sunshine solar air heating using two packed beds of latent storage energy. 2017 Oct 1. Solar Energy, 155, 1243-1253.
[27] Abuşka M, Şevik S, Kayapunar A. Experimental analysis of solar air collector with PCM-honeycomb combination under the natural convection. 2019 Jun 15. Solar Energy Materials and Solar Cells, 195, 299-308.

IOP Conf. Series: Materials Science and Engineering 1127 (2021) 012007 doi:10.1088/1757-899X/1127/1/012007

Recent Advancements in Design of Flat Plate Solar Collectors

K. Chopra [a,b*], Pratik Kumar Pathak[a], Mahendran Samykano[c], V. V Tyagi[b], A. K Pandey[d]

[a]School of Mechanical Engineering,Shri Mata Vaishno Devi University,Katra, Jammu & Kashmir, India-182320

[b]School of Energy Management, Mata Vaishno Devi University, Katra, Jammu & Kashmir, India-182320

[c]Faculty of Mechanical Engineering, University Malaysia Pahang, Malaysia- 26600

[d]Research Centre for Nano-Materials and Energy Technology (RCNMET), School of Science and Technology, Sunway University, No. 5, Jalan Universiti, Bandar Sunway, Petaling Jaya, 47500 Selangor Darul Ehsan, Malaysia.

E-mail: kapil.chopra@smvdu.ac.in

Abstract. Solar collector is an important technology for the effective utilisation of solar energy that the earth is blessed with. Flat plate solar collectors present a simple and easy to maintain design and thus are widely used for low and medium temperature applications. But being less efficient than alternatives, justifying the initial investment of flat plate solar collectors becomes difficult in the long run. This paper presents the efforts of researchers in the past some years to improve the efficiency of flat plate solar collectors through the improvement and optimization of the existing design. The range of research work covered gives a general idea of the variety of techniques being developed, analysed and tested to increase the efficiency of flat plate solar collector through means such as new absorber design, design of absorber tubes, new coatings on glass cover, and other means to reduce heat transfer losses, increase heat transfer from absorber to working fluid and absorbing and retaining direct as well as diffuse radiation. Design and efficiency improvement for better adoption of flat plate collectors in building facades has also been discussed. This paper will be beneficial for exploring the range of research avenues in the field of optimization and efficiency improvement of flat plate solar collectors.

1. Introduction

The ever-rising demand for energy with rapid industrialization has put a strain on limited fossil reserves as well as our environment. This has compelled us to explore all possible avenues of renewable energy generation. Among renewables, the potential of solar energy to meet the energy demands is enormous, that too, without harming the environment. Various technologies have been and are being developed to exploit this potential [1].

The difficulty associated with solar energy is that though it is in abundance during the day, it is less than scarce at night and this necessitates that the collection of energy be efficient when it is available and the storage be effective so that it can be utilized at night. This difficulty is overcome using solar thermal collectors. One of the types of solar thermal collector is the Flat Plat Solar Collector (FPSC) which though has lower efficiency compared to other collectors but is widely used owing to its simple construction and cheap maintenance cost [2].

A flat plate solar collector comprises of a casing which houses the absorber plate and absorber tubes and the associated manifold. Insulation and glazing cover of a layer or layers of glass sheets are provided to minimize the loss of heat [3]. Glazing is essentialto prevent loss due to irradiation as it allows short-range radiation through while blocking long-range radiation [4].

The parameters which govern the performance of FPSCs are numerous, like design of the absorber and absorber tube, thermo-physical properties of the working fluid, its inlet conditions, insulation, ambient conditions such as temperature, irradiation, etc. Volume flow rate of the working fluid is also an important parameter governingcollector efficiency [5].To enhance the performance of FPSCs various methods have been employed such as augmenting the absorptivity of the absorber, maximizing heat transfer between absorber plate and tubes, decreasing boundary layer in receiver tubes thus enhancing convective heat transfer and so on [6,7].

Many efforts towards improving the collector efficiency through optimization of design parameters have been made. Optimization through exergy analysis [8], multi-attribute optimization through Taguchi ANOVA technique, TOPSIS technique, etc. have been carried out [9].The optimal size of channels of flow for micro-channel absorber design has been obtained through exergy analysis [10]. Design of riser and header manifold has been optimized through a correlation model [11].

To enable mass adoption, the endeavor to make FPSCs friendly towards building integration has been pursued by various researchers. Several designs have been proposed and analysed to overcome the challenges involved in building façade integration of FPSCs namely architectural compatibility, inclusion of colour, safety for overheat, etc. [12, 13].

Though an old concept, evacuated flat plate collectors are gaining prevalence among researchers due to the fact that evacuated flat plate collectors combine the advantages of both the simplicity of flat plate collectors and better efficiencies of evacuated collectors due to minimized heat loss and thus are being explored as viable replacement of concentrating collectors that are currently used in Organic Rankine Cycle [14, 15, 16].

2. Advancements in FPSC Designs

Recent developments in the design of FPSCs based on the aforementioned factors have been discussed in detail in the following text. Techniques covered include ways to augment heat transfer from absorber to absorber tubes and ultimately to the working fluid, minimise heat loss from the absorber due to convection by the use of transparent insulation material or by evacuating the collector, boost capture of radiation by means of sun-tracking reflectors or coatings on the glass cover and some optimisation techniques to find optimal design dimensions for various components vital to performance of the collector.

Balaji K. et al. [17] investigated experimentally absorber tubes employing copper rod and copper tube thermal performance enhancers, figure 1. The convection effect was studied and found to be mixed mode but predominantly free convection though with diminished Grashof, Rayleigh and Richardson numbers in comparison to absorbers without thermal performance enhancers. For all Reynolds number, thermal enhancer with rod yielded greater efficiency compared to thermal enhancer with tube and also conventional absorber design collector. Also, the modified Richardson numbers were reported to be 37.5% and 25% lower for the rod and tube enhancer respectively as compared to conventional collectors.Two FPSCs, one with and one without the enhancers were used to power solar water heaters and they were experimentally probed by placing them in the sunlight with a slope angle of 13°. Temperature sensors were placed at various points and the solar radiation was measured using a pyranometer. Other ambient factors were measured by installing a weather station in the vicinity. Data acquisition systems were employed to record the measurements at frequent intervals.

Figure 1. Rod and tube thermal performance enhancers.

T.Jatau et al. [18] employed Constructal Method of design to optimise the geometry of the absorber numerically by seeking to minimise entropy generation subjected to the constraints of fixed absorber area and fixed volume of absorber tubes while allowing spacing between the tubes, the length of the tubes and the diameter of the tubes to vary.For minimum entropy generation,optimal parameters of absorber design were obtained and further its variance with modified Reynolds number was studied and the authors concluded that the increase in modified Reynolds number results in an increase of optimal length of the tube while it results in the decrease of optimal diameter and optimal spacing of the tube, figure 2. As they also observed the entropy generation to be minimum and conductance to be maximum for increased modified Reynolds number, they thereby optimized the parameters based on this. The design of the FPSC was simulated in SolidWorks and computational fluid dynamics analysis was performed in ANSYS-Fluent R 16.0 which gave solutions for equations of mass, energy and momentum conservation integrated over each control volume. External irreversibilities were taken as constant and internal irreversibilities were minimized to obtain the optimum geometry. Heat flux incident on the absorber was taken as 500 W/m^2 and fluid entry temperature was taken as 300K for the numerical simulation that was done for the optimisation.

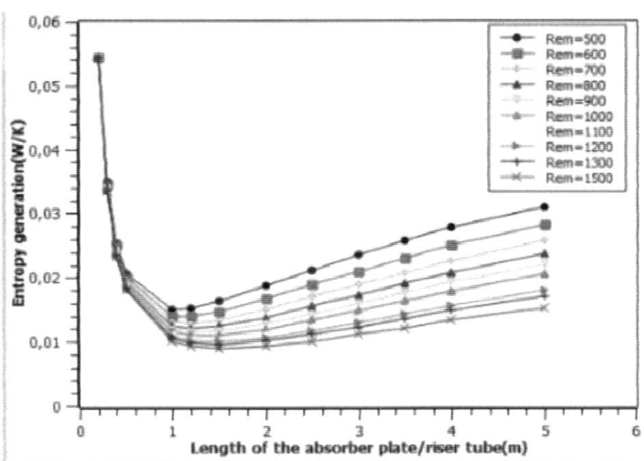

Figure 2. Trend of entropy generation with length of tube for different values of modified Reynolds number.

M. A. Oyinlola et al. [19] experimentally studied absorber plates with micro-channels for scaling effects and observed that Nusselt number was affected little by entrance effect and viscous losses.Conjugate heat transfer adversely affected Nusselt number and thus it was found to decrease with axial conductance number. Despite necessary conditions for neglecting entrance effects being met, an axial variation of local Nusselt number was seen and it became pronounced with an increase in Graetz number. Their experimental setup consisted of flow system with micro-channel test rig which acted as a simulation of a compact FPSC, figure 3. Coriolis mass flow meter, thermocouples and differential pressure sensors were used to capture various data.

IOP Conf. Series: Materials Science and Engineering 1127 (2021) 012007 doi:10.1088/1757-899X/1127/1/012007

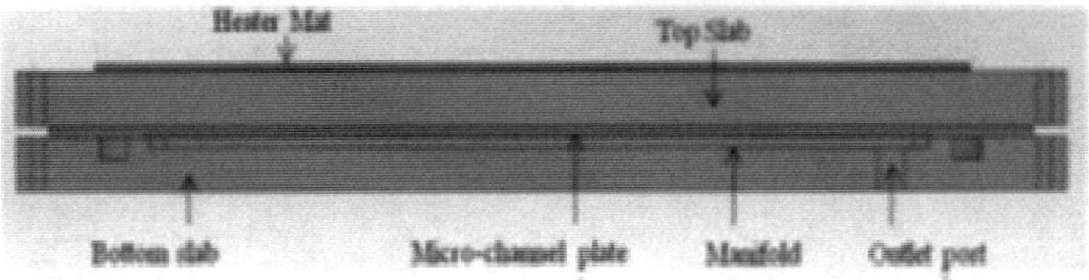

Figure 3. Micro-channel test rig cross-section.

R.W. Moss et al. [20] investigated numerous designs for an absorber for evacuated flat plate collector. The challenges encountered in designing an absorber for an evacuated collector included providing support for the glass cover as well as the necessity of the material being a low out-gassing one while providing better efficiency than conventional serpentine tube absorbers and at the same time the design must be easy to fabricate. Keeping this in view a flooded panel absorber was chosen and was fabricated from stainless steel sheets by hydroforming technique. CFDanalysis showed that halving the plate and having the inlet and outlet at diagonal corners could optimize the distribution of flow. FE analysis showed that the operational pressure could be withstood by a sheet of thickness 0.7 mm. Hydroforming pressure of 25 MPa was sufficient to mould the sheet of 0.7 mm thickness. Minimum safe height of the absorber in the evacuated collector was evaluated using distortion testing, figure 4. Authors have predicted 3% higher efficiency of the collector employing this design as opposed to conventional serpentine tube design.

Figure 4. Flooded panel absorber during distortion testing.

H. Bhowmik et al. [21] developed an FPSC with reflectors mounted on its two sides to concentrate diffuse as well as direct solar radiation on to the collector. The reflector panels were fitted such that their angle could be changed to track the sun. The authors report an improved efficiency of 10% over and above the conventional FPSC when tested with water heating prototype with the collector with reflectors at its core, figure 5. The collector with reflector invariably results in a higher outlet temperature of the fluid than the collector without reflector.The iron plate was used to make the collector and black paint was used as absorber coating and the final assembly was sealed off on top using transparent glass. Design of collector was done in a manner to obtain a parallel pattern of flow and constant rate of flow from entry to exit. Mercury glass was used for reflector panels. 25 l tank was utilized to supply water as working fluid.Using rotameters to measure and control the flow, flow rates of 0.1 l/min and 0.2 l/min were used for this experiment. The exit temperature of the fluid was seen to reduce with increase in volume flow rate. Thermometers were employed to measure fluid temperatures at entry and exit. Global average radiation was taken as 430 W/m^2. Collector surface area was 1.05 m^2 and reflector surface area was 1.85m^{2}and the mirror was assumed to have a reflectivity of 0.9.

Figure 5. Solar water heating system using FPSC with reflectors

I. Visa et al. [22] designed a new FPSC for better façade integration and architectural aesthetics with efficiency figures of 60.7% being obtained through optimisation of the isosceles trapeze-shaped design in terms of insulation, proper contact between absorber plate and absorber tubes, meander length of the tube, its diameter, etc., figure 6. The top surface of the glazing was spray-coated with a self-cleaning coating which improved performance. Certain characteristics of this design which make it apt for façade integration are that the geometry with an area of 0.63 m^2 makes it compatible for solar arrays. The absorber plate has been made into an assembly of 14 pieces and each can be coated with spectral selective coatings for a multi-colored design. An indoor testing rig with the capability to accommodate collectors with different shape and dimensions up to a maximum of 1 × 2 m^2 was used. A solar simulator with a maximum intensity of 1000 W/m^2 was employed. Pyranometer, temperature sensors, and linear actuators to rotate the frame holding the collector were incorporated in the rig. The flow of the working fluid was measured and controlled using a mass flow sensor. A standard mass flow rate of 0.012 kg/s was maintained and the water was taken as the working fluid in all iterations of the experiment.

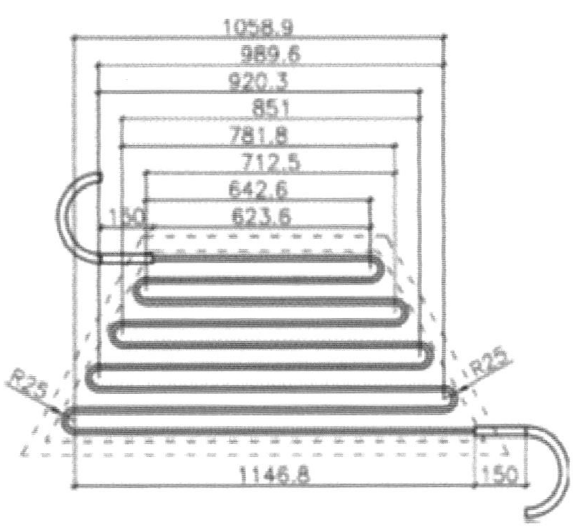

Figure 6. Optimised configuration of meander tubes.

I. Visa et al. [23] have further designed a FPSC with smaller area and triangular shape for improved façade integration by catering to the fact that facades of buildings have random shaped and small-sized sections. The area was taken as $0.083m^2$. Difficulty in using internal pipes was seen because of the small size of the collector design thus a flooded mode was used by making a cavity below the absorber plate. The distribution of flow was analyzed and optimized through simulation of the design in SolidWorks and its analysis in ANSYS. The optimal gap between glazing and absorber and optimal insulation width was found by applying the radiative mathematical model. Indoor testing of three collectors with different spectral selective coatings of orange black and green resulted in efficiencies of 35%, 55% and 42% respectively, figure 7. To validate the results of the simulation, experiments were performed on a collector with the cavity being fabricated using steel and the absorber plate with aluminum. These were mounted on a steel frame using fasteners and water was filled in the cavity up to a pressure of 3 bar at 20°C. A similar solar simulator as mentioned before was used to produce radiation of 800 W/m^2 to 900 W/m^2. Water with a specific flow rate of 0.02 kg/m^2s was used. The black absorber plate was obtained using commercially available black absorber coating, orange was obtained using vanadium oxide pigment and green was obtained with a mixture of hydrated copper sulphates and copper sulphides.

Figure 7. Solar thermal collector indoor testing rig.

Do Ango et al. [24]investigated polymer absorbers for FPSC through numerical simulation and found that the thickness of air gap influences the collector performance and the optimal thickness was obtained as 10 mm. Efficiency was seen to vary proportionally with the mass flow rate of the fluid but elevated mass flow rates reduce outlet temperature obtained. Efficiency was unaffected by intensity of radiation but it was found to affect the outlet temperature and an approximately linear relation was seen, figure 8. The temperature of the fluid at inlet influences the collector performance and it must be no less than the ambient for good collector performance. A portion from the central part of the polymer FPSC was simulated in computational fluid dynamics software CFD Star-CCM+ with dimensions 1 m × 90.5 mm. The choice of the central portion was done with the objective to prevent edge effects. The equations of a balance of mass, energy, and momentum were the governing equations. Radiation effects were simulated using a Surface to Surface model on the underlying principle of enclosure theory for diffuse gray surfaces. Taking the values of the parameters of air gap, solar radiation, fluid entry temperature, collector length and fluid mass flow rate from a leading authority on manufacturing information of solar collectors, a parametric study was thus performed.

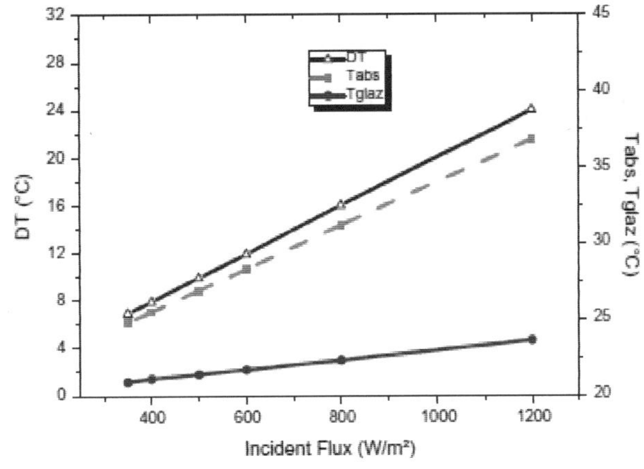

Figure 8. Variation of outlet temperature with incident flux

F. Giovannetti et al. [25] explored and analyzed various spectral selective and high transmittance glass coatings which are based on transparent conductive oxides, figure 9. Tin-doped indium oxide and aluminum-doped zinc oxide layers were analyzed. The spectral selective coating did not result in any enhancement of performance in single-glazed highly selective absorber collector; in fact, increased optical losses degraded the efficiency of the collector. But for other collectors, low emissivity coated glass resulted in improved performance. Thus single-glazed low selective absorber collector and double-glazed highly selective absorber benefit greatly from low emissivity, high transmittance spectral selective coating for the glass covers.

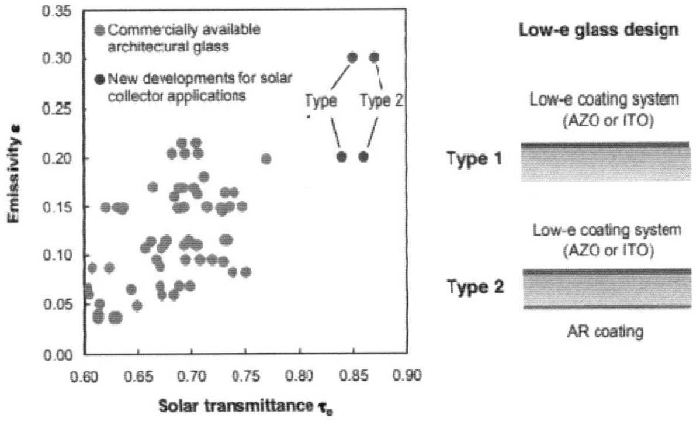

Figure 9. Emissivity versus transmittance plot.

H. Kessentini et al. [26] experimentally as well as numerically studied a design comprising of insulation of transparent material added between the absorber and the glass cover to minimize heat loss. They have added a system to protect the transparent insulation from overheat due to stagnation. The system uses a channel with a trap door which gets activated thermally. This has proven as an effective method of protection for transparent insulation material. The collector was then optimized through simulation and the optimized collector proved to be at par with existing designs in terms of performance while still being low cost. Experimental validation was done by performing tests on FPSC fabricated on the lines of conventional FPSCs with extra insulation of transparent material inserted between the glass cover and the absorber. Insulation material used was cellulose triacetate film arranged in the honeycomb structure. The thermally activated actuator was made of Nickel-Titanium shape memory alloy spring. The testing facility used had the capacity to perform experiments up to 8 bar and 150°C conditions of the working fluid. Temperatures were measured by employing thermocouples and resistance probes.

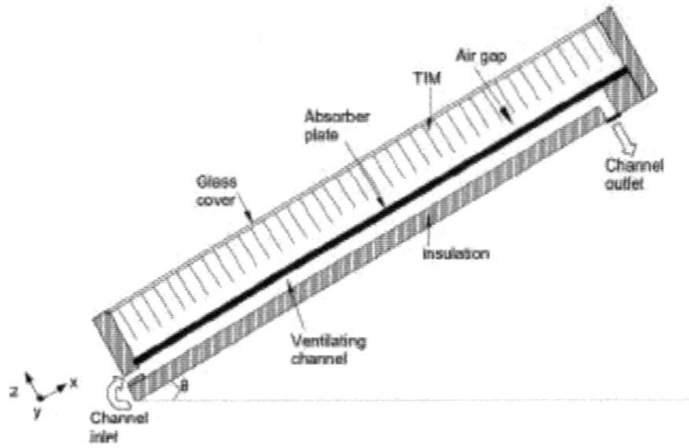

Figure 10. Cross-section of FPSC with transparent insulation.

Another recent and major development in the field of solar collectors in general and FPSCs in particular has been the use of nanofluids to enhance the heat transfer to the working fluid, which has not been covered in this paper. Nanofluids are a special type of fluid which can be used as working fluid in solar collector systems with and without heat pipes. They are synthesised by dispersing a small quantity of solid nanoparticles often less than 100 nm in a base of traditional working fluids such as water or ethylene glycol [27]. A rigotous treatment of the models for explaining the heat transfer augmentation has been presented by S.M.S Murshed et al [28]. S. K. Verma et al. [29], M. A. Alim et al. [30], A. Zamzamian et al. [31] have experimentally as well as analytically investigated various metal oxide based nanofluids for solar collector applications.

Table 1. Summary of few main papers.

Author	Type of investigation	Description	Performance
K. Balaji et al [17]	Experimental	Absorber tube with tube and rod thermal performance enhancer	Reynolds number, rod thermal performance enhancer yielded greater efficiency compared to tube design and also conventional absorber design collector.
T. Jatau et al. [18]	Theoretical	Optimisation through constructal method	Increase in modified Reynolds number results in an increase of optimal length of the tube while it results in the decrease of optimal diameter and optimal spacing of the tube.
R.W. Moss et al. [20]	Experimental and theoretical	Flooded panel design for evacuated FPSC	Authors have predicted 3% higher efficiency of the collector employing this design as opposed to conventional serpentine tube design.
I. Visa et al. [23]	Experimental	Facade integration of FPSCs	Three collectors with different spectral selective coatings of orange black and green resulted in efficiencies of 35%, 55% and 42% respectively

Do Ango et al. [24]	Theoretical	Numerical simulation of polymer absorbers	Efficiency was seen to vary proportionally with the mass flow rate of the fluid but elevated mass flow rates reduce outlet temperature obtained. Efficiency was unaffected by intensity of radiation but it was found to affect the outlet temperature and an approximately linear relation was seen.

3. Conclusion

This paper presented a general overview of design advancements in FPSCs in the past five years. Studies conducted showed a great scope for improvementin the design of FPSC to make it suitable for wider adoption by making it more efficient. Techniques covered in this paper include redesigning of absorber plate for efficient flow with the maximum transfer of heat from absorber to working fluid. The use of optimization techniques to evaluate the optimal dimension of absorber and manifolds prove to be effective. Coatings for glass covers, use of transparent insulation between absorber and cover may be used to minimize heat loss. This paper will be helpful in looking for future fields of research to pursue.

References

[1] M. A. Sabiha, R. Saidur, S. Mekhilef, O. Mahian. Progress and latest developments of evacuated tube solar collectors, *Renewable and Sustainable Energy Reviews*, **volume**51,pp. 1038–1054, (2015).

[2] K. Chopra, V. V. Tyagi, A. K. Pandey, A. Sari. Global advancement on experimental and thermal analysis of evacuated tube collector with and without hwat pipe systems and possible applications, *Applied Energy*, **volume** 228, pp. 351–389, (2018).

[3] Y. Tian, C. Y. Zhao. A review of solar collectors and thermal energy storage in solar thermal applications, *Applied Energy*, **volume**104, pp. 538–553, (2013).

[4] S. Suman, M. K. Khan, M. Pathak. Performance enhancement of solar collectors-A review, *Renewable and Sustainable Energy Reviews*, **volume**49, pp. 192–210, (2015).

[5] Z. Chen, S. Furbo, B. Perers, J. Fan, E. Andersen. Efficiencies of flat plate solar collectors at different flow rates, *Energy Procedia*, **volume**30, pp. 65–72, (2012).

[6] K. M. Pandey, R. Chaurasiya. Review on analysis and development of solar flat plate collector, *Renewable and Sustainable Energy Reviews*, **volume** 67, pp. 641–650, (2017).

[7] G. Colangelo, E. Favale, P. Miglietta, A. Risi. Innovation in flat solar thermal collectors: A review of the last ten years experimental results, *Renewable and Sustainable Energy Reviews*, **volume**57, pp. 1141–1159, (2016).

[8] S. Farahat, F. Sarhaddi, H. Ajam. Exergetic optimization of flat plate solar collectors, *Renewable Energy*, **volume** 34, pp. 1169–1174, (2009).

[9] F. A. Boyaghchi, H. Montazerinejad. Multi-objective optimisation of a novel combined cooling, heating and power system integrated with flat plate solar collectors using water/CuO nanofluid, *International Journal of Exergy*, **volume**21, pp. 202–238, (2016).

[10] R. W. Moss, G. S. F. Shire, P. Henshall, P. C. Eames, F. Arya, T. Hyde. Optimal passage size for solar collector microchannel and tube-on-plate absorbers, *Solar Energy*, **volume**153, pp. 718–731, (2017).

[11] J. Facão. Optimisation of flow distribution in flat plate solar thermal collectors with riser and header arrangements, *Solar Energy*, **volume**120, pp. 104–112, (2015).

[12] M. S. Buker, S. B. Riffat. Building integrated solar thermal collectors-A review, *Renewable and Sustainable Energy Reviews*, **volume**51, pp. 327–346, (2015).

[13] R. O'Hegarty, O. Kinnane, S. J. McCormack. Review and analysis of solar thermal facades, *Solar Energy*, **volume**135, pp. 408–422, (2016).

[14] R. W. Moss, P. Henshall, F. Arya, G. S. F. Shire, T. Hyde, P. C. Eames. Performance and operational effectiveness of evacuated flat plate solar collectors compares with conventional thermal, PVT and PV panels, *Applied Energy*, **volume** 216, pp. 588–601, (2018).

[15] F. Calise, M. D. d'Accadia, M. Vicidomini, M. Scarpellino. Design and simulation of a prototype of a small-scale solar CHP system based on evacuated flat plate solar collectors and Organic Rankine Cycle, *Energy Conversion and Management*, **volume**90, pp. 347–363, (2015).

[16] P. Henshall, P. Eames, F. Arya, T. Hyde, R. Moss, S. Shire. Constant temperature induced stresses in evacuated enclosures for high performance flat plate solar thermal collectors, *Solar Energy*, **volume**51, pp. 1038–1054, (2015).

[17] Balaji K., Ganesh K. P., Sakthivadivel D., Vigneswaran V. S., Iniyan S. Experimental investigation on flat plate solar collector using frictionally engaged thermal performance enhancer in the absorber tube, *Renewable Energy*, **volume**142, pp. 62–72, (2019).

[18] T. Jatau, T. Bello-ochende. Constructal design of flat plate solar collector, *Proceedings of the Romanian Academy*, **volume** special issue, pp. 160–165, (2018).

[19] M. A. Oyinlola, G. S. F. Shire, R. W. Moss. The significance of scaling effects in a solar absorber plates with micro-channels, *Applied Thermal Engineering*, **volume**90, pp. 499–508, (2015).

[20] R. W. Moss, G. S. F. Shire, P. Henshall, P. C. Eames, F. Arya, T. Hyde. Design and fabrication of a hydroformed absorber for an evacuated flat pate solar collector, *Applied Thermal Engineering*, **volume**138, pp. 456–464, (2018).

[21] H. Bhowmik, R. Amin. Efficiency improvement of flat plate solar collector using reflector, *Energy Reports*, **volume**3, pp. 119–123, (2017).

[22] I. Visa, A. Duta, M. Comsit, M. Moldovan, D. Ciobanu, R. Saulescu, B. Burduhos. Design and experimental optimisation of a novel flat plate solar thermal collector with trapezoidal shape for facades integration, *Applied Thermal Engineering*, **volume**90, pp. 432–443, (2015).

[23] I. Visa, M. Moldovan, A. Duta. Novel triangular flat plate solar thermal collector for facades integration, *Renewable Energy*, **volume** 143, pp. 252–262, (2019).

[24] A. C. M. Do Ango, M. Medale, C. Abid . Optimization of the design of a polymer flat plate solar collector, *Solar Energy*, **volume** 87, pp. 64–75, (2013).

[25] F. Giovannetti, S. Foste, N. Ehrmann, G. Rockendorf. High transmittance, low emissivity glass covers for flat plate collectors: Applications and performance, *Energy Procedia*, **volume**30, pp. 106–115, (2012).

[26] H. Kessentini, J. Castro, R. Capdevila, A. Oliva. Development of flat plate collectors with plastic transparent insulation and low-cost overheating protection system, *Applied Energy*, **volume**133, pp. 206–223, (2014).

[27] M. J.Muhammad, I. A. Muhammad, N. A. C Sidik, M. N. A. W. M. Yazid. Thermal performance enhancement of flat-plate and evacuated tube solar collectors using nanofluid: A review, *International Communications in Heat and Mass Transfer*, **volume**76, pp. 6–15, (2016).

[28] S.M.SMurshed, K. C. Leong, C. Yang. A Combined model for the effective thermal conductivity of nanofluids, *Applied Thermal Engineering*, **volume**29, pp. 2477–2483, (2009).

[29] S. K. Verma, A. K. Tiwari, D. S. Chauhan. Experimental evaluation of flat plate solar collector using nanofluids, *Energy Conversion and Management*, **volume** 134, pp. 103–115, (2017).

[30] M. A. Alim, Z. Abdin, R. Saidur, A. Hepbasli, M.A. Khairul, N. A. Rahim. Analyses of entropy generation and pressure drop for a conventional flat plate solar collector using different type of metal oxide nanofluids, *Energy and Buildings*, **volume**66, pp. 289–296, (2013).

[31] A. Zamzamian, M. KeyanpourRad, M. KianiNeyestani, M. T. Jamal-Adab. An experimental study of theeffect of Cu-synthesized/EG nanofluid on the efficiency flat plate solar collector, *Renewable Energy*, **volume**71, pp. 658–664, (2014).

IOP Conf. Series: Materials Science and Engineering 1127 (2021) 012008 doi:10.1088/1757-899X/1127/1/012008

Review on Economic analysis of Photovoltaic Thermal Systems

B Kalidasan*, R Divyabharathi#, A K Pandeyᵠ, R. Saidurᵠ

*Department of Mechanical Engineering, Bannari Amman Institute of Technology, Sathyamangalam, India-638401 email:kalidasancinna@gmail.com (B Kalidasan)

#Department of Agriculture Engineering, Bannari Amman Institute of Technology, Sathyamangalam, India-638401 email:divyabharathir@bitsathy.ac.in(R Divyabharathi)

ΨResearch Centre for Nano-Materials and Energy Technology (RCNMET), School of Science and Technology, Sunway University, No. 5, Jalan Universiti, Bandar Sunway, Petaling Jaya, 47500 Selangor Darul Ehsan, Malaysia

E-mail: adarsh.889@gmail.com

Abstract. Sun is the most promising renewable resource available to mankind for extracting maximum useful energy. One of the best ways to trap solar energy is to install photovoltaic thermal (PVT) systems whose main novelty is to produce power by photovoltaic conversion and also dissipate the heat evolved to recover thermal energy thereby enhancing the overall efficiency of the system. Implementing this approach in practical application system lags mainly due to the requirement of economic investment. This paper reviews on the economics involved in the installation, operation and maintenance of solar PVT systems. The economic viability of solar PVT systems is evaluated by considering five criteria (energy costs, life cycle costs, profit ratio, cost savings and pay-back period). Furthermore the performances of many PVT systems were compared and the economic evaluation has proved that the PVT system is economically feasible and has greater potential for incorporating in residential/commercial buildings especially for tropical countries like India.

Graphical Abstract

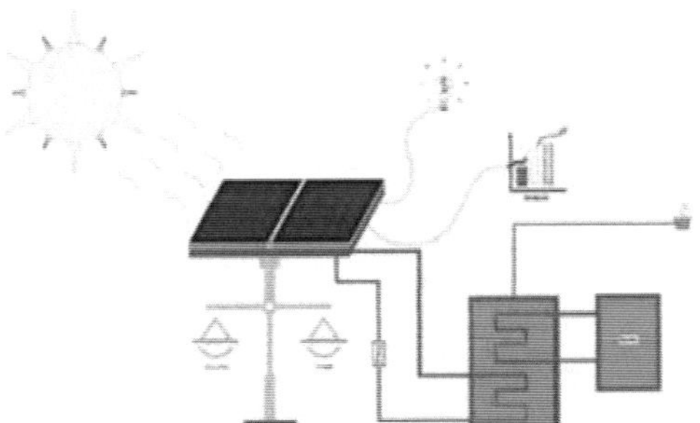

Economic Analysis of PVT

IOP Conf. Series: Materials Science and Engineering 1127 (2021) 012008 doi:10.1088/1757-899X/1127/1/012008

1. Introduction

Technological development and industrialization are the major factor for a developing country to compete in the global market. Industrialization depends on power consumption in terms of electricity for machine operations. Electric power generation using conventional energy source leads to global warming and environmental imbalance, as well many conventional energy are on the verge of extinction. Thus, on a global level people shift towards non-conventional energy source that matches their daily need with economic consideration. Solar energy being the most predominant and not polluting renewable source of energy tends to contribute towards energy scarcity. Solar Photovoltaic (PV) panels and Solar Thermal System (STS) are the basic classifications based on the application of solar energy utilization. Solar PV Panel converts solar radiation directly into electric power, whereas STS uses mechanical system following the rankine cycle principle to convert the solar radiation into electric power. For better performance and improved efficiency Solar PV panels and STS can be combined to a single system known as Solar Photovoltaic Thermal (PV-T) system [1]. Figure 01 shows the photovoltaic thermal system.

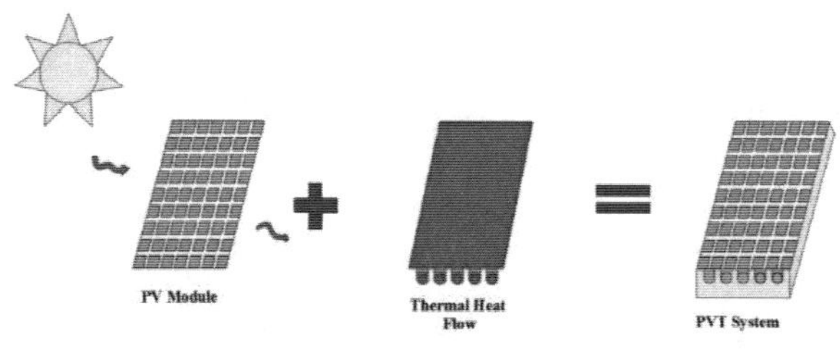

Figure 1. Photovoltaic Thermal System

Review [2] on the experimental research works that has carried out over the past ten years in the PVT System, he also reconnoitered in depth the liquid spectrum filter for PVT system. Direction to carry out research work in future in terms of materials, design, and application were listed in the review article. [3] Made a review study on the experimental and analytical work that has been carried out in terms of energy performance in Solar PVT system. Various PVT such as PVT-flat plate collector, PVT-compound parabolic concentrator, PVT-solar still and PVT greenhouse dryer were taken into account to review the thermal and electrical energy performance. [4] Communicated the recent works carried out in PVT system incorporating Phase change material. PCM based PVT systems would be more technically viable it the issue of thermal conductivity is addressed properly. From the review it was also noted that PCM based PVT system were 33-50% more efficient for storing heat than the conventional PVT System. Future works were focused on PVT microencapsulated PCM system which are on the verge of development. [5] Made a review on PVT system emphasizing the environmental benefits, life cycle assessment and environmental constraints while using PVT system. From the review it was noticed that most of the research work focused only on energy payback time CO_2 emission.

As brief in the literature [2-5] it's evident that PVT systems are quite efficient in terms of performance. Though many reviews have been carried out, still there is a gap economic analysis in terms of breakeven analysis, cost benefits and payback period of PVT system. This review manuscript has been divided into four sections to consolidate the economic analysis of Solar PVT system that has been carried by research. In the first section, basic necessity for the economic analysis of PVT and the novelty of this research paper is revealed. Second section gives a brief idea about the concept of PVT system and its working in terms of advancement in technology. The main core work of the paper is elaborated in the third section, the breakeven point, cost benefits, payback period and economic feasibility of using PVT in daily use is enlarged for both water and air working fluid in STS. Though there are numerous advantages the limitations still exist and thus the recommendations and future scopes are listed in fourth section.

2. Photovoltaic Thermal System

Hybrid PVT System combines PV and Solar thermal technologies; these innovative techniques are very encouraging for concurrent generation of both electric and thermal power. Hybrid PVT System contributes towards numerous advantages such as better performance, space saving, reduced heat dissipation etc. From research, it's been evident that due to high specific heat capacity of water compared to air, water based PVT system are mostly preferred [6].

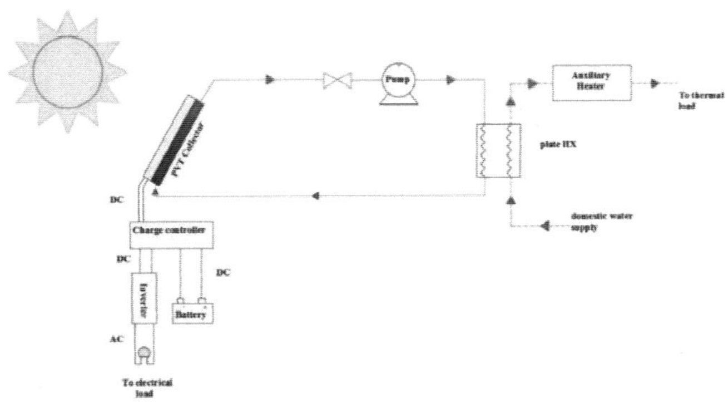

Figure 2. Working of Solar Photovoltaic Thermal System

The solar radiation from the sun is directly converted into electricity in the photovoltaic panel. Due to resistance temperature of the PV panel rises which decreases the efficiency of PV panel, combining thermal system with PV panel, dissipates the heat from PV panel and uses the excess heat for domestic purpose, saving the cost and space available. Working of Solar PVT system is elaborated in Figure 2.

3. Economic Analysis

The criteria under economic analysis of the PVT system aim at evaluating the paths of financial gain and losses involving from its capital, installation, operation and maintenance costs. Economic evaluation reviews the following areas of PVT systems employing both air and water as working fluid.

- Total cost of electricity generation from PVT system
- Profit rate of PVTs cost
- Cost savings of PVTs
- Payback period of PVTs
- Life cycle costs
- Energy costs

The economic evaluation is performed by assuming that electricity is produced continuously by the PVT system and also considering the electrical consumption of the fluid pumps, maintenance cost per annum and interest rate per annum.

3.1. Total cost of electricity generation from PVT system

For countries like India, with an average sunshine hours of 300 days, the total electricity generation is more than 334 GW [7] which is supplied mostly by coal, nuclear and renewable energy power plants. By utilizing renewable energy sources, the unit cost of electricity is reduced in addition to the reduction in emissions of GHGs.

A simulated equation for the total cost of electricity generation (C_{total}) in a PV/T collector heating system utilized for an indoor–outdoor pool, by evaluating the cost of installation (C_{inst}), inverter (C_{inv}), generator (C_{gen}) and subsidies (C_{sub}) which is given by [8] ,

$$C_{total} = C_{inst} + C_{inv} + C_{gen} - C_{sub} \tag{1}$$

In addition to these costs, in this study the costs of battery (C_{bat}), charge controller (C_{cc}) and pump (C_{pump}) are considered. Thus the equation can be rewritten as

$$C_{total} = C_{inst} + C_{inv} + C_{gen} + C_{bat} + C_{cc} + C_{pump} - C_{sub} \tag{2}$$

3.2. Profit rate, cost savings and Payback period of PVT system

The profit rate of PVT systems can be calculated by the following equation [9],

$$P = \sum_{i=1}^{n} \frac{ACCS_i}{(j+1)^i} - C_{total} \tag{3}$$

This equation is derived by considering the inflation costs per annum *j,* after *n* years. In order to improve this equation, the annual profit of economic costs generated through PVT systems is also considered for calculation in this study. The annual cumulative cost of savings in n years is given by [10],

$$CCS_n = \sum_{i=1}^{n} ACCS_i \, x \, (1+\beta)^{n-i} \tag{4}$$

$$ACCS_n = APC \; x \; Unit_{EC} \tag{5}$$

where, $ACCS_n$, APC and $Unit_{EC}$ represents the annual cumulative cost of savings over n years, annual profit of economic costs involved and unit cost of electricity respectively. β represents the coefficient for maintenance costs which is considered as 3% of the total costs of the PVT system (β = 0.03)

Considering the above mentioned costs of profit, unit and maintenance costs, the cumulative costs can be

$$CC_n = C_{total} \, x \, (1+\beta)^{n-1} \, x \, (1+\gamma)^{n-1} \tag{6}$$

where, γ is considered to be an arbitrary value of 3% (profit rate of production cost) based on the economics of the region.

Equation on net cash flow can be calculated by considering the cost of savings due to power production as input and operational, maintenance and other financial costs as output as expressed by [11],

$$NC_n = (cash\ input)_n - (cash\ output)_n \tag{7}$$

Thus the profit earned over n years is given by,

$$P_n = CCS_n - CC_n \tag{8}$$

$$P_n = \left[\sum_{i=1}^{n} ACCS_i \, x \, (1+\beta)^{n-i}\right] - C_{total} \, x \, (1+\beta)^{n-1} \, x \, (1+\gamma)^{n-1} \tag{9}$$

The Payback period costs is calculated from the following equation as presented by [12],

$$PBC = \frac{TIC}{AEC + AEES} \tag{10}$$

where, PBC, TIC, AEC and AEES represents the costs of pay back, total investment costs of PVTs, annual costs of electricity involved and the annual electrical energy savings by PVT system respectively.

3.3. Life cycle costs & Energy costs

To derive a feasible effective evaluation of PVT systems, more importance is given on calculating the energy costs and life cycle costs associated with the system.

The life cycle cost evaluation can be arrived by the equation as suggested by [10],

$$LC = C_{cap} + \sum_1^n C_{om} x \, R_{pw} + \sum_1^n C_{rep} \, x \, R_{pw} - C_{sal} x \, R_{pw} \tag{11}$$

where C_{cap}, C_{om}, C_{rep}, C_{sal} and R_{pw} represents the initial costs involved in purchase and installation, operational and maintenance involving inspection fee, taxes of property and insurance cost of replacement and repairs, salvage value which is 10% of the cost of equipment and current value of every factor involving the future sum of money and rate of interest respectively.

The initial cost of the system is based on the PVT system components involved, design and development in addition to the installation works. It is given by,

$$C_{cap} = (CE * UC) + CI \tag{12}$$

where, CE, UC and CI represents the costs of equipment, unit costs and installation costs respectively.

Cost of maintaining an equipment is proportional to the initial costs involved for the system as given by [13],

$$MC = \sum_{1}^{j} k_j * (C_{cap})_j \tag{13}$$

Based on the life cycle costs which involves the capital costs, maintenance and replacement/repairs costs involved in operation of the PVT system, the energy costs involved in PVT system can be calculated by,

$$EC = \frac{LC}{\sum_{i=1}^{n} E_{PVT}} \tag{14}$$

where, EC, LC and E_{PVT} represents the energy costs, life cycle costs and energy generated from PVT system respectively.

The overall comparison of the life cycle costs and payback period of PVT systems using various working fluids is given in the Table 1.

Table 1. Economic comparison of PVT system

Ref.	Fluids in the system	Output	Payback Period	Life Cycle Cost
[10]	Nano PCM (paraffin) + Nano-SiC-Water	0.125 \$ kWh of Electricity	4.4-5.3 year	288.4 USD
[14]	Water + PCM-lauric Acid	-	4 Years	357.8USD
[15]	Water	0.012 \$/kWh of Electricity	4.2 Years	292.0 USD
[16]	Capric Acid-Lauric Acid	20.45 USD per ton of refrigeration	3.405 Years	2122 USD

4. Recommendations & Scope

The review focuses significantly on the PV/T Systems involving air and water as working fluids. Various costs such as capital costs of system, electricity generation costs, profit rate, cost savings, payback period, life cycle costs and energy costs have been analyzed to evaluate the economic performance of the PV/T systems. Economic Analysis of PCM integrated PVT system shows better response then the ordinary PVT system, as they provide improved province for thermal energy storage. Thus the payback period was comparatively low, with better efficiency.

Researchers have conducted combined evaluation studies of enviro-economic analyses, energy-exergy analysis of PV/T systems. Furthermore, the literatures under techno-economic studies are limited. However no economic evaluations of PV/T system using air as working fluid have been conducted so far. Hence further studies on this area can improve the PV/T system's performance. In addition, performance evaluation through process optimization using numerical approach and simulation software can also be done. The rationale for this study is the result of economic evaluation through benefit-costs so that it paves the way for researchers, industrialists, engineers and stakeholders to find a feasible route for utilization of solar energy and producing maximum electricity at minimum system costs.

Acknowledgments
Authors gratefully acknowledge Bannari Amman Institute of Technology and Sunway University for providing facilities to conduct this research.

References

[1] Sultan, S.M. and M.E. Efzan, Review on recent Photovoltaic/Thermal (PV/T) technology advances and applications. *Solar Energy*, 2018. **173**: p. 939-954.
[2] Joshi, S.S. and A.S. Dhoble, Photovoltaic-Thermal systems (PVT): Technology review and future trends. *Renewable and Sustainable Energy Reviews*, 2018. **92**: p. 848-882.
[3] Sahota, L. and G. Tiwari, Review on series connected photovoltaic thermal (PVT) systems: analytical and experimental studies. *Solar Energy*, 2017. **150**: p. 96-127.
[4] Islam, M., et al., Recent progresses and achievements in photovoltaic-phase change material technology: a review with special treatment on photovoltaic thermal-phase change material systems. *Energy Conversion and Management*, 2016. **126**: p. 177-204.
[5] Lamnatou, C. and D. Chemisana, Photovoltaic/thermal (PVT) systems: A review with emphasis on environmental issues. *Renewable energy*, 2017. **105**: p. 270-287.
[6] Aste, N., C. Del Pero, and F. Leonforte, Water PVT collectors performance comparison. Energy Procedia, 2017. **105**: p. 961-966.
[7] (CEA), C.E.A., All India Installed Capacity of Power stations. 2019.
[8] Buonomano, A., et al., Dynamic simulation and thermo-economic analysis of a PhotoVoltaic/Thermal collector heating system for an indoor–outdoor swimming pool. *Energy conversion and management*, 2015. **99**: p. 176-192.
[9] Jia, Y., G. Alva, and G. Fang, Development and applications of photovoltaic–thermal systems: A review. *Renewable and Sustainable Energy Reviews*, 2019. **102**: p. 249-265.
[10] Al-Waeli, A.H., et al., Experimental investigation of using nano-PCM/nanofluid on a photovoltaic thermal system (PVT): Technical and economic study. *Thermal Science and Engineering Progress*, 2019. **11**: p. 213-230.
[11] Lari, M.O. and A.Z. Sahin, Design, performance and economic analysis of a nanofluid-based photovoltaic/thermal system for residential applications. *Energy conversion and management*, 2017. **149**: p. 467-484.
[12] Tse, K.-K., T.-T. Chow, and Y. Su, Performance evaluation and economic analysis of a full scale water-based photovoltaic/thermal (PV/T) system in an office building. *Energy and Buildings*, 2016. **122**: p. 42-52.
[13] Souliotis, M., et al., Experimental study and Life Cycle Assessment (LCA) of Hybrid Photovoltaic/Thermal (PV/T) solar systems for domestic applications. *Renewable energy*, 2018. **126**: p. 708-723.
[14] Hossain, M., et al., Two side serpentine flow based photovoltaic-thermal-phase change materials (PVT-PCM) system: Energy, exergy and economic analysis. *Renewable Energy*, 2019. **136**: p. 1320-1336.
[15] Kumar, S., Thermal–economic analysis of a hybrid photovoltaic thermal (PVT) active solar distillation system: Role of carbon credit. *Urban Climate*, 2013. **5**: p. 112-124.
[16] Ghorbani, B., M. Mehrpooya, and M.M.M. Sharifzadeh, Introducing a hybrid photovoltaic-thermal collector, ejector refrigeration cycle and phase change material storage energy system (Energy, exergy and economic analysis). *International Journal of Refrigeration*, 2019. **103**: p. 61-76.

IOP Conf. Series: Materials Science and Engineering 1127 (2021) 012009 doi:10.1088/1757-899X/1127/1/012009

Development of Eutectic Phase Change Materials for Solar Thermal Energy Storage

R. K. Sharma*, A. K. Ansu*, U. Laiker*

*Department of Mechanical Engineering, Manipal University Jaipur, India

E-mail: ravipvb@gmail.com

Abstract. This paper presents the findings on preparation and thermal analysis of binary eutectic mixture of polyethylene glycol 2000 and 6000. Their measured latent heat are 165.3 J/g and 220 J/g respectively. Eutectic mixtures of these materials were prepared by weight percentage composition with an increment of 10% i.e. compositions of PEG 2000 : PEG 6000 with 0:100, 10:90, 20:80 and so on till 100:0 were prepared. The thermal properties, melting temperatures and latent heats of fusion of these mixtures were measured by DSC analysis. The eutectic mixture with 50:50 composition was kept in a thermal cycling test rig to perform an accelerated thermal cycle test up to 1500 cycles. The melting temperature and latent heat of fusion of this thermally cycled mixture were obtained by DSC analysis on 0th, 100th, 500th, 1000th and 1500th cycle. There was a drop of 1.9 ºC in the melting temperature and 8% degradation in latent heat of fusion was observed in comparison to 0th cycle. Considering these values it can be deduced that the eutectic mixture is thermally stable.

1. Introduction

Thermal energy storage as sensible or latent heat is an effective way of using waste heat in industrial processes or excess energy available such as solar radiation. Using phase change materials for storage of sensible heat and latent heat gained importance after the energy crisis of 1973-1974 [1]. With rapid depletion of fossil fuels because of our over dependence on them and greenhouse emissions having adverse effect on our climate the effective utilization of energy became a key issue. Therefore new ways of integrating conventional sources of energy with recovering waste heat and excess energy came into the limelight and became focus of research. In the last three decades extensive research has been carried out in PCMs and their use in LHTES (latent heat thermal energy storage systems). PCMs find applications in the building industry, textiles, the automotive sector and solar energy installations. In recent years an increasing number of applications, including those in electronics and medicine has emerged. The traditional sectors, such as the construction industry, are being advanced by novel, more sophisticated TES materials for smart textiles and thermo regulated biomaterials etc. [1-5].

One way of storing heat is by using the latent heat of phase change of a substance, usually from solid to liquid, as it can provide high energy densities. Then, when this stored heat is needed, it can be released by leaving the material temperature decrease, becoming a solid again. Materials used for this purpose are known as phase change materials (PCM). PCM are being implemented in different systems, active or passive, and for several applications, cold storage, building comfort, medium and high temperatures [2, 6]. As the PCMs are used repeatedly over time their chemical stability is very important for prolonged periods. They should also be compatible with the encapsulation material. For the safety of the TES and to avoid any mishaps, the PCM should be non-toxic, non- flammable and non-explosive [3, 7].

Organic PCM show negligible supercooling during the freezing process and provide congruent melting without phase change segregation. Organic PCM are further classified as paraffin and non – paraffin. Hence organic materials are more suitable to be used as PCM and numerous research studies have been conducted focused on organic PCM's [8]. Hasan et al., [9] studied the thermal properties of myristic acid, palmitic acid and stearic acid, with melting temperatures between 50° and 70°C for 450 melt/freeze cycles. Sari et al., [5, 10] has conducted extensive research on fatty acids, namely, stearic, palmitic, myristic and lauric acid for 910 cycles. He has also studied the eutectic mixtures of these fatty acids such as eutectic mixture of lauric and palmitic acid, stearic and myristic acid [11].

Eutectic mixture are beneficial as the melting point can be tailored for PCM's having higher melting temperature which make their use difficult for solar applications. Polyethylene glycol (PEG) also known as polyoxyethylene (POE) or polyethylene oxide (PEO) is having the hydroxyl group at the end, they are soluble in water as well as in organic compounds. It is an important semi-crystalline polymer with a repeating unit of –CH2–CH2–O–. The melting point of PEG depends on its molecular weight and may vary from 4 °C to 70 °C with the heat of fusion in the range of 117 to 174 J/g. the melting temperatures increase with the corresponding increase in the molecular weight. Although PEG's of different molecular weight have been used along with other materials in binary or ternary mixtures as a PCM, there is not much research done on PEG and its eutectic mixtures as a PCM and its thermal stability till now. Sharma et al., [12] studied the thermal properties of PEG 6000 MW over 500 melt/freeze cycles. It was concluded that PEG 6000 proved to be thermally stable and its melting temperature changed in the range of 55 – 60 °C with a maximum deviation of 6.5%.

In view of the literature reviewed this project will be focused on thermal cycling test of eutectic mixture of PEG 2000 and PEG 6000 molecular weights.

2. Material Preparation

One step method was adopted to prepare the eutectic mixture of PEG 2000 and 6000. They were procured from Merck Company. The mixture compositions by weight % are shown in Table 1. They were mixed in a beaker kept on a hot plate above the melting point of both PCMs. This mixture was stirred using magnetic stirrer at 500 RPM for 30 minutes in order to prepare a homogeneous mixture.

Table 1. Mixture compositions by % weight.

Sample name	PEG 2000 MW (%)	PEG 6000 MW (%)
PEG 6K	0	100
U1	10	90
U2	20	80
U3	30	70
U4	40	60
U5	50	50
U6	60	40
U7	70	30
U8	80	20
U9	90	10
PEG 2K	100	0

3. Results and Discussion

3.1. *Thermal properties of eutectic PCMs*

The lower and the higher temperature range for the accelerated thermal cycle test in the tester for all the samples were kept as 30 °C and 120 °C respectively. The values of the melting temperature and

IOP Conf. Series: Materials Science and Engineering 1127 (2021) 012009 doi:10.1088/1757-899X/1127/1/012009

latent heat of fusion of the eutectic mixtures were measured by DSC at the flow rate of 10 ºC/min and the results are presented for various mixtures in Figures 1-8. For the thermally cycled U5 (50:50 composition by weight %), samples were collected and further tested at 500th and 1000th cycles. The values of melting temperature and latent heat of fusion for the eutectic mixtures have been presented in Table 2. The values of the melting temperature from these DSC curves are found on the intersection point of the tangent to the peak and the extrapolated baseline. This value obtained is actually the onset melting temperature of the material. The temperature corresponding to the peak is the maximum peak temperature. The latent of fusion is given by the area under the peak in the DSC curves. The values of latent heat are obtained in J/g.

Table 2. Melting temperatures and latent heat of fusions

Sample name	COMBINATION PEG 2K:PEG 6K (By weight %)	MELTING TEMPERATURE (ºC)	LATENT HEAT OF FUSION (J/g)
PEG 6K	0:100	59.5 ºC	220.0
U1	10:90	57.5 ºC	205.4
U2	20:80	56.5 ºC	208.2
U3	30:70	54.5 ºC	200.5
U4	40:60	54.0 ºC	196.6
U5	50:50	53.5 ºC	190.0
U6	60:40	52.0 ºC	184.7
U7	70:30	51.5 ºC	179.5
U8	80:20	51.0 ºC	174.0
U9	90:10	50.5 ºC	169.5
PEG 2K	100:0	50.0 ºC	165.3

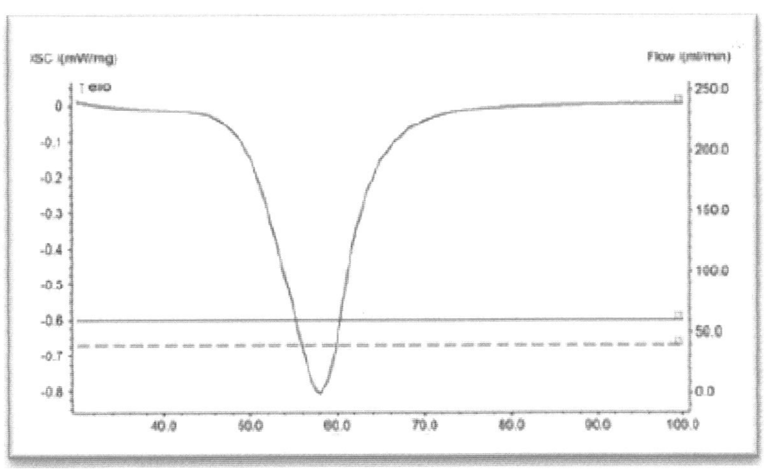

Figure 1. DSC curve for PEG 2000

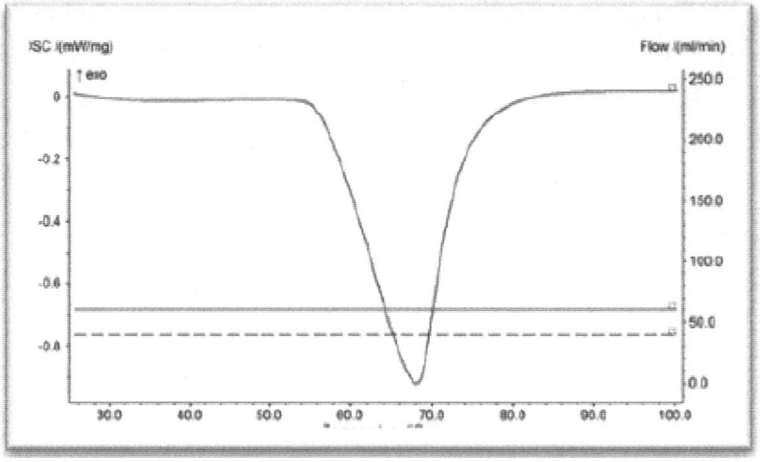

Figure 2. DSC curve for PEG 6000

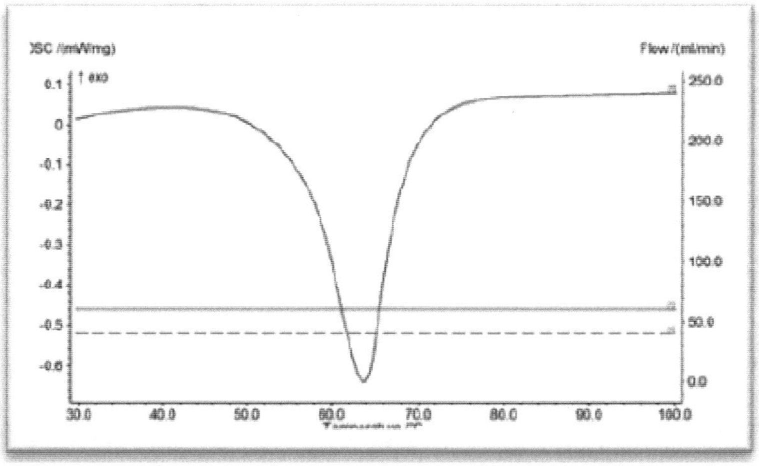

Figure 3. DSC curve of U2 eutectic mixture

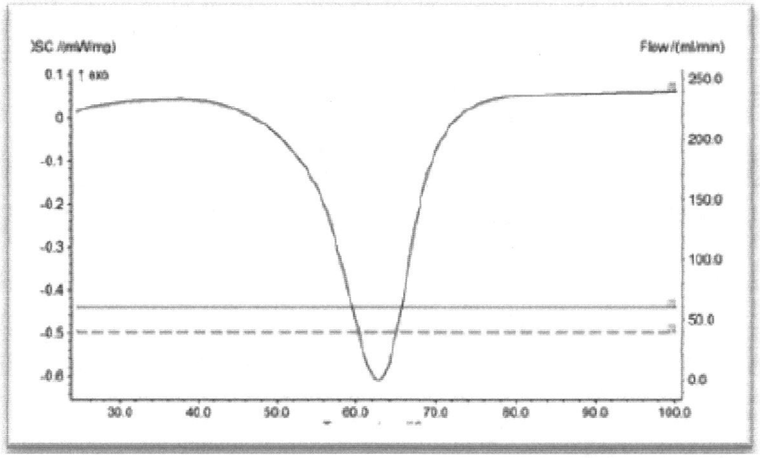

Figure 4. DSC curve of U3 eutectic mixture

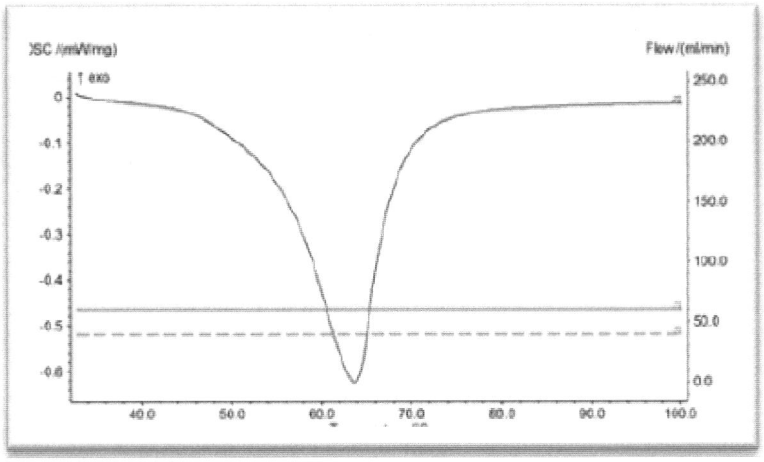

Figure 5. DSC curve of U4 eutectic mixture

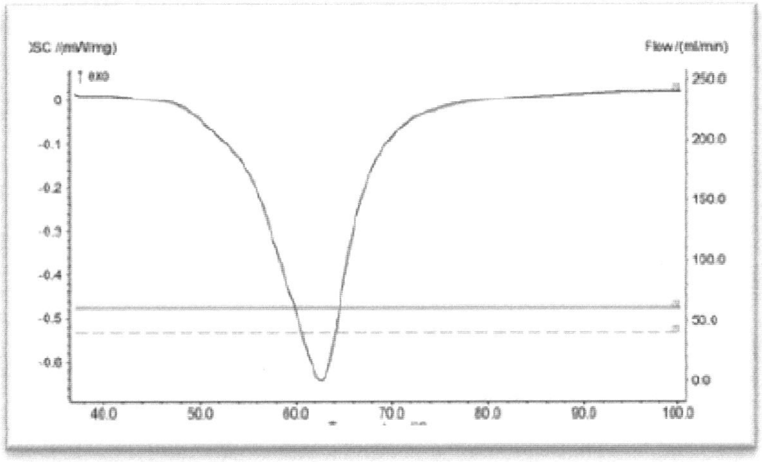

Figure 6. DSC curve of U5 eutectic mixture

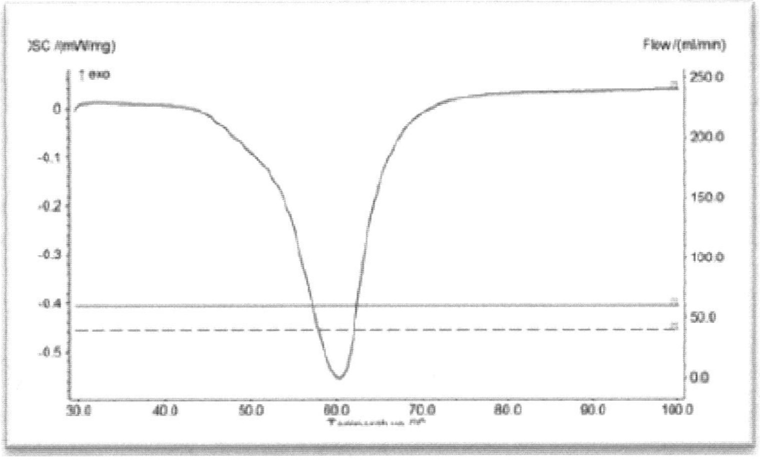

Figure 7. DSC curve of U6 eutectic mixture

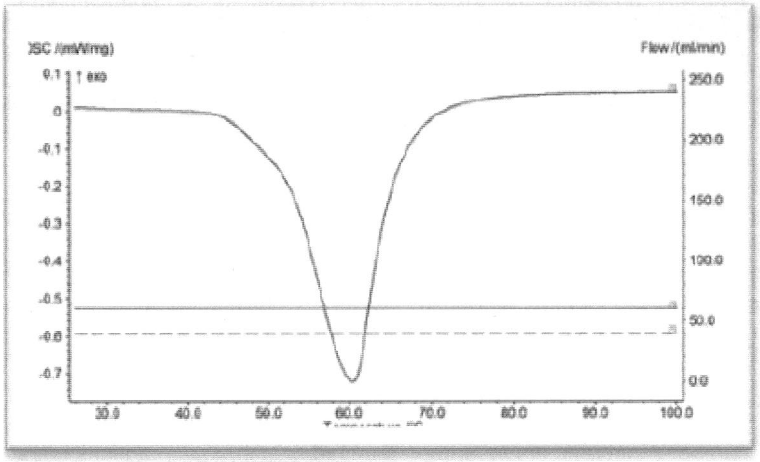

Figure 8. DSC curve of U7 eutectic mixture

For PEG 2000 the obtained value of melting temperature is 50°C, which lies in the melting temperature range of 50 – 55 °C, as quoted by Sigma Aldrich company and the measured latent of fusion is 165.3 J/g. for PEG 6000 the melting temperature found out from the DSC analysis was 59.5 °C, which again lies in between the temperature range of 59 - 64 °C as specified by Merck company and the latent heat of fusion was 220.0 J/g.

For the eutectic mixtures, the melting temperatures and latent heats of fusion lie in between the melting temperatures and latent heats of fusion of PEG 2000 and PEG 6000 as given in Table 1 and Fig 1-8.. The compositions by weight percentage of PEG 2000 and PEG 6000 have been arranged in ascending order with respect to the weight percentage of PEG 2000 in the said Table. Hence we see a downward trend in the values of the melting temperatures as the percentage of PEG 2000 increases in the mixtures. This is justified as the melting temperature of PEG 2000 is less than that of PEG 6000. The same is true for the values of latent heats of fusion. The maximum reduction of the melting temperature is seen from pure PEG 6000 to U1 (composition - 10% PEG 2K: 90% PEG 6K) of 2 °C. No sudden changes or exceptions in the values of the melting temperatures as well as the latent heats of fusion were observed as can be observed from the DSC curves as well.

The eutectic mixture, U5, (Fig. 6) of composition 50% PEG 2000 and 50 % PEG 6000 was selected for the accelerated thermal cycle test. When a substance undergoes repeated melt/freeze cycles consecutively it is known as an accelerated thermal cycle test.

The values of melting temperatures and latent heats of fusion of U5 measured by DSC at 0th, 100th, 500th, 1000th and 1500th cycle are given in Table 5. The values of U5 at the 0th cycle were taken as reference to evaluate and analyze the changes that took place after the material was thermally cycled. The un-cycled U5's (0th cycle) DSC analysis showed that it has a melting temperature of 53.5 °C and its latent heat of fusion was obtained as 190.0 J/g.

3.2. Accelerated thermal cycle test of selected eutectic PCM

An accelerated thermal cycle test for U5 was conducted on inhouse designed cycle tester. The minimum and maximum temperature in the tester was set as 30 °C and 80 °C. As can be seen from Table 2, the variation in the melting temperature of U5 after 1500 cycles is 1.9 °C, from 53.5 °C to 51.6 °C which amounts to 3.55 % change from the 0th cycle. This is the maximum variation. It can be noted there is a very minute increase in the melting temperature after 100 cycles from 53.5°C to 53.8 °C. The variation after 100th, 500th and 1000th cycle is 0.56%, 1.12% and 2.61% respectively as compared to 0th cycle. It can be noted that there is no substantial variation in the melting temperature after accelerated thermal cycle test.

For the latent heats of fusion, a regular decrease in the values is observed as is evident by data given in Table 2. There is slight degradation of latent of fusion after the first 100 cycles is 1.84% from 190

J/g to 186.5 J/g in comparison to fresh PCM (U5) at 0th cycle. After 500 cycles the latent heat is reduced by 2.84% and after 1000th cycle a 4.68 % reduction is observed. After 1000th cycle, there is a sharper decrease in the value of latent heat i.e. from 181.1 J/g to 174.3 J/g, a reduction of 3.75% from the 1000th cycle. This is the maximum variation in the value of latent heat. The overall variation from the 0th to the 1500th cycle is that of 8%.

Table 3. Melting temperatures and latent heats of fusion of thermally cycled U5

U5 - No. of cycles	Melting temperature (ºC)	Latent heat of fusion (J/g)
0	53.5	190.0
100	53.8	186.5
500	53.9	184.8
1000	52.1	181.1
1500	51.6	174.3

The binary system of PEG 2000 and PEG 6000 form eutectic mixtures with sufficiently high latent heats of fusion to be used for various solar thermal applications. The results of the accelerated thermal cycle test show that U5 with 50:50 composition is stable up to 1500 melt/freeze cycles with no substantial decrease in the thermal properties.

4. Conclusion
The objective was to develop a eutectic Phase Change Materials (PCM) mixtures of PEG 2000 and PEG 6000, which are organic materials with melting temperature range of 50 - 55 °C and 59 – 64 °C respectively. The melting temperatures of the binary system fall in between that of the constituents i.e. PEG 2000 and PEG 6000 and same is true for the values of latent heats of fusion. An accelerated thermal cycle test was conducted on the mixture of 50:50 composition by weight percentage (U5) between the temperature ranges of 30 - 120 °C to determine the thermal stability of the eutectic mixture. The results showed that the melting temperature of the PCM had a maximum variation of 3.55 % after 1500 melt/freeze cycles as compared to that of the 0th cycle. A regular degradation in the values of latent heat was observed and had a total variation of 8% after the material had undergone 1500 cycles. These changes are inconsequential to the potential of the PCM and hence the eutectic mixture can be said to be thermally stable after 1500 melt/freeze cycles. As per the result of this project, it can be assumed that, if 300 melt/freeze cycles are considered to occur in a year, then this material can be effectively used as a phase change material for thermal energy storage for a minimum of 5 years.

Acknowledgements
This research is financially supported by the Manipal University Jaipur under the grant of the Endowment Fund (EF/2017-18/QE04-04). Authors also express their sincere thanks to the Central Instrumentation Facilities at Jiwaji University Gwalior, India for characterization facilities availed.

References
[1] Sharma RK, Ganesan P, Tyagi VV, Metselaar HSC, Sandaran SC. Developments in organic solid–liquid phase change materials and their applications in thermal energy storage. *Energy Convers Manage.* 2015;**95**:193-228.
[2] Pielichowska K, Pielichowski K. Phase change materials for thermal energy storage. *Prog Mater Sci.* 2014;**65**:67-123.
[3] Sharma A, Tyagi VV, Chen CR, Buddhi D. Review on thermal energy storage with phase change materials and applications. *Renewable and Sustainable Energy Reviews.* 2009;**13**(2):318-45.

[4] Pielichowski K, Flejtuch K, Pielichowski J. Step-scan alternating DSC study of melting and crystallisation in poly(ethylene oxide). *Polymer*. 2004;**45**(4):1235-42.
[5] Sarı A, Sarı H, Önal A. Thermal properties and thermal reliability of eutectic mixtures of some fatty acids as latent heat storage materials. *Energy Convers Manage*. 2004;**45**(3):365-76.
[6] Zuo J, Li W, Weng L. Thermal performance of caprylic acid/1-dodecanol eutectic mixture as phase change material (PCM). *Energy and Buildings*. 2011;**43**(1):207-10.
[7] Ferrer G, Solé A, Barreneche C, Martorell I, Cabeza LF. Review on the methodology used in thermal stability characterization of phase change materials. *Renewable and Sustainable Energy Reviews*. 2015;**50**:665-85.
[8] Farid MM, Khudhair AM, Razack SAK, Al-Hallaj S. A review on phase change energy storage: materials and applications. *Energy Convers Manage*. 2004;**45**(9):1597-615.
[9] Hasan A, Sayigh AA. Some fatty acids as phase-change thermal energy storage materials. *Renewable Energy*. 1994;**4**(1):69-76.
[10] Sarı A, Kaygusuz K. Some fatty acids used for latent heat storage: thermal stability and corrosion of metals with respect to thermal cycling. *Renewable Energy*. 2003;**28**(6):939-48.
[11] Tunçbilek K, Sari A, Tarhan S, Ergüneş G, Kaygusuz K. Lauric and palmitic acids eutectic mixture as latent heat storage material for low temperature heating applications. *Energy*. 2005;**30**(5):677-92.
[12] Sharma RK, Ganesan P, Tyagi VV, Mahlia TMI. Accelerated thermal cycle and chemical stability testing of polyethylene glycol (PEG) 6000 for solar thermal energy storage. *Sol Energy Mater Sol Cells*. 2016;**147**:235-9.

Investigation on Thermal Properties of AL_2O_3 Based Phase Change Material Composite for Solar Thermal System Application

Mathew George*, A. K. Pandey †, Nasrudin Abd Rahim*, Syed Shahabuddin[T], V.V. Tyagi[#], R. Saidur†

*UM Power Energy Dedicated Advanced Centre, Wisma R&D, Level 4, Jalan Pantai Murni, University of Malaya, Kuala Lumpur 59990, Malaysia

† Research Centre for Nano-Materials and Energy Technology (RCNMET), School of Science and Technology, Sunway University, No. 5, Jalan Universiti, Bandar Sunway, Petaling Jaya, 47500 Selangor Darul Ehsan, Malaysia

[T]Department of Science, School of Technology, Pandit Deendayal Petroleum University, Knowledge Corridor, Raisan Village, Gandhinagar, -382007 Gujrat (State) India

[#]School of Energy Management, Shri Mata Vaishno Devi University, Katra 182320, Jammu & Kashmir, India

E-mail: adarsh.889@gmail.com

Abstract. Sustainable Development Goal and Paris Agreement are guidelines provided by the global community for a better future. One of the ways is through the utilization of solar energy. But solar energy is plagued by unpredictable energy delivery, which can be solved through latent heat energy storage system. In this study, a phase change material (A70) composite with 0.1% wt Al_2O_3 is prepared using probe sonication. The composite is characterized through Differential Scanning Colorimetry (DSC), Thermogravimetric Analyzer (TGA) and UV-VIS-NIR spectrometer. The DSC showed a 5.02% decrease in the value of latent heat while the composite was thermally stable for the temperature range 30 ^{0}C to 200 ^{0}C where majority of domestic solar application lies. The composite and pure A70 may be useful for thermal system application where indirect heat absorption like heat exchangers, heat sinks, photovoltaic thermal and concentrated photovoltaic thermal system.

1. Introduction

Solar energy is a readily available resource for successful implementation of Sustainable Development Goal 7 for reliable and affordable energy source and a method for reduction of carbon emission as agreed upon in Paris Agreement(2015)[1]. But solar energy is plagued by its unpredictability of delivering energy. To counter this unpredictability, a thermal energy storage is of utmost importance. Thermal energy storage can be classified into three namely: a) Sensible b) Latent and c) chemical energy storage[2]. The latent heat is a convenient form of thermal storage due to its capability to store energy at a temperature without considerable change in volume[3].

In thermal systems, there are several areas a phase change material may be utilized like heat exchangers, heat sink, solar water heater, photovoltaic thermal system, concentrated photovoltaic system, building cooling and heating load, greenhouse etc[2]. In concentrated photovoltaic thermal system, PCM spheres were used for improving the overall efficiency of the system by 15% compared to water cooling by Su et al. [4]. Nano-enhanced RT55 prepared with Al_2O_3 was implemented on a photovoltaic thermal system with an improvement of 13.2% in efficiency compared to without nano-enhanced PCM by S.A.Nada et al. [5]. Al-Waeli et al. [6] communicated a maximum temperature of 60.5 ^{0}C and an efficiency of 85.7 % for nano-enhanced paraffin wax with silicon carbide in a numerical study of photovoltaic thermal system. Nano-CuO with paraffin wax prepared through sonication was tested under 10 suns by M. Chen et al.[7]. The nano enhanced PCM reported 2.3times the maximum temperature compared to pure PCM. The nano enhanced PCM showing greater absorption of light.

From the studies it is evident, the presence of material additives to pure PCM has shown an improvement in its thermal characteristics. In this study, pure A70 phase change material and its composite with Al_2O_3 is investigated for thermal system application. The latent heat storage capacity is determined through Differential Scanning Calorimetry (DSC), the thermal stability is analysed through Thermogravimetric Analyser (TGA), and the light transmission through UV-VIS-NIR spectrometer. The three parameters measured will provide a pathway into determining the suitability of Al_2O_3 for direct absorbing solar thermal application.

2. Methods and Materials

2.1. Materials

PLUSICE A70 with melting point 70 ^{0}C and latent heat of 170 J/g was used as the PCM. It was purchased from the company Phase Change Material Products Ltd. Alumina particles were purchased from Sigma-Aldrich.

2.2. Methods

Paraffin based A70 PLUCISE with melting temperature 70 ^{0}C was used for the preparation of micro-enhanced PCM. The material additive used was Al_2O_3. The required weight of PCM was measured using Analytical Macrobalance (Model: TX323L, UNIBLOC) and the required weight percentage of material additive was measured using Analytical Microbalance (Model: EX224, OHAUS). The PCM was melted using a water bath at temperature 80 ^{0}C. The mixture was sonicated for 30 mins using Probe Sonicator (Model: FS-1200N). For example, A70 PCM with 0.1%Al_2O_3, 5 g of A70 was measured using Macrobalance and 5 mg of Al_2O_3 was measured using Microbalance. A70 was placed in a water bath of 80 ^{0}C. Once the PCM melted, five drops of Triton were added followed by the addition of 5 mg of Al_2O_3. This resultant mixture was probe sonicated for 30 mins. The probe sonication was done with water bath at 80 ^{0}C to keep the PCM in liquid form.

2.3. Characterization Technique

The latent heat and the melting of the pure A70 and its composites were determined using Linseis Differential Scanning Calorimeter Analyzer (Model: DSC-1000/C). The measurement was carried out under N_2 atmosphere from temperature 30 ^{0}C to 100 ^{0}C at a heating rate of 10 ^{0}C/min. The thermal stability investigation was carried out using Perkin Elmer TGA 4000 under an N_2 atmosphere at a heating rate of 10 ^{0}C/min. The sample was placed in an alumina crucible and the weight change was monitored from 30 ^{0}C to 600 ^{0}C.

3. Results and Discussion

3.1. Latent Heat

The pure A70 PCM and its composite with 0.1% Al_2O_3 is determined using DSC. The area under the peak represented the latent heat of the pure PCM and composite. The temperature at the peak of the

DSC curve represents the melting temperature of the pure PCM and its composite. This is represented in Fig.1. The area of the peak is calculated using Linseis software provided by with DSC instrument. The latent heat value for pure A70 PCM was determined to be around 172.17 J/g and melting point of pure A70 PCM is 70.3 ^{0}C. The peak area for composite with 0.1% Al_2O_3 is 163.53 J/g and its melting point temperature is 70.8 ^{0}C. A decrease in latent heat of 5.02% with respect to pure PCM is noted in composite. As the particle added do not change phase during the same phase transition temperature of the pure PCM[8], the reduction in latent heat may be attribute to this reason. A slight shift in melting temperature is also noted, from 70.3^{0}C to 70.8^{0}C for pure PCM to composite respectively.

3.2. Thermogravimetric Analysis (TGA)

The thermogravimetric analysis (TGA) of the composite and the pure PCM is represented in Fig.2. The rate of degradation over temperature is represented in Fig 3. The thermal degradation temperature of pure A70 PCM and its composite is evaluated at 3 % weight loss. It is noted that for pure PCM and composite the thermal degradation temperature are 273 ^{0}C and 273.16 ^{0}C respectively. The thermal degradation also occurs between 250 ^{0}C and 400 ^{0}C for pure PCM and composite respectively which is a characteristics of paraffin wax[9].

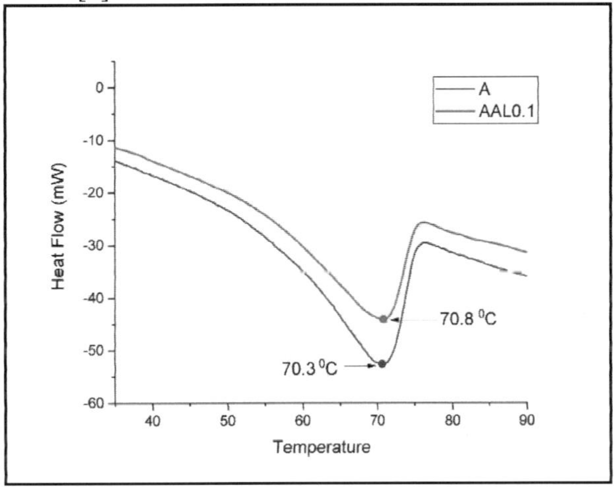

Figure 1. DSC curve of Pure A70 PCM (A) and 0.1% composite with Al_2O_3 (AAL0.1) with their respective melting point temperature.

This shows that very little change occurs to the thermal stability properties due to the addition of the Al_2O_3 particle to the A70 PCM. Maximum rate of degradation of pure A70 PCM is at -10.73 %wt/min at a temperature of 385.33 while that of 0.1% Al_2O_3 is around -11.48 %wt/min at 389.67 ^{0}C. A slight increase in the maximum rate of degradation as well as temperature is noticed.

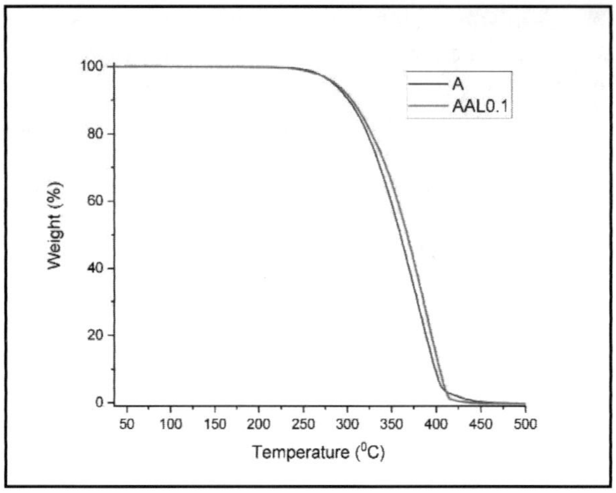

Figure 2. Thermogravimetric Analysis (TGA) curve for pure A70 and A70 composite with 0.1% wt. Al_2O_3.

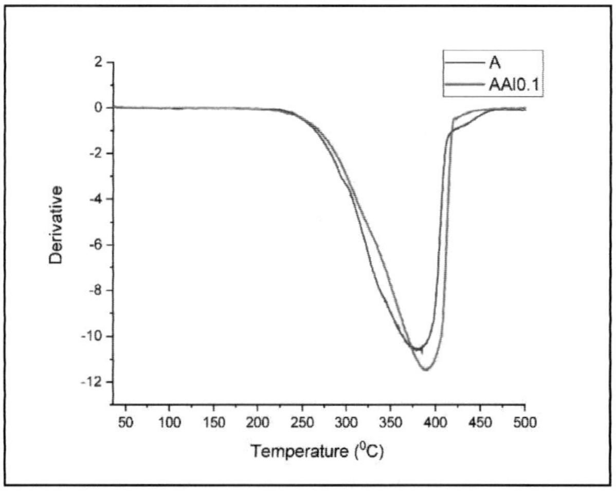

Figure 3. The derivative (%wt/min) over temperature curve representing pure A70 (A) and A70 composite with 0.1% wt. Al_2O_3.

3.3. Light Transmission

The light transmission of pure A70 PCM and 0.1%wt of Al_2O_3 composite with A70 is determined through UV-VIS spectrometer and is represented in Fig.4. We can define the range 200-300nm as ultraviolet, 300-700nm as visible range and 700-800nm as near infrared. The transmission of pure A70 is lower compared to the composite except for the ultraviolet region. A considerable increase in transmission percentage can be seen in the composite in the visible and near infrared region. Due to the lower transmission of pure A70 compared to its composite with Al_2O_3, A70 is a better alternative in terms of direct solar radiation absorbing material. A low transmission percentage is a material better equipped for solar thermal application.

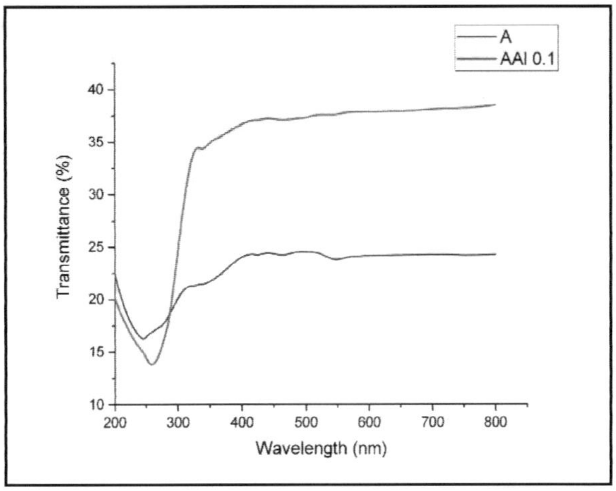

Figure 4. Light transmission of A70 (A) and its composite with 0.1% wt Al_2O_3 (AAl 0.1) in the ultraviolet, visible and near infrared region.

4. Conclusion

Phase change material A70 composite with Al_2O_3 is successfully prepared using probe sonication method. The prepared composite showed a decrement of 5.02% in the latent heat value using DSC. The TGA curve showed that thermal degradation of pure A70 and the composite occurred between the temperature 250 ^{0}C to 450 ^{0}C. Hence, very little change in the thermal degradation temperature due to the preparation technique. But there was an increase in the maximum degradation rate of the composite from -10.73%wt/min of pure A70 to -11.48%wt/min of the composite. The composite may not be suitable for direct solar absorption application as the composite had higher light transmission percentage compared to pure A70. This increase in transmission percentage, decreases the light absorbed by the composite and hence the energy. To understand the suitability of the composite in other thermal system application further properties of the composite need to be performed. In the case of heat exchangers, heat sinks and building application, the thermal conductivity plays an important role. In the future studies, further weight composition needs to be prepared, thermal conductivity of the material, and thermal reliability needs to be investigated. These investigations may shed more knowledge on the suitability of A70 with Al_2O_3 on other thermal system application like heat exchanger, heat sinks, and concentrated photovoltaic thermal systems.

Acknowledgements

The authors thank the technical and financial assistance of Sunway University collaborative research fund: MRU 2019 (STR-RMF-MRU-004-2019) and (UMPEDAC-2018); HICoE, Ministry of Education.

References

[1] George M, Pandey AK, Abd Rahim N, Tyagi VV, Shahabuddin S, Saidur R. Concentrated photovoltaic thermal systems: A component-by-component view on the developments in the design, heat transfer medium and applications. *Energy Convers Manag* 2019;186. doi:10.1016/j.enconman.2019.02.052.

[2] Leong KY, Rosdzimin M, Rahman A, Gurunathan BA. Nano-enhanced phase change materials : A review of thermo-physical properties , applications and challenges. *J Energy Storage* 2019;21:18–31. doi:10.1016/j.est.2018.11.008.

[3] Nourani M, Hamdami N, Keramat J, Moheb A, Shahedi M. Thermal behavior of paraffin-nano-Al2O3 stabilized by sodium stearoyl lactylate as a stable phase change material with high thermal conductivity. *Renew Energy* 2016;88:474–82. doi:10.1016/j.renene.2015.11.043.

[4] Su Y, Zhang Y, Shu L. Experimental study of using phase change material cooling in a solar tracking concentrated photovoltaic-thermal system. *Sol Energy* 2018;159:777–85. doi:10.1016/j.solener.2017.11.045.

[5] Nada SA, El-Nagar DH, Hussein HMS. Improving the thermal regulation and efficiency enhancement of PCM-Integrated PV modules using nano particles. *Energy Convers Manag* 2018;166:735–43. doi:10.1016/j.enconman.2018.04.035.

[6] Al-Waeli AHA, Chaichan MT, Sopian K, Kazem HA, Mahood HB, Khadom AA. Modeling and experimental validation of a PVT system using nanofluid coolant and nano-PCM. *Sol Energy* 2019;177:178–91. doi:10.1016/j.solener.2018.11.016.

[7] Chen M, He Y, Ye Q, Zhang Z, Hu Y. Solar thermal conversion and thermal energy storage of CuO/Paraffin phase change composites. *Int J Heat Mass Transf* 2019;130:1133–40. doi:10.1016/j.ijheatmasstransfer.2018.11.026.

[8] Mayilvelnathan V, Valan Arasu A. Characterisation and thermophysical properties of graphene nanoparticles dispersed erythritol PCM for medium temperature thermal energy storage applications. *Thermochim Acta* 2019. doi:10.1016/j.tca.2019.03.037.

[9] Tang G, Wang X, Song G, Ma S, Yin Z. Preparation and characterization of flame retardant form-stable phase change materials composed by EPDM, paraffin and nano magnesium hydroxide. *Energy* 2010;35:2179–83. doi:10.1016/j.energy.2010.02.002.

IOP Conf. Series: Materials Science and Engineering 1127 (2021) 012011 doi:10.1088/1757-899X/1127/1/012011

Dissemination of Sustainable Cooking: A Detailed Review on Solar Cooking System

Hemant Raj Singh[*†1], Dilip Sharma [†2], Shyam Lal Soni [†3]

[*]Manipal University Jaipur, India 303007

[†1]Malaviya National Institute of Technology, Jaipur, India 302017

[†2]Malaviya National Institute of Technology, Jaipur, India 302017

[†3]Malaviya National Institute of Technology, Jaipur, India 302017

E-mail: hemant.singh@jaipur.manipal.edu, hemantrajs@gmail.com

Abstract. Cooking sector is considered as one of the major energy consuming sector in developing and underdeveloped countries. From the primitive age, humankind largely rely on biomass for their cooking energy need, even though there has been a tremendous upgradation in cooking methods. Due to hazardous effects and energy inefficiencies of conventional biomass cooking system, unconventional cooking energy (method) is becoming popular day by day. Dependency on ever depleting fossil fuel has forced us to switchover to renewable energy options and amongst all; solar energy seems best option due to its inherent reasons. In this study, efforts have been made to review the entire solar cooking systems incorporating with and without thermal energy storage available hitherto. The study reveals the different thermal performance parameters of solar cooking system and attempts to highlight the adaption factors of solar cooking system in the designated market.

1. Introduction

Energy consumption reflects the living standard of any country. Global energy demand is increasing very fast with the upgradation in living standard and development of the country. World energy consumption is growing with a fast rate as shown in Fig.1. [1].

Scarcity of fossil fuels and their harmful effects on environment are the two major challenges for today's world. As these fossil fuels are the major source of energy being utilized by the human, hence the energy demand depends on the availability of fossil fuels. Hence, with the increase in energy demand, risk the survival and sustainability of life support system. In this regard depletion of fossil fuel reserves, energy scarcity, global warming, and increase in pollutants in atmosphere are matters of great concern today.

In the era of global warming, climate justice has become the need of hour. Renewable energy is gaining popularity day by day by not only seeking attention of its researchers and end users, but also reasonably affects its policy makers, as the energy demand is growing very fast and supply is limited.

The thermal application of solar energy is seeking attention very rapidly in sustainable cooking by knowing the fact of inefficiencies of conventional cooking and ever rising cost of fossil fuels. Apart from that cooking energy also contributes 36% in the world's total primary energy consumption. Rural folks usually use wood, charcoal, agricultural residues and dung cakes as a cooking fuel.

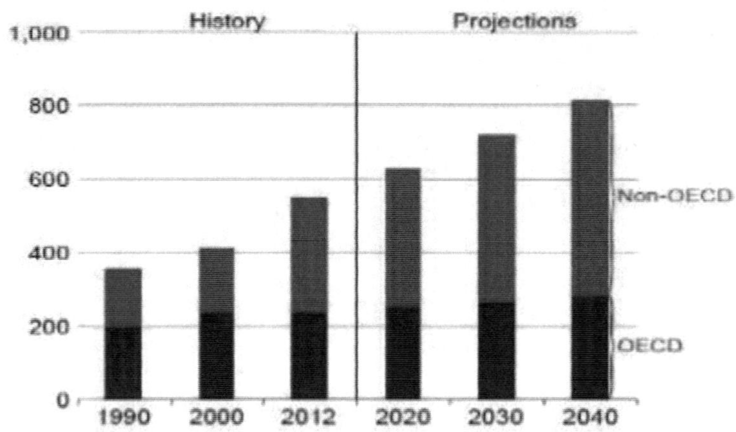

Figure 1. Energy consumption of the world for the period 1990– 2040 (in quadrillion Btu) [1]

86.1% of the rural households uses fuel wood and cow-dung for cooking [3]. In order to get, this cooking fuel around 16 million hectares of forest are destroyed annually [4]. These cooking fuels also inherently associated with indoor air pollution which results hazardous effect on rural folks.

Table 1.World energy consumption by country / region (in quadrillion Btu) [1]

		2012	**2020**	**2025**	**2030**	**2035**	**2040**
OECD	America	118	126	128	131	134	138
	Europe	81	85	87	90	93	96
	Asia	39	43	45	46	47	48
NON OECD	Europe	51	52	55	56	58	58
	Asia	175	223	246	270	295	322
	Africa	22	26	30	34	38	44
	Middle East	32	41	45	51	57	62
	America	31	33	37	40	43	47
Total		**549**	**629**	**673**	**718**	**765**	**815**

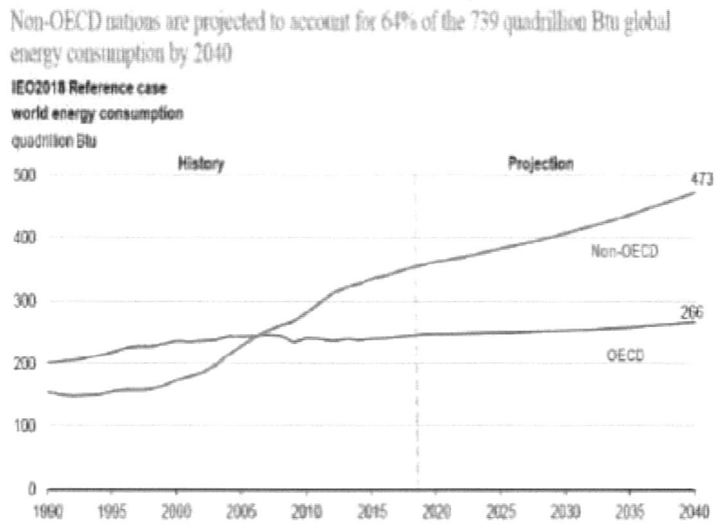

Figure 2. World energy consumption by energy source (in quadrillion Btu) [2]

This paper will perform a state-of-the-art review about solar cooking system and the objective was kept in mind while performing this study is to identify availability of different solar cooking system and provide a framework for researchers and policy makers in solar cooking areas to increase its adaptability in concerning area.

2. Solar Cooker

Definition: Solar cooker term usually associated with the device or combinations of devices that are utilised for cooking processes like boiling, frying, baking and grilling processes, which involve its prime heat energy by solar radiation directly or indirectly.

2.1. History:

History of solar cooking is very old, it had been started in 1767, when Nicholas de-Saussure was first attempted as usage of solar energy for fruit cooking [16]. In 1876, in India, Adams developed a solar cooker which consists of an octagonal oven with 8 mirrors and it was designed for 7 soldiers. Secretary of the A.S.I., Abbot, was the first recorded inventor of solar cookers in which the solar heat receiver was outside the house but the cooking stove was inside. Heat energy was transferred from solar receiver to stove by thermic fluid [17]. In 1940s, in the U.S.A, Telkes investigated different types of solar cookers, which also involved the heat storage type [18]. In 1945, in India, Ghosh's solar box cooker was commercialized, which was first of its kind [19]. In1973, in the USA, Kerr fabricated several solar cookers, by using simple low cost and high thermal inertia material, which involve focusing i.e. concentrating type and non-focusing i.e. box type solar cookers [20]. In 1980s, the governments of India and China expanded national promotion of box cookers. A lot of researches have been carried out in this era. Brace Research Institute, McGill U., Canada, researched and field tested solar cookers. Prof. S.S. Nandwani in Costa Rica researched solar cookers. Box cookers were distributed to 20,000 Afghan refugees in Pakistan by SERVE (Serving emergency Relief and Vocational Enterprises). M.H. Gurley Larson wrote first U.S. solar cookbook, Solar Cooking

Naturally, in 83 Metcalf. In 84 Metcalf published his 9-page instructions for building solar box cookers. ULOG was started by Ulrich and Lisel Oehler to promote box and parabolic cookers in many countries. With Kerr–Cole instructions, solar cookers were built in Kitui, Kenya in'85[21]. In 1987, in India, Mullick presented two TPP (Thermal Performance parameters) i.e. two figure of merits (F1 and F2) for box-type solar cookers [22]. In 2000, Funk presented universally adopted international standard i.e. cooking power (W) for thermal assessment of solar cooker. It is more generic that means applicable for box type and concentrating type solar cooker and useful for comparison of dissimilar designs also. It is also recognized by ASAE (American Society of Agriculture Engineers) [23].

2.2. Classification:

There are many type of solar cookers found in the literature. Although there are no fixed criteria for classification of solar cooker, however some researchers have classified. In this review study, an attempt has been made to classify all the solar cooker available hitherto with the help of a hierarchy chart as shown in figure 3.

The first broad classification is based on mode of heat transfer to cooking vessel from cooking system and are classified as direct type and indirect type. In direct type of solar cooking system, solar radiation directly take part in cooking process. In indirect type of solar cooking system, solar radiation gives its heat to a particular fluid i.e. thermic fluid, then this thermic fluid transfers the heat to cooking vessel. Direct type of solar cookers are again classified as concentrating and non- concentrating type. Concentrating type further classified on the basis of nature of concentrator i.e. opaque and non- opaque. In non-opaque type solar cooker, sun rays transmits through concentrator and focus on a single point i.e. cooking vessel. In opaque type solar cooker, solar radiation reflected by its concentrator to an appropriate point i.e. cooking vessel. Paraboloid, Scheffler, and Fresnel type are the most common examples of opaque type concentrating solar cooker. Different designs of Box type solar cookers are the most common examples of non-concentrating type solar cooker.

Indirect type solar cookers are of concentrating type only, as intense heat is required to heat the thermic fluid which transport its heat to cooking vessel. Indirect type is again classified as single generation and multi generation system. Single generation system is designed for cooking only while in case of multi generation system excess heat would utilised for another process apart from cooking. Single and multi- generation are both again classified on the basis of their focus i.e. point and line focus. Parabolic Trough Collector (PTC) is most common example of line focus solar cooker.

3. Performance Assessment

Performance assessment parameters was first available for box type then concentrating type solar cooker. Several advances are done on box type solar cooker by researcher especially in tropical country like India due to better availability of solar radiation [5-10]. These countries are having better solar radiation e.g. many Indian regions receive 5–7 kWh/m^2 as average daily solar radiation. Number of clear sky days per year in India is greater than 275 [24]. This implies solar cooking seems high adaptability capacity in these countries and can provide a feasible alternate in the households.

In 2000, Funk presented TPP like Cooking power (P), Standardized cooking power (P_s), Adjusted cooking power (P_a) and Heat loss coefficient (U_L) [12]. In 1987, Mullick presented TPP like First figure of merit (F_1), Second figure of merit (F_2), Heat loss factor ($F' U_L$) and Optical efficiency factor ($F' \eta_o$) [13]. In 1994, in Germany, ECSCR (European Committee on Solar Cooking Research) testing standard presented different factors like thermal testing regime, safety factors, ease of cooking, cooking pot access and estimated durability [14]. In 2002, Shaw presented several factors (associated with assessment) in Ph.D. thesis. These factors are factors due to environment e.g. "wind, ambient temperature, insolation, and precipitation; controlled factors e.g. cooking vessels, tracking, time, thermal loading, data collection, and recording; thermal figures of merit" e.g. "cooking power, standard cooking power, standard stagnation temperature, standard sensible heating time, heating time, and unattended cooking time and utility figures of merit" e.g. "cooking power per kilogram and aperture area per kilogram" [15].

IOP Conf. Series: Materials Science and Engineering 1127 (2021) 012011 doi:10.1088/1757-899X/1127/1/012011

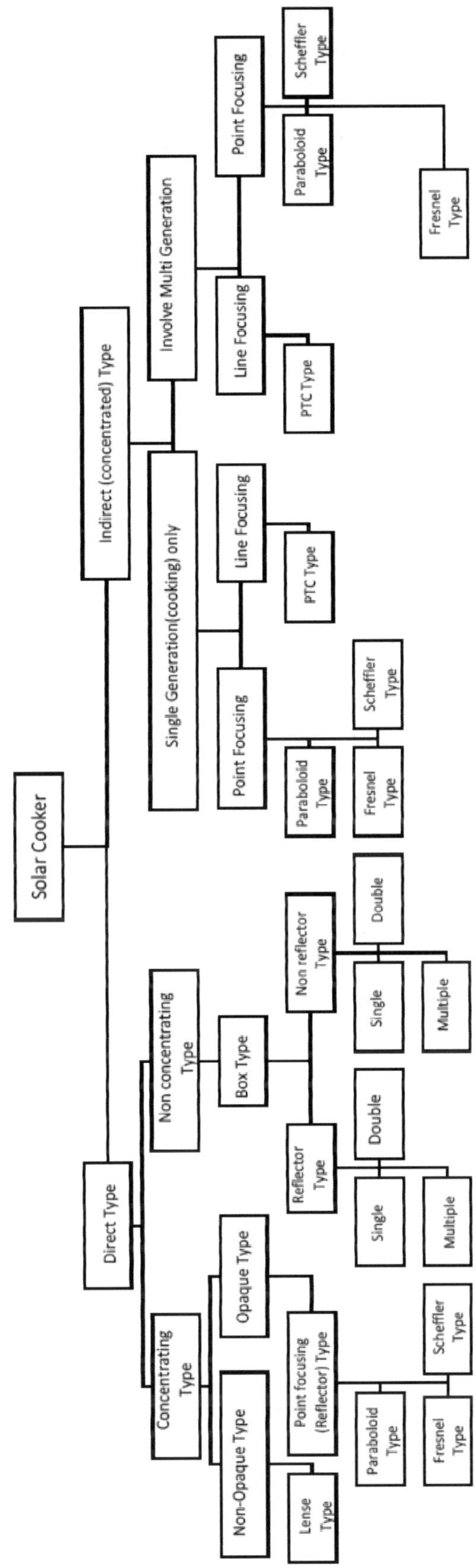

Figure 3. Detailed Classification of Solar Cooker

4. Conclusion

From the past several decades sustainable/solar cooking gradually receiving new advances in energy deprivation. However, paucity of actualisation of solar cooking is still there. Adoption of solar cooker is still far below their potential.

There can be several reasons for the lack of dissemination of solar cooker in the designated market. Study reveals some factors are more prominent e.g. intermittent nature of sunshine, suitable cooking place, suitable cooking pots, cost, ethnical and geographical issues, different region different cooking pattern/practice and micro level cooking are more preferred as compared to community cooking.

Rural area is also a concerning area for researchers as they constitute a significant portion of country's population and quenching of energy deprivation in that area has to coincide with the requirements of the rural folks for their overall socio-economic development.

Finally, the study recommends for researcher and policy makers to incorporate some factors like suitable cooking place, suitable cooking pots, cost effectiveness, different region different cooking pattern/practice and adaptability of micro level cooking over macro level in the solar cooking system to reduce the reluctance to acceptance or risk of dissemination in the designated market.

References

[1] EIA, International Energy Outlook 2016, https://www.eia.gov/forecasts/ieo/pdf/0484(2016).Pdf, 2016.

[2] EIA, International Energy Outlook 2018, https://www.eia.gov/pressroom/presentations/Capuano_07242018.pdf,2018.

[3] Pohekar, S. D., Dinesh Kumar, and M. Ramachandran. "Dissemination of Cooking Energy Alternatives in India - A Review." *Renewable and Sustainable Energy Reviews*, 2005; 9 (4): 379–93. https://doi.org/10.1016/j.rser.2004.05.001.

[4] Umanand, L., and J. R. Prasanna. "Modeling and Design of a Solar Thermal System for Hybrid Cooking Application." *Applied Energy*, 2011;88: 1740–1755

[5] Garg H.P, Thanvi K.P. Simple hot box type solar cooker. *Res Indus*, 1976;21(3):184-6

[6] Nahar N.M. Performance and testing of an improved hot box solar cooker. *Energy Convers Manage*, 1990.;30(1):9-16.

[7] Kumar S, Kandpal TC, Mullick SC. Solar cooker use in Delhi: Pilot survey of urban households. *Int J Ambient Energy*, 1997;18(2):77.

[8] Balasubramanium V, Philips SK, Singhal AK. Performance testing of box type solar cookers with differing vessel combinations. *SESI J (India)*, 1990;4(1):1-25.

[9] Thulasi Das TC, Karmarkar S, Rao DP. Solar box cooker: Part I—Modelling. *Solar Energy*, 1994;52(3):265-72.

[10] Thulasi Das TC, Karmarkar S, Rao DP. Solar box cooker: Part II—Analysis and simulation. *Solar Energy*, 1994;52(3):273-82.

[11] Kalogirou Solar Energy Engineering Processes and Systems: Academic Press,2013.

[12] Funk, P. A., "Evaluating the International Standard Procedure for Testing Solar Cookers and Reporting Performance," *Sol. Energy*, 2000; 68(1), pp.1–7

[13] Mullick, S. C., Kandpal, T. C., and Saxena, A. K., "Thermal Test Procedure for Box-Type Solar Cookers," *Sol. Energy*, 1987; 39(4), pp. 353–360.

[14] ECSCR, "Solar Cooker Test Procedure," Version 3, European Committee on Solar Cooking Research, Eschborn, Germany, 1994.

[15] Shaw, S., "Development of a Comparative Framework for Evaluating the Performance of Solar Cooking Devices," Ph.D. thesis, Rensselaer Polytechnic Institute, Troy, NY, 2002.

[16] Hottel HC, Woertz BB. The performance of flat plate solar heat collectors. *Trans ASME*, 1942; 64:91–104.

[17] Telkes M. Solar cooking ovens Princeton, New Jersey: Solar Energy Laboratory, Curtiss–Wright Corporation, 1955.

[18] Duffle JA, Lof GOG, Beckman B., Laboratory and field studies of plastic reflector solar cookers.

Solar Energy,1962;6(8).
[19] Lof GOG. Recent investigations in the use of solar energy for cooking. *Solar Energy*,1963;7(3),
[20] Kerr BP, The expanding world of solar box cookers. Taylor, AZ: Kerr-Cole Solar Box Cookers, (1991).
[21] Saxena, Abhishek, Varun, S. P. Pandey, and G. Srivastav. "A Thermodynamic Review on Solar Box Type Cookers." *Renewable and Sustainable Energy Reviews*, 2011;15(6): 3301–18. https://doi.org/10.1016/j.rser.2011.04.017
[22] Funk PA. Evaluating the international standard procedure for testing solar cookers and reporting performance. Solar Energy, 2000;68(1):1–7
[23] ASAE S580 JAN-03. "Testing and Reporting Solar Cooker Performance", 2003.
[24] Mani A, Rangarajan S. Solar radiation over India. New Delhi: Allied Publishers; 1982.
[25] Singh, H.R. "A Step towards Energy Efficiency in Solar Thermal Energy: Solarized Trigeneration-A Review." Poster Presentation. Rapperswil: International Solar Energy Society (ISES) ,2018, https://www.youtube.com/watch?v=yeU5HVDX870,
[26] Duffie, John A., and William A. Beckman.,Solar Engineering of Thermal Processes: 2013; Fourth Edition. Solar Engineering of Thermal Processes: 4ed., 53. https://doi.org/10.1002/9781118671603,

Application of Phase Change Materials in Solar Water Heating Systems for Thermal Energy Storage

Dheeraj Kumar Nagilla[1], V. V. Tyagi[1,*], Kumaran Kadirgama[2], K. Chopra[1,3], A. K. Pandey[4], Richa Kothari[5]

[1]School of Energy Management, Shri Mata Vaishno Devi University, Katra, Jammu & Kashmir, India-182320

[2]Faculty of Mechanical Engineering, University Malaysia Pahang, Malaysia- 26600

[2]Faculty of Mechanical & Manufacturing Engineering, Universiti Malaysia Pahang, 26600 Pekan, Pahang, Malaysia.

[3]School of Mechanical Engineering, Shri Mata Vaishno Devi University, Katra, Jammu & Kashmir, India-182320

[4]Research Centre for Nano-Materials and Energy Technology (RCNMET), School of Science and Technology, Sunway University, No. 5, JalanUniversiti, Bandar Sunway, Petaling Jaya, 47500 Selangor Darul Ehsan, Malaysia

[5]Department of Environmental Sciences, Central University of Jammu, Rahya Suchani, Samba Jammu and Kashmir 181143

E-mail: v.tyagi@smvdu.ac.in, vtyagi16@gmail.com

Abstract. One of the major drawbacks of solar water heating systems is unable to supply hot water during night time or off sunshine hours. The integration of phase change material with solar water heating systems is cost effective and efficient solution to overcome this major problem associated with solar water heating systems. The phase change material integrated with solar water heating system stores thermal energy during sun shine hours and this stored energy can be recovered during off shine hours or night time to produce hot water. The phase change material can be integrated with water tank of collector, evacuated tubes, external water tank for solar collector and flat plate collector by adding layers at the bottom of absorber plate. The integration of phase change material with SWH system not only overcome the drawbacks of SWH system but also enhance the efficiency of conventional SWH system. Many investigations for the application of TES materials integrated SWH system have been carried out and found a significant enhancement in the performance. This paper presents a comprehensive review of recent advances in the applications of PCM with SWH system for TES.

1. Introduction

As a renewable energy source, solar energy is free, environmentally friendly and available in abundant. Solar energy can be converted into thermal energy using solar collectors. Solar collectors are used in many industrial, commercial and domestic applications such as solar water heating, solar space heating, solar drying, solar desalination [1]. Also, there is a consumption of 75% of the total energy for the production of hot water in the domestic, commercial and industrial sectors. There is a huge demand for the production of hot water using solar collectors. Solar water heating (SWH) system is popular as it covers about 80% of the worldwide market of solar systems [2]. However, for the production of hot water, solar energy is relatively being used only to a small extent due to the intermittent availability of solar radiation and uneven load profiles of domestic hot water consumption. There must be deployment of viable technology to mitigate the impact of these factors. Thermal energy storage is a technology that stores the thermal energy in the form of heat during daytime and recovers heat during nigh time or cold weather conditions. The integration of phase change materials (PCMs) with SWH systems can improve the system's performance by storing solar energy during daytime and recovering the stored heat during night-time [3,4]. In the PCM integrated SWH system, PCM acts as heat-absorbing units and SWH system heat the water. Also, the integration of PCM with SWH helps to increase solar fraction. The rise and drop of water temperature in SWH system throughout the day can be controlled by improving the PCM's characteristics and its configuration. The integration of PCM with novel properties acts as a solution to overcome the drawbacks of SWH system and helps to promote the SWH system in the market worldwide as well as reduce the SWH system costs. This review paper presents a detailed work of the application of the PCM integrated SWH systems. This manuscript provides the ongoing research work opportunities to enhance the performance of SWH system with the use of PCM's materials

2. Phase Change Material

More than 150 potential PCMs are being used as thermal energy storage materials in solar thermal energy systems [5]. Thermal energy storage materials are classified into sensible heat thermal energy storage (SHTES) materials, latent heat thermal energy storage (LHTES) materials and chemical heat storage materials. SHTES materials are those materials that store thermal energy with a rise in its temperature and vice-versa. LHTES materials store heat by changing their phase and vice-versa. Some of the desirable characteristics required for efficient usage of TES are constant operating temperature, high energy density, high heat transfer rate [6]. Of the above-mentioned characteristics, LHTES shows most of them. Therefore, this method of energy storage has attracted a large number of applications. LHTES materials are being used in various applications such as solar thermal energy systems, space heating and cooling, and peak load reduction. When compared to sensible heat storage (SHS) method, the latent heat storage (LHS) method provides enhanced energy storage density with a small temperature difference. Phase change material (PCM) is a type of LHTES material stores and releases thermal energy in the form of heat by changing its phase under nearly constant temperature. There is an availability of a large number of PCMs in required temperature range (-5°C to 190°C). Therefore, PCMs are used in many applications such as solar energy storage, industrial waste heat recovery, heat pumps, spacecraft thermal control applications, thermal storage in satellites, solar heating systems, energy-efficient buildings, heating and cooling of buildings, building air conditioning, central air-conditioning systems, indoor temperature controlling, underfloor heating system, building envelope, refrigeration system, temperature-adaptable greenhouses, electronics cooling, protective clothing, textiles, thermo-regulating fibres, preservation of food. Some of the novel thermo-physical properties of PCM are such as favourable phase equilibrium, high density, small volume change, low vapour pressure, no supercooling, sufficient crystallization rate, long-term chemical stability, compatibility with materials of construction, non-toxic, non-fire hazard, abundantly available, cost-effective. PCMs are being selected on the basis of their thermo-physical properties, kinetic properties, chemical properties, and economics. There are three categories of PCM which are solid-solid phase-changing materials, solid-liquid phase-changing materials, solid-gas phase-changing materials. Of all the three categories, there is wide use of solid-liquid PCMs. For any effective application, the required thermal properties of PCM are high latent heat of fusion, high thermal conductivity, less volume change during phase transition, little or no supercooling, low vapour pressure in the melt, chemically inert and stable, self-nucleating, no phase segregation. Solid-liquid PCMs consists three groups of which are organic, inorganic and eutectic solid-liquid PCMs. Organic solid-liquid PCMs includes paraffin, fatty acids, and sugar alcohols [7,8]. Inorganic solid-liquid PCMs includes salt hydrates, salt solutions, and metals. Eutectic solid-liquid PCMs are developed according to the requirement for desired operating temperature of an application. Some of the characteristics of the inorganic group of solid-liquid PCMs which makes it less reliable than organic solid-liquid group of PCMs are super-cooling, phase segregation and nucleation [9]. The advantage of organic solid-liquid PCMs are non-corrosive, chemically stable, exhibit little or no sub-cooling, compatible with most building materials, high latent heat per unit weight, low vapour pressure. Organic solid-liquid PCMs also possesses some of the drawbacks such as low thermal conductivity, high changes in volume on phase change, flammability. Among the organic solid-liquid PCMs, paraffin has been widely used due to its advantageous characteristics such as high enthalpy of phase change, small segregation of components, and small changes in structure during repeated phase transitions, negligible supercooling, low vapour pressure, self-nucleating behaviour, and low cost. The melting temperature range of paraffin is 25-100°C [10,11,12]. Capric acid, lauric acid, myristic acid, stearic acid, palmitic acid are some of the fatty acids comes under the same class of organic solid-liquid PCMs. Where stearic acid and palmitic acid have superior properties such as easy availability, congruent melting/ freezing, good thermal and chemical stability, non-toxicity, and suitable phase change temperature. In spite of these desirable properties of the mentioned fatty acids, high phase change temperature, corrosiveness, bad odour (especially sublimating during heating), poor thermal stability, and thermal conductivity limit are some disadvantages. Therefore, PCMs possess many advantages and disadvantages due to which their application is in limited use. Many researchers are working on material characteristics of PCMs to enhance their performance for the efficient use of PCMs. The novel properties of PCM act as a solution to remove the drawbacks of SWH system [13,14,15].

3. Application of PCM in SWH Systems

There are various research studies are available on the applications of phase change materials integrated SWH system. The studies revealed that thermal performance of SWH system can be enhanced by integration of PCM storage unit with high latent heat and a large area for heat transfer. Al-Kayiem et al. [16] experimentally investigated the performance of flat plate solar collector (FPSC) based SWH systems by integrating nanocomposite PCMs at various inclination angles with SWH system as shown in the Fig. 1. As a nanocomposite PCM, a mixture of paraffin wax and 20-nm copper nanoparticles of 1wt% is considered in this study as shown in the Fig. 2. They considered three cases comprising of SWH without PCM, SWH with PCM and SWH with Cu-Paraffin wax nanocomposite and performed different observations at 10°, 20°, 30° inclination angles of FPSC, respectively. They found that the water flow rate of 0.5kg/min and inclination angle of 10° is the optimal operation conditions for the high performance of the system. The results indicated that the obtained water tank temperatures of SWH without PCM, with paraffin wax as PCM and with Cu-paraffin wax as PCM were 35.1°C, 40.1°C and 40.7°C respectively. The efficiency of SWH (at 10° inclination angle) without PCM, with paraffin wax, with Cu-paraffin wax were 47.6%, 51.1%, 52%, respectively. They found a significant enhancement in the efficiency of SWH system with paraffin wax used as PCM compared to SWH system without PCM. Further they found that there was no significant enhancement in the efficiency of SWH system was observed when Cu-paraffin wax used as PCM. They recommended that further investigation should be done at various flow rates. They also found an enhancement in the thermal conductivity of Cu-paraffin wax nanocomposite compared to paraffin wax. There is an efficiency rise of 6.9% and 8.4% when solar collector integrated with paraffin wax and Cu-paraffin wax nanocomposite as PCM respectively. During morning time, they found that the SWH with paraffin wax and SWH with Cu-paraffin wax produced hot water of 40.1°C, 40.7°C.

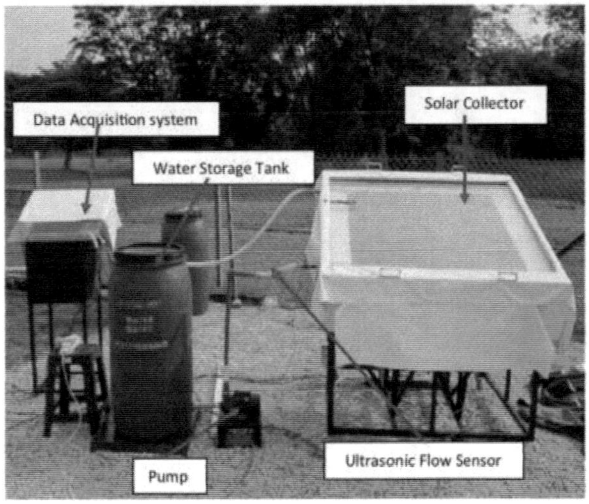

Figure 1. Solar water heating system [16].

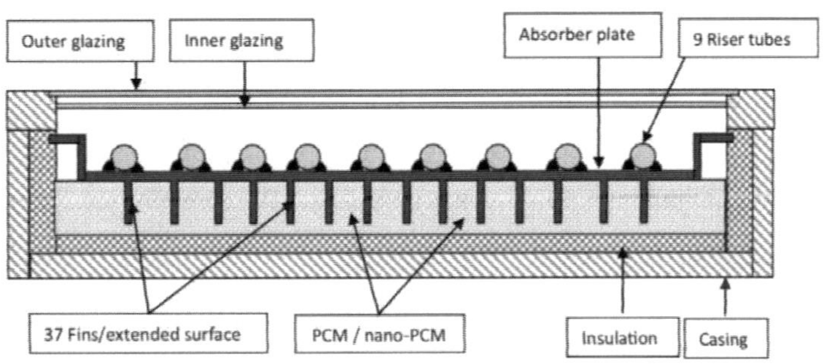

Figure 2. Flat plate collector filled with Cu-paraffin nanocomposite [16].

Due to the large demand for hot water in residential buildings, more attention is being paid to solar water heating systems. In China, there is a wide use of centralized solar water heating systems due the consumption of huge amounts of hot water in multi-story buildings. In the same context, Zhou et al. [17] experimentally investigated the performance of a centralized solar water heating system by retrofitting phase change materials (PCMs) as a thermal energy storage (TES) unit as shown in the Fig.3. The TES unit comprised of paraffin wax which was packed in stainless steel ball. Also, expanded graphite was added in order to enhance the rate of heat transfer. They selected two PCMs with a melting temperature of 55°C and 60°C according to the operating temperature range of SWH system. At different values of solar radiations, they investigated the heat storage, heat-releasing and insulation properties of water tank equipped with paraffin heat storage unit. Depending on the users demand, the solar fraction of the PCM-SWH system is also calculated. The heat storage time for paraffin with melting temperatures 55°C and 60°C was about 2-4.5 hours. When SWH integrated with a PCM with melting temperature of 55°C, the rate of heat release and solar fraction were 78% and 75%, respectively. When SWH integrated with a PCM with melting temperature of 60°C, the rate of heat release and solar fraction were 83% and 80%, respectively. They found a performance enhancement in a centralized solar water heating system integrated with paraffin as PCM compared to the system without PCM. By Analysing the experimental results, they recommended using paraffin with 60°C melting temperature in centralized solar water heating systems. They also stated that there is an effective rise in the heat storage rate of paraffin when it is changed from PVC unit to stainless steel ball unit along with expanded graphite.

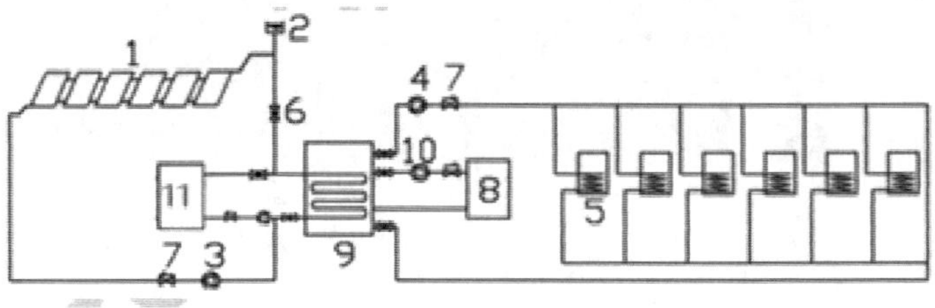

Figure 3. Centralized solar hot water heating system in residential building [17].

Felinski et al. [18] investigated the performance of heat pipe based evacuated tube solar collector (HP-ETSC) by integrating it with paraffin wax as phase change material as shown in the Fig.4. They conducted experiments to investigate the impact of paraffin wax on the system's performance. They found that operating time of the HP-ETSC system was increased by integration of paraffin wax as PCM. Also, due to the integration of paraffin wax, they found less heat loss from the system due to the decrease in the mean temperature of the system and the temperature of the water can be brought up to 45°C. Depending on the water flow rate during discharge cycle, the amount of useful heat from the paraffin-HPETSC was increased by 45-79%. PCM was integrated with SWH system to increase the density of stored heat which was obtained from the conversion of solar energy into thermal energy. PCMs are being integrated within the systems such as by inserting PCM in the water tanks of the SWH system, by filling PCM in the Evacuated collector tubes, adding PCM layers in the flat plate collectors; and or filled PCM in an additional external tank. Many researchers are working to develop simple and compact collector-storage systems in order to save space and reduce the heat losses followed by increased system's thermal efficiency. Such a compact collector –storage SWH system, simultaneously absorbs and stores the solar energy. But one of the drawbacks of this collector-storage SWH system is loss of heat during the night time when solar irradiance is unavailable. Various researchers found solutions such as by changing the water tank's shape, inserting baffled plates, insulating with transparent or non-transparent material. They found that the system's thermal efficiency of 62.6% was obtained at conditions when the paraffin's weight is 13.82 kg, collector inclination angle is 60°, and solar radiation is 900 W/m^2. They also performed an experiment on the same type of SWH system without paraffin (PCM) to investigate the impact of paraffin wax on SWH system. They conducted simultaneous experiments on ETC/Storage system and ETC system without storage. They followed two measurement procedures which were (i) heat reception-heating medium passing through the heat collection unit, (ii) without heat reception-heating medium bypassing the heat collection unit as shown in the Fig.5. In the both ETC and ETC/S systems, they maintained a flow rate of $0.02kg/sm^2$ during the charging process of paraffin wax for both the tests (with and without heat reception). Whereas, during the discharge cycle in order to investigate the amount of heat recovered from the paraffin during night time, they conducted a series of experiments containing with and without heat reception as shown in the Fig.3. Since at higher operating temperature, the rate of heat loss was more, therefore working temperature of 45°C was required to obtain useful heat for water heating for domestic purposes. As shown in Fig. 6, during the charge cycle for up to 250 minutes, the rate of increase in the heating medium's mean temperature and the final mean temperature values were higher in case of ETC system compared to ETC/S system. It was observed that ETC/S could able to generate hot water up to 45°C only after 200 minutes in comparison to heat reception due to the presence of paraffin as PCM in the system. Therefore, the proposed ETC/S can be utilized to obtain the desired temperature of about 45°C after 200 minutes. They observed that after the completion of 250 minutes of study, there was a reduction in the heat losses by 31% and 32% in case of without and with heat reception modes of ETC/S system, when compared to the ETC system without storage. Compared to ETC system without PCM, depending on the flow rates, the amount of useful heat received was 45-79% higher than ETC/S system.

Figure 4. Evacuated tube collector-storage system [18].

Abokersh et al. [19] developed a compact SWH system comprised of U-pipe ETC integrated with paraffin (ALEX WAX 600) as a storage unit. The characteristics of paraffin (ALEX WAX 600) organic-PCM are 60°C melting temperature, and 0.21W/mK thermal conductivity. To enhance the rate of heat transfer of paraffin in the collector, a fin of $0.1251\,m^2$ area was placed inside the collector. In this study, they considered two system comprising (i) finned U-pipe ETC based SWH system and (ii) un-finned U-pipe ETC based SWH system as shown in the Fig. 7.

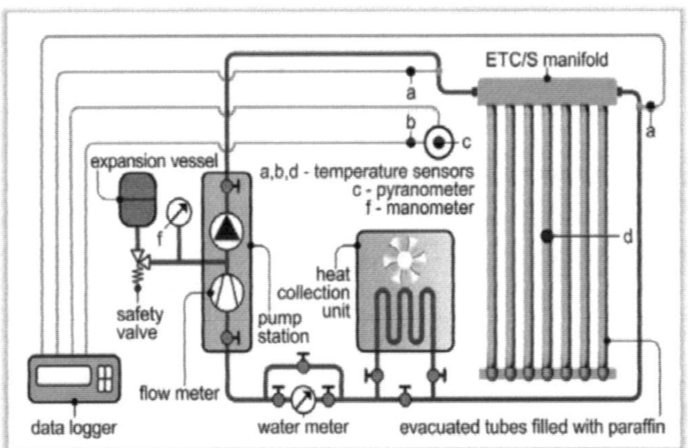

Figure 5. Schematic diagram of the collector-storage SWH system [18].

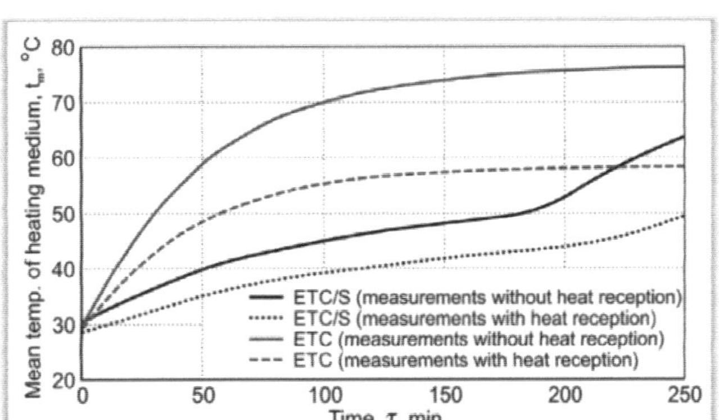

Figure 6. Mean temperature of heating medium during charge cycle [18].

They also compared the performance of these two systems with a typical forced recirculation solar water heating (FSWH) system under same operating conditions. They found that due to lower thermal inertia, there was an enhancement in the performance of two compact systems over a typical SWH system operating under forced recirculation mode tested at various operating conditions. Also, the arrangement of fin enhanced the rate of heat transfer of paraffin which further increased the stability of compact SWH system. They also found that the amount of energy discharge from the finned and un-finned U-pipe ETC based SWH system was 35.8% and 47.7% higher, respectively in comparison to the FSWH system. They stated that efficiency of finned, un-finned and typical SWH system were 85.7%, 71.8%, and 40.5% respectively. The solar fraction of finned, un-finned and typical SWH system were 40.1%, 50.5%, and 32% respectively.

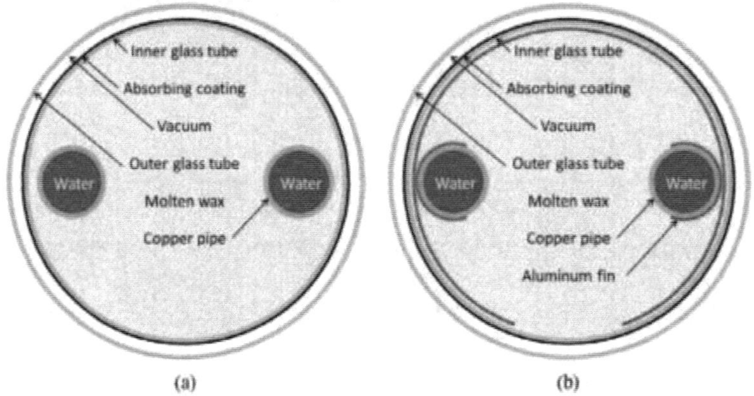

Figure 7. (a) un-finned and (b) finned based U-pipe ETC [19].

Kanimozhi et al. [20] developed an experimental setup of an SWH system integrated with TES system and used paraffin wax as LHS material and honey wax as SHS material to enhance the heat transfer rate and energy storage of TES system as shown in the Fig. 8. In this study, TES system is a cylindrical tank fixed where copper tubes were filled with the paraffin and honey waxes as shown in the Fig. 9. It was found that this TES system is beneficial for both isothermal and non-isothermal storage processes. This TES system improved the heat transfer rate of water passing from the water tank of SWH system to the waxes in the cylindrical unit. They examined the charging and discharging processes of PCMs and observed enhancement in heat absorption, heat transfer, and heat rejection. They found an enhancement of 40% in the heat absorption, heat transfer, and heat rejection processes.

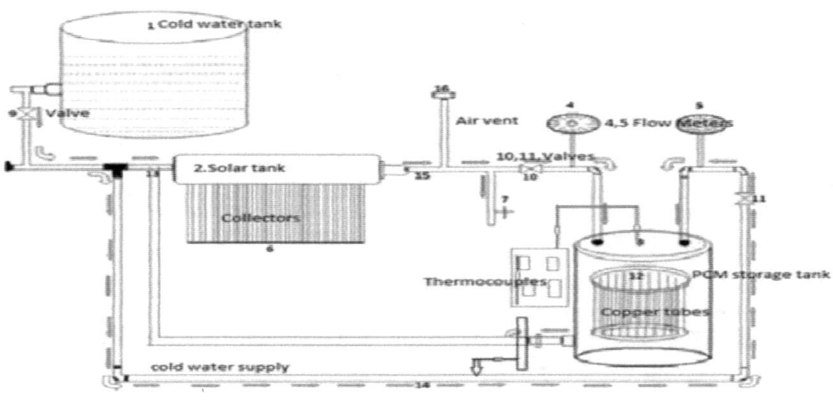

Figure 8. Schematic diagram of experimental setup [20].

They also developed a test rig to measure the system's performance and took total year measurements at all climatic conditions and continued for a couple of successive years. They collected measurements of average efficiency of collector, efficiency of collector, water outlet temperature, coefficient of performance (COP) and made a comparative study of a system without and with PCM. They found that there was an enhancement in the performance of the SWH system with PCM compared to the system without PCM. Compared to the system without PCM, the system with PCM can able to exhibit enhance performance, achieved higher efficiency, hold water at required temperatures, overcome the impact of solar radiation fluctuations. During daytime, they found 30% lesser fluctuations in collector efficiency in the system with PCM compared to the system without PCM.

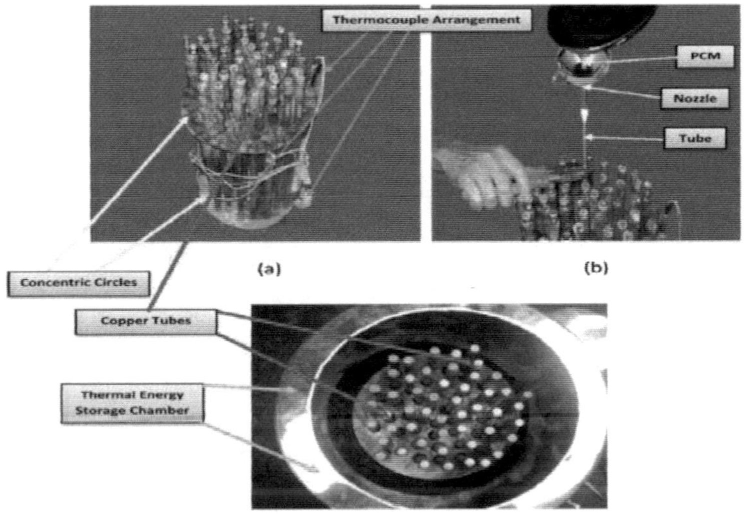

Figure 9. (a) Copper tubes fixed in concentric circles (b) pouring PCM in copper tubes (c) Arrangement of copper tubes in TES system [20].

During summer nights, the system with PCM was able to achieve 50°C water outlet temperature but it was lesser than system without PCM. During winter nights, the COP of system with PCM was 3.0 which made to achieve water outlet temperature of 50°C in less time compared to system without PCM.

Figure 10. Collector-storage based SWH system [21].

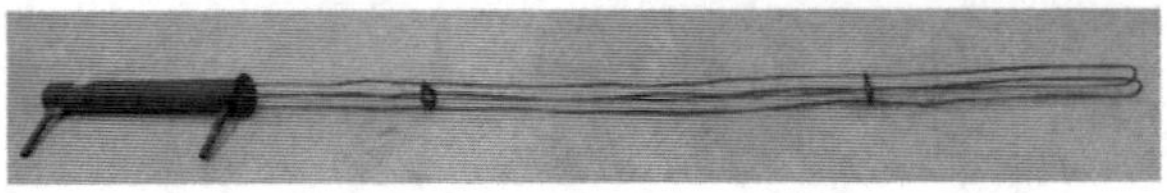

Figure 11. Oscillating heat pipe (OHP) [21].

4. Conclusion

The SWH system absorbs solar energy and converts into thermal energy in the form of heat which further utilized to produce hot water. Since solar energy is free available and environmentally friendly which having significant potential for the use to hot water through SWH system. Thermal energy storage (TES) materials are available in the temperature range of 50-70°C temperature ranges which can integrate with the SWH system to store solar energy during daytime and recovered it during night time to produce hot water. This manuscript offers an outline of the latest investigations for SWS system and its performance by the integrating different solar collectors with the PCM unit. This type of new innovative systems has vast possibilities for domestic and industrial sector to reduce energy consumption for hot water production. A collector-storage water heating system is extremely endorsed for low/medium temperature applications as they can store solar energy in the form of latent heat during daytime and can provide heat at night time or unavailability of sun radiation to produce hot water. It was found that the PCM's based Solar Water Heating systems having high potential to replace existing system to enhancing the thermal efficiency of SWH system.

Acknowledgements

One of the author (Kumaran Kadirgama) gratefully acknowledges the Universiti Malaysia Pahang Internal grant (RDU1803176).

References

[1] Sabiha, M. A., Saidur, R., Mekhilef, S., & Mahian, O. (2015). Progress and latest developments of evacuated tube solar collectors. *Renewable and Sustainable Energy Reviews*, *51*, 1038-1054.

[2] Shukla, R., Sumathy, K., Erickson, P., & Gong, J. (2013). Recent advances in the solar water heating systems: A review. *Renewable and Sustainable Energy Reviews*, *19*, 173-190.

[3] Sadhishkumar, S., & Balusamy, T. (2014). Performance improvement in solar water heating systems—A review. *Renewable and Sustainable Energy Reviews*, *37*, 191-198.

[4] Gautam, A., Chamoli, S., Kumar, A., & Singh, S. (2017). A review on technical improvements, economic feasibility and world scenario of solar water heating system. *Renewable and Sustainable Energy Reviews*, *68*, 541-562.

[5] Zalba, B., Marın, J. M., Cabeza, L. F., & Mehling, H. (2003). Review on thermal energy storage with phase change: materials, heat transfer analysis and applications. *Applied thermal engineering*, *23*(3), 251-283.

[6] Sharma, A., Tyagi, V. V., Chen, C. R., & Buddhi, D. (2009). Review on thermal energy storage with phase change materials and applications. *Renewable and Sustainable energy reviews*, *13*(2), 318-345.

[7] Sharma, R. K., Ganesan, P., Tyagi, V. V., Metselaar, H. S. C., & Sandaran, S. C. (2015). Developments in organic solid–liquid phase change materials and their applications in thermal energy storage. *Energy Conversion and Management*, *95*, 193-228.

[8] Da Cunha, J. P., & Eames, P. (2016). Thermal energy storage for low and medium temperature applications using phase change materials–a review. *Applied Energy*, *177*, 227-238.

[9] Lin, Y., Alva, G., & Fang, G. (2018). Review on thermal performances and applications of thermal energy storage systems with inorganic phase change materials. *Energy*.

[10] Vadhera, J., Sura, A., Nandan, G., & Dwivedi, G. (2018). Study of Phase Change materials and its domestic application. *Materials Today: Proceedings*, *5*(2), 3411-3417.

[11] Sharif, M. A., Al-Abidi, A. A., Mat, S., Sopian, K., Ruslan, M. H., Sulaiman, M. Y., & Rosli, M. A. M. (2015). Review of the application of phase change material for heating and domestic hot water systems. *Renewable and Sustainable Energy Reviews*, *42*, 557-568.

[12] Seddegh, S., Wang, X., Henderson, A. D., & Xing, Z. (2015). Solar domestic hot water systems using latent heat energy storage medium: a review. *Renewable and Sustainable energy reviews*, *49*, 517-533.

[13] Wang, Z., Qiu, F., Yang, W., & Zhao, X. (2015). Applications of solar water heating system with phase change material. *Renewable and Sustainable Energy Reviews*, *52*, 645-652.

[14] Kee, S. Y., Munusamy, Y., & Ong, K. S. (2018). Review of solar water heaters incorporating solid-liquid organic phase change materials as thermal storage. *Applied Thermal Engineering*, *131*, 455-471.

[15] Zayed, M. E., Zhao, J., Elsheikh, A. H., Hammad, F. A., Ma, L., Du, Y., ... & Shalaby, S. M. (2019). Applications of cascaded phase change materials in solar water collector storage tanks: A review. *Solar Energy Materials and Solar Cells*, *199*, 24-49.

[16] Al-Kayiem, H. H., & Lin, S. C. (2014). Performance evaluation of a solar water heater integrated with a PCM nanocomposite TES at various inclinations. *Solar Energy*, *109*, 82-92.

[17] Zhou, Z., Liu, J., Wang, C., Huang, X., Gao, F., Zhang, S., & Yu, B. (2018). Research on the application of phase-change heat storage in centralized solar hot water system. *Journal of cleaner production*, *198*, 1262-1275.

[18] Feliński, P., & Sekret, R. (2016). Experimental study of evacuated tube collector/storage system containing paraffin as a PCM. *Energy*, *114*, 1063-1072.

[19] Abokersh, M. H., El-Morsi, M., Sharaf, O., & Abdelrahman, W. (2017). On-demand operation of a compact solar water heater based on U-pipe evacuated tube solar collector combined with phase change material. *Solar Energy*, *155*, 1130-1147.

[20] Kanimozhi, B., Bapu, B. R., & Pranesh, V. (2017). Thermal energy storage system operating with phase change materials for solar water heating applications: DOE modelling. *Applied Thermal Engineering*, *123*, 614-624.

[21] Wu, W., Dai, S., Liu, Z., Dou, Y., Hua, J., Li, M., ... & Wang, X. (2018). Experimental study on the performance of a novel solar water heating system with and without PCM. *Solar Energy*, *171*, 604-612.

A Brief Review on Conducting Polymer Nanocomposite Based Epoxy Coatings for Marine Applications

A.M Fitriya [a,b], Syed Shahabuddin [c*], Nanthini Sridewi [b*], M.Norsyarizad [b*] and A.K. Pandey [a]

[a] Research Centre for Nano-Materials and Energy Technology (RCNMET), School of Science and Technology, Sunway University, 47500 Selangor Darul Ehsan, Malaysia.

[b] Department of Maritime Technology, Faculty of Science and Defence Technology, National Defence University of Malaysia, 57000, Kuala Lumpur, Malaysia

[c] Department of Science, School of Technology, Pandit Deendayal Petroleum University" Knowledge Corridor, Raisan Village, Gandhinagar – 382007 Gujarat, India

E-mail: fitriyamansur@gmail.com; syedshahab.hyd@gmail.com; nanthini@upnm.edu.my

Abstract. Recently, fouling has become the main concern in the marine, naval and shipping industries. Furthermore, fouling has caused an increase in ship drag and fuel consumption whilst reducing the hydrodynamic performance of ships. Marine structures that partly or fully submerged such as oilrigs, submarine pipes and marine buoys are also impacted. Therefore, research and numerous studies had conducted to propose the best antifouling coating. The main objective of this review is to provide an overview of polymer-based surface coatings for protection from marine biofouling organisms. The physicochemical and mechanical properties of polymers had found to make these materials promising as polymeric coating that can be used in marine applications. Furthermore, this review will discuss the utilization of nanocomposites as additives, modifiers and nanofillers in order to enhance the properties of the polymer for antifouling coating.

1. Introduction

Antifouling (AF) known as the ability of specifically designed coatings to remove or prevent biofouling by any number of organisms on wetted surfaces. For example, AF paints are usually used to prevent the settlement of marine organisms on ship hulls including smaller vessels most of which are made of steel, polymeric resin and wood. Figure 1 shows the development in antifouling research by number of publications in respective years.

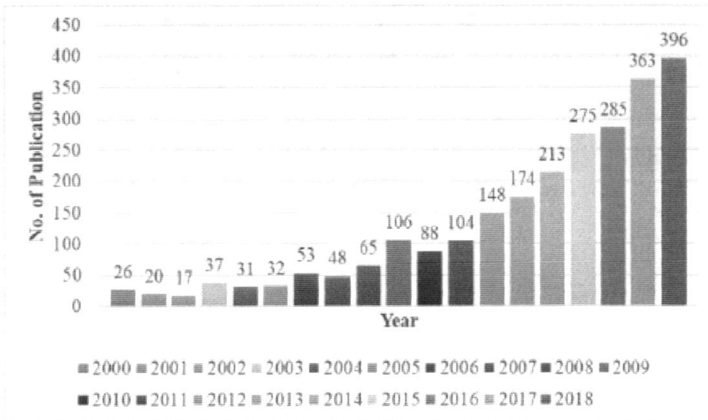

Figure 1. Statistics of Antifouling Research Paper By Year (Source: Web of Science)

In addition, a few requirements have to be fulfilled to produce an effective coating including adhesion to the substrate, mechanical and functional properties, wear resistance, hydrophobicity and hydrophilicity, antibacterial and chemical resistance [1]. Temperature is also included as one of the important conditions required for coating application. Therefore, figure 2 shows the types of coatings based on their usage and application in marine paint industry.

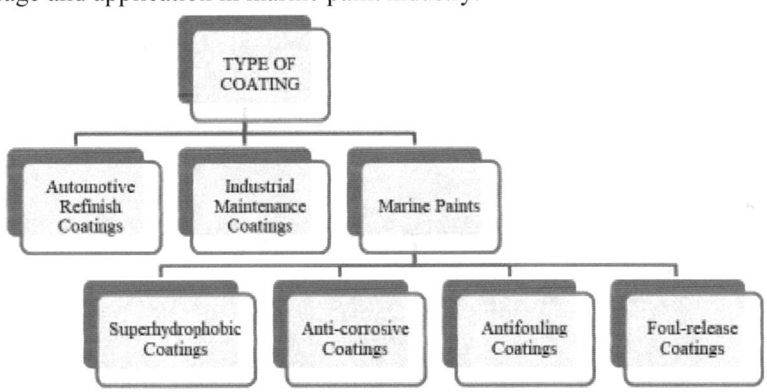

Figure 2. The types of coatings as per ACA and US Census Bureau Current Industrial Report

Table 1. Antifouling Technologies in the Last 10 Years

METHOD	STRATEGIES	REF
Physical and biological	Biomimetic	[2]
Chemical, biological and physical	Chemical: Traditional and modern methods. Biological: Enzymes. Physical: Elctrolysis & radiation, modification of surface topography and changing the zeta potential.	[3]
Biological	Tara Tannin (TT)	[4]

Physical and biological	Surface topography & chemical properties & biomimetic using natural products, cells & enzymes.	[5]
Chemical	Tin free self-polishing copolymer, polymer coating	[6]
Chemical	Tributyltin (TBT), copper, organic additives and polymers.	[7]
Physical	Microstructures of *Cancer pagurus* by using Poly(methyl methacrylate) (PMMA).	[8]
Biochemical	Biodegradable polymer based polyurethane.	[9]
Biochemical	Biodegradable polyurethane with low surface energy polydimethylsiloxane.	[10]

Other than that, polymers have been a part of our lives, from the early ages - such as in tar, shellac, tortoise shells and horns until more recently in synthetic offerings like polyolefins, epoxies and engineering resins [11]. Year by year, many researchers find interest in the study of polymer coatings. This has shown in Table 1. In addition, polymers have the ability to be used as additives in antifouling paints besides protecting metals against corrosion [12]. Figure 3 shows the percentage of publications on polymer nanocomposites that each subject area and category comprises.

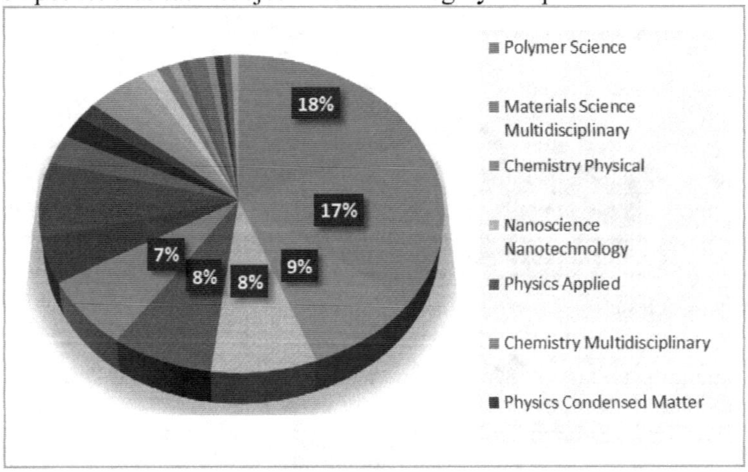

Figure 3. Statistics Polymer Nanocomposite Publication by Subject Area and Categories (Source: Web of Science)

On the other hand, studies on polymer coatings focus more on the development of nontoxic and environment-friendly coatings against corrosion and fouling [13]. Interaction between polymer coatings can be halt by combining them with low surface energy polymers [13]. In addition, many of the protective coatings contain a few types of long-chain polymers of which the nature of polymers is altered by the multicomponent processing of their inorganic or organic precursors that influence the material structure and functions [1]. Table 2 summarises the studies that have observed an improvement in polymer coating with the addition of nanoparticles.

Table 2. Improvement of Polymer Coatings on Value Addition of Nanoparticles [1]

POLYMER MATRIX	NANOPARTICLE EMBEDDED	%WT LOAD /RATIO	REF
Polyurethane	Molybdenum disulfide	20-55.6	[14]
Fluorinated polysiloxane	Steric acid-modified ZnO	13:7	[15]
Teflon tailings	Tetrafluoroethylene and hexafluoropropylene	-	[16]
Fluoride latex	Phosphating material	40 vol%	[17]
Epoxy resin	Fatty acids and epoxidized oleic acid	5	[18]

2. CONDUCTING POLYMER NANOCOMPOSITE BASED EPOXY AS ANTIFOULING COATING

2.1. Epoxy Based Coating

Epoxy resin that is used as a coating material has been shown to have good mechanical and chemical properties that are relatively simple and safe [19]. In addition, epoxy is desirable for its easy processing, high safety, excellent solubility, good chemical resistance and toughness, low shrinkage on cure, good electrical insulating properties and corrosion resistance with strong adhesion to a lot of substrates [19]. Nanofillers were added into the epoxies to improve their antifouling properties [20]. Furthermore, blending two polymers through the incorporation of epoxy resin with additional fillers and modifiers into the base polymer matrix had seen to enhance the coating properties as in the figure 4.

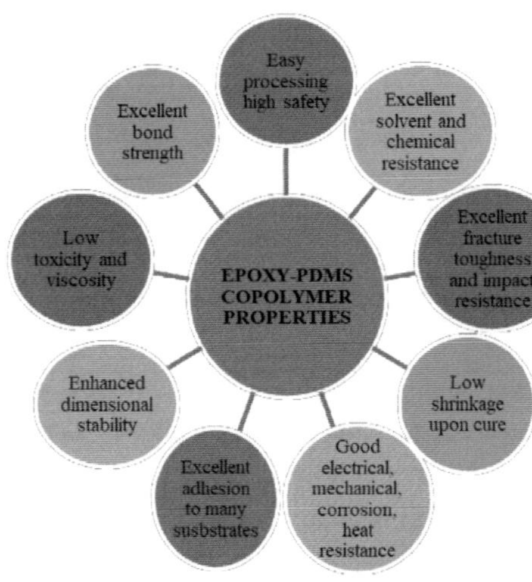

Figure 4. Properties of Epoxy-PDMS filled and unfilled systems [1]

Dispersion of polysiloxane into an epoxy resin matrix has been achieved using suspension polymerization technique by utilizing the core-shell microsphere analogy [1]. Figure 5 shows siloxane core and epoxy shell lead to the creation of elastomeric microspheres. Based on the results of SEM and X- ray microanalysis, the microhardness and the gradient of domain size were homogenized to present mixable and fully compatible microstructures [1].

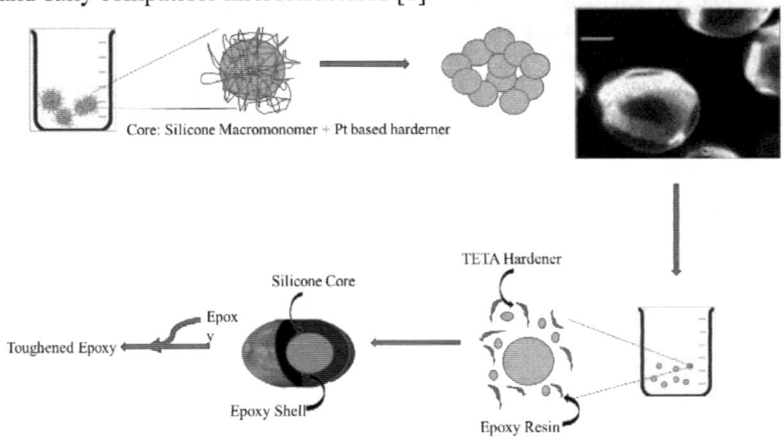

Figure 5. The Synthesis Route of Epoxy-PDMS Core-shell Microspheres [1]

2.2. Polyaniline (PANI) as Antifouling Coating
Polyaniline has good potential for use as an antifouling agent to provide a coating that is non-toxic, low density and has stable electrical conductivity [21]. Fazli-shokouhi et al. has conducted research on PANI combined with graphene oxide nanosheets making PANI-GON nanocomposite by using in-situ polymerization method to form an epoxy based paint coating that showed significant antifouling properties [22]. Based on figure 6, the samples had shown the lowest and highest antifouling behaviour for which the reason was the ability of inorganic nanoparticles like graphene oxide to disperse and fill the micropores of epoxy [22].

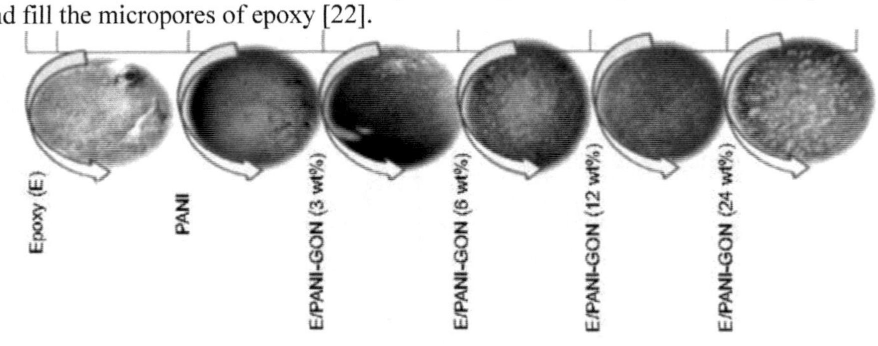

Figure 6. Coating samples after immersion in simulated seawater for 3 months [22]

Furthermore, the pure epoxy coating on metal is colourless and the micropores that are not filled with any nanocomposites can allow the droplets of electrolyte to diffuse easily into the metal surface while the micropores of the epoxy coating that are filled with nanocomposite prevent the diffusion of electrolyte droplets from reaching the metal surface [22].

Other than that, with the good conductivity of PANI coupled with the excellent photocatalytic effect of zinc oxide in a polymer nanocomposites has shown ability to reduce the adhesion of marine bacteria [23]. Mooss et al. studied the embedding of polymer nanocomposite of PANI-ZnO in the basis of thermoplastic polyurethane that had good hydrolytic stability as well as chemical and microbial resistance as shown in Figure 7 [23].

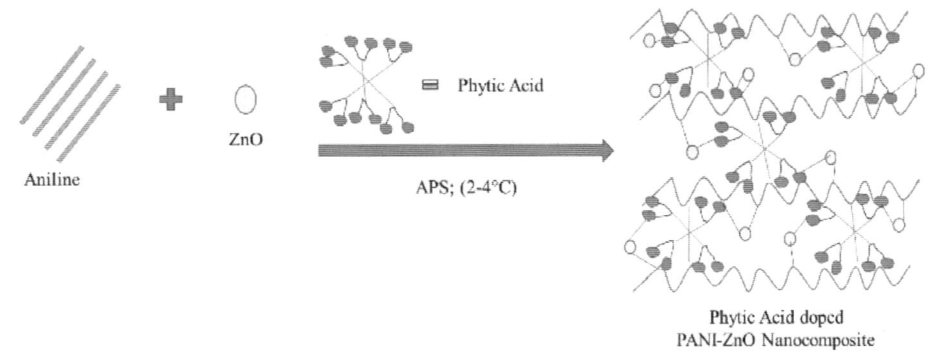

Figure 7. Steps involved in the synthesis of PANI-ZnO Nanocomposite using Phytic Acid as Dopant [23]

3. Conclusion and Remarks

This review article has shown the development of antifouling coating that utilize a polymer basis for marine application. Nanocomposites, for antifouling coating, were use as additives in the coating in order to enhance the properties of antifouling and reduce the adhesion behaviour of the biofouling organism attached to the coating surface materials. However, there is little research on the use of polymer nanocomposite for marine antifouling coating. Therefore, further investigation focusing on conducting polymer nanocomposite for marine coating is need, as the marine field requires more surveys for surface coating applications.

References

[1] S. Verma, S. Mohanty, and S. K. Nayak, "A review on protective polymeric coatings for marine applications," *J. Coatings Technol. Res.*, vol. 16, no. 2, pp. 307–338, 2019.
[2] K. R. Stokes *et al.*, "Designing biomimetic antifouling surfaces," *Philos. Trans. R. Soc. A Math. Phys. Eng. Sci.*, vol. 368, no. 1929, pp. 4729–4754, 2010.
[3] C. A. O. Shan, W. Jiadao, C. Haosheng, and C. Darong, "Progress of marine biofouling and antifouling technologies," vol. 56, no. 7, pp. 598–612, 2011.
[4] N. Bellotti, B. Amo, and R. Romagnoli, "Progress in Organic Coatings Tara tannin a natural product with antifouling coating application," *Prog. Org. Coatings*, vol. 74, no. 3, pp. 411–417, 2012.
[5] J. E. Gittens, T. J. Smith, R. Suleiman, and R. Akid, "Current and emerging environmentally-friendly systems for fouling control in the marine environment," *Biotechnol. Adv.*, vol. 31, no. 8, pp. 1738–1753, 2013.
[6] L. P. Liu and B. R. Hou, "Development of environmental friendly antifouling coatings," *Asian J. Chem.*, vol. 26, no. 13, pp. 4009–4010, 2014.
[7] J. Telegdi, L. Trif, and L. Románszki, "Smart anti- biofouling composite coatings for naval applications," *Smart Compos. Coatings Membr. Transp. Struct. Environ. Energy Appl.*, pp. 123–155, 2015.
[8] P. O'Neill, A. Barrett, T. Sullivan, F. Regan, and D. Brabazon, "Rapid Prototyped Biomimetic Antifouling Surfaces for Marine Applications," *Mater. Today Proc.*, vol. 3, no. 2, pp. 527–532, 2016.
[9] C. Ma, W. Zhang, G. Zhang, and P. Y. Qian, "Environmentally Friendly Antifouling Coatings Based on Biodegradable Polymer and Natural Antifoulant," *ACS Sustain. Chem. Eng.*, vol. 5, no. 7, pp. 6304–6309, 2017.
[10] Y. Kang *et al.*, "Preparation of novel marine antifouling polyurethane coating materials," *Polym. Bull.*, vol. 75, no. 11, pp. 5143–5162, 2018.
[11] K. I. Winey and R. A. Vaia, "P olymer," vol. 32, no. April 2007, pp. 314–322, 2019.
[12] A. Fiorini, K. Leite, D. Miranda, and C. Bressy, "Using Conducting Polymers as Active Agents for Marine Antifouling Paints," vol. 18, no. 6, pp. 1129–1139, 2015.

[13] S. Verma, S. Das, S. Mohanty, and S. K. Nayak, "Development of multifunctional polydimethylsiloxane (PDMS)-epoxy-zinc oxide nanocomposite coatings for marine applications," *Polym. Adv. Technol.*, no. January, pp. 1–26, 2019.

[14] Y. Tang *et al.*, "Fabrication of superhydrophobic polyurethane/MoS2 nanocomposite coatings with wear-resistance," *Colloids Surfaces A Physicochem. Eng. Asp.*, vol. 459, pp. 261–266, 2014.

[15] Y. Qing, C. Yang, C. Hu, Y. Zheng, and C. Liu, "A facile method to prepare superhydrophobic fluorinated polysiloxane/ZnO nanocomposite coatings with corrosion resistance," *Appl. Surf. Sci.*, vol. 326, pp. 48–54, 2015.

[16] A. Satyaprasad, V. Jain, and S. K. Nema, "Deposition of superhydrophobic nanostructured Teflon-like coating using expanding plasma arc," *Appl. Surf. Sci.*, vol. 253, no. 12, pp. 5462–5466, 2007.

[17] L. Zhu and Y. Jin, "A novel method to fabricate water-soluble hydrophobic agent and super-hydrophobic film on pretreated metals," *Appl. Surf. Sci.*, vol. 253, no. 7, pp. 3432–3439, 2007.

[18] A. M. Atta, H. A. Al-Lohedan, A. O. Ezzat, and S. A. Al-Hussain, "Characterization of superhydrophobic epoxy coatings embedded by modified calcium carbonate nanoparticles," *Prog. Org. Coatings*, vol. 101, pp. 577–586, 2016.

[19] S. A. Kumar, T. Balakrishnan, M. Alagar, and Z. Denchev, "Development and characterization of silicone/phosphorus modified epoxy materials and their application as anticorrosion and antifouling coatings," *Prog. Org. Coatings*, vol. 55, no. 3, pp. 207–217, 2006.

[20] P. Saravanan, D. Duraibabu, K. Jayamoorthy, S. Suresh, and S. A. Kumar, "Twin Applications of Tetra-Functional Epoxy Monomers for Anticorrosion and Antifouling Studies," *Silicon*, vol. 10, no. 2, pp. 555–565, 2018.

[21] X. H. Wang *et al.*, "Polyaniline as marine antifouling and corrosion-prevention agent," *Synth. Met.*, vol. 102, no. 1–3, pp. 1377–1380, 1999.

[22] S. Fazli-shokouhi, F. Nasirpouri, and M. Khatamian, "Polyaniline-modified graphene oxide nanocomposites in epoxy coatings for enhancing the anticorrosion and antifouling properties," *J. Coatings Technol. Res.*, 2019.

[23] V. A. Mooss, F. Hamza, S. S. Zinjarde, and A. A. Athawale, "Polyurethane films modified with polyaniline-zinc oxide nanocomposites for biofouling mitigation," *Chem. Eng. J.*, no. November, pp. 1–11, 2018.

IOP Conf. Series: Materials Science and Engineering 1127 (2021) 012014 doi:10.1088/1757-899X/1127/1/012014

Influence of SDBS Surfactant on Stability, Thermal Conductivity and Viscosity of h-BN/EG Based Nanofluids

Syed Shahabuddin[a*], Syed Nadeem Abbas Shah[b,c] , Mohd Faizul Mohd Sabri[b*] and A.K. Pandey[d]

[a] Department of Science, School of Technology | Pandit Deendayal Petroleum University, Knowledge Corridor, Raisan Village, Gandhinagar, 382007 Gujarat, India

[b] Department of Mechanical Engineering, Faculty of Engineering, University of Malaya, 50603, Kuala Lumpur Malaysia

[c] Department of Mechanical Engineering (Main campus Lahore), University of Engineering and Technology Lahore, 54890, Pakistan.

[d] Research Centre for Nano-Materials and Energy Technology (RCNMET), School of Science and Technology, Sunway University, 47500 Selangor Darul Ehsan, Malaysia.

E-mail: syedshahab.hyd@gmail.com; faizul@um.edu.my

Abstract. In present work, an experimental study was carried out to investigate the influence of anionic surfactant (SDBS) on stability, thermal conductivity and viscosity of h-BN/EG based nanofluids. The nanofluids were prepared via two-step method and characterized by using zeta-sizer, thermal property analyser and viscometer. The volume concentration of h-BN (0.05%) remains constant whereas SDBS volume concentration sweep from 0.05% to 2%. Current results indicate, the zeta potential increases with surfactant addition, but overall, zeta potential shows inverse relation with SDBS concentration. As the value reduced from -57.5mV to -40.5mV when SDBS concentration increase from 0.05% to 2%. While the particle size increase (396nm to 817.9nm) linearly with SDBS concentration due adsorption of surfactant molecules. On the other hand, thermal conductivity over temperature range 25°C-50°C shows maximum enhancement ~6.57% at 25°C for 1 vol% SDBS. However, dynamic viscosity increase abruptly after 0.5 vol% of SDBS. Therefore, 0.5 vol% could be treated as the optimum SDBS concentration with reduced viscosity (~2.85%) and increased thermal conductivity (~3.26%) as compared to base fluid at 25°C. Potential of such optimum combination of nanofluids (0.05vol% h-BN with 0.5 vol% SDBS) may be manipulated in circulating fluid thermal management systems.

1. Introduction

Many commercial devices such as electric vehicle lithium-ion batteries, power transformer and PEM fuel cells need proper thermal management for durable and efficient operation [1-3]. Nanofluids are widely adopted in heat transfer systems of such devices in order to dissipate heat at higher rates. However, in these devices thermal management systems, the electrical conductivity of nanofluids plays a vital role. Any increase in electrical conductivity might suppress their performance. For instance, electrical conductivity reduces the break down voltage of transformer cooling system. Therefore, nanomaterials with promising dielectric properties are first choice for preparation of nanofluids. Another possibility to enhance the convective heat transfer rates for cooling of small scale is the topological alteration of nano-channels [4]. The former is presumed to be cost effective than later. As the surface modification of nano-channels require design changes and wall nano-coatings. That is why nanofluids are getting much attention in heat transfer applications.

Recently, a geothermal coupled multi-generation systems has been introduced which employs nanofluids for hydrogen and power production [5]. The performance of nanofluids in circulating fluid heat transfer systems primarily influenced by colloidal stability, thermal conductivity and rheological properties. Surfactants have been utilized to prepare stable nanofluids. However, the foaming is the obvious phenomenon with surfactant inclusion due to reduced coalescence of bubbles especially at higher temperatures [6,7]. Owing to such limitation researchers opted the covalent functionalization approach to prepare stable colloidal suspension. Subsequently, it was reported that the covalent functionalization damage the intrinsic structure of nanomaterials leading to suppression of material intrinsic properties [8]. Conclusively, both stabilization techniques have pros and cons.

Li et al. proposed the use of gemini surfactant with large spacer length to stabilize metal (Ag, Au) based nanofluids in aqueous medium using one step method [9]. Borode et al. presented a detailed review on surfactant aided dispersion of carbon nanomaterials in base fluid network [10]. Yekeen et al. reported that the nanoparticles-surfactants mixture can lower the oil-water interfacial tension (IFT) over tested temperature range (303K to 353K) for all nanoparticle and surfactants considered in the study [11]. Cabaleiro et al. studied the effect of chemical conversion of graphene oxide (GO) to reduced graphene oxide (rGO) on dynamic viscosity and surface tension. It was found surface tension reduced ~3% whereas dynamic viscosity showed reduction for rGO as compared GO base nanofluids. However, viscosity showed ~70% and 130% enhancement as compared to base fluid for rGO and GO respectively [12].

Recently, Wu et al. introduced novel platinum/hydrocarbon nanofluids with polymer stabilizer for hypersonic aircrafts thermal management. They observed 2% reduction in dynamic viscosity as compared to methylcyclohexane (MCH) base fluid along with upgraded cooling[13]. The surfactants have obvious capability to lower the interfacial tension (IFT) as demonstrated by Tagavifar et al. [14]. Carrillo-Berdugo et al. outlined a new theoretical approach to achieve colloidal stability and termed as three step method. Following three step method supposed to work by adjusting polar and dispersive component of base fluid as per nanoparticle dispersion requirements. The better colloidal stability will be expected by minimizing the interfacial tension (IFT) at solid-liquid interface. The computed the 0.0017 vol% of Triton X-100 surfactant as the optimum which can impart minimum solid-liquid IFT for SWCNT-TX100-water. Maximum ~13.9% increase in dynamic viscosity was observed with thermal conductivity improvement ~79.5% and specific heat enhancement ~8.6%. It was also urged that the dilution in surfactant concentration do not much affect the viscosity so the increment might be ascribed to nanoparticles only [15]. Guo et al, based on MD simulation studied the effect of surfactant adsorption on heat transfer at solid-liquid interface. It was concluded, the surfactant adsorption stimulated the heat conduction path by lower the interfacial thermal resistance as compared to bared solvent. Consequently, heat flux increased.

The surfactant affinity to solid as well as solvent plays an essential role in such improvement of interfacial heat transfer [16]. Shah et al. demonstrated that the aqueous: ethylene glycol based CuO nanofluids with PVA surfactant had small size aggregates when compared without surfactants. While the large size aggregate nanofluids showed less viscosity as compared to surfactant contained nanofluids at similar conditions [17]. Zhou et al. explored the potential of SDS surfactant in reducing the viscosity of water. They also showed a synergistic effect of titania nanoparticles with SDS which further reduced the water viscosity. The following potential was attributed to the lubrication effect of nanoparticles as well the analogous micelle [18].

Moreover, when the fluid circulate through pump in forced convection systems, the deposit in pump may mitigate the working life due to corrosion. So the knowledge of nanofluids corrosion properties is also vital. Abdeen et al. reported the effect CNT-water nanofluids with gum Arabic (GA) surfactant on corrosion evaluation of stainless steel (SS) substrate. They found the small concentration of CNT (0.1 wt. %) had slow corrosion rates on SS surface as compared to high concentration (0.5 wt. %). The high corrosion rates at high concentration was ascribed to formation of anodic active sites on SS surface [19]. Stable insulating mineral oil (MO) based AlN and TiO_2 nanofluds were prepared using olein acid (OA) with preserved thermos-physical properties. They showed ~16% and ~11% increment in thermal conductivity for AlN and TiO_2 nanofluids respectively. Conversely, low particle loading (0.01wt. % and 0.1 wt. %) showed viscosity reduction while high concentration (0.5wt. %) augmented dynamic viscosity upto ~22% as compared to MO. Moreover, results indicated that OA treatment reduced the dynamic viscosity of AlN nanofluids when compared without OA treatment. The reduction in viscosity was ascribed to shear force needed for bond breaking between nanoparticles and MO. The addition of OA weakens the bond resulting in reduced viscosity [20].

Barai et al. showed 83.444% augmentation in thermal conductivity for reduced graphene oxide and Fe_3O_4 nanocomposite based aqueous nanofluids. Also, the heat transfer coefficient increased which was assigned to improvement in thermal conductivity of nanofluis. In addition, ~10.83% more pressure drop in nanofluids flow was observed as compared to water which was due to increase in viscosity [21]. Jorge et al. modified MWCNTs surface by combining plasma polymer (PP) coating and vapour functionalization techniques. The (3-aminopropyl) triethoxysilane (APTS) treated MWCNTs found stable in ethanol but not in water. However, tetraethylenepentamine (TEPA) treatment made MWCNTs stable in water at specific pH values (2, 5.5). This was happened due to TEPA adsorption onto PP coating [22]. Hernaiz et al. analysed the nanofluids contact angle and regarded it as thermo-physical property as it plays a key role in phase change heat transfer processes. Contact angle was studied for SS surfaces in order to mimic the industrial application. It was concluded the limiting contact angle (for zero volume) resembled the nanofluids thermo-physical property. This could be further employed to investigate the relation of nanofluids droplets in contact with any substrate under certain conditions. Their work elucidated no significant change in density, viscosity and thermal conductivity of Au, GO and alumina based aqueous nanofluids at 0.1 vol% particle loading [23]. Vallejo et al. developed antifreeze glycolated water-based nanofluids based on functionalized graphene nanoplatelets for wind turbine thermal management using SDBS as stabilizer. Their results showed ~7.3% thermal conductivity augmentation along with viscosity enhancement up to 20% [24].

Comprehensive literature review with diversified findings suggested, surfactants have been widely utilized for stable colloidal dispersion. However, the critical micelle concentration is also important to avoid excess usage which might mitigate the thermal and rheological performance of nanofluids. Also, based on previous research findings, surfactant can reduce the interfacial tension. Such potential of surfactants could be manipulated to reduce the dynamic viscosity of colloidal suspensions. On the other hand, the lubricating properties of nanoparticles have also been regarded as beneficial in order to improve rheological characteristics of nanofluids.

The present work is motivated from updated literature to find an appropriate solution for heat transfer fluid which can reduce the viscosity along with improvement in stability and thermal conductivity. Two dimensional hexagonal boron nitride (h-BN) is found to be appropriate choice with low coefficient of friction, layered morphology with better lubrication effect and good dielectric and thermal properties. Therefore, in present work hexagonal boron nitride (h-BN) nanoparticles are employed to prepare ethylene glycol based nanofluids using sodium dodecyl benzene sulphonate (SDBS) as stabilizer. The main objective of the present research is to explore the potential of SDBS surfactant on stability, viscosity and thermal conductivity.

2. Experimental

2.1. Materials

All the chemicals in the experimental study were used without modification. Two dimensional h-BN N70 with average particle size 70nm, hexagonal crystal structure, chemical formula (BN), coefficient of friction 0.15-0.7, thermal conductivity 33.5 (W/m.K), density 2.3 (g/cm^3), purity 99% and white colour was purchased from lower friction (M.K Impex, Canada). Sodium dodecyl benzene sulphonate (SDBS) with chemical formula ($C_{18}H_{29}NaO_3S$), molecular weight 348.48 g/mol and density 1.06 (g/cm^3) was purchased from Fisher scientific UK. Ethylene glycol (EG) with chemical formula $(CH_2OH)_2$, molecular weight 62.07 g/mol and density 1.11 (g/cm^3) was purchased form Sigma Aldrich.

2.2. Nanofluids Preparation

Two-step method is employed to prepare nanofluids. The weight of nanoparticle (h-BN) and SDBS was computed from the density and total volume of the solution. The weight corresponding to volume concentration of hBN (0.05%) was dispersed into 50ml of EG. The suspension was subjected to magnetic stirring for 15 minutes (40 °C, 250 rpm) then desired weight of SDBS corresponding to volume concentrations (0.05%, 0.5%, 1%, 2%) was added followed by magnetic stirring for next 15 minutes. Finally, the dispersed suspension was sonicated (bath) for 90 minutes to get stable colloidal dispersion.

2.3. Nanofluids Characterizations

The stability and particle size distribution were measured using the electrophoretic and dynamic light scattering techniques respectively at 25 °C with litesizer500 (Anton Paar). Supernatant liquid from parent samples was used to dilute the samples for particle size distribution measurements. To prepare supernatant a table top centrifuge was employed (5000rpm for 3 minutes). KD2 thermal property analyser (Decagon devices Inc., Pullman Washington, USA) was used to estimate thermal conductivity at 25 °C, 30 °C, 40 °C and 50 °C. The thermal conductivity was measured up to 50°C because beyond this temperature limit the micro-convection cause less reproducibility of results [20]. Sample (s) temperature was maintained with temperature controlled circulating water bath within ±0.5 °C. A 15 minutes idle time was allowed to attain thermal equilibrium before starting each measurement. The SV-10 sine wave vibro viscometer was employed to measure dynamic viscosity (measurement range 0.3-10000 mPa.s, A&D instruments ltd) [15]. The calibration was performed with DI water. The measured value was equated with literature to find measurement error which was appeared to be ~3.2%. This error was accounted for all tested samples. All viscosity measurements were taken at ambient temperature (25°C) with ± 1 °C variation.

All tested samples measurements were triplicate and an average values was used in the analysis.

3. Results and Discussions

The long term homogeneous dispersion of nanoparticles within the base fluid is essential to preserve thermo-physical properties over time. Therefore, the stability evaluation is the first step before gauging the heat transfer potential of nanofluids. Figure 1 (a) represents the particle size distribution of 0.05 vol% h-BN particles loading and effect of SDBS surfactant volume concentrations (0.05-2%). For the pristine nanofluids the measured mean particle size is 365±15.3nm with polydispersity index (PDI) of 24±2.1%. Conversely, inclusion of SDBS surfactant shows the mean particle size increases endlessly with studied concentration of surfactant. The particle size increases from 365±15.3nm to 817.9±64.8nm when volume concentration of surfactant sweep from 0.05% to 2%. However, the PDI values remains within acceptable limit where the dispersion can be considered more likely to be mono-disperse. The increment in particle is the apparent indication that the surfactant molecules are adsorbed successfully on the surface of h-BN nanoparticles.

Though it's a good sign that surfactant molecules are attached to the nanoparticles surface but the increment in particle size is also alarming. As the particle size becomes bigger, the tendency of sedimentation increases under the action of gravity. In addition, the highest particle size at 2 vol% of SDBS might also be owing to micelles formation. Because, there exist a CMC limit for surfactants use. Any additional surfactant amount may form micelles leading to sedimentation as they become bigger.

In order to verify the adsorption of surfactant molecules the surface charge of colloidal suspension is estimated by measuring the zeta potential values. Figure 1 (b) shows the mean zeta potential values for pristine as well as surfactant contained nanofluids. It can be seen from the Figure 1 (b), addition of surfactant improves the surface charge of colloidal suspension. The zeta potential shows maximum -57.5±3.6 mV at 0.05 vol% SDBS. This values start decreasing as the volume concentrations of changes to 2% and reach on a value of -40.5±6.2 mV. For all tested SDBS volume concentrations, the mean zeta potential remains above the threshold limiting value (±30mV). But this cannot be considered as the ever stable colloidal suspension. Because, the increment in mean particle size is already known with increasing concentration of SDBS. So a synergy can be made between mean particle size and zeta potential before drawing final conclusion. The increment in particle size and decrease in zeta potential with SDBS concentration are well agreed to fact of sedimentation. The bigger size leads to quick sedimentation under gravity.

On the other hand, large number of particles in same volume of nanofluids also give rise to coagulation process. The coagulation finally transforms to flocs which in turns sediment quickly. Therefore, from the results it could be postulated, for nanofluids stability, only zeta potential value is not a reliable indicator. One must consider the particle size of colloids to evaluate the stability of nanofluids. The mechanism of colloidal dispersion in this work arises from well-known steric effect. The adsorbed surfactant molecules on particle surface repels each other causing long term dispersion. Theoretically, such colloidal dispersion is governed by DLVO theory which works on the balance of repulsive and attractive forces among the particles [25] . So in present work the concentration of surfactant up to 0.5 vol% can be regarded as optimum as far as stability is concerned. However, to declare optimum nanofluids remaining thermos-physical properties also need to evaluate first. Following stability evaluation, subsequent, thermal conductivity and dynamic viscosity are measured.

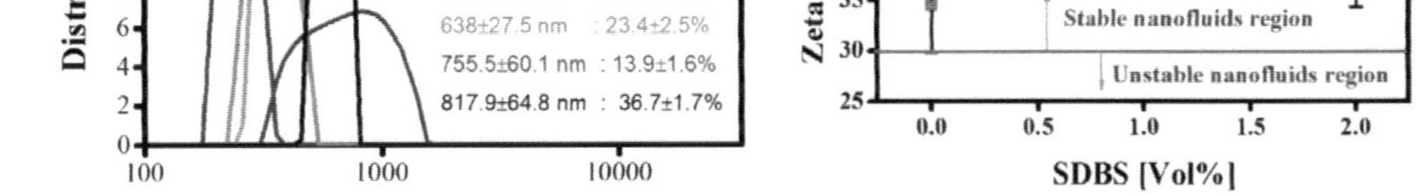

Figure 1. a) Particle size distribution, polydispersity index (PDI) of h-BN/EG nanofluids and b) mean zeta potential

The heat transfer performance of thermos-fluids depends on the thermal conductivity and dynamic viscosity. Figure 2 (a) compares the thermal conductivity of base fluid (EG) and pristine nanofluids (0.05 vol% h-BN) from 25°C to 50°C. Apparently, both, base fluid and pristine nanofluids shows increase in thermal conductivity with increasing temperature. However, maximum enhancement is observed ~4.7% at 30°C as compared to base fluid. But, the increment in thermal conductivity is obvious at all tested temperature. When temperature rises to 40°C, augmentation becomes ~3.43% and subsequently reaches to ~3.68% at 50°C. While at 25°C, increment appears to be ~3.68%. The behaviour of pristine h-BN nanofluids is quite complex at various tested temperatures. This behaviour may be correlated to stability of pristine nanofluids. At 25°C and 30°C, the augmentation could be attributed to the better static stability as evident by the zeta potential. However, from 40°C to 50°C, the increment may be ascribed to intensify Brownian motion of the particle.

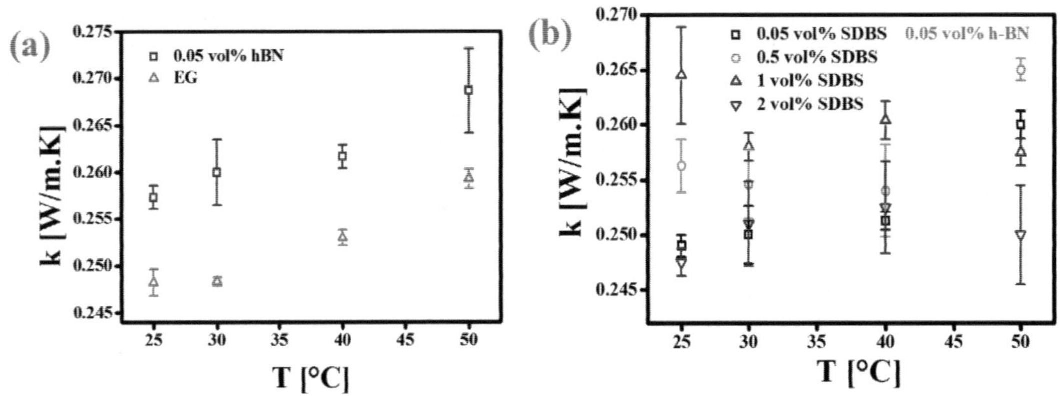

Figure 2. Thermal conductivity of a) Pristine nanofluids and base fluid, b) Surfactant contained nanofluids at various temperatures

Likewise, to previous studies, in current research, the potential of h-BN nanoparticles intrinsic thermal conductivity is not prevented. As, the pristine nanofluids do not show thermal conductivity improvement comparable to h-BN nanoparticles (33.5 W/m.K). This is obvious due to interfacial thermal resistance at the solid-liquid interface which reduces the effective thermal conductivity. Contrariwise, surfactant contained h-BN/EG nanofluids thermal conductivity is presented in Figure 2(b). From 0.05-1 vol% SDBS, the thermal conductivity increases from 0.32% to 6.57% at 25°C. It is noteworthy, suddenly thermal conductivity becomes low when SDBS volume concentration change from 1 to 2%. This is clear indication the CMC of surfactant has reached and large micelles are formed which ultimately sediment. Such results are consistent with zeta potential values at higher concentration of surfactant. Similar trend of thermal conductivity is observed at other tested temperatures. But, at 50°C, the thermal conductivity reduction effect becomes more prominent starting from 1% SDBS. This may be due to foam formation at high temperature which is an obvious phenomenon with surfactant use. Here, the 0.5 vol% SDBS could be regarded as the CMC concentration of SDBS in present nanofuids. There has been many phenomena's ascribed to thermal conductivity augmentation such as Brownian motion, liquid layering, clustering, synergistic effect of nanoparticles etc. However, the chaotic motion of nanoparticles may enhance the thermal conductivity as well as convective heat transfer coefficient. The increase in thermal conductivity reduces the thickness of boundary layer in circulating fluid flow leading to enhanced convective heat transfer coefficient [21]. The detailed summary on thermal conductivity results is shown in Table 1. Still, conclusion cannot be drawn about optimum nanofluids combination. As, the pressure drop in circulating fluid systems is the obvious outcome form the use of nanofluids which depends on dynamic viscosity.

Thermal conductivity results suggested the potential of SDBS contained nanofluids is dominant at 25°C. Therefore dynamic viscosity is measured at 25°C for surfactant contained nanofluids. Figure 3 demonstrated the effect of surfactant on nanofluids as well as on base fluid. The addition of 0.05 vol% SDBS in base fluid shows viscosity reduction ~3.96%, subsequent increment in concentration increases the viscosity. When the concentration of SDBS changes to 2%, the viscosity shows increment up to 9.41% as compared to base fluid. Further, the addition of fixed volume concentration of h-BN (0.05%) in a mixture of surfactant and base fluid, synergetic effect of viscosity reduction is observed. The synergistic effect of h-BN and SDBS persist up to 0.5 vol% of surfactant. Afterward, again viscosity starts increasing. Based on viscosity reduction, it could be postulated, the 0.5 vol% SDBS is optimum CMC. Similar effect of surfactant and nanoparticles on dynamic viscosity has already been demonstrated by Zhou et al. They also showed a synergistic effect of titania nanoparticles with SDS which further reduced the water viscosity. The following potential was attributed to the lubrication effect of nanoparticles as well the analogous micelle [18]. Therefore, the low coefficient of friction possessed by h-BN and interfacial tension lowering capability of SDBS plays a vital role lowering the dynamic viscosity in present research. However, the present work utilized a vibro-viscometer which cannot truly depicts the shear flow behaviour as well as temperature sweep effect for studied nanofluids. But the results are interesting and also well in agreement to reported literature. The detailed summary on dynamic viscosity results is shown in Table 2.

Table 1. Thermal conductivity enhancement/reduction results summary

SDBS (Vol%)	h-BN 0.05 vol%			
	25 °C	30 °C	40 °C	50 °C
	k Enhancement/Reduction [%]			
0	3.68	4.70	3.43	3.61
0.05	0.32	0.67	-0.69	0.27
0.5	3.26	2.52	0.4	2.2
1	6.57	3.89	2.92	-0.69
2	-0.28	1.08	-0.2	-3.59

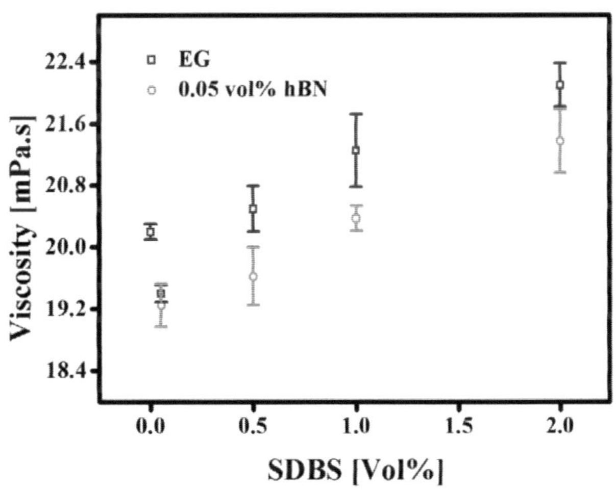

Figure 3. Dynamic viscosity of base fluid and nanofluids at 25°C

Table 2. Dynamic viscosity enhancement/reduction results summary

SDBS (Vol%)	h-BN 0 vol%	h-BN 0.05 vol%
	Viscosity Enhancement/Reduction as compared to base fluid [%]	
0.05	-3.96	-4.70
0.5	1.49	-2.85
1	5.20	0.87
2	9.41	5.82

4. Conclusions

The stability, thermal conductivity and dynamic viscosity is studied experimentally. The properties are measured with fixed volume concentration of h-BN and varying the volume concentration of SDBS (0.05-2%). The stability of pristine and surfactant contained nanofluids revealed, the addition of SDBS significantly improves the dispersion stability. The zeta potential increases from -34.5±4.64 mV to -53.63±5.6 mV when 0.05 vol% SDBS added to pristine nanofluids. This value is sufficiently above the threshold limit (±30mV) of zeta potential which confirms the effect of surfactant on stability. However, the particle also shows increment from 365±15.3 nm to 638±27.5 nm at same volume concentration of SDBS. While, the thermal conductivity shows maximum improvement ~6.57% with 1 vol% SDBS at 25°C. Base on overall evaluation, the 0.5 vol% SDBS is regarded as the optimum surfactant concentration which shows ~3.26% thermal conductivity enhancement, ~2.85% dynamic viscosity reduction along with substantial improvement in stability at 25°C. The present findings strongly suggested the further evaluation on nanofluids rheology in the presence of surfactants. As surfactant might reduce the viscosity leading to less pumping cost.

Acknowledgements

The authors acknowledge the financial support from University of Engineering and Technology Lahore, Pakistan under faculty development program. Moreover, the Mechanical Engineering department, University of Malaya is highly acknowledged for the research facilities.

References

[1] Nasir FM, Abdullah MZ, Ismail MA. Experimental Investigation of Water-Cooled Heat Pipes in the Thermal Management of Lithium-Ion EV Batteries. Arab J Sci Eng 2019. doi:10.1007/s13369-019-03851-5.

[2] Islam R, Shabani B. Prediction of electrical conductivity of TiO 2 water and ethylene glycol-based nanofluids for cooling application in low temperature PEM fuel cells. Energy Procedia 2019;160:550–7. doi:10.1016/j.egypro.2019.02.205.

[3] Beheshti A, Shanbedi M, Zeinali S. Heat transfer and rheological properties of transformer oil-oxidized MWCNT nanofluid. J Therm Anal Calorim 2014;118:1451–60. doi:10.1007/s10973-014-4048-0.

[4] Chakraborty P, Ma T, Cao L, Wang Y. Significantly enhanced convective heat transfer through surface modification in nanochannels. Int J Heat Mass Transf 2019;136:702–8. doi:10.1016/j.ijheatmasstransfer.2019.03.053.

[5] Ratlamwala TAH, Waseem S, Salman Y, Bham AA. Geothermal and solar energy–based multigeneration system for a district. Int J Energy Res 2019:1–22. doi:10.1002/er.4480.

[6] Raza MQ, Kumar N, Raj R. Effect of foamability on pool boiling critical heat flux with nanofluids. Soft Matter 2019;15:5308–18. doi:10.1039/c8sm02565g.

[7] Sarsam WS, Kazi SN, Badarudin A. A review of studies on using nanofluids in flat-plate solar collectors. Sol Energy 2015;122:1245–65. doi:10.1016/j.solener.2015.10.032.

[8] Georgakilas V, Otyepka M, Bourlinos AB, Chandra V, Kim N, Kemp KC, et al. Functionalization of graphene: Covalent and non-covalent approach. Chem Rev 2012;112:6156–214. doi:10.1021/cr3000412.
[9] Li D, Fang W, Feng Y, Geng Q, Song M. Stability properties of water-based gold and silver nanofluids stabilized by cationic gemini surfactants. J Taiwan Inst Chem Eng 2019;97:458–65. doi:10.1016/j.jtice.2019.02.017.
[10] Borode AO, Ahmed NA, Olubambi PA. Surfactant-aided dispersion of carbon nanomaterials in aqueous solution. Phys Fluids 2019;31:071301. doi:10.1063/1.5105380.
[11] Yekeen N, Padmanabhan E, Idris AK. Synergistic effects of nanoparticles and surfactants on n-decane-water interfacial tension and bulk foam stability at high temperature. J Pet Sci Eng 2019;179:814–30. doi:10.1016/j.petrol.2019.04.109.
[12] Cabaleiro D, Estellé P, Navas H, Desforges A, Vigolo B. Dynamic Viscosity and Surface Tension of Stable Graphene Oxide and Reduced Graphene Oxide Aqueous Nanofluids. J Nanofluids 2018;7:1081–8. doi:10.1166/jon.2018.1539.
[13] Wu X, Ye D, Jin S, He G, Guo Y, Fang W. Cracking of platinum/hydrocarbon nanofluids with hyperbranched polymer as stabilizer and initiator. Fuel 2019;255:115782. doi:10.1016/j.fuel.2019.115782.
[14] Tagavifar M, Balhoff M, Mohanty K, Pope GA. Dynamics of Low-Interfacial-Tension Imbibition in Oil-Wet Carbonates 2019:1–16.
[15] Carrillo-Berdugo I, Zorrilla D, Sánchez-Márquez J, Aguilar T, Gallardo JJ, Gómez-Villarejo R, et al. Interface-inspired formulation and molecular-level perspectives on heat conduction and energy storage of nanofluids. Sci Rep 2019;9:1–13. doi:10.1038/s41598-019-44054-0.
[16] Guo Y, Surblys D, Kawagoe Y, Matsubara H, Liu X, Ohara T. A molecular dynamics study on the effect of surfactant adsorption on heat transfer at a solid-liquid interface. Int J Heat Mass Transf 2019;135:115–23. doi:10.1016/j.ijheatmasstransfer.2019.01.131.
[17] Shah J, Ranjan M, Sooraj KP, Sonvane Y, Gupta SK. Surfactant prevented growth and enhanced thermophysical properties of CuO nanofluid. J Mol Liq 2019;283:550–7. doi:10.1016/j.molliq.2019.03.127.
[18] Zhou J, Hu M, Jing D. The synergistic effect between surfactant and nanoparticle on the viscosity of water-based fluids. Chem Phys Lett 2019;727:1–5. doi:10.1016/j.cplett.2019.04.052.
[19] Abdeen DH, Atieh MA, Merzougui B, Khalfaoui W. Corrosion Evaluation of 316L Stainless Steel in CNT-Water Nanofluid: Effect of CNTs Loading. Materials (Basel) 2019;12:1634. doi:10.3390/ma12101634.
[20] TAHA-TIJERINA J, CADENA-DE LA PEÑA N, CUE-SAMPEDRO R, RIVERA-SOLORIO C. Thermo-physical evaluation of dielectric mineral oil-based nitride and oxide nanofluids for thermal transport applications. J Therm Sci Technol 2019;14:JTST0007–JTST0007. doi:10.1299/jtst.2019jtst0007.
[21] Barai DP, Bhanvase BA, Saharan VK. Reduced Graphene Oxide-Fe 3 O 4 Nanocomposite Based Nanofluids: Study on Ultrasonic Assisted Synthesis, Thermal Conductivity, Rheology, and Convective Heat Transfer. Ind Eng Chem Res 2019;58:8349–69. doi:10.1021/acs.iecr.8b05733.
[22] Jorge L, Coulombe S, Girard-Lauriault PL. Tetraethylenepentamine and (3-aminopropyl)triethoxysilane adsorbed on multi-walled carbon nanotubes for stable water and ethanol nanofluids. Thin Solid Films 2019;682:50–6. doi:10.1016/j.tsf.2019.05.007.
[23] Hernaiz M, Alonso V, Estellé P, Wu Z, Sundén B, Doretti L, et al. The contact angle of nanofluids as thermophysical property. J Colloid Interface Sci 2019;547:393–406. doi:10.1016/j.jcis.2019.04.007.
[24] Vallejo JP, Álvarez-Regueiro E, Cabaleiro D, Fernández-Seara J, Fernández J, Lugo L. Functionalized graphene nanoplatelet nanofluids based on a commercial industrial antifreeze for the thermal performance enhancement of wind turbines. Appl Therm Eng 2019;152:113–

25. doi:10.1016/j.applthermaleng.2019.02.046.
[25] Korada, Viswanatha Sharma and Hamid NHB. Engineering Applications of Nanotechnology: From Energy to Drug Delivery. Springer International Publishing; 2017.

IOP Conf. Series: Materials Science and Engineering 1127 (2021) 012015 doi:10.1088/1757-899X/1127/1/012015

Comparative analysis of the performance of cabinet solar dryer and open sun drying for Banana slices

Than Tun Naing, Cho Thandar Soe

Mechanical Department, Yangon Technological University, Myanmar,

E-mail: konaing@s.okayama-u.ac.jp

Abstract. Sun-drying is the traditional method of preservation fruits, crops and vegetables. It is also an alternative way to prevent the shortage of natural resources. But traditional sun drying causes damage to food quality and cannot prevent dust, any other insets and birds. For this reason, this paper represents the design calculation of cabinet solar dryer for 1kg of banana slices and the comparative investigation of cabinet solar dryer and open sun drying. The thickness of a banana slice is taken 4mm. After 14 hours of drying time in two days experiment, moisture content of banana slices inside of the cabinet dryer is 14% meanwhile open sun drying is still 40%. As a result, the desired moisture content of banana slice can be received quickly in cabinet solar dryer. And then, it can easily construct with locally cheap and environmentally acceptable, non-hazard materials. Cabinet solar dryer can also reduce the loss of food due to high temperature, insects, and birds. Hence, it is more effective than open sun drying.

1. Introduction

Drying is a method of dehydration of food products which means reducing the moisture content from the food to improve its shelf life by preventing bacterial growth [1]. Drying is the essential process to keep fruits, crops and paddy without spoilage for the desired time. Open sun drying is the cheapest and easiest way, but there was a lot of limitations such as the birth of microorganisms and mould due to its unstable temperature and weather conditions. In traditional sun drying, due to its unstable temperature, the drying materials loss their vitamins that are required for human health, their colour and texture. During off sunshine hours and cloudy days, it may encourage the birth of mould. The required temperature for drying food is the range from 30°C to 60°C [2]. In traditional sun drying, it needs labours to look after the drying materials for the prevention of damage by insects, birds, other herbivores, rain, dust and wind. In large scale preservation of food business, electrical drying machines are used. But farmers and households can't afford to use this type of machines because of their large power consumption and expensive cost. In our country, Myanmar, a lot of agricultural products are spoiled every year by adverse climate conditions during the harvesting period. The solar dryer can maintain the required temperature of food. It reduces the spoilage of food by touching the direct sunlight, unlike open sun drying. Drying time is shorter than open sun drying. It prevents the damage of food by external enemies. Unlike electrical drying mechanisms, it is more economical because it has a low cost for power consumption. Some types of dryers have no cost for power consumption especially natural convection type and direct type solar dryers. For these reasons, a solar dryer is a suitable solution to solve these problems for farmers and small business. The paper described that the comparison between sun drying and solar drying and obtained that solar dryer generates high temperature, lower relative humidity, lower product moisture content and reduce spoilage during the drying process than sunshine [3]. The drying time reduces about 20% and produces a better quality of drying products in the case of using a cabinet solar dryer and compare with open sun drying [4].

There are different types of solar dryers that are basically classified as natural convection type solar dryer, force convection type solar dryer, direct type solar dryer, indirect type solar dryer and mixed-mode solar dryer by various methods depending on the usage of mechanisms for airflow and how to exposure to sunlight whether directly or not [5] .In this present work, design considerations and comparative study of moisture content, relative humidity, the temperature of cabinet solar dryer for 1kg of the Banana slices with open sun drying. The recommended thickness of the banana slice to be dried is 4-5 mm [6]. In this paper, 4 mm of Banana slices were used.

2. Design Consideration

The amount of moisture to be removed, m_w (kg)

$$m_w = m_p (M_i - M_f)/(100 - M_f) \quad (1)$$

where M_i =Initial moisture content of the material
M_f = final moisture content of material
m_p = initial mass of material(kg)

- Moisture content dry basis, M

$$M = M_f / (1 - M_f) \quad (2)$$

- Water activity, a_w

$$a_w = 1 - \exp[-\exp\{0.914 + 0.5369 \ln(M)\}] \quad (3)$$

- Equilibrium relative humidity, ERH

$$ERH = a_w \times 100 \quad (4)$$

- Average drying rate, m_{dr} (kg of water/h)

$$m_{dr} = m_w / t_d \quad (5)$$

where t_d=drying time (hour)

- The mass flow rate of air, m_a (kg of dry air /h)

$$m_a = m_{dr} / (w_f - w_i) \quad (6)$$

where, w_i = initial humidity ratio (kJ/kg of dry air)
w_f = final humidity ratio (kJ/kg of dry air)

- The total heat energy required to remove moisture, E (MJ)

$$E = m_a (h_f - h_i) \times t_d \quad (7)$$

where, h_i =initial enthalpy of moist air (kJ/kg)
h_f=final enthalpy of moist air (kJ/kg)

- Collector Area, Ac (m^2)

$$A_c = E/I \times \eta \ (m^2) \quad (8)$$

where I = incident solar radiation (MJ/m^2)
η =collector efficiency

- Volumetric airflow rate, V_a (m^3/s)

$$V_a = m_a / \rho \quad (9)$$

where, ρ =density of air (kg/m3)

- Vent Area, A_v (m^2)

$$A_v = V_a / v \quad (10)$$

where, v =wind speed (m/s)

2.1. Desired data for design calculation

Initial moisture content (M_i)	92.3%
Final desired moisture content (M_f)	14%
Ambient air temperature	30 °C
Ambient relative humidity	70%
Collector efficiency (η)	30%
Wind speed (v)	1m/s
Daily average incident solar radiation in September for Myanmar (I)	15.22 MJ/m^2
Desired drying temperature (T_f)	56°C
Drying time(t_d)	7hr
The thickness of a banana slice	4 mm
Tilted angle (inclination angle of glass collector to cabinet	15°

2.2. Results of Design Calculation

Table 1 shows the results of design calculation for cabinet solar dryer for 1 kg of banana slices. Design calculation based on the weather condition of Yangon Technological University, Myanmar. Figure 1 and Figure 2 describe the schematic diagrams of the cabinet solar dryer according to the design calculation results.

Table 1. Results of design calculation

Parameter	**Value**
the initial mass of material, m_p	1kg
amount of moisture to be removed, m_w	0.91 kg of water
moisture content dry basis, M	0.15
water activity, a_w	0.59
equilibrium relative humidity, ERH	59%
average drying rate, m_{dr}	0.13 kg/h
initial humidity ratio, W_i	0.0185 kg of water/kg of dry air
h_i=initial enthalpy	77 kJ/kg of dry air
h_f=final enthalpy	104 kJ/kg of dry air
final humidity ratio, W_f	0.0256 kg of water/kg of dry air
mas of airflow rate, m_a	18.403 kg/h
the total heat energy to evaporate water, E	3.4782 MJ
collector area, A_c	0.761 m^2
collector length, l_c	1m
collector width, b_c	0.761 m
The density of air, ρ_a	1.2 kg/m^3
volumetric airflow rate, V_a	0.0245 m^3/s
air vent area, A_v	0.0245 m^2

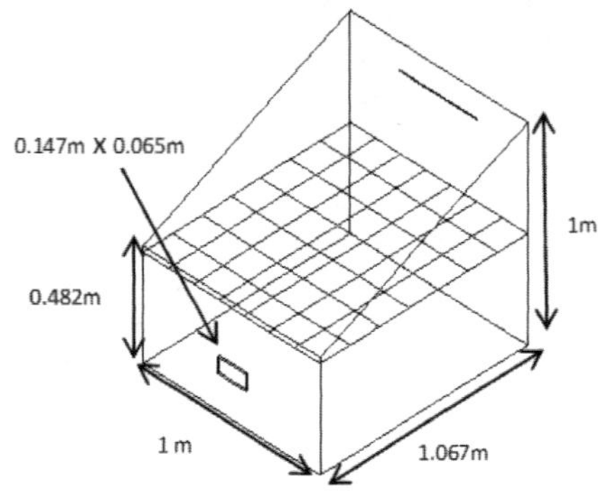

Figure 1. Schematic diagram of cabinet solar dryer

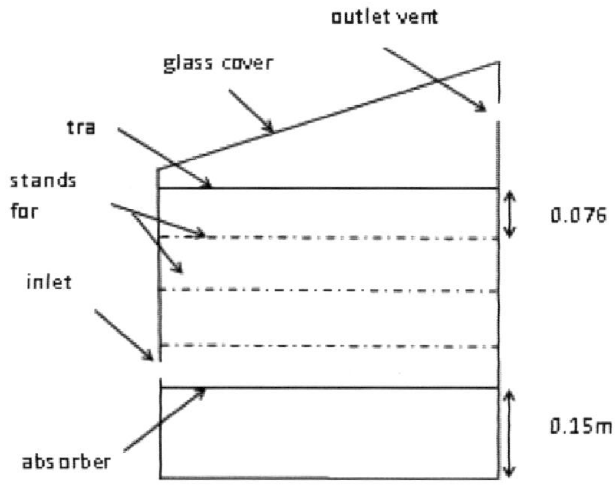

Figure 2. Schematic diagram of cabinet solar dryer (Side View)

2.3. Components of Cabinet Solar Dryer

The top part of the cabinet solar dryer is covered by a glass plate. Most of the researchers suggested to use the glass thickness range from 4-5 mm and it should be a good transparent for solar radiation [1, 6, 7]. In this paper, 4mm thickness, 1 m x 1.067 m area of transparent glass is used. Some paper described the common materials used for the cover plate are plastic and polycarbonate. The tilt angle of the glass cover is 15° due south.

Absorber Plate is used to absorb the solar heat energy. Black painted of 1 mm thick aluminium plate is placed about 0.15 m height from the bottom of the dryer. It absorbs heat passing through the glass and re-emits heat to the passed inlet air inside the cabinet. Common materials for absorber plates are copper, galvanized steel sheet, aluminium sheet.

An insulating material is used to prevent leakage the emitted heat energy from the absorber plate to the surround of the cabinet 2.54 cm thickness Styrofoam sheets are used to minimize the heat transfer from the walls of the drying chamber to the surroundings. All sides of the drying chamber and bottom of the absorber plate are covered by these Styrofoam sheets. Some other common insulating materials are thermocole sheets, fibreglass, foam, and sawdust.

The drying chamber is mostly made up of readily available materials especially plywood. Steel, iron, copper, plywood and aluminium can be also used. Plywood is a poorer heat conductor and cheaper than aluminium. But aluminium sheets are selected in this chamber to prevent the condition of being swollen during the rainy season.

The tray is used to spreading drying materials over it inside the drying chamber. The 1 cm x 1 cm iron mesh tray is used to allow the airflow inside the cabinet. And also, the measuring devices such as solar power meter, infrared thermometer, thermocouple, datalogger, hygrometer, air velocity meter are used to measure and collect the data during the experiment.

3. Experiment Procedure

Cabinet solar dryer is set up at the unshaded place in YTU campus faced south. Two hygrometers are used to measure the relative humidity of the environment and inside of the cabinet. The inside temperature of cabinet and ambient temperature is measured by two thermocouples that connected to a data logger. The infrared thermometer is used to measure the absorber plate temperature. To measured solar irradiance, solar power meter (TES1333) is used. Figure 3 shows the inside view of cabinet solar dryer. Figure 4 shows the experimental set up for open sun drying for 1 kg of banana slices. Experiments were made at the same weather condition, at the same place and at the same time.

Figure 3. Inside view of cabinet solar dryer

Figure 4. Experimental set up for open sun drying

4. Experimental Results and Discussions

Solar radiation of the experimental site is measured hourly by solar power meter during drying time. Variation of solar radiation on each day is shown in Figure. 5 and Figure. 6. On the first day of the drying process, solar radiation gradually increases from 735.8 W/m^2 to 1007 W/m^2 between 9 am to 1 pm and then decreases to 63.6 W/m^2 at 4 pm. It was found that the solar radiation sharply decreases from 1 pm to 2 pm because of weather condition on that day. At the second day of the drying process, solar radiation slightly decreases from 304 W/m^2 and after that sharply increases to 1100 W/m^2 about noon and then dropdown. To be summarized, maximum radiation is obtained about (1000 to1100) W/m^2 during 12 pm to 1 pm and it was found that the fluctuation of solar radiation was very high and it depends on weather condition.

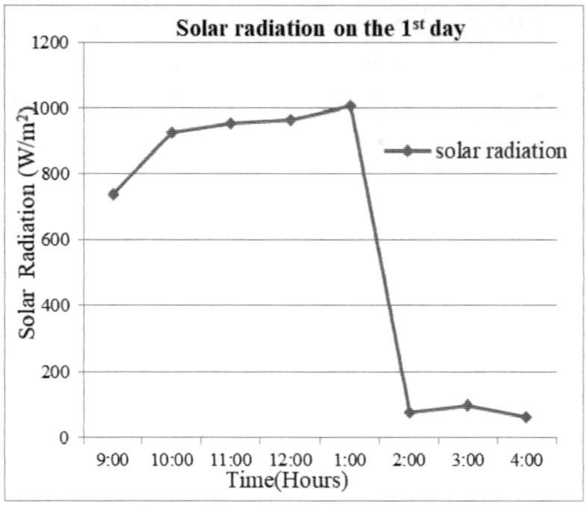

Figure 5. Variation of solar radiation on the 1st day

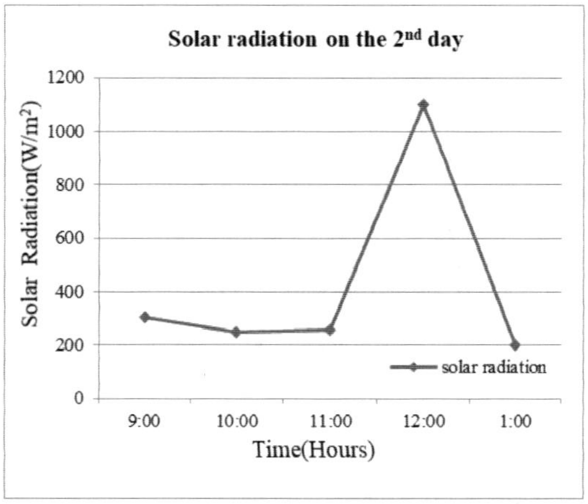

Figure 6. Variation of solar radiation on the 2nd day

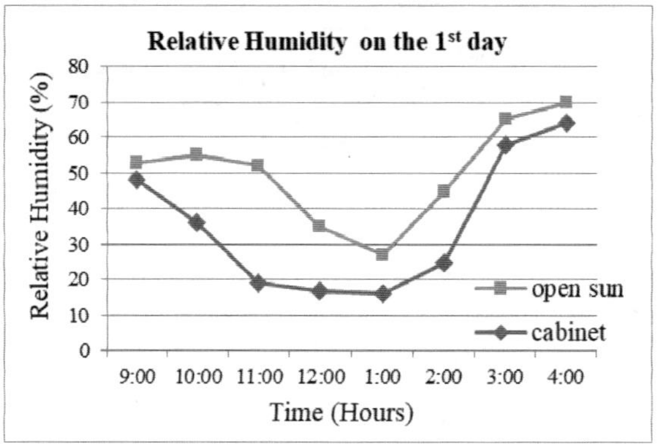

Figure 7. Relative humidity on the 1st day

Variation of the relative humidity of cabinet solar dryer and open sundry are shown in Figure. 7 and Figure. 8 for the first and second day of the experiment. The relative humidity of open sun drying is always higher than cabinet solar dryer in all time of the experiment in two days. Because of this reason, the cabinet solar dryer is more effective in drying than open sun drying process. According to these results, it was found that the relative humidity is a little bit high in the morning and afternoon and the lowest is at the afternoon both two days of the experiments.

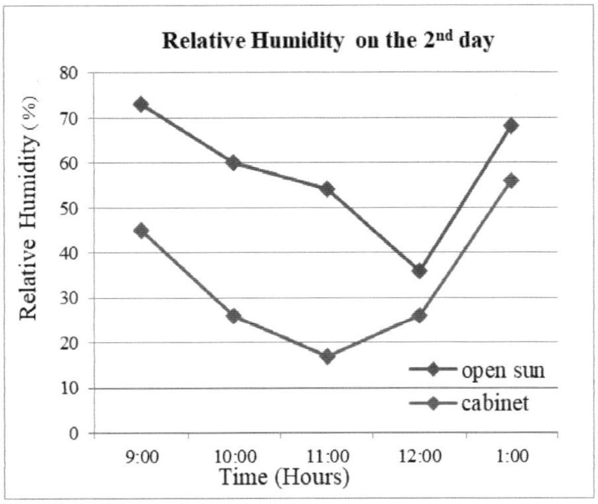

Figure 8. Relative humidity on the 2nd day

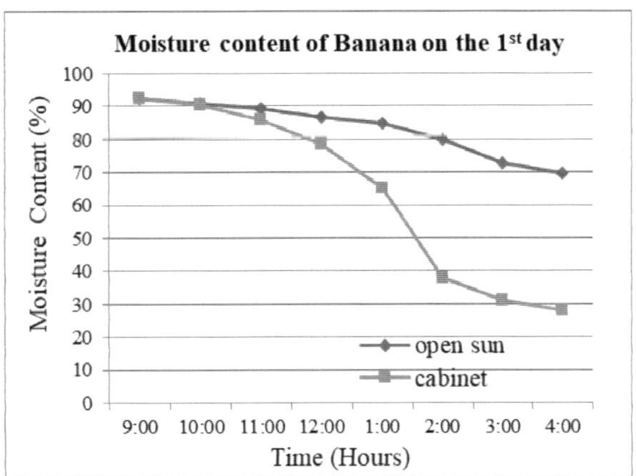

Figure 9. Moisture content of Banana on the 1st day

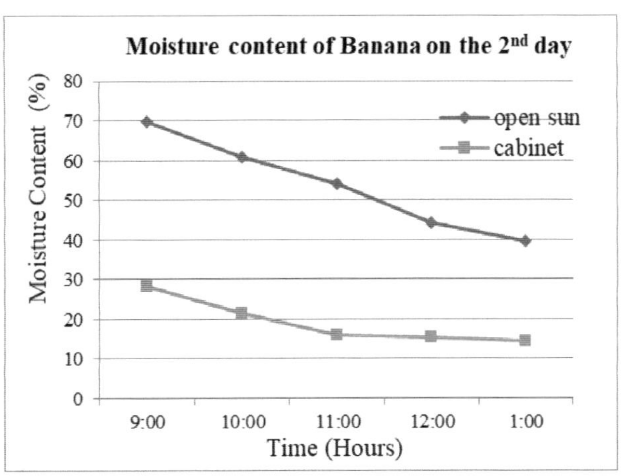

Figure 10. Moisture content of Banana on the 2nd day

Variations of moisture content of banana slices for each day are shown in Figure. 9 and Figure. 10. In cabinet solar dryer, moisture content was fast removed than open sun drying at all of the drying time on the first day. After 7 hours of the first day, it can reduce the moisture content of banana about 93 % to 29% in cabinet solar dryer meanwhile only reduce 93% to 70% in open sun drying. At the end of the experiment of two days, the moisture content of Banana was 14% in cabinet solar dryer meanwhile only 40% in open sun dryer. It can be concluded that the removing of moisture content from the food is quite effective in cabinet solar dryer.

Figure 11 and Figure 12 describe the temperature distributions of the absorber plate, inside the cabinet and open sun in outside. All of the three temperatures of the absorber plate, inside the cabinet and open sun outside, was maximum at 1 pm on the first day of the experiment. At the day, the temperature of all three conditions decreased at noon because of the weather condition. The temperature of the absorber plate is highest among them because the black painted absorber plate absorbed heat energy from the direct sunlight and emits heat energy inside of the cabinet solar dryer. It also clear that the temperature of the inside of the cabinet more than the outside temperature. It can dry the food faster inside of the cabinet solar dryer. From the figure 12, nearly stable maximum temperature could be obtained from 10 am to 12 noon on the second day of the experiment.

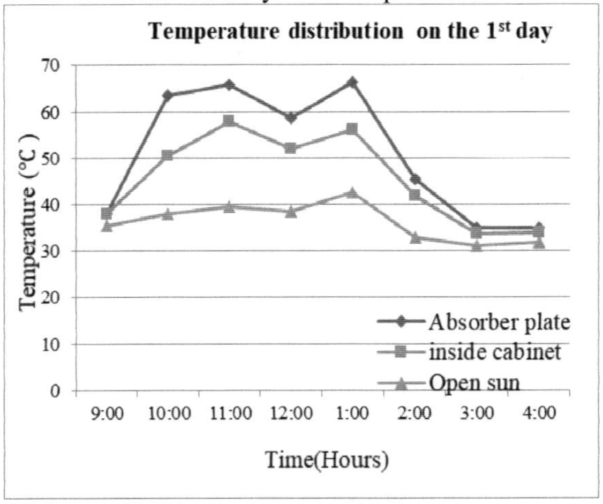

Figure 11. Temperature distributions of Banana on the 1st day

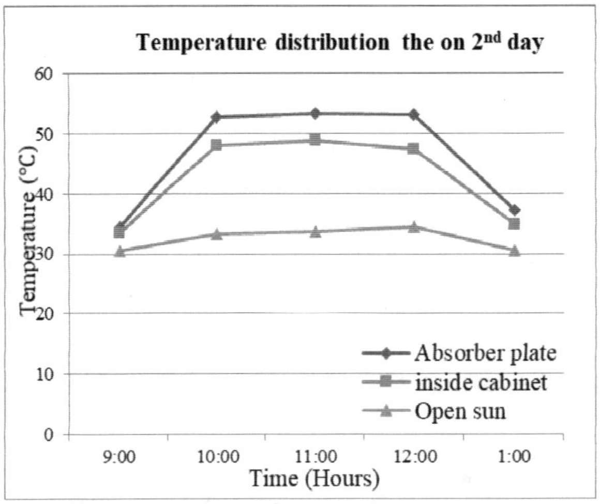

Figure 12. Temperature distributions of Banana on the 2nd day

5. Conclusions

Cabinet solar dryer can reduce moisture content than open sun drying under the same weather condition even the cloudy hours without the aid of other heat energy source because the absorber plate absorbs heat during sunny hours and emits it. During 14 hours of drying time, cabinet dryer can reduce moisture from initial moisture content of 93% to final desired moisture content of 14% while the open sun can reduce 40%. For this reason, the desired moisture content of the food could be reached quickly in cabinet solar dryer without the use of any extra energy or electricity. And then, it can easily construct with locally cheap and environmentally acceptable, non-hazard materials. Cabinet solar dryer can also reduce the loss of food due to high temperature, insects, and birds. Hence, it is more effective than open sun drying.

Acknowledgements

The authors would like to thank the JICA EEHE project for supporting us. The authors also would like to thank the financial and technical assistance provided by AUN-SEED/Net Japan ASEAN Collaborative Education Programme (JACEP) under the CEP 2019.

References

[1] Garg H. Solar energy: fundamentals and applications: Tata McGraw-Hill Education; 2000.
[2] Raju RVS, Reddy RM, Reddy ES. Design and fabrication of efficient solar dryer. *International Journal of Engineering Research and Applications*. 2013;**3**(6):1445-58.
[3] Umogbai V, Iorter H. Design, construction and performance evaluation of a passive solar dryer for maize cobs. *African Journal of Food Science and Technology*. 2013;**4**(5):110-5.
[4] Rajeshwari N, Ramalingam A. Low cost material used to construct effective box type solar dryer. *Archives of Applied Science Research*. 2012;**4**(3):1476-82.
[5] El-Sebaii A, Shalaby S. Solar drying of agricultural products: A review. *Renewable and Sustainable Energy Reviews*. 2012;**16**(1):37-43.
[6] Hegde VN, Hosur VS, Rathod SK, Harsoor PA, Narayana KB. Design, fabrication and performance evaluation of solar dryer for banana. *Energy, sustainability and Society*. 2015;**5**(1):1-12.
[7] Geramitchioski T, Mitrevski V, Vilos I. Design and construction of a mobile solar dryer. *Journal on Processing and Energy in Agriculture*. 2011;**15**(3):197-200.

IOP Conf. Series: Materials Science and Engineering 1127 (2021) 012016 doi:10.1088/1757-899X/1127/1/012016

Design and performance testing of a bread box type low cost solar water heater for rural area

Than Than Htike*, Mi Sandar Mon †, Saw Ba Tin#

*Department of Mechanical Engineering,Yangon Technological University,Yangon, Myanmar

†Department of Mechanical Engineering,Yangon Technological University,Yangon, Myanmar

#Department of Mechanical Engineering,Pyay Technological University,Pyay, Myanmar

E-mail: drthanthan.mech@gmail.com

Abstract. This paper deals with the design of integral passive low cost solar water heating system of "Bread Box" type for rural area. The designed capacity is 200lit and aims to cover the need of hot water at evening/night and morning for one to three people. This type of heater consists of horizontal cylindrical Mild-Steel tank painted flat black and is enclosed in a glazed (glass) insulated rectangular wooden box whose lid and front are opened as reflector (mercury-backed mirror) during the day and closed as insulation at night. In this research, bread box solar water heater was designed and constructed with local materials and then, this heater was tested. In this design, the diameter of collector-storage tank is 0.55 m, its length is 0.797 m and collector area is 1.377m^2. The total glazing area of this heater is 1.859 m^2 and box size is 1.341m x 0.884m x 0.76m.

1. Introduction

In developing nations, heating water for bathing and cooking is often the most expensive and time-consuming process in the household energy budget. And then, their traditional fuel sources such as wood, gas and electricity are unsustainable as they are costly to households and contribute to build up of greenhouse gases in the atmosphere. One potential solution to this problem is the use of low-cost residential solar water heater such as integral collector-storage (ICS) heater also called "Bread Box or Batch Solar Water Heater" for households used especially for the rural area. In which, at the winter time hot water is necessary for early morning and at night. By using the renewable energy, it can reduce electrical energy and hence, energy saving and cost saving as well.

The main objective of this paper is to design, construct with locally available material and test the "Bread Box" type low cost solar water heater for hygiene and health and to contribute global climate protection by using small-scale renewable energy application.

2. Solar radiation

In designing solar energy systems, it is usually necessary to go beyond basic ideas to more advanced concepts in the science of heat energy. This section is devoted to discussion of solar radiation, the energy that arrives at the earth's surface from the sun. Total solar radiation intensity (insolation) received on a surface is illustrated as shown in Figure 1.

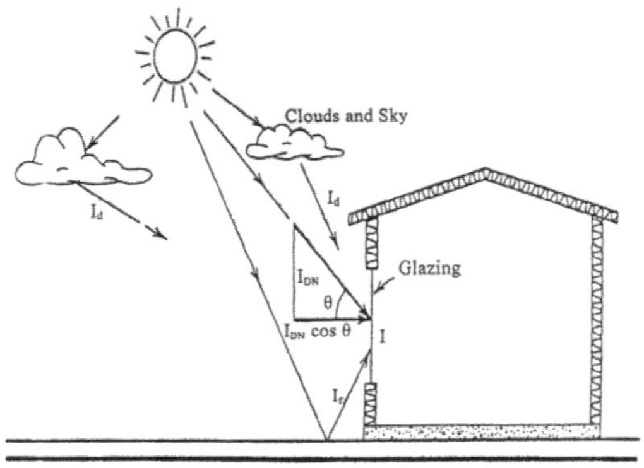

Figure 1. Illustration of the solar radiation intensity [1]

2.1. Direct and diffuse solar radiation

The intensity of solar radiation that is available on the surface of the earth depends on reflection of the atmosphere and the earth's outer surface, scattering in the earth's atmosphere and absorption by the atmosphere. The solar radiation intensity on the surface of the earth's;

$$I_{DN} = [G_{o,H} + G_{o,V}]\,\bar{K}_T \quad [2] \qquad (1)$$

Where,
I_{DN} = Direct beam solar radiation intensity
$G_{o,H}$ = Extraterrestrial radiation onto a horizontal surface
$G_{o,V}$ = Extraterrestrial radiation onto a vertical surface
$\bar{K}_T$ = Mean daily clearness index

Diffuse solar radiation intensity is calculated by:

$$\frac{I_d}{I_{DN,avg}} = 0.175, \text{ (for } \bar{K}_T \geq 0.722 \text{ and } \omega_s \geq 81.4) \ [3] \qquad (2)$$

Where,
I_d = Diffuse solar radiation intensity
$I_{DN,avg}$ = Average direct beam solar radiation intensity (daily)
$\bar{K}_T$ = Mean daily clearness index,
ω_s = sunset hour angle

2.2. Reflected short-wave solar radiation

The amount of solar radiation reflected from the reflector onto a surface can be calculated by equation 3 [2].

$$I_r = (I_{DN} + I_d) \times \rho_s \times F_{vf} \qquad (3)$$

Where,
ρ_s = Reflectivity of a surface
F_{vf} = View factor from the reflector

2.3. Total solar radiation

The total solar radiation intensity is the sum of the direct beam solar radiation intensity (I_{DN}) and diffuse radiation (I_d) and reflected radiation (I_r) from the external reflectors.

$$I_{Total} = I_{DN} + I_d + I_r \quad (4)$$

The design data are based on 25th December, 2017 and the day length is from 8:00 AM to 4:00 PM. In this paper, the solar radiation intensity is calculated by using equation 1 to 4. The calculated hourly total solar radiation intensity is shown in Figure 2. The calculated available solar energy is Q_{solar} is 48.558 MJ/m^2.

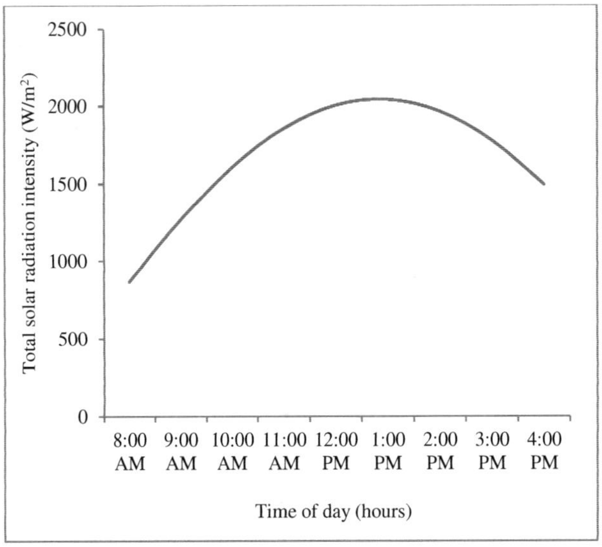

Figure 2. Variation of hourly total solar radiations intensity

2.4. Average transmittance of glass cover

In solar applications, the transmission of radiation is through a slab or film of material. Transmittance of a single glass cover is allowed for both reflection and absorption losses. For the average transmittance of the two components is calculated by equation 5 [4].

$$\tau = \frac{\tau_a}{2}\left[\left(\frac{1-r_{\perp}}{1+r_{\perp}} \times \frac{1-r_{\perp}^2}{1-(r_{\perp}\tau_a)^2}\right)+\left(\frac{1-r_{II}}{1+r_{II}} \times \frac{1-r_{II}^2}{1-(r_{II}\tau_a)^2}\right)\right] \quad (5)$$

Where,

τ_a = Absorption losses

K = The extinction coefficient

$r_{\perp}$ = Perpendicular component of polarized radiation

r_{II} = Parallel component of unpolarised radiation

Transmittance of the glass cover can be calculated by using equation 5 [4]. The variation of hourly average transmittance of glass cover is shown in Figure 3.

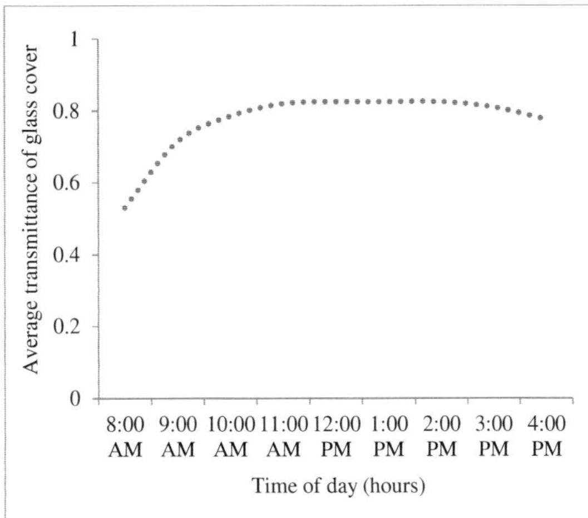

Figure 3. Variation of hourly average transmittance of glass cover.

3. Design of Bread Box solar water heater

This paper aims to design "Bread Box" solar water heater. Design is considered for households use of two persons per day. The capacity of the heater is 200 litre. The source water temperature is 25 °C and expected maximum design temperature is 70 °C. This heater is designed as shown in Figure 4.

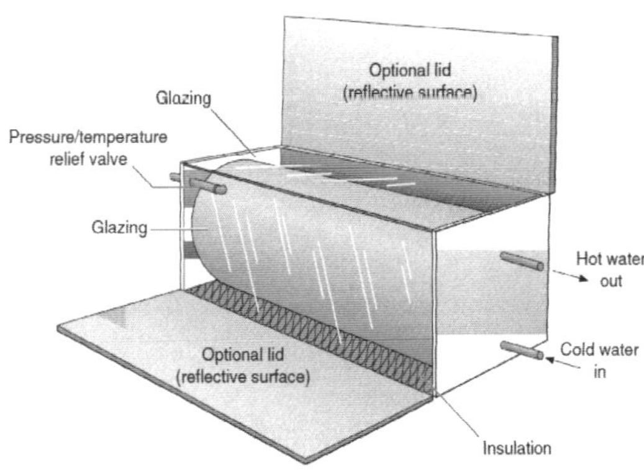

Figure 4. Bread box solar water heater

It consists of a single cylindrical storage mild-steel tank with painted flat black to absorb heat from the sun and then transfer to water stored within, thus, it do dual operation as collector-storage. The collector-storage tank is placed horizontally in an insulated rectangular wooden box to reduce heat loss and its lid and front are opened to collect sufficient solar radiation. Both apertures are covered with glazing material (glass) that is allowed to transmit short-wave solar radiation through glass and prevented long-wave thermal radiation came back from collector as convection and radiation losses. Two insulated externals reflectors (outside surface is made with wood and inside is mercury back mirror) are installed at lid and front of box. These two reflectors are opened at morning and closed at evening. They offer the advantages of increasing collector area (increasing solar gain), and then they can reduce night time heat loss, maintain higher temperatures through the night and protect heater overheat at noon time by closing reflectors. A bread box heater (integral passive solar water heater), allows cold water to flow in from the bottom of the tank and hot water to flow out of the top.

3.1. Design parameter

The design calculation is made by using the following design parameters.

Locality = PYAY (height 31.4 m above sea level)
Latitude, l = 18° 48' 0" N (18.8° N)
Longitude, LL = 95° 13' 0" E (95.217° E)
Design time = 25th December, 2017 from 8:00 AM to 4:00 PM

3.2. Required energy

The total required energy to heat the water can be calculated by using equation 6 and 7.

$$\text{Required energy, } Q_w = m\, C_w\, \Delta T \quad (6)$$

$$= \rho\, V\, C_w\, (T_{set} - T_{source}) \quad (7)$$

Where, C_w= Specific heat capacity of water, 4.187 kJ/kg °C
ρ = Density of water, 1000 kg/m^3

The parameters for water heating which are to be considered are:

i. Volume of water, V(m^3) required in a given time period
ii. The temperature of the cold water
iii. Require delivery temperature

3.3. Sizing of collector/storage tank design

Available solar energy/area is calculated with equation 8 [5].

$$Q_{input} = Q_{solar} \times \overline{\tau\alpha} \quad (8)$$

Where,
τ = Average Transmittance of glass cover, 0.78
α = Absorptance of collector, 0.56 [5]

$$\text{Capacity of water, } V = \pi/4\, D^2\, L \quad (9)$$

$$\text{Length of tank, } L = 4V/\pi D^2 \quad (10)$$

$$= (4 \times 200 \times 10^{-3})/\pi D^2$$

$$\text{Collector area, } A_c = \pi\, D\, L \quad (11)$$

Note 1. All surface area of tank absorbed energy equally.
2. To simplify the calculation, authors assume that energy losses are not significant.

3.4. Glazing area or aperture area design

The glazing must be adequately large and well enough insulated to collect sufficient solar radiation.

$$A_g/\text{surface} \geq A_c/2 \text{ (as glazing have two surfaces)}$$

$$L \times W \geq A_c/2 \quad (12)$$

where, L = Length of glazing
W = Width of glazing

3.5. Calculation of water tank delivery temperature

The temperature rise of water tank delivery temperature (Table 1) can be obtained by using equation 13 and 14. By calculation result, the maximum desire temperature is 71°C at 1:00 PM.

$$Q_{solar} \times \overline{\tau\alpha} \times A_c = Q_w/\text{Time Taken} \quad (13)$$

$$= [\rho V C_w (T_{set} - T_{source})]/\text{Time Taken} \quad (14)$$

Table 1. Calculated results of the water tank temperature

Time of Day (hours)	Water Tank Delivery Temperature (˚C)
8:00 AM	28.26˚C
9:00 AM	34.54˚C
10:00 AM	43.03˚C
11:00 AM	52.94˚C
12:00 PM	62.65˚C
1:00 PM	71.04˚C

3.6. Calculated results for Bread Box solar water heater

This research is designed to heat 200 litres water from 25˚C to 70˚C. The design month is based on 25th December,2017 and the day length is from 8:00 AM to 4:00 PM. Firstly, the direct-beam solar radiation, diffuse solar radiation and reflected short-wave solar radiation intensity are calculated. The results of total direct-beam solar radiation intensity is shown in Figure 2. Then, the required energy to heat water can be calculated by using equation 6 and 7. Transmittance of the glass cover is also calculated. Then the required collector area can be obtained. Then the required collector area can be obtained. After that collector area, size of collector/storage tank, glazing area and size of insulated box are designed.

For the desire capacity 200 litres, the various sets of diameter, length and area of collector/storage tank and glazing can be obtained according to the equation 10 to 12. From these values, the diameter, length and collector area of tank is chosen to satisfy the required collector area. The calculated results for bread box solar water heater is as shown in table 2.

From these results, the collector area 1.377 m^2 and the glazing area 1.858 m^2 are satisfied with Rule of Thumb [6].

Table 2. detailed dimension of a Bread Box solar water heater

Components	Materials	Dimensions		
Collector	Mild-Steel Tank (50 gallon)	Diameter: 0.55 m	Length: 0.797 m	Area: 1.377 m^2
Glazing	Window Glass (0.0023m thick)	Length: 1.2192 m	Width: 0.76 m	Area: 1.858 m^2 (two sides)
Reflector	Mirror (mercury backed)	Length: 1.2192 m	Width: 0.76 m	Area: 1.858 m^2 (two reflectors)
Insulated Box	Wood, Sawdust	Length: 1.341 m	Width: 0.884 m	Height: 0.76 m

4. Fabrication of a Bread Box solar water heater

By using the above calculated results from table 2, the bread box solar water heater is constructed with locally available materials. The following materials are used in fabricating a bread box solar water heater;

Mild-Steel Tank (200 lit capacity) with urethane interior coating (heated water is for domestic use, not for drinking), Flat-Black Paint, GI Pipes, Copper or CPVC Pipes, Sockets and Valves, Wood, Sawdust, Window Glass, Mercury-Backed Mirrors (Reflectors) and Tools.

Fabrication is done by the following procedure.

Step 1: Wash the inside surface of Mild-Steel tank with mixture of caustic, detergent and water by rolling three or more time and then, cleaning and unrusting the outer (exterior) surface of the tank by using hand-grinder with abrasive paper, shown in Figure 5. Check water leakage of the tank.

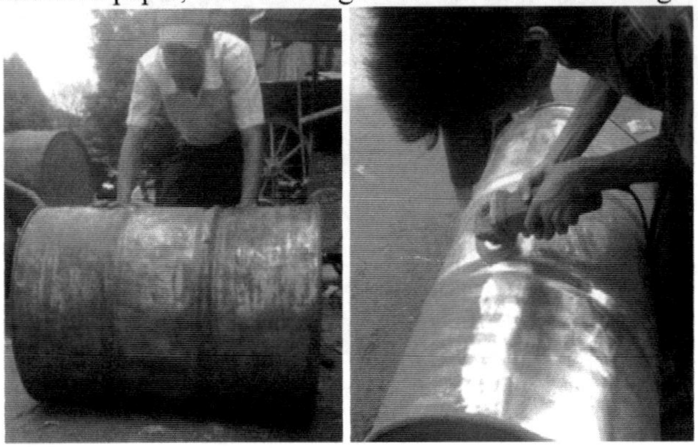

Figure 5. Cleaning Inside and outside surface of tank

Step 2: After the tank have been cleaned, dry the tank and then, paint the exterior surface of tank with a flat-black paint

Step 3: Drill inlet hole at the bottom and outlet hole at the two-third of the collector-storage tank (Figure 6). And then, drill the drain hole and air vent at the bottom and at the top of the collector-storage tank, shown in Figure 6. After that, weld copper pipes at the inlet, outlet and drain holes by using welding machine. Check water leakage at these joint (of welding).

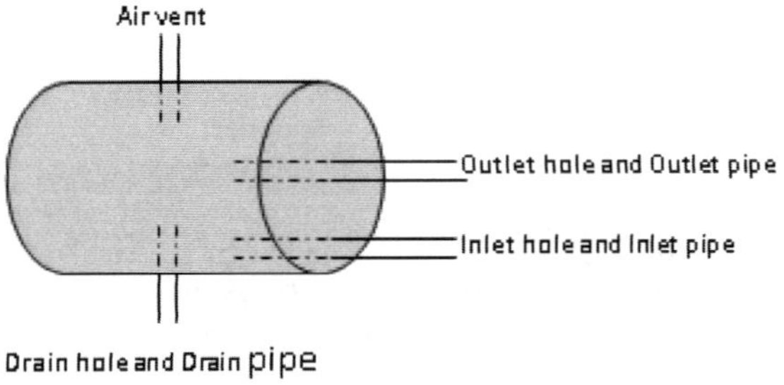

Figure 6. Drilling and Pluming

Step 4: Construct insulated rectangular wooden box as shown in Figure 7 (a), with required dimensions. The front and lid of this insulated box are opened to cover with glazing. Fill inside surface of this box with sawdust at the bottom, back and side (left and right) and recover with wooden of this (sawdust) filling, shown in Figure 7 (b). And then, paint inside surface of this box with metallic shiny coating

Figure 7. (a) Insulated Rectangular Wooden Box, (b) Sawdust Filling

Step 5: Place collector-storage tank (horizontally) in the insulated rectangular wooden box. And then, cover with window glass at the lid and front of this box. Check corners and joints of this box to prevent thermal losses. Construct reflectors with the dimension of the glazing and then, install at the lid and front of the box.

Step 6: After passing through step 1 to 5, construction of bread box solar water heater was finished.

5. Performance testing of a fabricated Bread Box solar water heater

To construct this bread box solar water heater, it took over one week. After the heater has been constructed, performance test was conducted in Pyay Technological University campus. To test this heater, firstly, it is necessary to look for south facing sunny spot. Then, cold water is filled from the bottom of the collector-storage tank and its reflectors are opened at morning time (sunrise) and closed at evening time (sunset), shown in Figure 8 (a) and (b).

Figure 8. (a) Opened Reflectors at Morning (Sunrise) (b)Closed Reflectors at Evening (Sunset)

The hot water temperature of this constructed heater are measured with three layers such that maximum temperature ($T_{maximum}$) from the top of the tank or air vent, delivery temperature ($T_{delivery}$) from outlet pipe and bottom temperature (T_{bottom}) from the inlet pipe and then, the ambient-water temperature is measured from the surrounding. The average results of tested water temperature from 1st November to 8th November, 2017 are shown in Figure 9.

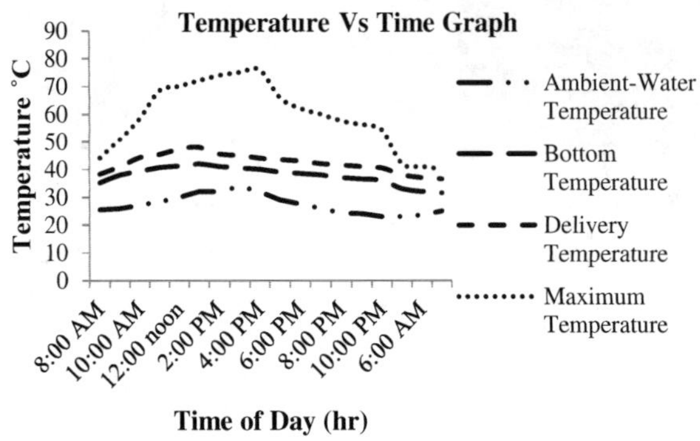

Figure 9. Variation of hourly temperature

According to the tested results of this bread box solar water heater, the maximum design temperature 70 °C of hot water reached at 12 noon and then, this maximum temperature was lowered to 40 °C at 7 o'clock (AM) after passing through night.

6. Cost analysis of a Bread Box solar water heater

The cost when the unit is manufactured commercially is about (150,000 MMK) and its expected life is about 20 years. When using electric power for heating 200 litre of water up to 70°C when the inlet temperature of water is 25°C, the energy required is 37.683 MJ.

One unit of electricity = 1KWh (3.6 MJ)

37.683 MJ = 10.47 KWh

It means that this amount of energy must put into the water. Therefore, for heating 200 litres of water through 45°C requires 1 KW rated electric immersion heated running for nearly 10 hours daily, which cost is about (263 MMK)/day or (7,875 MMK)/month or (94,500 MMK)/year. Therefore, the payback period for the solar water heater is only about one and a half year.

7. Conclusion

Renewable energy includes hydropower, solar energy, wind energy, wave energy and tidal energy. Like other renewable energy systems, solar water heater minimizes the environmental impacts of enjoying a comfortable, modern lifestyle. Solar water heater helps to reduce our dependence on imported fuel.

This bread box solar water heater would be provided numerous benefits to households for rural area. Many households could reduce their fuel costs by eliminating or reducing their need for wood, gas or electricity to heat water and also could improve their health and hygiene because of their low cost, ease of installation, durability and reliability.

The designed solar water heater costs around (150,000 MMK) and it saves about 10 KWh of energy per day and payback period of the heater is just over one year. It expected life is about 20 years and therefore it saves large amount of energy and households' energy budged.

Acknowledgement

The authors would like to thank the financial and technical assistance provided by AUN-SEED/Net Japan ASEAN Collaborative Education Programme (JACEP) under the CEP 2019.

References

[1] HTUN ZM. Design and Performance Analysis of Solar Water Heating System for 435 Liters Water Tank: Yangon Technological University; 2006.

[2] Version 1ME. Cooling and Heating Load Calculations, Estimation of Solar Radiation. IIT Kharagpur.

[3] DUFLIE JA. Solar engineering of thermal processes. *A Wiley-Interscience Publication.*

1991:770-81.
[4] BAKIR O. Experiment Investigation of a Spherical Solar Collector: Middle East Technical University; 2006.
[5] Kreider JF, Kreith F. Solar heating and cooling: engineering, practical design, and economics. *NASA STI/Recon Technical Report A*. 1975;**76**:35225.
[6] Bainbridge DA. The integral passive solar water heater book. *The Passive Solar Institute*. 1981:2-79.

IOP Conf. Series: Materials Science and Engineering 1127 (2021) 012017 doi:10.1088/1757-899X/1127/1/012017

Japan Solar PV manufacturing in the past and future

Keiichi N. Ishihara*

*Graduate School of Energy Science, Kyoto University, Yoshida Honmachi, Sakyo-ku, Kyoto, Japan

E-mail: ishihara@energy.kyoto-u.ac.jp

Abstract. Over the past 10 years, the number of solar power generation systems installed around the world has increased rapidly, partly due to the strengthening of measures against global warming. Chinese companies meet most of the demand. While the installation of PV in developed countries is declining, the demand in China, the Middle East, and Southeast Asia are expected to grow continuously. Meanwhile, Chinese exports to the United States declined sharply due to trade friction between the two countries. In this work, the Japan market mechanism of solar PV was investigated to consider how they can survive in the future. Firstly, the historical trend of Japanese solar PV production in the domestic market is reviewed. It is indicated that the Japanese consumer has selected Japanese product even it is expensive. Because they can sell electricity by a reasonable price under the FiT scheme and spend more money on construction to protect from natural disasters. Recently, both Japanese and Chinese companies start to produce solar PV cells outside the country. The production price becomes competitive in the future. The hot production countries in southeast Asia are Malaysia and Vietnam.

1. Introduction

The demand for solar PV has accelerated in the last decade. In Japan, energy security is the most concern in the energy policy, since Japan produces limited fossil energy. Japanese government subsidized PV installation in the household between 1980 and 2006. Japanese solar PV manufacturing had led the world market. From around 2000, the wind power had introduced alternatively, since it is more cost-effective than PV. The Japanese share was top before the Chinese started mass production since 2006. In 2012, the Feed-in Tarif introduced for all kinds of renewable electricity generation. The electricity production by solar PV in Japan has increased five times. However, in 2018, Chinese companies supplied about 58% of world demand, and following Malaysia, Taiwan, Vietnam, Korea. However, when we look at the Japan market, Japanese brand occupies about half. The relation between energy policy and the market of solar PV has investigated by several authors [1, 2]. They report that the domestic policy affects the market and sometimes destroy the domestic firms. The recent trade conflict between China and the US will add another point of view and modify the historical trend of the solar PV market.

In this work, the difference between Japan and the world solar PV market was analyzed. After that, the future Japan market will be discussed, considering the US-China trade conflict.

2. Characteristics of Japan solar PV

2.1. Price of Module

The price of the module made in China is the cheapest in the world, which was about 600USD/kW, closely following made in Germany. It cost about 1,000USD/kW for made in the US and Korea. While it made in Japan was about 1300USD/kW, in 2014 [3]. But the Japanese manufactures have occupied 80% for the house and 50% for non-house in the market share, as shown in Fig.1 [4]. The reason can be identified three; a) high Feed-in Tarif, b) high construction cost, and c) close relation to the hosing maker. We will discuss the details in the following section.

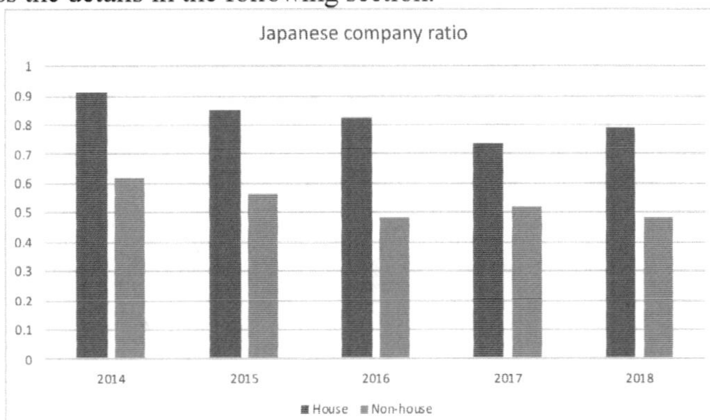

Figure 1. Market share of solar PV by Japanese companies

2.2. The effect of Feed-in Tarif

Japan introduced Feed-in Tarif for renewable energy since 2012 (Fig.2). In the beginning, it was relatively high comparing the selling price in the house, 25JPY/kWh. So, the user does not consider the module price seriously.

2.3. The construction cost

Many house vendors are connecting PV manufactures. One of the reasons is that the holding company, such as Panasonic and Panahome, provides both [5]. They trust each other. Furthermore, the shape of the roof is not simple in Japan because of the small land. So, the PV module should be flexible to fit any shape of the roofs. Japanese PV producer provides a variety of products to fit Japanese houses. The Japanese market share of rooftop PV in Japan is more than 80%.

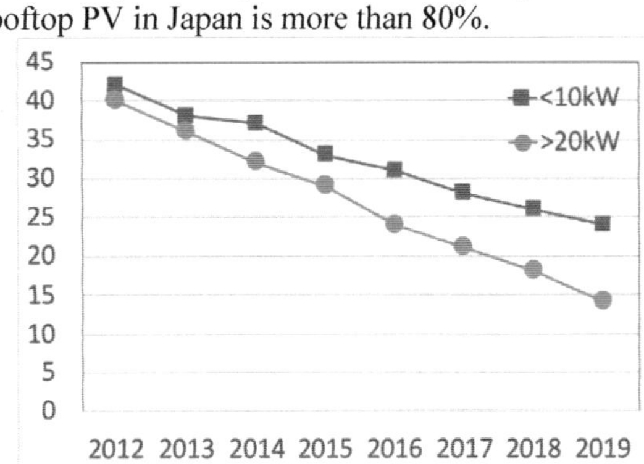

Figure 2. FiT price of solar PV in Japan. (unit: JPY/kWh)

3. Future of solar PV manufacturing

Because of the recent trade conflict between China and the US, Chinese producer starts to make their products outside of China and export from them. Before this movement, Japanese producers have started to produce to reduce labor cost. Notably, in southeast Asian countries, Malaysia and Vietnam will be the production center of solar PV. Fig. 3 shows the historical trend of solar PV exports from Malaysia and Vietnam. In the case of Malaysia, the major countries are the US, Hongkong, Japan, and Germany. While, in the case of Vietnam, it suddenly increased in 2017, and they export it to the Netherland and the US. Also, Japanese companies produce the module and panel using the imported cells. Fig. 4 shows the shipping of the solar module in Japan. Though market share by Japanese companies is more than 50%, the modules are mainly imported. Because the Japanese company produces solar PV outside Japan. As a result, the quality and module of solar PV become similar. In 2019, the system price is about 2,000 USD/kWh in both cases.

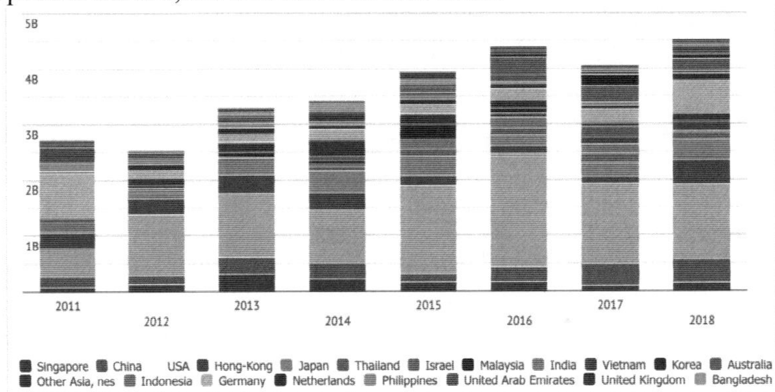

(a) Malaysia

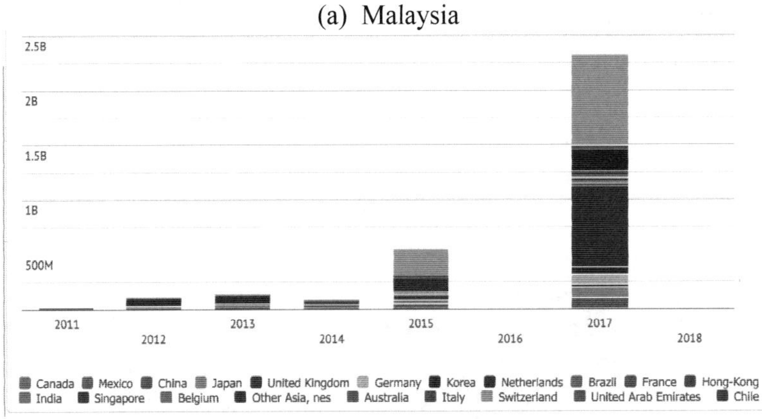

(b) Vietnam

Figure 3. The trend of Solar PV exports (unit: USD) [6].

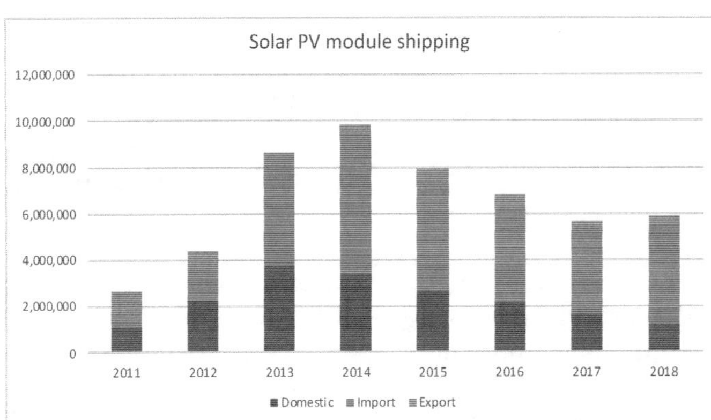

Figure 4. Solar PV shipping in Japan [4] (unit: kW)

4. Discussion

Thanks to FIT policies, solar PV power generation has increased rapidly in recent years. On the other hand, in 2007, the Japanese solar cell industry occupied 47% of the world market, but despite rising domestic demand, much of it was imported from China. The Japanese producers try to survive, focusing on the rooftop solar market. They believe in avoiding the severe competition by dividing the market between the inexpensive import module for a large solar farm and the high-performance module made in Japan well in terms of the application. Behind this, there is a strong trust between home builders and solar cell producers. The recent trade friction between the US and China forces Chinese manufacturers to actively promoting their products to Japan. However, the complex cost structure in the Japanese solar PV market alienates from international venders. The reason is that Japanese consumer believes the quality of Japanese products, shows less concern with the module price. There are two reasons. One is the high FIT price. The other is that the ratio of the module is small in the total cost because of high construction costs. Since Japan has many disasters such as typhoons, earthquakes, we need a solid foundation, which raises the cost. Besides, the construction period is more extended than in Germany. German companies succeeded in shortening the period by using a specialized tool while it is impossible to introduce the technology to Japan since the land is tight and the surrounding road is narrow.

Japanese manufactures of solar PV have shifted their production place to outside japan, seeking low labor and energy cost. On the other hand, the Chinese company also starts to produce PV outside China to avoid the trade conflict. Also, there is less difference in quality manufacturing in the same place and environment. Notably, they will be concentrated in Malaysia and Vietnam in SE Asia. In the future, as the price competition will intensify, for Japanese solar cell manufacturers to survive, it is necessary to provide even higher quality panels at low prices.

5. Summary

In this paper, the historical trend of Japanese solar PV production in the domestic market is reviewed. It is indicated that the Japanese consumer has selected Japanese product even it is expensive. Because they can sell electricity by a reasonable price under the FiT scheme and spend more money on construction to protect from natural disasters. Recently, both Japanese and Chinese companies start to produce solar PV cells outside. The production price becomes competitive in the future. The hot production countries in southeast Asia are Malaysia and Vietnam.

Acknowledgments

The authors would like to thank the financial and technical assistance provided by AUN-SEED/Net Japan ASEAN Collaborative Education Programme (JACEP) under the CEP 2019.

References

[1] Rammer C, Gottschalk S, Peneder M, Wörter M, Stucki T, Arvanitis S. Does energy policy hurt international competitiveness of firms? A comparative study for Germany, Switzerland and Austria. *Energy Policy*. 2017;**109**:154-80.

[2] Yu HJJ, Popiolek N, Geoffron P. Solar photovoltaic energy policy and globalization: a multiperspective approach with case studies of Germany, Japan, and China. *Progress in Photovoltaics: Research and Applications*. 2016;**24**(4):458-76.

[3] Japan Renewable Energy Foundation. The comparison of Solar PV Cost between Japan and Germany (in Japanese). 2016.

[4] Japan Photovoltaic Energy Foundation. Japan Photovoltaic Energy Association [last accessed Nov. 5, 2019. Available from: http://www.jpea.gr.jp/]

[5] Eco Life. The comparison of solar PV price (in Japanese) [last accessed Nov. 5, 2019. Available from: http://standard-project.net/solar/hikaku_brands.html]

[6] TrendEconomy. TrendEconomy [last accessed Nov. 5, 2019. Available from: https://trendeconomy.com/data/.]

Barrier Analysis for Deployment of Renewable Energy Mini-grids in Myanmar Using AHP

Masako Numata[a,*], Masahiro Sugiyama[a], Gento Mogi[b]

[a]Institute for Future Initiatives, The University of Tokyo, 7-3-1 Hongo, Bunkyo-ku, Tokyo 113-0033 Japan,

[b]Department of Technology Management for Innovation (TMI), School of Engineering, The University ofTokyo, 7-3-1 Hongo, Bunkyo-ku, Tokyo 113-0033 Japan

E-mail: m.matsuo-numata@05.alumni.u-tokyo.ac.jp

Abstract. Energy access is still a challenge for many countries, as demonstrated by Sustainable Development Goal (SDG) 7. Though the government of Myanmar set a target of 100% electrification by 2030, currently only less than half of households are connected to national grids. To accelerate electrification, decentralized approaches should be considered more. Mini-grids are an effective option that can fill the gap between a solar home system and a national grid. However, many of the existing mini-grids in Myanmar are powered by diesel generators. In rural areas, diesel fuel is much more expensive than in urban areas because of transportation cost. Under this condition, mini-grids powered by solar photovoltaics and batteries are already cost-competitive with diesel generators. Nevertheless, the deployment of mini-grids powered by renewable energy is still slow. In this study, we analysed barriers to the deployment and the prioritization of these barriers. We conducted a questionnaire survey with stakeholders using the analytic hierarchy process (AHP) to identify the prioritization of each barrier factor. To see tendencies, we used k-means for clustering results. The results showed that opinions were divided among stakeholders. There is no single silver bullet for the mini-grids and overcoming the barriers needs steady work.

1. Introduction

Among the United Nations SDGs, "Goal 7: Affordable and Clean Energy" states that access to electricity is still an issue worldwide [1]. The global population without access to electricity finally dropped to less than 1 billion, 840 million, in 2017 [2]. At the same time with improving access to electricity, it should be achieved by renewable energy rather than by fossil fuels. Sub-Saharan Africa attracts the most attention because of the large population that lives without access to electricity. However, in Asia, Myanmar has nearly the same electrification rate as Sub-Saharan Africa [3]. The Myanmar government has set a target of 100% electrification by 2030 [4] and the number of households connected to the national grid has grown from 34% in 2016 [5] to 42% in 2018 [6]. Still, more than half of total households lack access to electricity. As of 2017, only 24.7% of households in the rural area were connected to the national grid, in contrast with 42.2% in urban area [7]. The rural electrification is an urgent matter for universal access in Myanmar.

Mini-grids have recently begun to attract attention as being a bridge between household electrification methods such as solar lanterns and home systems and large-scale national grids [8,9].

According to the estimation by IRENA, the national grid would supply the 37% of additional power source to achieve universal energy access in developing Asian countries and then mini-grids would supply 44% of it [10]. In Myanmar, the national grid accounts for about 13% of the source of electrification in rural areas and mini-grids account for almost same with 13% though the national grid covers over 80% of the electrification in urban areas [7]. Mini-grids could contribute more for rural electrification.

In rural areas, roads are often not well-developed, and fuel prices are higher than urban area because of the transportation cost. In areas where diesel fuel is expensive, solar photovoltaics (PV) backed up with batteries are estimated to be cost-competitive with diesel generators as a power source for mini-grids [11]. However, in Myanmar, diesel generators are still dominant as a power source of mini-grids. It supplies power in 13,000 villages, whereas micro-hydropower, biomass, and solar energy are used as power sources in 2,400, 1,200, and 150 villages, respectively [12]. Various international aid agencies promote the introduction of mini- grids combining solar power and storage batteries [13,14], but mini-grids powered by renewable energy are still not expanding rapidly enough.

Myanmar is committed to "implement mitigation actions in line with sustainable development needs" in its Intended Nationally Determined Contributions (INDCs) to the Paris Agreement [15]. Also, Myanmar is a resource-rich country in terms of renewable energy, as estimated that 27GW for the potential capacity of solar power, 0.23GW for small and medium hydropower, and 100GW for large-scale hydropower [16,17]. Those abundant resources should be utilized for mini-grids as well as large scale generation [18].

In a prior study [19], we constructed a barrier typology based on discussions with stakeholders (international organizations, private companies, non-governmental organizations, and field researchers) and a bibliographic survey. In this study, we conducted a questionnaire survey of stakeholders based on an analytic hierarchy process (AHP) to analyse the priority of each barrier. The results were clustered using K-means methodology to see tendencies.

2. Barrier factors

We identified the barrier factors below in prior study [19] and chose 4 of them from each category to analyse prioritization in this study. The factors without citations were based on stakeholders' interview.

2.1. *Social/cultural barriers*

S1: Negative externalities caused by international organizations
Existing local mini-grid businesses were mostly non- commercial and social welfare purpose, but the introduction of business models has changed the mindsets of operators and/or customers. Then it caused to break trust relationship between operators and customers.

S2: Ethnic or language difference
Residential areas with minor ethnic groups overlap with off- grid areas. Language and cultural differences hinder project implementation.

S3: Education gap
The educational gap hinders financing for local companies from international organizations that provide lower capital costs. The language barrier (non-English speakers) is part of the reason.

S4: Perception of inferior quality [20,21,22]
Especially in the early stages, it is difficult to offer a 24-hour, seven-days-a-week service.

2.2. *Regulatory barriers*

R1: Lack of regulatory framework [23,24,25]

There have not been regulations for mini-grids yet though the off-grid mini-grid legal system has been developed under the DRD and Deutsche Gesellschaft für Internationale Zusammenarbeit (GIZ), and was submitted to the Ministry Of Electricity And Energy (MOEE) [26].

R2: Lack of technical standards[24,27,22,28]
Without technical standards or codes, it is difficult to maintain a certain level of quality for mini-grids. Moreover, rules for industrial waste, tar, and lead-acid should be established for sustainable development.

R3: Institutional capacity[29,30,22,18,25]
Institutions are attached to their current work, and it is difficult to coordinate priorities between ministries and/or other institutions.

R4: Threat of grid extension [30,31,22,32,33]
Mini-grid operators do not know what happens after the national grid reaches their customers' villages.

2.3. *Technical barriers*

T1: Technology gap
Indigenous technologies were sometimes different from international standards in many aspects, but it should not be flatly dismissed.

T2: Lack of interoperability with national grid [22]
Mini-grids might be designed without connections to the national grid due to the absence of technical rules.

T3: Intermittency [27,22,25]
The energy supply fluctuates over the day or season (typical for intermittent renewable energy sources).

T4: Operation and maintenance (O&M) [34,22,29] Appropriate operation and maintenance often fail to be maintained.

2.4. *Economic barriers*

E1: Small market size [24,35,30,25]
The Myanmar market is in its initial state, despite the rapid development of the international market.

E2: Low demand [24,35,30,29]
Creating demand in addition to basic use, such as for lighting, is still a challenge for operators.

E3: Tariff structure: cost-revenue gap [30,27,22,29,32,33] The design of the tariff structure affects the business model.

E4: Fee collection uncertainty [21,30,36,32]
Operators must ensure that customers pay for the electricity, sometimes using new technologies such as Pay As You Go.

2.5. *Financial barriers*

F1: Access to financing [34,23,29,27,25,28]
Due to the lack of familiarity with project financing through a financial institution, obtaining loans is difficult. Moreover, immature stock and debt markets limit the options for financing arrangements.

F2: High cost of capital [24,23,34,22,25,28]
Even if funds are arranged, financing c
osts are high. In the case of loans, interest rates are higher and loan fees are costly.

F3: Insufficient customers' capital [24,34,22]
Customers' financing methods are also limited. Microfinance is relatively new, and unofficial money lenders are expensive.

F4: Currency risk [34,37]
If financing is based on foreign currency, companies are exposed to exchange rate risks because their revenue and expenses are in different currencies.

3. Methodology

3.1. Analytic hierarchy process

The Analytic hierarchy process (AHP) is a decision-making method developed by Saaty in the 1970s and has been widely used since then [38,39,40]. A problem is hierarchically decomposed into key factors and each factor is compared in pairwise. Then the ranking of each factor is defined. The procedure is as follows[41];
(1) Model the problem considering the hierarchy of the key factors
(2) Determine the importance of the elements by comparing pairs based on knowledge and emotion, and score them on a five-point scale from 1 "equal importance" to 9 "extreme importance"
(3) Calculate the priority of each factor based on the scores

The result obtained is a square matrix (we chose 4 factors so here 4×4) with a diagonal component of 1. It is necessary to check if the obtained results are consistent; this is done using the random index (0.89 for a matrix of $n = 4$) [42]. The consistency ratio is set to 0.1 as suggested in many papers [41,43,44] and the only answers whose consistency score are lower than 0.1 are used for the analysis.

There are various multi-criteria decision-making methods that can be applied to energy planning[45,46], but we chose AHP because it is used widely and is easy to understand. AHP has been applied to barrier analysis for renewable energy development in Nepal [47]; for small-scale power sources in Sri Lanka [48]; for the adaptation of renewable energy in India [25]; for cooking stoves and biogas fermenters in rural Thailand [49]; for energy efficiency in small-scale industries in India [50]; and for cleaner production by small and medium-sized enterprises (SMEs) in China [51]. Also, it has been applied to other energy topics; the prioritization of the energy technologies to allocate research budget [52]; of decentralized power in Iran [53] and Jordan [54]; to develop energy in rural China [55]; and to select suitable locations for wind power generation [44].

In this study, we applied AHP to analyse the main barrier to the dissemination of mini-grids powered by renewable energy in Myanmar. A questionnaire survey was conducted during the period between September 2018 and February 2019. Questionnaires were sent by email or passed in person. Table 1 shows the list of respondents. Energy-related stakeholders were selected from among various occupations. We sent 53 and received 42 answers to analyse.

Table 1. Number of respondents of the survey

No. of respondents (individuals)	Sent	Answered
NGO (international +local)	8	8
Government	8	7
Private Company	25	15
Media	2	2
Academia	6	6
International organization	4	4
Total	**53**	**42**

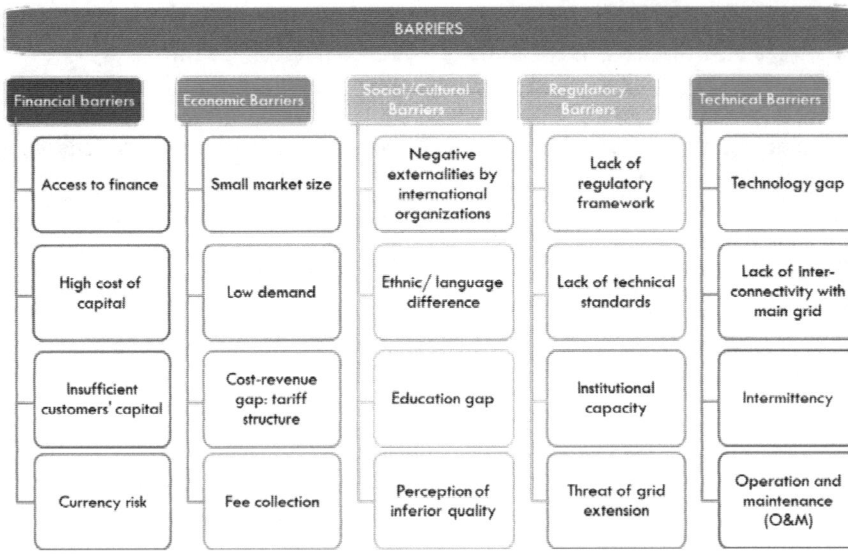

Figure 1. Hierarchy of barriers

3.2. K-means

K-means is a non-hierarchical clustering algorithm. The values are partitioned to the nearest cluster which has the nearest mean value within the cluster. The number of clusters k is given. The elbow method was used to investigate the number of clusters in this study [56].

4. Results

In AHP, a problem is hierarchized as shown in Figure 1. We identified four factors for each category prior to the questionnaire survey. Respondents were asked to compare one factor to another factor within the same category and continued to other categories. Only answers passed consistency check were used to analyse.

The number of clusters was examined using the elbow method. However, since most of the barrier categories did not display a clear elbow shape, the number of clusters was set to three based on the balance with the number of valid answers. For all barrier categories, the number of clusters was set as the same in a way that was easy to understand. Within each category, the scoring results obtained from the respondents were clustered into three groups and analysed. After the consistency check using the consistency ratio 0.1, the number of valid answers was 17 for regulatory category, 13 for the social and cultural category, 12 for the economic category and for the technical category, and 8 for the financial category of 42 responses.

4.1. Social/cultural barriers

Figure 2 shows the clustering results of priorities of the social and cultural barrier factors. The valid responses were divided into three clusters using k-means, and the weight of each barrier factor within the same cluster was averaged. The vertical axis in the figure shows the averaged value of prioritization score for each factor. In the social and cultural category, Clusters 1 and 2 have six valid responses each. "S4: Perception of inferior quality" was considered the most important factor in Cluster 1, and "S3: Education gap" the most important in Cluster 2. These factors are the most important by far in their Clusters (S4 has a weight of 0.63 in Cluster 1 and S3 has a weight of 0.60 in Cluster 2). Other barrier factors, S1 and S2 are relatively less prioritized in those two majority clusters. Minority group, cluster 3 shows a different tendency.

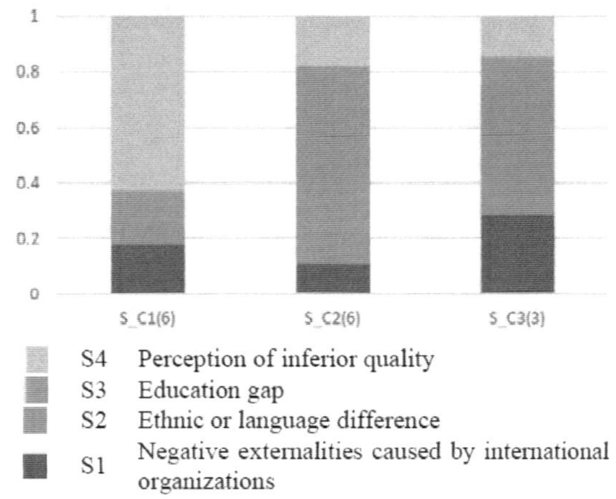

Note: S_C1(6) indicates that Cluster 1 of the social category contains six valid responses. Same with S_C2(6) and S_C3(3).

Figure 2. Clustering results of priorities of the social/cultural barrier factors

4.2. Regulatory barriers

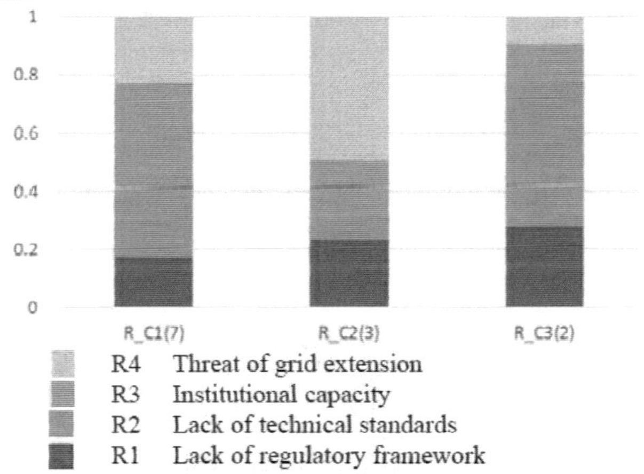

Note: R_C1 (7) indicates that Cluster 1 of the regulatory category contains seven valid responses regarding regulatory barriers. Same with R_C2(3) and R_C3(2).

Figure 3. Clustering results of priorities of regulatory barrier factors

Figure 3 shows the clustering results of priorities of the regulatory barrier factors. Cluster 1 is the majority and "R3: Institutional capacity" was the most prioritized. In Myanmar, electrification by the extension of the national grid is under the jurisdiction of the MOEE, and electrification by off-grid components such as SHS and mini-grids are under the jurisdiction of the Department of Rural Development of the Ministry of Agriculture, Livestock and Irrigation (DRD). The respondents commented that the coordination between the ministries was not smooth.

4.3. Technical barriers

Figure 4 shows the results of the technical barriers. Cluster 1 includes five valid answers, slightly more than other clusters which have the same number of answers. Cluster 1 prioritizes "T1: Technology gap" among factors. The tendency of prioritization is different among clusters. However, "T3: intermittency" is relatively less prioritized in all clusters.

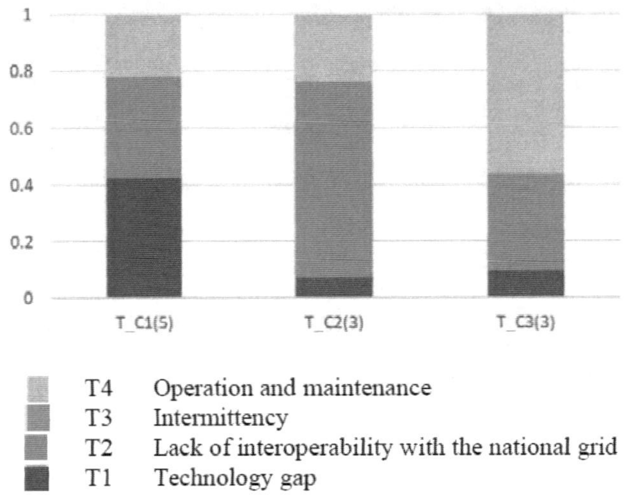

Note: T_C1 (5) indicates that Cluster 1 of technical category contains five valid responses. Same with T_C2(3) and T_C3(3).

Figure 4. Clustering results of priorities of technical barrier factors

4.4. Economic barriers

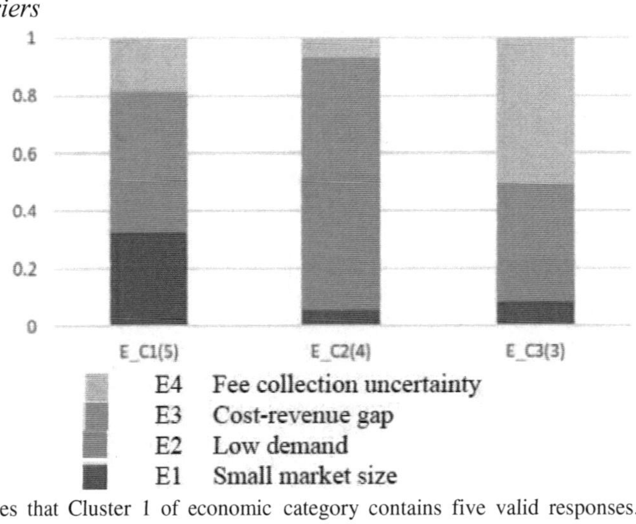

Note: E_C1 (5) indicates that Cluster 1 of economic category contains five valid responses. Same with E_C2(4) and E_C3(3).

Figure 5. Clustering results of priorities of economic barrier factors

Figure 5 shows the clustering results of the priorities of economic barrier factors. Cluster 1 prioritizes "E1: Small market size" (weight of 0.33) and "E3: Cost-revenue gap" (weight of 0.31). Cluster 2 prioritizes "E2: Low demand" (weight of 0.46) and E3 (weight of 0.41). E3 is evaluated as relatively important in common by the top two clusters.

4.5. Financial barriers

As a result of the consistency check, many responses in the financial category were inconsistent. This result indicates that it is difficult to decide the main barrier factor in the financial category.

5. Discussion and Conclusions

Results were different by categories. Some categories showed high agreement and other categories not. Clear prioritization was not observed unlike other countries; e.g., the threat of the national grid extension in India [22]. The results indicate that there are many kinds of barriers to the deployment of renewable energy mini-grids in Myanmar, and Myanmar is still at a stage where the many barriers remain still to be tackled.

In the social and cultural category, the opinions were divided among respondents and "S4: Perception of inferior quality" and "S3: Education gap" were evaluated as the greatest barrier factor. The solar and hydropower are intermittent by dry/rainy seasons, and day/night in case of solar. Solar PV needs back up such as batteries or diesel generator to secure its power supply. Micro hydropower needs to be designed carefully with a decrease in flow during the dry season in mind. It is important that consumers understand the characteristics of renewable energies. It leads to decrease dissatisfaction when electricity shortage occurs.

The education gap has an impact on mini-grids business. The school enrolment rates are different between the urban and rural area in Myanmar. The high school enrolment rate is 60% in the urban area though it is 39% in the rural area in 2017 [7]. International donor organizations have entered Myanmar and they have provided soft loans. However, well- developed documents e.g., excel-based financial plan, are needed to access these low-interest loans and it is difficult for existing developers/operators to prepare. The soft loan provider should consider to whom they would like to provide and take measures to make their loan easier to use such as consultation service.

In the regulatory category, a considerable degree of agreement about institutional capacity was observed. The power sources generating 30 MW or more, and any power generation connected to the national grid are under the jurisdiction of the MOEE according to The Electricity Law 2014 [57]. Power generation under 30 MW without connection to the national grid are under the jurisdiction of the state and regional governments. However, the off-grid electrification of the National Electrification Project (NEP) funded by the World Bank [58] is being promoted by the DRD. Off-grid components are supposed to be installed where the national grid will not reach in certain years. Therefore the coordination between ministries is needed. However, it seems not going smoothly. The regulatory guideline was drafted by DRD and circulated in the second half of 2018 [26] and has not published yet.

Under the technical category, the technology gap was evaluated as most important by the majority group. Before its democratization of 2011, Myanmar had been closed to foreign countries and the indigenous technology in small- scale hydropower and biomass has been developed. It sometimes differs from current international practices. However, there have been many cases that high-level technology was installed and discarded after a breakdown without repair. In a rural area, it takes a few days to arrange engineer from an urban area and to obtain repair parts. Therefore, those difficulties discourage to keep using. When technology is installed, it should be considered if it is an affordable level for operation and maintenance in the area.

In the economic category, the cost-revenue gap was evaluated as relatively important by the two major clusters. The importance of the cost-revenue gap shows how difficult to balance tariff and customers' ability to pay. Unlike diesel generator, renewable energy mini-grids are capital intensive, and developers are intended to be paid back by tariffs. The rural consumers' ability to pay is often limited and consumers may refrain from using electricity if a tariff is felt as too expensive. It is necessary to balance the consumers' ability to pay and the recovery of cost. Also, the range of acceptable tariff depends on the reference price. If villagers previously relied on diesel power, the tariff of solar-powered mini-grids could be felt cheaper. If villagers refer the subsidized national grid tariff, villagers answered that they would not like to pay tenfold tariff.

There are some limitations in this study. Ideally, it is desirable to define prioritization among categories following to prioritization of the factors. However, since the paper-based and e-mail based questionnaire survey was conducted, it was difficult to conduct superior prioritization of categories which needs the calculated results of prioritized factors. We placed emphasis to collect opinions of various stakeholders. We used the consistency ratio of 0.1 which is suggested in most of the extant literature, but some of the prior studies mentioned the range of consistency ratio from 0.1 to 0.2 [59,60]. Sensitivity to consistency ratio could be analyzed for further research.

Acknowledgements
This research was supported by the Economic Research Institute for ASEAN and East Asia (ERIA).

The work in progress of this research was reported in the ERIA research project report 'Shaping Energy Policies to Achieve the Sustainable Development Goals in Myanmar and the Greater Mekong Subregion' in September 2019 [61].

References

[1] United Nations, "Sustainable Development Goals," 2015. [Online]. Available: http://www.un.org/sustainabledevelopment/sustainable-development-goals/. [Accessed: 29-May-2017].

[2] IEA, IRENA, UNSD, WB, and WHO, "Tracking SDG7: The Energy Progress Report (2019)," Washington DC, 2019.

[3] Climatescope, "1Q 2019 OFF-GRID AND MINI-GRID MARKET OUTLOOK," *Medium*, 2019. [Online]. Available: https://medium.com/climatescope/1q-2019-off-grid-and-mini-grid-market-outlook-ec4a400ced93. [Accessed: 17-Apr-2019].

[4] Ministry of Electricity and Energy Electricity Supply Enterprise The Republic of the Union of Myanmar, "NEP plan," *Ministry of Electricity and Energy The Republic of the Union of Myanmar*, 2019. [Online]. Available: http://www.moee.gov.mm/en/ignite/page/80. [Accessed: 17-May-2019].

[5] Frontier Myanmar Research Ltd., "Myanmar Energy Brief," *Myanmar Energy Monitor*, 2019, no. 191, Yangon, p. 3.

[6] D. Billen and G. Bianchi, "Decentralised Energy Market Assessment in Myanmar," Yangon, 2019.

[7] Central Statistical Organization, UNDP, and World Bank, "Myanmar Living Conditions Survey 2017: Key Indicators Report," Nay Pyi Taw and Yangon, Myanmar, 2018.

[8] D. Schnitzer, D. S. Lounsbury, J. P. Carvallo, R. Deshmukh, J. Apt, and D. M. Kammen, *Microgrids for Rural Electrification*. United Nations Foundation, 2014.

[9] BNEF, "2Q 2017 Frontier Power Market Outlook Micro-grids on the horizon," 2017.

[10] IRENA, "REthinking Energy 2017: Accelerating the global energy transformation," Abu Dhabi, 2017.

[11] M. Numata, M. Sugiyama, G. Mogi, Wunna Swe, and V. Anbumozhi, "Technoeconomic Assessment of Microgrids in Myanmar," Jakarta, ERIA-DP-2018-05, 2018.

[12] C. Greacen, "Myanmar Mini-Grid Overview," 2017. [Online]. Available: https://www.esmap.org/sites/default/files/Greacen ESMAP Myanmar minigrid situation_Optimized.pdf. [Accessed: 01-Jul-2017].

[13] ADB, "Developing Renewable Energy Mini-Grids in Myanmar," Manila, Philippines, 2018.

[14] [14] Frontier Myanmar Research Ltd., "Second batch of NEP mini-grid proposals finalised," *Myanmar Enery Monitor*, 2018. [Online]. Available: http://energy.frontiermyanmar.com/news/power-and-electricity/second-batch-nep-mini-grid-proposals-finalised. [Accessed: 27-Apr-2018].

[15] The Republic of the Union of Myanmar, *Myanmar's Intended Nationally Determined Contribution-INDC*. 2015.

[16] ADB, "Myanmar: Energy Sector Assessment, Strategy, and Road Map," Manila, Philippines, 2016.

[17] ADB, "Renewable Energy Developments and potential in the Greater Mekong Subregion," Manila, Philippines, 2015.

[18] D. Del Barrio-Alvarez and M. Sugiyama, "Toward a sustainable electricity policy in Myanmar: Recommendations for policy makers and development partners," Tokyo (Japan), 2018.

[19] H. Yoshikawa and V. Anbumozhi (Eds), "Electricity Futures in the Greater Mekong Subregion: Towards Sustainability, Inclusive Development, and Conflict Resolution," Economic Research Institute for ASEAN and East Asia, Jakarta, 2018.

[20] S. C. Bhattacharyya, *Mini-Grids for Rural Electrification of Developing Countries: Analysis and Case Studies from South Asia*. London: Springer, 2014.
[21] M. Franz, N. Peterschmidt, M. Rohrer, and B. Kondev, "Mini-grid Policy Toolkit," Eschborn, 2014.
[22] S. D. Comello, S. J. Reichelstein, A. Sahoo, and T. S. Schmidt, "Enabling Mini-Grid Development in Rural India," *World Dev.*, 2017, vol. 93, pp. 94–107.
[23] C. Greacen, "Role of Mini-grids for Electrification in Myanmar - SWOT Analysis and Roadmap for Scale up," Washington, DC, 2017.
[24] J. P. Painuly, "Barriers to renewable energy penetration: A framework for analysis," *Renew. Energy*, 2001, vol. 24, no. 1, pp. 73–89.
[25] S. Luthra, S. Kumar, D. Garg, and A. Haleem, "Barriers to renewable/sustainable energy technologies adoption: Indian perspective," *Renew. Sustain. Energy Rev.*, 2015, vol. 41, pp. 762–776.
[26] P. Du Pont, "Decentralizing power:The Role of State and Region Governments in Myanmar's Energy Sector," Yangon, 2019.
[27] T. S. Schmidt, N. U. Blum, and R. S. Wakeling, "Attracting private investments into rural electrification - A case study on renewable energy based village grids in Indonesia," *Energy Sustain. Dev.*, 2013, vol. 17, no. 6, pp. 581–595.
[28] UNCDF/UNDP, "CleanStart - Microfinance Opportunities for a Clean Energy Future," 2012.
[29] H. Ahlborg and L. Hammar, "Drivers and barriers to rural electrification in Tanzania and Mozambique - Grid-extension, off-grid, and renewable energy technologies," *Renew. Energy*, 2014, vol. 61, pp. 117–124.
[30] S. C. Bhattacharyya, "To regulate or not to regulate off-grid electricity access in developing countries," *Energy Policy*, 2013, vol. 63, pp. 494–503.
[31] T. Kobayakawa and T. C. Kandpal, "Photovoltaic micro-grid in a remote village in India: Survey based identification of socio-economic and other characteristics affecting connectivity with micro-grid," *Energy Sustain. Dev.*, 2014, vol. 18, no. 1, pp. 28–35.
[32] F. Hasan, "ARE Newsletter April 2018 : Off-grid Investments : Financing & Funding In focus," *The Alliance for Rural Electrification*, 2018. [Online]. Available: https://www.ruralelec.org/newsletter/are-newsletter-april-2018-grid-investments-financing-funding#guest-editorial-1638. [Accessed: 26-Apr-2018].
[33] [33] B. Tenenbaum, C. Greacen, T. Siyambalapitiya, and J. Knuckles, *From the Bottom Up*. World Bank, 2014.
[34] [34] D. Gershenson, M. Tilleard, J. Cusack, D. Cooper, A. Monk, and D. Kammen, "Increasing Private Capital Investment into Energy Access: The Case for Mini-grid Pooling Facilities," 2015.
[35] D. Palit and A. Chaurey, "Off-Grid Rural Electrification Experiences from South Asia," *Green Energy Technol.*, 2011, vol. 15, pp. 266–276.
[36] N. U. Blum, R. Sryantoro Wakeling, and T. S. Schmidt, "Rural electrification through village grids - Assessing the cost competitiveness of isolated renewable energy technologies in Indonesia," *Renew. Sustain. Energy Rev.*, 2013, vol. 22, pp. 482–496.
[37] BloombergNEF, "Emerging Markets Outlook 2018. Energy transition in the world's fastest growing economies," 2018.
[38] R. W. Saaty, "The analytic hierarchy process-what it is and how it is used," *Math. Model*, 1987, vol. 9, no. 3–5, pp. 161–176.
[39] T. L. Saaty, "An Exposition of the AHP in Reply to the Paper 'Remarks on the Analytic Hierarchy Process,'" *Manage. Sci.*, 1990, vol. 36, no. 3, pp. 259–268.
[40] T. L. Saaty, "Decision making with the AHP,Why is the principal eigenvector necessary," *Eur. J. Oper. Res.*, 2003, vol. 145, pp. 85–91.
[41] T. L. Saaty, "How to make a decision: the analytic hierarchy process.," *Eur. J. Oper. Res*, 1994, vol. 74, no. 1, pp. 426–447.

[42] T. L. Saaty, "The Analytic Network Process," in *Decision Making with the Analytic Network Process International Series in Operations Research & Management Science*, Boston, MA: Springer, 2013, pp. 1–26.
[43] K. Soma, "How to involve stakeholders in fisheries management - A country case study in Trinidad and Tobago," *Mar. Policy*, 2003, vol. 27, no. 1, pp. 47–58.
[44] H. Aras, Ş. Erdoğmuş, and E. Koç, "Multi-criteria selection for a wind observation station location using analytic hierarchy process," *Renew. Energy*, 2004, vol. 29, no. 8, pp. 1383–1392.
[45] S. C. Bhattacharyya and D. Palit (Eds), *Mini-Grids for Rural Electrification of Developing Countries: Technical Aspects of Mini-Grids for Rural Electrification*. Springer, 2014.
[46] P. G. A. Kumara, "Report of Off-Grid Hydro Power Assessment of Myanmar," 2015. [Online]. Available: https://energypedia.info/images/3/3e/REVIEW_REPORT_OF_OFF-GRID_HYDRO_POWER_Jan_23_2015.pdf. [Accessed: 01-Jul-2018].
[47] L. P. Ghimire and Y. Kim, "An analysis on barriers to renewable energy development in the context of Nepal using AHP," *Renew. Energy*, 2018, vol. 129, pp. 446–456.
[48] P. D. C. Wijayatunga, K. Siriwardena, W. J. L. S. Fernando, R. M. Shrestha, and R. A. Attalage, "Strategies to overcome barriers for cleaner generation technologies in small developing power systems: Sri Lanka case study," *Energy Convers. Manag.*, 2006, vol. 47, no. 9–10, pp. 1179–1191.
[49] B. Limmeechokchai and S. Chawana, "Sustainable energy development strategies in the rural Thailand: The case of the improved cooking stove and the small biogas digester," *Renew. Sustain. Energy Rev.*, 2007, vol. 11, no. 5, pp. 818–837.
[50] N. Nagesha and P. Balachandra, "Barriers to energy efficiency in small industry clusters: Multi-criteria-based prioritization using the analytic hierarchy process," *Energy*, 2006, vol. 31, no. 12, pp. 1633–1647.
[51] H. Shi, S. Z. Peng, Y. Liu, and P. Zhong, "Barriers to the implementation of cleaner production in Chinese SMEs: government, industry and expert stakeholders' perspectives," *J. Clean. Prod.*, 2008, vol. 16, no. 7, pp. 842–852.
[52] S. K. Lee, G. Mogi, and J. W. Kim, "Decision support for prioritizing energy technologies against high oil prices: A fuzzy analytic hierarchy process approach," *J. Loss Prev. Process Ind.*, 2009, vol. 22, no. 6, pp. 915–920.
[53] A. Zangeneh, S. Jadid, and A. Rahimi-Kian, "A hierarchical decision making model for the prioritization of distributed generation technologies: A case study for Iran," *Energy Policy*, 2009, vol. 37, no. 12, pp. 5752–5763.
[54] M. Kablan, "Prioritization of decentralized electricity options available for rural areas in Jordan," *Energy Convers. Manag.*, 1997, vol. 38, no. 14, pp. 1515–1521.
[55] W. Xiaohua and F. Zhenmin, "Sustainable development of rural energy and its appraising system in China," *Renew. Sustain. Energy Rev.*, 2002, vol. 6, no. 4, pp. 395–404.
[56] S. Raschka, "Python Machine Learning - Code Examples Chapter 11 - Working with Unlabeled Data – Clustering Analysis," *GitHub, Inc.*, 2017. [Online]. Available: https://github.com/rasbt/python-machine-learning-book/blob/master/code/ch11/ch11.ipynb. [Accessed: 07-May-2019].
[57] Pyidaungsu Hluttaw, *Electricity Law*, no. l. Republic of the Union of Myanmar, 2014, pp. 1–15.
[58] Ministry of Electricity and Energy Myanmar, "NEP Plan," 2018. [Online]. Available: http://www.moee.gov.mm/en/ignite/page/80. [Accessed: 20-Apr-2018].
[59] T. L. Saaty and K. P. Kearns, *Analytical planning: The organization of system*, First. New York: Pergamon Press, 1985.
[60] A. M. Cox, J. Alwang, and T. G. Johnson, "Local preferences for economic development outcomes: Analytical hierarchy procedure," *Growth Change*, 2000; vol. 31, no. 3, pp. 341–366.
[61] H. Yoshikawa and V. Anbumozhi (Eds.), "Shaping Energy Policies to Achieve the Sustainable

Development Goals in Myanmar and the Greater Mekong Subregion," Jakarta, 2019.

IOP Conf. Series: Materials Science and Engineering 1127 (2021) 012019 doi:10.1088/1757-899X/1127/1/012019

A review paper related to the various variable's measurements on the relationship between environmental performance and financial performance

Qichun Wu *, Fumitaka Furuoka *, Kiew Ling Pui †

* Asia-Europe Institute, University of Malaya, 50603 Kuala Lumpur, Malaysia

†Faculty of Economics & Administration, University of Malaya, 50603 Kuala Lumpur, Malaysia

E-mail: axliewu.b@gmail.com

Abstract. The present-day environmental concerns have provided an incentive for firms to reduce environmental pollution by the reduction in resource consumption which in turn reduces the carbon emissions and other pollutants. To achieve the stakeholder requirement, the firm needs to improve the environmental management system, utilize environmentally friendly activities and strategies for efficient utilization of materials. Different results (positive, negative, no influence) have been documented in previous articles. This study focusses on the measurement of variables and the result of these papers. An inconsistent result has been found between Corporate Environmental Performance (CEP) and Corporate Financial Performance (CFP) relationship, for a sample of 63 empirical studies. In general, environmental performance is positively related to corporate financial performance. The result indicates the variety of environmental variables influencing the results. It is demonstrating that financial benefits (both accounting-based, market-based and others) are more from environmental management and environmental pollution than environmental disclosure. But there is no obvious evidence to demonstrate different financial variables can lead to different correlations.

1. Introduction

Nowadays air pollution and global environmental problem are major issues all over the world. Firms are making more contribution to reduce its impact on air pollution and global environmental issues [1]. Several studies provide evidence on the relationship between corporate environment performance (CEP) and company's financial performance (CFP). However, a puzzle remains as to the precise effect of CEP and CFP on the organizations.

The increased environmental concern provides an incentive for the company to reduce environmental pollution through pro-active ways [2,3]. From the resource-based review, companies need to reduce the resources consumption to reduce the carbon emissions and other pollutions, thus increase CFP [4,5,6,7,8 9]. Another way that can improve CEP is to improve the corporate environmental management system and use friendly environmental activities and strategies leading to efficient material utilization [10,11,12]. Different variables have been measured in previous studies. Most of these papers measured financial performance by accounting-based such as return on assets (ROA) [13, 14], return on equity (ROE) [5,14]; market-based such as stock return or Tobin's Q [15,6,16,17]. There are some studies which use both, accounting-based and market-based variables

[18,19]. The environmental performance measured variables can be categorised into three groups. The first group is the environment pollution caused by the activities of the firm that had effects on the environment [21]. It includes some variables such as carbon emissions and waste [21,22]. As a minimum, in stakeholder theory corporate business has to reduce their environmental pollution to satisfy related stakeholders, at least retain environmental legitimacy [23]. The second group is environmental disclosure, the firm needs to disclose environmental information to interested parties to make decisions. They release the information under voluntary and mandatory rules, both as an annual report and sustainability reports [24]. For example, firms in the world registered under the Carbon Disclosure Project (CDP) is a convenient platform for studying cross-country carbon emission [25]. The third group is environmental management, the corporates improve the management structures and systems to adopt sustainability development [26]. Now firms have access to more approaches and objectives for enhancing the environmental risk management to avoid punishment by environmental protection agency. Such as carbon risk management can mitigate the penalty of firms [27].

Different results (positive, negative, no influence) have been documented in previous articles. We focus on the measurement of variables and the result of these papers. An inconsistent result has been found between Corporate Environmental Performance (CEP) and Corporate Financial Performance (CFP) relationship, for a sample of 63 empirical studies. We analyse whether different variables measured in CEP and CFP can determine empirical outcomes. The result shows that the variable environmental management has a positive influence on corporate financial performance.

The result indicates how various variables influence corporate financial performance. Besides, the results are also influenced by different theories and analysis methods. Some of the results suggest that financial benefits (accounting-based, market-based, and business operational performance) come from environmental management and environmental pollution more than environmental disclosure.

2. Environmental performance and financial performance

2.1. Definition of environmental performance

Corporate environmental performance is taking more attention in the last few decades, but there is not a clear definition. The ISO 14031 states that environmental operational performance as 'outcomes are benefits from the organizations being performed' [28]. There is two dimensions of CEP, environmental management performance (EMP) and environmental operational performance (EOP) [29, 30]. Both dimensions are capturing a different aspect of environmental performance [28,31,32]. This paper will use three dimensions to analysis corporate environmental performance. CEP is related to financial elements, environmental costs, resource consumption and allocation for the prevention, reduction and avoidance of environment impact [33].

Firm's activities influence on the natural environment [20], can be grouped into two categories: input- based measurement (resources consumption and energy input) and output-based measurement (GHG emission and waste) [34].

2.2. Definition of financial performance

Financial performance measures both accounting- based (profit) and market-based (stock prices) [35]. Financial performance is defined as "interaction among an organization's attributes, actions, and environment based on the economic outcomes" [36]. Financial performance is a multivariate construct [28] that consists of stock liquidity, profit, growth and stock market performance [37]. Most research on the relationship between CEP and CFP had been on profit and stock market performance due to different theoretical effects and empirical evidence across different study contexts [11,38,39,40]. On the other hand, existing research studies in environmental or theoretical concepts are not focused on stock market liquidity and growth [28].

Return on equity (ROE) is a profitability ratio which provides stockholders' evaluation and performance of the market. Return on assets (ROA), Return on investment (ROI), and Return on invested capital (ROIC) not only affects the equity capital but also the borrowing capital through trade payables and investors. Tobin's q−1 and the natural logarithm of Tobin's q is a ratio known as the market price of the firm to the book value of total assets [9,41,42].

3. Different measurement of environmental performance and financial performance

Corporate environmental performance on corporate financial performance has taken more awareness of the researchers during the last three decades. But this construct is difficult to measure. Some studies find that CEP can be composed of two dimensions, called environmental management performance (EMP) connecting to environmental operational performance (EOP) [28]. EMP composed of five sub-dimensions which are environmental policy, objective, processes, organizational structure, and monitoring [29,30]. All these sub-dimensions can be captured in several environmental indicators [28,31,32].

3.1. Different dimensions of Corporate environmental performance

a. Environmental pollution

The important part of environmental pollution is carbon emission and air pollution. Carbon emission releases its trace gas in Earth's atmosphere. Carbon emissions are the greenhouse gas emissions as a contributor to climate change due to global warming or the greenhouse effect. The industrial revolution began the blazing of carbon (fossil fuels) which has grown the concentration of carbon dioxide (CO2). Their levels in the atmosphere have rapidly increased global warming. Environmental pollution evaluates environmental impacts in terms of physical units and monetary value [43]. It comprises of many variables that can be categorised into two groups as input variables (resource consumption and total energy input) and output variables (GHG emissions and waste) [34].

b. Environmental disclosure

According to Ministry of the Environment Government of Japan (June 2007, p.7), 'environmental disclosure is, regardless of its name or disclosure media, to promote communication of organizations, to fulfil its accountability regarding environmental efforts in their activities, and to provide useful information to decision making of interested parties.'

In the previous studies, most of the variables utilized as environmental disclosure can be summarised as information releases regarding toxic emissions, environmental awards, environmental accidents and crises, and environmental investment announcement [43]. Azar Shahgholian (2019) introduced a new variable environmental disclosure quality. The author observed that environmental disclosure quality is more important than the toxic emissions [34]. Besides, legitimacy theory, stakeholder theory and voluntary theory are the most popular theories that have been used in prior studies related to social and environmental disclosure [44,45,46,47]. Carbon disclosure belongs to environmental disclosure. Legitimacy theory claims that carbon disclosure is a function of political, social and stakeholder pressures. In response to public pressure, firms with poor carbon performance need stronger incentives in an attempt to disclose more environmental information to the public [6]. Carbon disclosure is a sub-set of environmental disclosure [48], this information that related companies' past and forecast carbon emission level, the carbon emission level will associate with the firm's risk and opportunities, and the firm's action to manage this risk in the past or future [49,50].

Environmental disclosure can significantly enhance a firm's legitimacy. Organizations can adopt environmental disclosure to make the changes in performance and the firm's image [51]. The empirical findings of the previous study show that carbon disclosure and performance are different from the carbon study findings on environmental disclosure. The idea of carbon disclosure is a wider discourse than environmental and sustainability disclosure [6]. A changing climate creates pervasive risks; thus, a comprehensive strategy and actions will need to be adopted by a firm to deal with climate change. Even though all firms emit carbon, but many firms do not have environmental issues. On the one hand, most environmental disclosure, for example, toxic emissions and chemical waste are nondiscretionary, whereas carbon disclosure is voluntary. Hence, there is uncertainty about the nature of the relationship between carbon information and market reactions.

c. Environmental management

Environmental management is a new green technology or strategy that can improve the firm's financial performance. Such as, implementing a new efficient system to improve the green management of the production process. Environmental and research expenditure to reduce environmental-related risks. Regulatory risk is the risk of a change of policies and regulations that are likely to exert a significant effect on the financial performance and capital cost of a firm through compliance costs and/or trading emission credits. Risk is directly associated with climate change, such as drought, floods, storms and sea-levels rise. And last, corporate carbon risk is the risk of market competition due to environmental disclosure [52].

3.2. Dissimilarity dimensions of financial performance

This study from four dimensions to illustrate the corporate financial performance, mentioned as follows:

a. Accounting-based financial performance

Accounting-based financial performance is based on the financial statement and reflect the firm's valuation from the statement of financial position

b. Market-based financial performance

Market-based financial performance reflects the valuation by investors from the fluctuation in market value.

c. Firm cost

The firm cost has two aspects, first is the cost of the products [53], and the other one is the cost of financial debts [54,55].

d. Business operation performance

Business operation performance is evaluated from the process of the firm business operation, to reflect how well the firm's business is functioning.

4. Types of variables

4.1. Environmental performance variables measurement

This section describes the types of CEP measures employed by previous studies. Corporate environmental performance includes three dimensions, as tabulated in table 1: environmental pollution, environmental disclosure and environmental management. Environmental pollution includes GHG emission reduction [4,5,14,18], environmental rating score [18,22,56] and pollutant release [57,58]. Researchers have used environmental disclosure for the measurement of CEP. Environmental disclosure includes carbon emission disclosure [13,59] and others measure using environmental disclosure score [18,60] for specific measurement. Other researchers use third dimension environmental management to measure corporate environmental performance. Environmental management is when companies use environmental strategies and practices to improve the firm's environmental performance. Such as environmental management [61,62,63], CSR concern [64] and firm register in ISO 14001 certificate [19,30,65].

Table 1. Environmental variables measurement

Environmental pollution	*GHG emission reduction*
	Environmental rating score
	Pollutant release
Environmental disclosure	*Environmental disclosure score*
	Carbon emission disclosure
Environmental management	*Environmental management*
	CSR concern
	ISO 14001 certificate

4.2. Financial performance variables measurement

This part explains the different variables of financial performances from a different perspective, namely: accounting-based financial performance, market-based financial performance, cost and business operational performance as tabulated in Table 2.

There are some financial indicator to measure accounting-based financial performance, return on assets (ROA) [8,60,65], return on equity (ROE) [66,67], return on sales (ROS) [5,68], return on investment (ROI) [69,70,71] and Earning [58,72,73].

Table 2. Financial variables measurement

Accounting based financial performance	*ROA,*
	ROE
	ROS
	ROI
	Earning
Cost	*Production cost*
	Financial debt cost
Market based financial performance	*Tobin's Q*
	Market value
	Market value added
Business operation performance	*Financial risk*
	Future cash flow
	Capital turnover
	Sales growth

Market-based financial performance is measured using stock exchange price. Stock market price have been directly used by researchers like market value [60] and market values added [59]. While some other researches use stock price by company replacement costs to calculate Tobin's Q [9,30, 61]. In contrast, financial performance measures the cost of the firm. Financial performance includes production cost [53,74] and financial debt cost [54,55]. And other variables associated with business operation performance are financial risk [22], future cash flow [75], capital turnover [8]and sales growth [10,71].

5. Discussion and Conclusion

After highlighting the corporate environmental performance and corporate financial performance, there are different categories of measurement of corporate environmental performance and corporate financial performance as shown in Fig.1. In Fig.1, CEP and CFP categories are combined from 63 empirical studies and analysed the correlation through vote counting.

There are 110 correlations of different environmental performance on financial performance. There are 44 correlations (40%) using environmental pollution, 18 correlations (16%) using environmental disclosure and 48 correlations (44%) using environmental management. The four main financial categories are 61 correlations (56%) using accounting-based performance, 11 correlations (10%) using firm cost, 29 correlations (26%) using market-based financial performance and 9 correlations (8%) using business operation performance. For the measurement of environmental performance, the majority of researchers are using environmental performance and corporate environmental management, only 16% of relationship using environmental disclosure.

According to the different combinations can be inferred like under the environmental pollution measure, there are 22 positive results (50%), 12 negative results (27%), 9 non-relationship results and 1 U-shape result (2%). Under environmental disclosure measure respective, 7 positive results (39%), 3 negative result (17%), 8 non-relationship (44%). In the case of environmental management measures, 36 positive result (75%), 4 negative results (8%), 7 non- relationship (15%) and 1 U-shape (2%). This study demonstrates, using the firm's environmental management variable has the highest relative frequencies of positive results. But there are no obvious correlations for different corporate financial relationship. It reveals that firms need to take more consideration to improve environmental management tools or skills, these have a great impact on firm financial performance. Sustainability development and environmentally friendly production will more and more important in future company governance.

The limitation of this research is that the relationship between CEP and CFP may be influenced if analysed from a different database. These articles from a different area, from developed and developing countries, and different economics context may also influence the relationship.

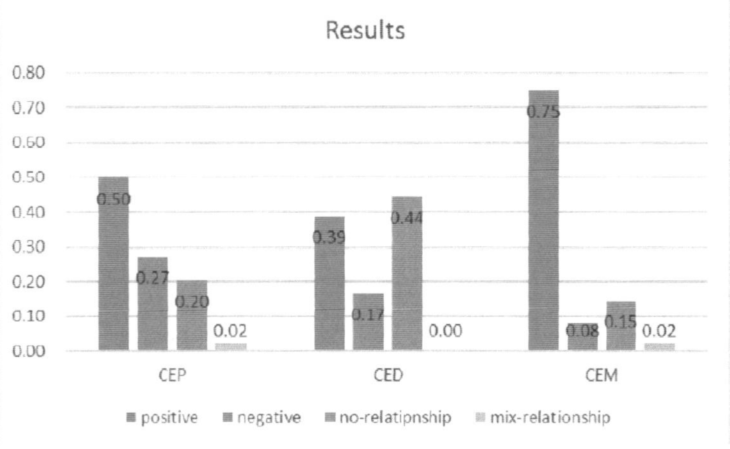

Figure 1. The relationship between Corporate Financial Performance and Corporate Environmental Performance

References

[1] Molina-Azorin, J. F., & Fetters, M. D. (2018). In This Issue. Journal of Mixed Methods Research, 12(3), 251– 253. doi:10.1177/1558689818781885

[2] Porter, M. E., & Reinhardt, F. L. (2007). A strategic approach to climate. Harvard Business Review, 85(10), 22-26.

[3] Ki-Hoon Lee, Byung Min. Green R&D for eco- innovation and its impact on carbon emissions and firm performance. Journal of Cleaner Production 108 (2015) 534-542.

[4] Carmen Fernandez-Cuesta, Paula Castro, María T. Tascon, Francisco J. Castano, The effect of environmental performance on financial debt. European evidence, Journal of Cleaner Production 207 (2019) 379-390

[5] Fortune Ganda & Khazamula Samson Milondzo (2018). The Impact of Carbon Emissions on Corporate Financial Performance: Evidence from the South African Firms. Sustainability 2018, 10, 2398; doi:10.3390/su10072398

[6] Yu He, Qingliang Tang & Kaitian Wang (2017): Carbon performance versus financial performance, China Journal of Accounting Studies, DOI: 10.1080/21697213.2016.1251768

[7] Isabel Gallego-Alvarez, Liliane Segura, Jennifer Martínez-Ferrero. Carbon emission reduction: the impact on the financial and operational performance of international companies. Journal of Cleaner Production 103 (2015) 149-159.

[8] Hidemichi Fujii and Kazuyuki Iwata and Shinji Kaneko and Shunsuke Managi. Corporate environmental and economic performances of Japanese manufacturing firms: Empirical study for sustainable development. Munich Personal RePEc Archive (2012).

[9] Hiroki Iwata, Keisuke Okada. How does environmental performance affect financial performance? Evidence from Japanese manufacturing firm. Ecological Economics 70 (2011) 1691–1700.

[10] Menguc B, Ozanne LK. 2005. Challenges ofthe “green imperative”: a natural resource-based approach to the environmental orientation–business performance relationship. J. Bus. Res.

58(4):430–38
[11] Marc Orlitzky, Frank L. Schmidt, Sara L. Rynes.(2003). Corporate Social and Financial Performance: A Meta- analysis. Organization Studies.24(3): 403–441
[12] Lee, K.-H., Min, B., 2014. Globalization and carbon constrained global economy: a fad or a trend? J. Asia-Pacific Bus. 15 (2), 105-121
[13] Zahra Borghei, Philomena Leung & James Guthrie. Voluntary greenhouse gas emission disclosure impacts on accounting-based performance: Australian evidence. Australasian Journal of Environmental Management (2018).
[14] Matthias Damert, Arijit Paul, Rupert J. Baumgartner. Exploring the determinants and long-term performance outcomes of corporate carbon strategies. Journal of Cleaner Production 160 (2017) 123-138.
[15] Feng Shen, Yunwen Ma2, Run Wang, Ningning Pan, Zhiyi Meng, (2019)Does environmental performance affect financial performance? Evidence from Chinese listed companies in heavily polluting industries, Quality& Quantity.
[16] Gilley KM, Worrell DL, Davidson WN 3rd, El-Jelly A. 2000. Corporate environmental initiatives and antic- ipated firm performance: the differential effects of process-driven versus product-driven greening initia- tives. J. Manag. 26(6):1199–216
[17] Li Chang, Wenjing Li and Xiaoyan Lu. Government Engagement, Environmental Policy, and Environmental Performance: Evidence from the Most Polluting Chinese Listed Firms. Bus. Strat. Env. 24, 1–19 (2015). DOI: 10.1002/bse.1802
[18] Mikael Petitjean, Eco-friendly policies and financial performance:Was the financial crisis a game changer for large US companies? Energy Economics 80 (2019) 502–511.
[19] Miroshnychenko I, Barontini R, Testa F. 2017. Green practices and financial performance: a global outlook. J. Clean. Prod. 147:340–51
[20] Wall, B. T., Stephens, F. B., Constantin-Teodosiu, D., Marimuthu, K., Macdonald, I. A., & Greenhaff, P. L. (2011). Chronic oral ingestion ofl-carnitine and carbohydrate increases muscle carnitine content and alters muscle fuel metabolism during exercise in humans. The Journal of Physiology, 589(4), 963–973. doi:10.1113/jphysiol.2010.201343
[21] CÓRDOVA, C. R., ZORIO-GRIMA, A., & GARCÍA- BENAU, M. (2018). NUEVAS FORMAS DE REPORTING CORPORATIVO: INFORMACIÓN SOBRE LA HUELLA DE CARBONO EN ESPAÑA. Revista de Administração de Empresas, 58(6), 537–550. doi:10.1590/s0034-759020180603
[22] Yasir Shahab, Collins G. Ntim, Yugang Chen, Farid Ullah, Hai- Xia Li1, Zhiwei Ye, Chief executive officer attributes, sustainable performance, environmental performance, and environmental reporting: New insights from upper echelons perspective. Bus Strat Env. 2019;1–16. DOI: 10.1002/bse.2345
[23] Qin, Y., Harrison, J., & Chen, L. (2019). A framework for the practice of corporate environmental responsibility in China. Journal of Cleaner Production, 235, 426–452. doi:10.1016/j.jclepro.2019.06.245
[24] Frost, G., 'The Introduction of Mandatory Environmental Reporting Guidelines: Australian Evidence', Abacus, Vol. 43, No. 2, 2007.
[25] Unerman, J., Chapman, C., 2014. Academic contributions to enhancing accounting for sustainable development. Acc. Organ. Soc. 39 (6), 385–394.
[26] Schultze, W., & Trommer, R. (2012). The concept of environmental performance and its measurement in empirical studies. Journal of Management Control, 22, 375-412.
[27] Peter M. Clarkson, Yue Li, Matthew Pinnuck & Gordon D. Richardson (2015) The Valuation Relevance of Greenhouse Gas Emissions under the European Union Carbon Emissions Trading Scheme, European Accounting Review, 24:3, 551-580, DOI: 10.1080/09638180.2014.927782
[28] Christoph Trumpp and Thomas Guenther. (2015). Too Little or too much? Exploring U-shaped Relationships between Corporate Environmental Performance and Corporate Financial

Performance. Bus. Strat. Env. (2015). DOI: 10.1002/bse.1900
[29] Milne, M. J., & Gray, R. (2012). W(h)ither Ecology? The Triple Bottom Line, the Global Reporting Initiative, and Corporate Sustainability Reporting. Journal of Business Ethics, 118(1), 13–29. doi:10.1007/s10551-012-1543-8
[30] Wahba H. 2008. Does the market value corporate environmental responsibility? An empirical examination. Corp. Soc. Responsib. Environ. Manag. 15(2):89–99.
[31] Xie, S., & Hayase, K. (2007). Corporate environmental performance evaluation: A measurement model and a new concept. Business Strategy and the Environment, 16, 148–168.
[32] Clemens, B., & Bakstran, L. (2010). A framework of theoretical lenses and strategic purposes to describe relationships among firm environmental strategy, financial performance, and environmental performance. Management Research Review, 33, 393–405.
[33] Voicu D. Dragomir. (2018) How do we measure corporate environmental performance? A critical review. Journal of Cleaner Production 196 (2018) 1124-1157.
[34] Azar Shahgholian. Unpacking the relationship between environmental profile and financial profile; literature review toward methodological best practice. Journal of Cleaner Production 233 (2019) 181-196.
[35] Dietrich Earnhart. (2018) The Effect of Corporate Environmental Performance on Corporate Financial Performance. Annu. Rev. Resour. Econ. 2018. 10:425– 44.
[36] Combs, J. G., Russell Crook, T., & Shook, C. L. (2005). The dimensionality of organizational performance and its implica- tions for strategic management research. In
[37] D. J. Ketchen (Ed.), Research methodology in strategy and management (pp. 259–286). Amsterdam: Elsevier.
[38] Hamann, P. M., Schiemann, F., Bellora, L., & Guenther, T. W. (2013). Exploring the dimensions of organizational performance: A construct validity study. Organizational Research Methods, 16, 67–87.
[39] Delmas M, Hoffmann VH, Kuss M. 2011. Under the tip of the iceberg: absorptive capacity, environmental strategy, and competitive advantage. Bus. Soc. 50(1):116–54
[40] Dixon-Fowler, H.R., Slater, D.J., Johnson, J.L., Ellstrand, A.E., Romi, A.M., 2013. Beyond 'does it pay to be green?' A meta-analysis of moderators of the CEP–CFP relationship. J. Bus. Ethics 112(2), 353–366.
[41] Endrikat, J., Guenther, E., & Hoppe, H. (2014). Making sense of conflicting empirical findings: A meta-analytic review of the relationship between corporate environmental and financial performance. European Management Journal, 32(5), 735–751.
[42] Konar S, Cohen MA. 2001. Does the market value environmental performance? Rev. Econ. Stat. 83(2):281–89
[43] Hirsch, B.T., Seaks, T.G., 1993. Functional form in regression models ofTobin's q. Review of Economics and Statistics 75 (2), 381–385.
[44] Elisabeth Albertini. (2013). Does Environmental Management Improve Financial Performance? A Meta- Analytical Review. Organization & Environment 26(4) 431 –457. DOI: 10.1177/1086026613510301
[45] Bouten, L., Everaert, P., Roberts, R.W., 2012. How a two-step approach discloses different determinants of voluntary social and environmental reporting. J. Business Finance Account. 39 (5–6), 567–605.
[46] Stephen Brammer, Chris Brooks, and Stephen Pavelin. (2006). Corporate Social Performance and Stock Returns: UK Evidence from Disaggregate Measures. Financial Management • Autumn 2006 • pages 97 - 116
[47] Deegan, C. (2002). Introduction: The legitimizing effect of social and environmental dis- closures: A theoretical foundation. Accounting, Auditing & Accountability Journal, 15(3), 282–311.
[48] Owen, D. (2008). Chronicles of wasted time? A personal reflection on the current state of, and

future prospects for social and environmental accounting research. Accounting, Auditing & Accountability Journal, 21(2), 240–267.

[49] Andrew, J., Cortese, C., 2011. Accounting for climate change and the self-regulation of carbon disclosures. Account. Forum 35 (3), 130–138.

[50] Najah, M. M. S. (2012). Carbon risk management, carbon disclosure and stock market effects: An international perspective. University of Southern Queensland.

[51] Noor Raida Abd Rahmana, Siti Zaleha Abdul Rasid, Rohaida Basiruddin. Exploring the relationship between carbon performance, carbon reporting and firm performance: A conceptual paper. Procedia - Social Social and Behavioral Sciences 164 (2014) 118 – 125

[52] Patten, D.M., 1992. Intra-industry environmental disclosures in response to the Alaskan oil spill: a note on legitimacy theory. Acc. Organ. Soc. 17 (5), 471– 475.

[53] Labatt, S., & White, R. R. (2007). Carbon finance: The financial implications of climate change. Hoboken, NJ: Wiley.

[54] Pereira-Moliner J, Font X, Tar´ı J, Molina-Azorin J, Lopez-Gamero M, Pertusa-Ortega E. 2015. The Holy Grail: environmental management, competitive advantage and business performance in the Spanish hotel industry. Int. J. Contemp. Hosp. Manag. 27(5):714–38

[55] Zhifang Zhou, Tao Zhang, Kang Wen, Huixiang Zeng, Xiaohong Chen. (2018) Carbon risk, cost of debt financing and the moderation effect of media attention: Evidence from Chinese companies operating in high- carbon industries. Bus Strat Env. 2018;1 – 14.DOI: 10.1002/bse.2056

[56] Juhyun Jung, Kathleen Herbohn, Peter Clarkson. Carbon Risk, Carbon Risk Awareness and the Cost of Debt Financing. J Bus Ethics (2016). DOI 10.1007/s10551- 016-3207-6

[57] Marilyn T. Lucas and Thomas G. Noordewier, (2016) Environmental Management Practices and Firm Financial Performance: The Moderating Effect of Industry Pollution-Related Factors, Intern. Journal of Production Economics.

[58] Noor Muhammad, Frank Scrimgeour, Krishna Redd, Sazali Abidin. The relationship between environmental performance and financial performance in periods of growth and contraction: evidence from Australian publicly listed companies. Journal of Cleaner Production 102 (2015) 324-332.

[59] Cordeiro JJ, Sarkis J. 1997. Environmental proactivism and firm performance: evidence from security analyst earnings forecasts. Bus. Strategy Environ. 6(2):104–14

[60] Fortune Ganda. The effect of carbon performance on corporate financial performance in a growing economy. (2016) SOCIAL RESPONSIBILITY JOURNAL. LARELLE CHAPPLE, PETER M. CLARKSON AND DANIEL L. GOLD. The Cost of Carbon: Capital Market Effects of the Proposed Emission Trading Scheme (ETS) ABACUS, Vol. 49, No. 1, 2013 doi: 10.1111/abac.12006

[61] Yuriko Nakao,Akihiro Amano, Kanichiro Matsumura, Kiminori Genba and Makiko Nakano. Relationship Between Environmental Performance and Financial Performance: An Empirical Analysis of Japanese Corporations. Bus. Strat. Env. 16, 106–118 (2007) DOI: 10.1002/bse.476

[62] Gonz´alez-Benito J, Gonz´alez-Benito O.2005. Environmental proactivity and business performance: an empirical analysis. Omega 33(1):1–15

[63] Carmona-Moreno E, C´espedes-Lorente J, De Burgos- Jim´enez J. 2004. Environmental strategies in Spanish hotels: contextual factors and performance. Serv. Ind. J. 24(3):101–30

[64] Isabel- María García- Sánchez & Jennifer Martínez- Ferrero (2018), Chief executive officer ability, corporate social responsibility, and financial performance: The moderating role of the environment, Bus Strat Env. 2018;1– 14. DOI: 10.1002/bse.2263

[65] Heras-Saizarbitoria I,Molina-Azor´ın JF,DickGPM.2011. ISO 14001 certification and financial performance: selection-effect versus treatment-effect. J. Clean. Prod. 19(1):1–12

[66] Zeng S, Meng X, Yin H, Tam C, Sun L. 2010. Impact of cleaner production on business performance. J. Clean. Prod. 18(10):975–83

[67] Dietrich Earnhart & Lubomir Lizal. Effect of Pollution Control on Corporate Financial Performance in a Transition Economy. Eur. Env. 17, 247–266 (2007). DOI: 10.1002/eet.447

[68] Mengying Feng, Wantao Yu, Xingyu Wang, CheeYew Wong, Maozeng Xu, Zhi Xiao, Green supply chain management and financial performance: The mediating roles of operational and environmental performance. Bus Strat Env. 2018;1–14. DOI: 10.1002/bse.2033

[69] Yavuz Agan, Cemil Kuzey, Mehmet Fatih Acar, Atif Açıkgoz. The relationships between corporate social responsibility, environmental supplier development, and firm performance. Journal of Cleaner Production (2014), http://dx.doi.org/10.1016/j.jclepro.2014.08.090

[70] Arag ´on-Correa J, Rubio-Lopez E. 2007. Proactive corporate environmental strategies: myths and misunder- standings. Long Range Plan. 40(3):357–81

[71] Montabon F, Sroufe R, Narasimhan R. 2007. An examination of corporate reporting, environmental management practices and firm performance. J. Oper. Manag. 25(5):998–1014

[72] De Burgos Jime´nez, J., &Ce´spedes Lorente, J. J. (2001). Environmental performance as an operation objective. International Journal of Operations and Production Management, 21(12), 553–572.

[73] Judge WQ, Douglas TJ. 1998. Performance implications of incorporating natural environmental issues into the strategic planning process: an empirical assessment. J. Manag. Stud. 35(2):241–62

[74] Sharma S, Vredenburg H. 1998. Proactive corporate environmental strategy and the development of compet- itively valuable organizational capabilities. Strateg. Manag. J. 10(2):729–53

[75] Marlene Plumlee, Darrell Brown, Rachel M. Hayes, R. Scott Marshall. Voluntary environmental disclosure quality and firm value: Further evidence. J. Account. Public Policy (2015),

IOP Conf. Series: Materials Science and Engineering 1127 (2021) 012020 doi:10.1088/1757-899X/1127/1/012020

An Improved Mathematical Model Derivation Based on Circuitry Approach for Crystalline Silicon Photovoltaic Module

K.L. Teong, Rodney H.G. Tan, Landon Y.H. Hoo

*UCSI University, Malaysia

E-mail: rodneytan@ucsiuniversity.edu.my

Abstract. This paper revises the five-parameter model, an approach that contributes the fundamental knowledge on performing accurate mathematical modelling for most of the commercially-available photovoltaic cells. However, such model can be further improved in order to increase the modelling accuracy while reducing the required modelling parameters at the same time. Improvement is made through formula derivation for photo-generated current and diode saturation current, while the remaining parameters (series resistance, shunt resistance, and diode ideality factor) remained un-managed for the parameterisation procedure. The derivations presented in this paper are direct, yet such procedures are frequently overlooked by others. The comparison between conventional and proposed modelling methods is presented. The methodology of this work is presented through MATLAB script demonstrated on the measured field data for 70W mono- and poly- crystalline silicon photovoltaic modules. The performed methodology justifies that formula derivation can be applied for photo-generated and diode saturation current parameters that originated from five-parameter model. Such approach simplifies the works of parameterising the theoretical model matches with the experimental I-V curve. Generally, the work of this paper contributes to the mathematical modelling of photovoltaic module, simulation of experimental I-V curve result, and interactive education tool.

1. Introduction

Over the year, the role of renewable energy becomes increasingly important through the power generation aspect as it enhances the flexibility to energy sourcing while reducing the dependence on fossil fuel-based energy generation [1]. The photovoltaic (PV) system, which is categorised under renewable energy, refers to the solar power generation technology which sources energy from the sunlight and converts into electricity form. Solar energy considered as a promising alternative to replace the non-renewable energy source due to its abundant existence in nature. Through such aspect, solar energy possesses high growing rate amongst all of the renewable energy including ocean wave, wind, hydro, and biomass [2]. This scenario also reflects the increase of PV system worldwide [3]. Regardless of their relatively high installation and operative cost of PV system during its initial development stage, such aspect is progressively overcome by lower-cost-focused research over the years [3-5].

Most of the conducted research applies the similar approach on modelling the PV system through mathematical or circuitry means, with the sole concern on allowing the designer to optimize the performance of the PV system [6]. Through mathematical modelling perspective, the most prominent work is presented by [7], which conclude the five required parameters to model the I-V curve of PV cell. This five-parameter model is crucial as the parameters of photo-generated current (I_{ph}), diode saturation current (I_o), diode ideality factor (A), shunt (R_{sh}) and series (R_s) resistance are required for I-

V curve analysis but these are usually not provided in manufacture's datasheet [7, 8]. However, the five-parameter model can be further simplified to ease the parameterisation of I-V curve. Presented in [1], the five-parameter model is simplified by deriving I_{ph}, I_o and R_s parameter, and thus the I-V curve is parameterised with remaining A and R_{sh} parameter. However, the methodology shown in [1] lacks convincing statements as two different derived equations for I_o parameter are used at the same time. Such methodology is indirectly debated by [9] stating that both derived equations presented the duplicated function. Besides, the modelling accuracy for the derived five-parameter model can be improved as well, through another derivation approach that will be proposed in this paper. As a continuation of the research outcomes, the I-V curve of PV module could be parameterised with the remaining A, R_s and R_{sh} parameters through the algorithm which presented in [10], after obtaining the I_{ph} and I_o parameter through derivation. Nonetheless, applied works of this paper also describe the reason for amending the algorithmic parameterisation approach applied in [10].

To converge the researches result and adopted into the work of this paper, it serves the purpose to model a PV module through the mathematical approach. The I-V characteristic curve of PV module is described by the lumped parameter equivalence, in which the total model parameter reflects the internal mechanism of PV module [11]. The objective of the work is to develop a more accurate general mathematical model for crystalline silicon PV module in order to evaluate the long-term performance. Nonetheless, the comparison between proposed and conventional model is presented in this paper, justifying that improvement of modelling accuracy can be achieved through the application of the proposed model.

2. Derivation of Mathematical Model

Many modelling methods exist to describe the behaviour of the PV system. The available modelling method could be categorised through either circuitry or mathematical mean. Among these two, the electrical equivalent circuit-based model is considered the popular method to simulate the behaviours of the PV cell. This statement is supported by the reason which circuitry model possesses the advantages of available electrical simulation software, such as PSpice and Proteus. With a proper model applied, the performance of PV cell, module, or even array can be predicted accurately through simulations [12]. As per today, the available circuitry modelling for PV module categorised into three, which are the double-diode model, single-diode R_s and R_p model [8]. The commonly used model is the single-diode R_p model, as shown in Figure 1.

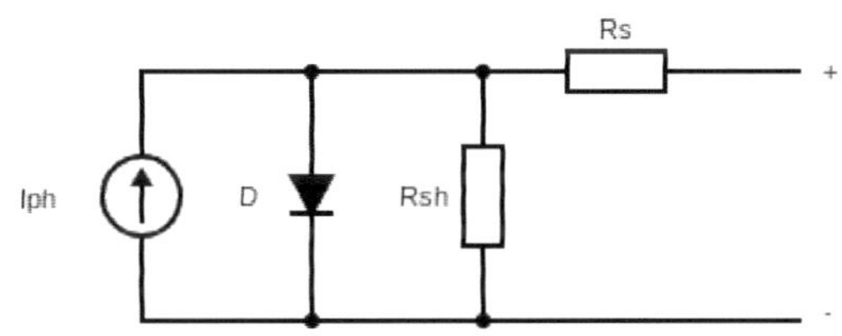

Figure 1. Single-diode equivalent circuit R_p model of PV cell

Due to simplicity, the single-diode R_p model is widely used to model a PV module [13], same reason is applied to justify the application of this model into this paper. As shown in Figure 1, the single-diode R_p model consisted of the current source connected in parallel to a PN junction diode and a shunt resistor, while connected in series to a resistor that leads to the terminal output. The current source models the irradiance current generated by the photon carriers that sourced from the sun; the diode models the electron-hole recombination between the PN junctions, thus representing the diode saturation current of PV cell. Lastly the shunt and series resistance models the leakage current of the diode and internal resistance of the PV module respectively.

Applied with Kirchhoff's Current Law, the mathematical model for single diode R_p model can be derived to determine the PV cell's terminal current through Equation (1), where I_{ph} is the photo-generated current which varies linearly to the irradiance level of the sunlight; I_D is the current flowing through the parallel diode, which resulting the non-linear characteristic of I-V curve; and $I_{R_{sh}}$ is the leakage current at shunt resistor.

$$I = I_{ph} - I_D - I_{R_{sh}} \tag{1}$$

With the substitution of Shockley diode equation and also relevant variables, the complete mathematical model of photovoltaic module is obtained as Equation (2), where I_o is the diode saturation current; R_s is the series resistance; R_{sh} is the shunt resistance; N_s is number of PV cells connected in series; V_T is the thermal voltage; V and I respectively represents the voltage and current of output terminal. The thermal voltage can be calculated using Equation (3), where A is the diode ideality factor of PV cell, k is the Boltzmann constant of 1.38×10^{-23} ; T is the PV cell's temperature; q is the electron charge constant of 1.6×10^{-19} .

$$I = I_{ph} - I_0\left(e^{\frac{V+IR_S}{N_S V_t}} - 1\right) - \frac{V+IR_S}{R_{sh}} \tag{2}$$

$$V_T = \frac{A \cdot k \cdot T}{q} \tag{3}$$

Equation (2) represents the relationship of the PV module's terminal output current respect to different terminal output voltage value. This equation applies the physic of PN junction and generally accepted to represent the behaviour of PV cell, especially for crystalline silicon [14]. Thus, a conventional single-diode R_p circuitry model is mapped to the mathematical model of Equation (2), which the equation is the foundation leads to the five-parameter model.

2.1. *Five-parameter Model*

In [7] and [15], the five-parameter is obtained by deriving the I-V characteristic curve general equation of Equation (2) under three conditions of short-circuit current condition I_{sc}, open-circuit condition V_{oc} and maximum output power of current I_{mp} and voltage V_{mp}. With such approach, both [7] and [15] conclude that a minimum of five parameters are required to model the I-V characteristic curve of PV cell, which also known as the behaviours of PV cell. The five parameters I_{ph}, I_o, A, R_{sh}, R_s represents the photo-generated current, diode saturation current, diode ideality factor, shunt and series resistance respectively. It is proven in [7] and [15] that each of these five parameters does influence the characteristic of the I-V curve. The solid justification of five parameter model towards the characteristic of I-V curve is recognized by most of the publication [1, 12, 16] and further improvement for PV system is proceeded after the contribution of the five-parameter model. However, the five-parameter model can be further simplified, through the formula derivation for photo-generated current, I_{ph} and diode saturation current, I_o.

2.2. *Derivation for Photo-generated Current*

For simplicity, many researchers [1, 7, 9, 12] apply the concept of photo-generated current, I_{ph} is equivalent to short-circuit current, I_{sc} as presented in Equation (4). Besides, some researchers derive the photo-generated current to increase modelling accuracy, such as [3, 4, 16, 17]. The prominent work is presented in [16]. The photo-generated current can be calculated through Equation (5), which is the derivation from Equation (2) at open-circuit voltage condition. However, the decision of deriving the formula to obtain photo-generated current parameter would be questionable as the photo-generated current should be mapped with short-circuit current condition based on the general assumption made by researchers [1, 7, 9, 12].

$$I_{ph} \approx I_{sc} \tag{4}$$

$$I_{ph} = I_0\left[e^{\frac{V_{OC}}{N_S V_T}} - 1\right] + \frac{V_{oc}}{R_p} \tag{5}$$

In [17], a similar method is applied, in which the derivation of photo-generated current is derived based on short-circuit current condition. As in general, the assumption is made where the photo-generated current is almost equal to short-circuit current [1, 7, 9, 12], and thus the derivation for photo-generated current should be applied on short-circuit current condition of single-diode R_p model, as shown in Figure 2.

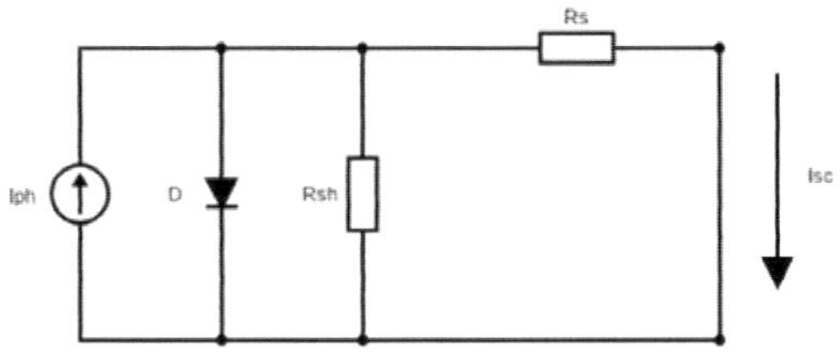

Figure 2. Single-diode R_p model under short-circuit condition

Under the short-circuit condition, the output terminal voltage and current value would be zero and I_{sc} respectively. With these parameters substituted into Equation (2), the result is obtained as Equation (6). Thus, the derivation for the photo-generated current of Equation (7) is obtained by rearranging Equation (6) in form of photo-generated current.

$$I_{sc} = I_{ph} - I_0\left(e^{\frac{I_{sc}R_s}{N_s V_t}} - 1\right) - \frac{I_{sc}R_s}{R_{sh}} \tag{6}$$

$$I_{ph} = I_{sc} + I_0\left(e^{\frac{I_{sc}R_s}{N_s V_t}} - 1\right) + \frac{I_{sc}R_s}{R_{sh}} \tag{7}$$

2.3. Derivation for Diode Saturation Current

The same concept is applied which parameterisation of the five-parameter model can be simplified by reducing required parameters. Two popular equations are used among the researchers on deriving diode saturation current parameter. It is shown as Equation (8) [1, 3, 12, 18] and Equation (9) [1, 7, 9, 12], where E_g is the bandgap energy for PV cell which can be calculated from Equation (10).

Shown in [1], which applied both of Equation (8) and Equation (9) to model the I-V curve of PV cell at operating condition, the MATLAB script provided in the publication shows that both of the equations have repeated function of determining the diode saturation current parameter at different PV cell's temperature. This statement is supported by [9] stating that Equation (9) is the ratio of diode saturation current at different temperature. Thus, the application of Equation (8) had already catered the requirement to model PV cell or module at specific temperature. The works of applying the derived diode saturation current formula can be traced from [16] and [5], which respectively shown in Equation (11) and Equation (12). However, the derivation methodology is remained debatable as the short-circuit current parameter I_{sc}, open-circuit voltage parameter V_{oc}, and series resistance R_s exist at the same equation, which such condition is physically impossible to exist on the circuitry model.

$$I_0 = \frac{I_{sc}}{\left[e^{\frac{V_{oc}}{N_s V_T}} - 1\right]} \tag{8}$$

$$I_0 = I_{0,\mathrm{ref}} \times \left(\frac{T}{T_{ref}}\right)^{\frac{3}{n}} e^{\left[\frac{1}{k}\left(\frac{E_{g,ref}}{T_{ref}} - \frac{E_g}{T}\right)\right]} \tag{9}$$

$$\frac{E_g}{E_{g,ref}} = 1 - 0.0002677(\mathrm{T} - T_{ref}) \tag{10}$$

$$I_0 = \frac{I_{sc}\left(1 + \frac{R_s}{R_p}\right) - \frac{V_{oc}}{R_p}}{e^{\frac{V_{oc}}{N_s V_T}}} \tag{11}$$

$$I_0 \approx (I_{sc} - \frac{V_{oc}+I_{sc}R_s}{R_{sh}})e^{\frac{-V_{oc}}{V_T}} \quad (12)$$

The derivation of diode saturation current formula is based on the open-circuit condition of the single-diode R_p model, as shown in Figure 3. Under open-circuit condition, the output terminal voltage and current value would be V_{oc} and zero respectively. With these parameters substituted into Equation (2), the result is obtained as Equation (13). Thus, the derivation for the diode-saturation current of Equation (14) is obtained by rearranging Equation (13) in form of diode-saturation current.

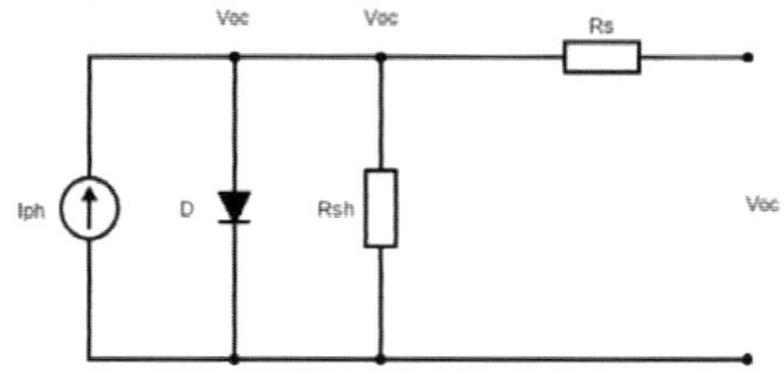

Figure 3. Single-diode R_p model under open-circuit condition

$$0 = I_{sc} - I_0\left(e^{\frac{V_{oc}}{N_s V_T}} - 1\right) - \frac{V_{oc}}{R_{sh}} \quad (13)$$

$$I_0 = \frac{I_{sc} - \frac{V_{oc}}{R_{sh}}}{\left[e^{\frac{V_{oc}}{N_s V_T}} - 1\right]} \quad (14)$$

Comparing Equation (2) and Equation (13), the photo-generated current variable is changed to short-circuit current variable during the derivation. Such assumption is required as the derivation procedure for Equation (6) requires the diode saturation current parameter while Equation (13) requires the photo-generated current parameter. Assumption of photo-generated current equal to short-circuit current is required to avoid such recursive loop.

2.4. Algorithmic parameterisation for A, R_s and R_{sh}

The five-parameter model is simplified through the derivation of photo-generated and diode saturation current parameters. This simplified five-parameter model can also use to determine the behaviours of PV cell or module, as presented through the work of [3, 5, 9, 10]. All of these works have the similarities, which the algorithmic parameterisation for A, R_s and R_{sh} parameters requires unique mathematical derivation approaches. However, the mathematical approach from [7] is applied for the work presented in this paper and the algorithmic parameterisation of the simplified five-parameter model is implemented by taking inspiration from [3, 5, 9, 10] without justifying their mathematical approach.

Both [10] and [18] give a strong reference for the work of this paper. [18] uses the optimization algorithm for the five parameter model while [10] applies algorithmic parameterisation through the simplified five-parameter model of the remaining A, R_s and R_{sh} parameters. [10] states that both ideality factor A and shunt resistance R_{sh} have an effect on the series resistance R_s. With such idea, the algorithmic parameterisation in [10] is executed through the sequence of A, to R_s, and to R_{sh}. The algorithmic parameterisation continues until minimum error is achieved. To apply this algorithm into the work published in this paper, the algorithm is amended into the sequence of R_s, to A, and to R_{sh}. This algorithm represents the idea that characteristic change of I-V curve under the influence of R_s parameter is adjusted with the following changes of A and R_{sh} parameters. This statement is justified through Equation (2), which the changes of R_s parameter will affect both diode saturation and shunt resistance leakage current, and the counter-feedback can be applied respectively with the change of A and R_{sh} parameters.

The algorithmic parameterisation methodology shown in Figure 4 is applied on both conventional and proposed model through MATLAB script, as shown in Figure 5.

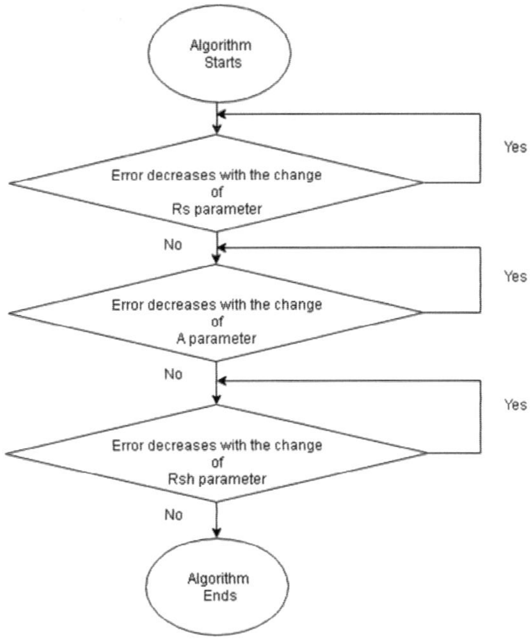

Figure 4. Proposed algorithmic parameterisation through three-parameter model

```
I0 = (Isc)/(exp(Voc/vt)-1);                                   % Equation (8)
%I0   = (Isc - (Voc/Rsh) ) / ( exp ( Voc / (vt)) -1 );         % Equation (14)
IL = Isc;                                                     % Equation (4)
%IL = Isc +  I0*( exp ( (Isc*Rs)/(vt) ) -1 ) + (Isc*Rs)/Rsh;   %Equation (7)

i = 0;                      % Set initial current i=0
I_tmp = zeros(1,length(V));

for idx = 1:length(V)
    I_tmp(idx)= IL - I0*(exp((V(idx)+(i*Rs))/vt)-1)-((V(idx)+(i*Rs))/Rsh); %Equation (2)
    i = I_tmp(idx);             % Update Current Data Point
end
```

Figure 5. Pseudo code implementation for conventional or proposed mathematical model

Applying the simplified five-parameter model, the algorithm can be applied to characterise the commercially available PV module today. This algorithm is tested with the field measurement data obtained for both mono- and poly- crystalline silicon PV module. Comparison is made between the field measurement data with the I-V curve generated by the mathematical model of Equation (2). The algorithmic parameterisation result of mono- or poly- crystalline silicon PV module consider obtained when the minimum possible Mean Absolute Percentage Error (MAPE) achieved. MAPE is chosen due to its robustness and suitability of measuring the data point distance between two graphical models, as stated in [19]. Equation (15) shows the MAPE equation, which I is the data point for terminal output current value of PV module; and n is the total number of data point to model the I-V curve.

$$MAPE = \sum_{i=1}^{n} \frac{|I_{Measured} - I_{calculated}|}{I_{Measured}} \times \frac{100}{n}\% \qquad (15)$$

3. Result and Discussion

The application of the simplified five-parameter model is tested on mono- and poly- crystalline silicon PV module of unknown specification. The measured specifications for mono- and poly- crystalline silicon PV modules are presented in [20] and the data is summarised into Table 1. The number of PV cells should be included in Equation (2) as the parameterisation procedure executed in this paper is applied on PV module instead of PV cell.

Table 1. Measured specification of mono- and poly- crystalline PV module

Measured Specification of PV Module	Mono-crystalline	Poly-crystalline
Measured maximum power (Pmax)	69.52W	69.54W
Maximum power voltage (Vmp)	15.48V	15.08V
Maximum power current (Imp)	4.49A	4.61A
Open-circuit voltage (Voc)	20.27V	19.99V
Short-circuit current (Isc)	5.00A	5.10A
Number of cell in series	36	36

The field measurement data is obtained for both of the PV module models and imported through MATLAB script to compare with the mathematical modelling of Equation (2) individually. The original MATLAB script and field measurement data are obtained from [20] as well, with some amendment. The final parameterisation result is obtained when the mathematical model could match to the measured I-V data with lowest MAPE value as possible. The conventional model refers photo-generated current I_{ph} and diode saturation current I_o parameter are based on Equation (4) and Equation (8) respectively, while the proposed three-parameter model refers the photo-generated current I_{ph} and diode saturation current I_o parameter are based on Equation (7) and Equation (14) respectively. Comparison between the conventional and proposed model is made through the same field measured data, indicating that both data are influenced by irradiance and temperature factors under the same level, thus these effects towards the PV module is not investigated for the work presented in this paper. The comparison of both models is presented in Figure 6 to Figure 9. The algorithmic parameterisation is applied based on Figure 4 to match both measured and modelled I-V curve until the MAPE value from Equation (15) no longer be reduced. The essential analytic parameters are then extracted from the modelled I-V curve, summarised as Table 2 and Table 3. All of the improved MAPE results throughout the iteration are recorded and plotted, as shown in Figure 10.

Interpreting the result obtained for Figure 6 to Figure 9, the proposed model shows the better result on matching both measured and modelled I-V curve, especially at the near open-circuit condition. This result is apparent at the mono-crystalline silicon PV module as referred to Figure 6 and Figure 7. This statement is supported by [2] stating that there is denser data distribution after the maximum power point of P-V curve. However, the near open-circuit voltage of the I-V curve presented in Figure 6 and Figure 7 does not show a smooth curve like Figure 8 and Figure 9. This is due to the inaccurate series resistance value is introduced at the result of Figure 6 and Figure 7, as this parameter will result in a significant effect on the slope of the near open-circuit voltage data point. However, these series resistance parameter can no longer be managed as it is the best result obtained from the algorithm shown in Figure 4. Such inaccuracy is obtained because the error induced from the non-smooth region of I-V curve generated at near open-circuit voltage data point is compensated at another region of I-V curve, such as the I-V curve slope at maximum power point or short-circuit current region. However, a hypothesis can be made which reducing the step-size on R_s parameterisation might able to eliminate this phenomenon.

Table 2. Parameterisation result comparison between mono- crystalline silicon PV modules applied with both conventional and proposed three-parameter modelling

Parameterisation Result	Mono-crystalline silicon PV Module	
	Conventional Model	Proposed Model
Diode Ideality Factor(A)	1.52	1.53
Series Resistance (R_s)	0.28 Ω	0.27 Ω
Shunt Resistance (R_{sh})	248 Ω	124 Ω

Mean Absolute Percentage Error (MAPE)	1.2654%	0.9167%
Total number of iteration	84	212

Table 3. Parameterisation result comparison between poly- crystalline silicon PV modules applied with both conventional and proposed three-parameter modelling

Parameterisation Result	Poly-crystalline silicon PV Module	
	Conventional Model	Proposed Model
Diode Ideality Factor(A)	1.30	1.27
Series Resistance (R_s)	0.34 Ω	0.35 Ω
Shunt Resistance (R_{sh})	248 Ω	164 Ω
Mean Absolute Percentage Error (MAPE)	1.2786%	1.0041%
Total number of iteration	176	196

Interpreting the result summarised in Table 2 and 3, the proposed model able to further reduce the MAPE error for both mono- and poly- crystalline silicon PV module models. Although the conventional and proposed model extracts the three parameters into different results, the proposed model considered to possess higher modelling accuracy as it able reaches to a lower MAPE value. For both of the crystalline silicon PV module model, the proposed model requires a higher number of iteration to complete the algorithmic parameterisation compared to the conventional model, which the result is shown in Figure 10. This reflects that the proposed model is able to continue the algorithmic parameterisation which already terminated by the conventional model at the pre-mature state of the parameterisation for the I-V curve fitting. As the trade-off of higher modelling accuracy, the parameterisation of the proposed model will show higher MAPE value at the initial stage of parameterisation compared to conventional model. However, the proposed model able to achieve lower MAPE value as the iteration of parameterisation proceeds.

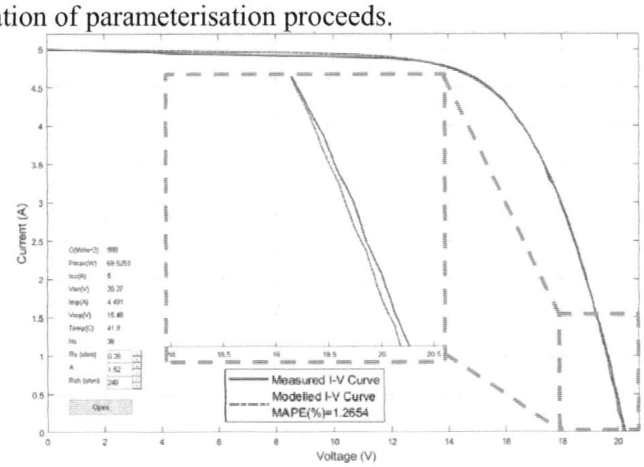

Figure 6. Algorithmic parameterisation for I-V characteristic curve fitting applied on mono-crystalline silicon PV module (Conventional Model)

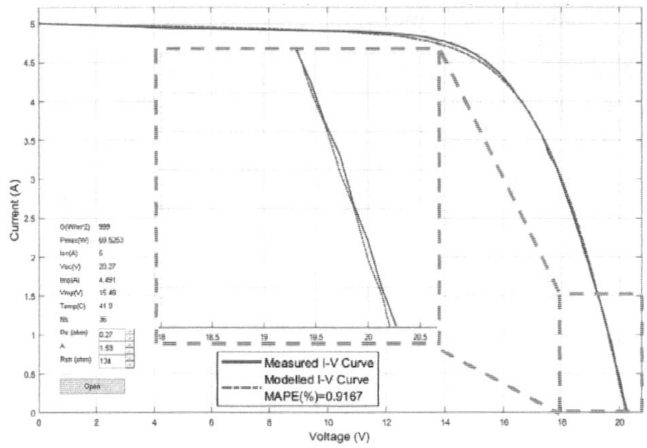

Figure 7. Algorithmic parameterisation for I-V characteristic curve fitting applied on mono-crystalline silicon PV module (Proposed Model)

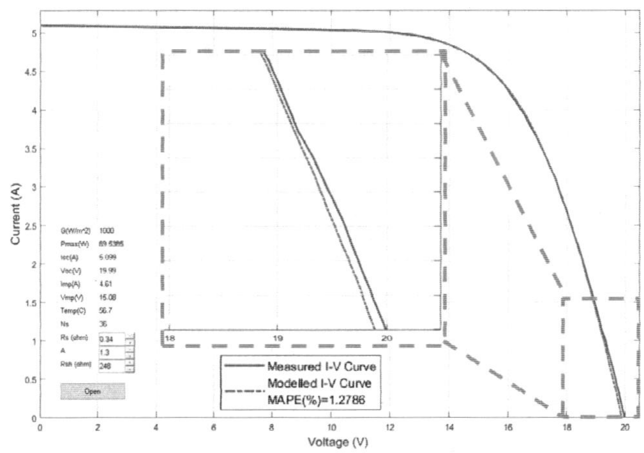

Figure 8. Algorithmic parameterisation for I-V characteristic curve fitting applied on poly-crystalline silicon PV module (Conventional Model)

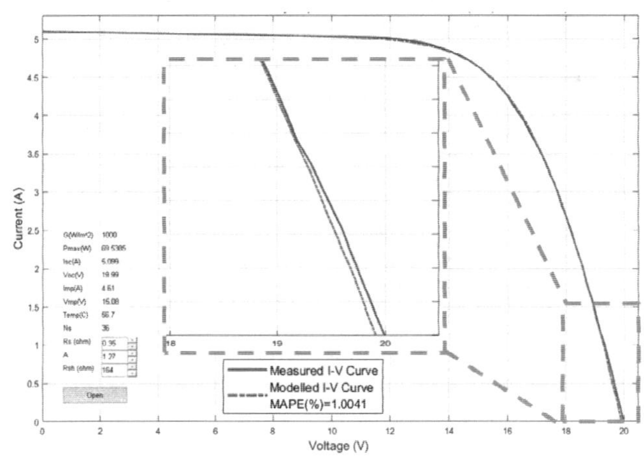

Figure 9. Algorithmic parameterisation for I-V characteristic curve fitting applied on poly-crystalline silicon PV module (Proposed Model)

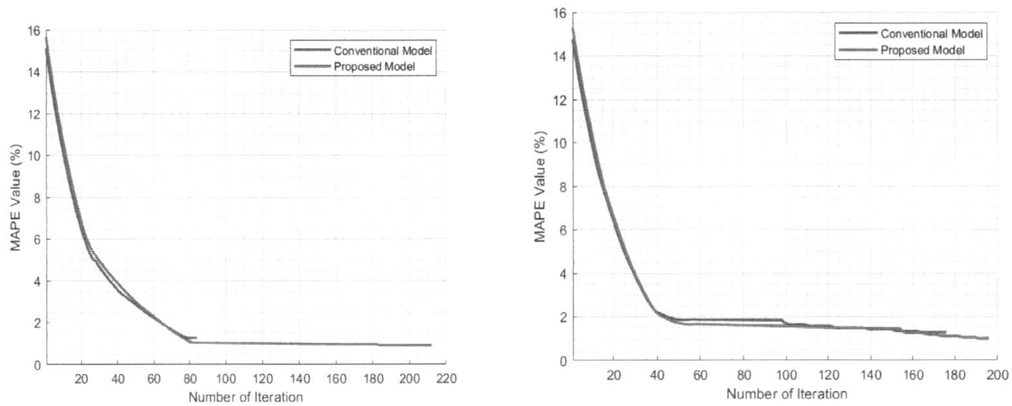

Figure 10. Comparison of iterative MAPE succession between conventional and proposed model applied in (a) mono-crystalline silicon PV module; (b) poly-crystalline silicon PV module

Since the formula derivation of the proposed model is introduced with a higher degree of resemblance towards the single-diode R_p model, thus a finer adjustment of the three parameters are required in order to obtain the modelled I-V curve with lowest possible MAPE value, and this explains the higher number of iterations required for the proposed model. Comparing Equation (8) and Equation (14), the derivations of these two formulas indirectly state that the single-diode R_p model possesses higher modelling accuracy compared to the single-diode R_s model. Equation (8) omitted the open-circuit voltage to shunt resistance ratio as compared to Equation (14), this related to the assumption of single-diode R_s circuit modelling on considering shunt resistance with infinite value. Considering this aspect, it is predicted that a much accurate algorithmic parameterisation result can be obtained by applying the same derivation methodology on the double-diode model.

To summarize the result, the proposed three-parameter model is proven to be valid on performing the algorithmic parameterisation curve fitting between the measured and modelled I-V curve. The proposed three-parameter model increases the modelling accuracy and simplified the parameterisation procedure as compared to the conventional five-parameter model. By introducing a higher resemblance of circuitry elements to derive the proposed model, the error between measured and modelled I-V curve could be reduced. Overall, the proposed model does improve the modelling accuracy for crystalline silicon PV module model.

4. Conclusion

To conclude the work presented in this paper, the commonly used five-parameter model consisted of I_{ph}, I_o , A, R_s and R_{sh} parameters can be simplified into a five-parameter model that consisted of diode ideality factor (A), series (R_s) and shunt (R_{sh}) resistance with derived photo-generated (I_{ph}) and diode saturation (I_o) current parameters. It proves that the proposed three-parameter model does improve the modelling accuracy of crystalline silicon PV module's I-V characteristic curve. The modelling accuracy is determined by the minimum difference can be achieved between the field measured and mathematically modelled I-V characteristic curve. Comparison between the proposed three-parameter and conventional five-parameter model is made, showing that the proposed three-parameter model able to improve the modelling accuracy of I-V characteristic curve through a higher number of iteration for algorithmic parameterisation. A better overview of the connection between the circuitry and mathematical modelling of PV cell/module is presented. The work presented also indirectly proves that the single-diode R_p model possesses higher modelling accuracy than the single-diode R_s model, based on the derivation methodology of the proposed and conventional model. The MATLAB scripts entitled

PV Module's I-V Characteristic Curve Fitting Tools used for the data presentation in this paper are made available by the author at Mathworks official MATLAB Central File Exchange [21].

Acknowledgements
The authors would like to acknowledge the project code FRGS/1/2015/TK07/UCSI/02/1 and conference financial support from UCSI University research management unit CERVIE so that the research and publication of this work are made possible.

References
[1] González-Longatt FM. Model of photovoltaic module in Matlab. *Ii Cibelec*. 2005;**2005**:1-5.
[2] Tan RH, Teow MY. Comprehensive Modelling of Photovoltaic Module Characteristic Curve in MATLAB/Simulink. 2016.
[3] Sera D, Teodorescu R, Rodriguez P, editors. PV panel model based on datasheet values. 2007 IEEE international symposium on industrial electronics; 2007: IEEE.
[4] Ishibashi K-i, Kimura Y, Niwano M. An extensively valid and stable method for derivation of all parameters of a solar cell from a single current-voltage characteristic. *J Appl Phys*. 2008;**103**(9):094507.
[5] Chupong C, Plangklang B. Forecasting power output of PV grid connected system in Thailand without using solar radiation measurement. *Energy Procedia*. 2011;**9**:230-7.
[6] Orioli A, Di Gangi A. A procedure to calculate the five-parameter model of crystalline silicon photovoltaic modules on the basis of the tabular performance data. *Applied energy*. 2013;**102**:1160-77.
[7] De Soto W, Klein SA, Beckman WA. Improvement and validation of a model for photovoltaic array performance. *Solar energy*. 2006;**80**(1):78-88.
[8] Chin VJ, Salam Z, Ishaque K. Cell modelling and model parameters estimation techniques for photovoltaic simulator application: A review. *Applied Energy*. 2015;**154**:500-19.
[9] Chenni R, Makhlouf M, Kerbache T, Bouzid A. A detailed modeling method for photovoltaic cells. *Energy*. 2007;**32**(9):1724-30.
[10] Carrero C, Ramírez D, Rodríguez J, Platero C. Accurate and fast convergence method for parameter estimation of PV generators based on three main points of the I–V curve. *Renewable Energy*. 2011;**36**(11):2972-7.
[11] Ortiz-Conde A, Sánchez FJG, Muci J. New method to extract the model parameters of solar cells from the explicit analytic solutions of their illuminated I–V characteristics. *Sol Energy Mater Sol Cells*. 2006;**90**(3):352-61.
[12] Kou Q, Klein S, Beckman W. A method for estimating the long-term performance of direct-coupled PV pumping systems. *Solar Energy*. 1998;**64**(1-3):33-40.
[13] Ibrahim H, Anani N. Variations of PV module parameters with irradiance and temperature. *Energy Procedia*. 2017;**134**:276-85.
[14] Gow J, Manning C. Development of a photovoltaic array model for use in power-electronics simulation studies. *IEE Proceedings-Electric Power Applications*. 1999;**146**(2):193-200.
[15] Kennerud KL. Analysis of performance degradation in CdS solar cells. *IEEE Transactions on aerospace and electronic systems*. 1969(6):912-7.
[16] De Blas M, Torres J, Prieto E, Garcıa A. Selecting a suitable model for characterizing photovoltaic devices. *Renewable energy*. 2002;**25**(3):371-80.
[17] Arab AH, Chenlo F, Benghanem M. Loss-of-load probability of photovoltaic water pumping systems. *Solar energy*. 2004;**76**(6):713-23.
[18] ALQahtani AH, editor A simplified and accurate photovoltaic module parameters extraction approach using matlab. 2012 IEEE International Symposium on Industrial Electronics; 2012: IEEE.
[19] De Myttenaere A, Golden B, Le Grand B, Rossi F. Mean absolute percentage error for regression models. *Neurocomputing*. 2016;**192**:38-48.
[20] Tan R. PV Panel Characterization Tool [Online]. MATLAB Central File Exchange; 2018.

[21] Teong R. PV Module's I-V Characteristic Curve Fitting Tools [Online]. MATLAB Central File Exchange; 2019.

Investigation on wind energy for grid connection in Bangladesh: Case study

N A Madlool [1,*], M S Hossain [2,5,*], M J Alshukri [1], Jeyraj Selvaraj[2], Ahmed A Taher [1], Vaskar Das Gupta [3], Prasun Mazumder [3], Nasrudin Abd Rahim[2, 4]

[1] Dept. of Mechanical Engineering, University of Kufa, 21 Kufa, Iraq

[2]Higher Institution Centre of Excellence (HICoE), UM Power Energy Dedicated Advanced Centre (UMPEDAC), Level 4, Wisma R&D, University of Malaya, Jalan Pantai Baharu, 59990 Kuala Lumpur, Malaysia.

[3]Dept. of Electrical& Electronic Engineering, International Islamic University Chittagong, Chittagong, Bangladesh

[4]Renewable Energy Research Group, King Abdulaziz University, Jeddah 21589, Saudi Arabia.

[5]College of Environmental Science and Engineering, Peking University, Beijing 100971, China

E-mail: naseer.alkhalidi@uokufa.edu.iq and engi.shouquat@um.edu.my

Abstract. The increasing demand for energy in Bangladesh as well as all over the world is continuously growing unabated because of the increase of the world population and the countries' fast-developing industrialization programs. In this critical stage of energy crisis, renewable energy is a highly welcome reliable energy source wherein wind energy is the cheapest replenishable and available energy source of the time. Regarding this particular energy source, a feasibility study of the wind energy in Chittagong has been done by analyzing the wind speed data. It is estimated that the maximum practical power output generated can be 124.53 MW at Parky beach. The number of wind turbines, operation and maintenance cost and pay-back period of those sites are also calculated.

1. Introduction

Bangladesh is geographically situated between 20°34' and 26°38' north latitude and 88°01' and 92°41' east longitude with an area of 147,570 sq.km. It has 724 km of the coastal line along the Bay of Bengal [1]. The average population density is about 2,917.6 capita/sq. mile [2]. The electricity demand is nearly 6,000 MW but the country can only currently generate 4,500 MW of electricity [3] from which only 49% of the total population can gain access to the facility which translates into the fact that the per-capita energy usage is 180 kWh which is one of the lowest in the region. To meet this demand it is quite usual to use fossil fuels, like, coal, gas, and diesel which are used as energy sources. However, there are no alternative reliable sources of fossil fuels that are renewable [4]. In Bangladesh, researches within the sphere of wind energy began only a few years ago and it was also found that some southern districts of Bangladesh, especially in the coastal regions, have a very good potential location of wind energy extraction where the seasonal trade-winds blow over the country from March to October for a period of 8 months each year. The wind is enhanced when it enters the V-shaped coastal regions of Bangladesh because as this trade wind strikes the coastal belt of Bangladesh, it carries a lot of kinetic energies [6]. These coastal areas have better opportunities for wind power exploitation in the form of electricity generation which can supply electricity to many production areas [7] in which case it also has the potential to be a source of decentralized energy for Bangladesh, hence, simultaneously solving the critical emerging energy problems. There is no available adequate information on wind speeds over the country, particularly, on wind speeds at hub levels of wind turbines in Bangladesh.

In 2005, for the first time in Bangladesh, Bangladesh Power Development Board (BPDB) had implemented a pilot project of 0.90 MW capacity of the Grid-Connected Wind Energy (GCWE) in the Muhuri Dam areas. The installation, commissioning and erection works of 4 units of the 225 kW GCWE turbine was progressing very fast and after the successful completion of this 0.90 MW pilot project it was further provided with 1 MW unit capacity wind turbines as this site was found to be most suitable for another 100 MW wind energy power production [8]. Nevertheless, this project by the BPDB had not been blessed with success because of the wrong measurement of the wind speeds. An earlier project, located in the southwestern part of Kutubdia Island and fully exposed to the sea, had been completed for a wind-power electricity generation plant.

The wind speeds of the coastal regions of Chittagong have been considered in the present paper and the data obtained shows a prospective source of wind energy in the coastal areas of Chittagong. The data analysis and study of the area show that the seasonal or monthly average wind speed of above 6 m/s is very strong in the coastal regions of Chittagong. During the earlier part of the windy months, the hourly wind speed variations are low, but as time progresses the wind speed remains high during the rest of the months. As energy depends on the 3rd power of speed (v^3), the available energy can be higher in the coastal wind in Bangladesh than for locations having the same annual wind speed with a low-speed variation [9], hence, the possibility of producing power using the wind energy in Chittagong is immensely bright and quite timely.

2. Wind Turbine Systems

A wind turbine is a rotating machine that converts the kinetic energy of the wind into mechanical energy. If the mechanical energy is directly used by machinery, such as a pump or flour grinder, the machine is usually called a wind mill. If the mechanical energy is converted into electrical energy, the machine can then may be called by other names, such as wind generator or wind turbine or wind power unit (WPU) or wind energy converter (WEC) [10]. Figure. 1(A) shows the wind-turbine major components [11, 15].

2.1. Components of a typical Wind Turbine

A wind turbine usually has seven main components, namely, i) the nacelle, ii) the rotor, iii) the gearbox, iv) the generator, v) the control and protection system, vi) the tower and, vii) the foundation.

2.1.1. Nacelle It is the box-like structure located behind the rotor blades.

2.1.2. Rotor The rotor is an elegantly shaped blade that captures the wind energy and converts its kinetic energy into mechanical energy through connected shafts.

2.1.3. Gearbox The gearbox changes the rotational velocity of the shaft to suit the generator [14]. The gearbox increases the slow speed of the main shaft to a speed suitable for the generator to function.

2.1.4. Generator The generator is a device that converts mechanical energy to the system into electrical energy. The generator is typicaly of the induction type, operating at the near-fixed speeds.

2.1.5. Control and Protection System The computer-based central control panel of the wind turbine is typically mounted inside the tower (if tubular). The control system monitors the gearbox and the generator temperatures, wind speed and so on. The protection system protects the turbine from abnormal conditions, such as extremely high wind speeds during stormy periods.

2.1.6. Tower The tower is the main shaft that connects the rotor to the foundation [12, 13]. It also raises the rotor high in the air to capture the strong winds.

2.1.7. Foundation The foundation supports the wind turbine and ensures that it is well established and solidly founded onto the ground all of which usually consist of solid concrete.

With regard to the axis of rotation of the rotor shafts, the wind turbines are separated into two types [13]:

(i) Horizontal.

Horizontal Axis wind Turbines (HAWT).

Figure 1(B) shows a Horizontal Axis Wind Turbine [11, 15].

(ii) Vertical Axis Wind Turbines (VAWT) [14].

Figure. 1(C) shows a Vertical Axis Wind Turbine [11, 15].

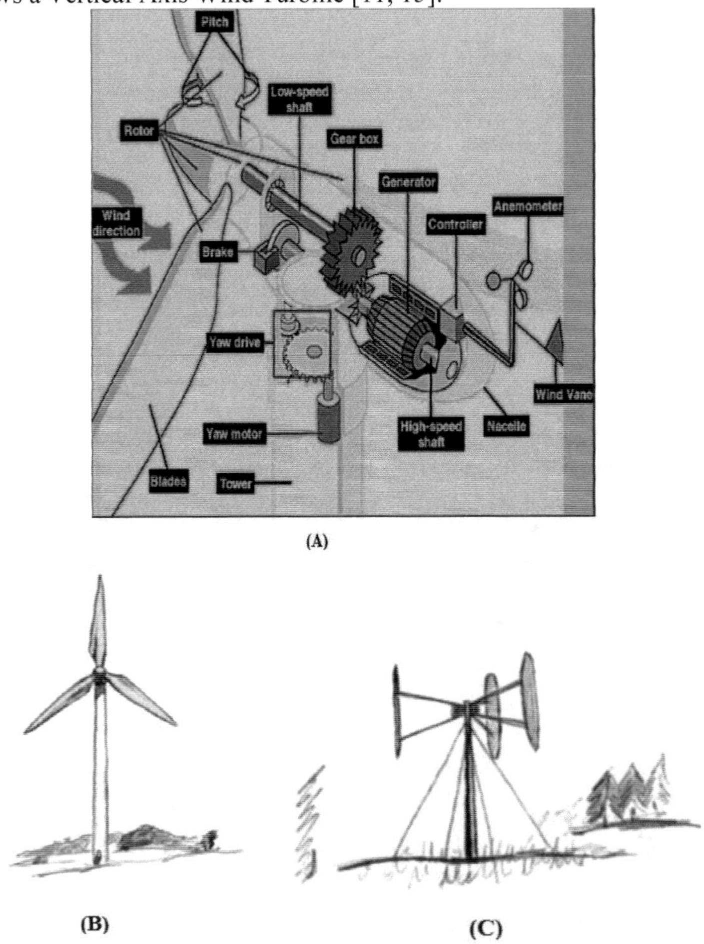

Figure 1. (A) Wind Turbine Major Components, (B) Horizontal Axis Wind Turbine (HAWT), (C) Vertical Axis Wind Turbine (VAWT) [12, 16].

3. Site selection

Around one or two months before and after the Monsoon season, high-speed winds are mainly available in Bangladesh. During the months starting from late October to the middle of February, the wind speed

remains either calm or too low. The peak wind speed occurs during the months of June and July [16]. However, the wind speed of the coastal area holds a good prospect to make the best use of wind energy. Six locations were selected for this investigation and the locations were, namely;
i) Patenga Sea-beach, ii) Parky Beach, iii) Batali Hill, iv) Cox's Bazar, v) IIUC City Campus at Chawkbazar and, vi) Shah Amanat Bridge.
The wind speed data of Patenga Sea-beach, Batali Hill, IIUC City Campus and Shah Amanat Bridges were measured with an Anemometer. A comparison had been made between the measured wind speed data and the collected wind speed data from different research centres.

3.1. Patenga Sea-beach

The wind speeds at this site had been measured at a height of 1.62 m in October and the variation of wind speeds for 1.30 hr duration is as shown in Figure 2(A) and (B). From Figure 2(A) it can be seen that the speed of the wind varies with time. The highest value of the wind speed is above 7.5 m/s and the lowest value is around 2.8 m/s while the mean wind speed at this site is around 5.15m/s. Figure 2(B) illustrates a power versus wind speed graph in which as the wind speed increases the power output is also increased. The theoretical power output of the wind at this site at 1.62 m height is around 298 W/m^2 of wind area.

The speed versus time (months) and power (watts) Vs speed (m/s) graphs of wind speed data of one year at 20 m height collected from the Bangladesh Meteorological Departments (BMD) are shown below respectively as in Figures 2(C) & (D) where Figure 2(C) shows that during the months of May, June, July, August and September, the wind speed is high and in other months it is lower. Figure 2(D) shows that the power output increases as the wind speed increases and by taking the mean wind speed, the theoretical power output is about 389.82 W/m^2

3.2. Batali Hill

The wind speed at this site was measured at 1.62 m height in October and the variation of the wind speed for a duration of 1.00 hr is as shown in Figures 2(E) & 3(A) where Figure 2(E) shows that the speed of the wind varies with the time. The highest value of the wind speed was above 5 m/s and the lowest value was around 0.6m/s, while the mean wind speed was around 3.35 m/s. Figure 3(A) shows a power versus wind speed graph in which when there is no wind, the power output is correspondingly zero and as the wind velocity increases, the power output also increases. The theoretical power output at this site at 1.62 m height is around 98 W/m^2 of wind area.

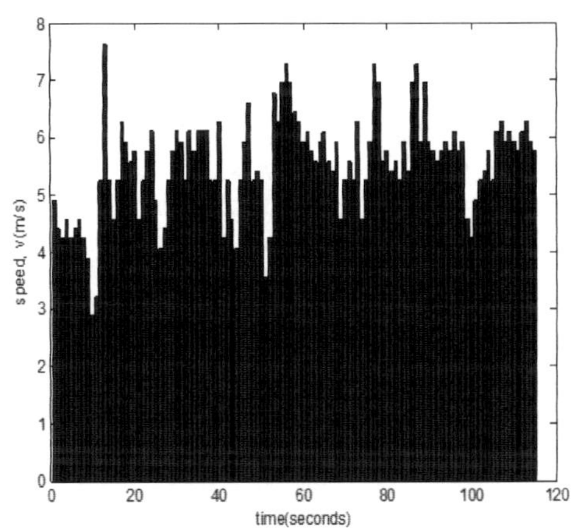

(A)

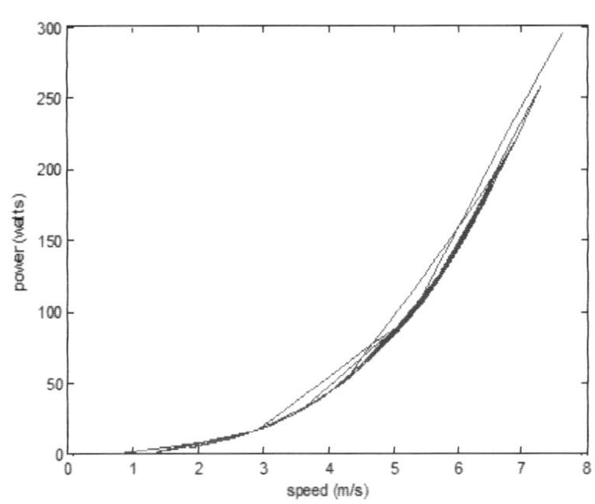

(B)

IOP Conf. Series: Materials Science and Engineering 1127 (2021) 012021 doi:10.1088/1757-899X/1127/1/012021

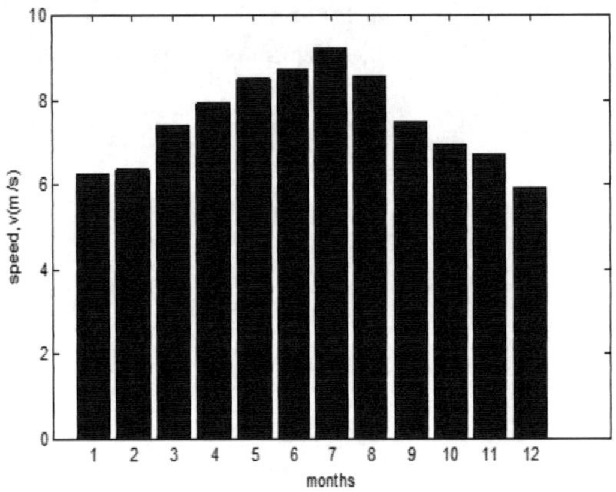

(C)

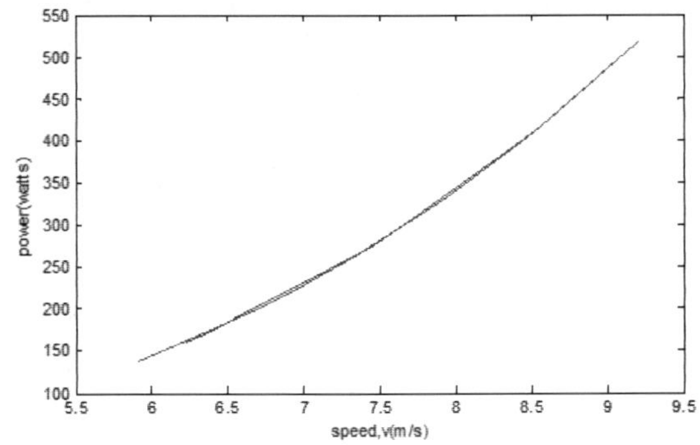

(D)

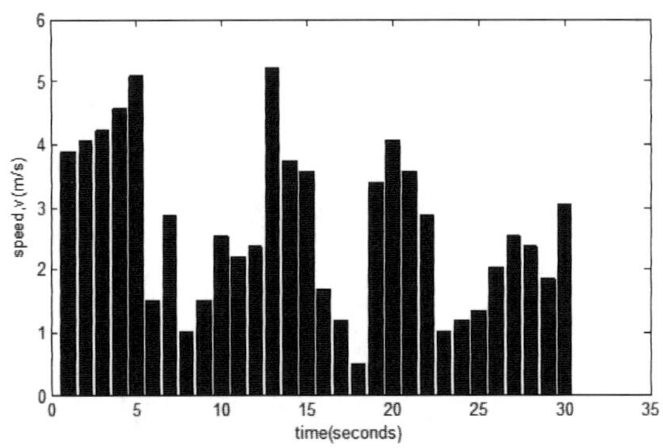

(E)

Figure 2. (A) V vs. Time (seconds) graph, (B) Power (watts) vs. Speed (m/s) graph, (C) V vs. Time graph, (D) P vs. Speed (m/s) graph, (E) V vs. Time (seconds) graph.

3.3. IIUC City Campus

Here, the wind speed was measured at 1.62 m height in December and the variation of the wind speed for a duration of 1.00 hr is as shown in Figure 3(B) & (C) where Figure 3(B) shows that the speed of the wind varies with the time. The highest value of the wind speed was above 6.5m/s and the lowest values were around 0.7m/s while the mean wind speed at this site was around 3.6m/s. Figure 3(C) shows a power versus wind speed graph where it illustrates that when there is no wind the power output is correspondingly zero and as the wind speed increases, the power output also increases. The theoretical power output of this site at 1.62 m height is around 192 W/m^2 of wind area.

3.4. New Shah Amanat Bridge at Karnafuly

The wind speed here was measured at 1.62 m height in January and the variation of the wind speed for a duration of 1.00 hr is as shown in Figures 3(D) and (E) where Figure 3(D) illustrates that the speed of the wind varies with the time. The highest value of the wind speed was above 5.6 m/s and the lowest value was around 0.8 m/s while the mean wind speed at this site was around 3.2 m/s. Figure 3(E) shows a power versus wind speed graph where it illustrates that when there is no wind the power output is zero and as the wind speed increases the power output also increases. The theoretical power output at this site at 1.62 m height is around 102 W/m^2.

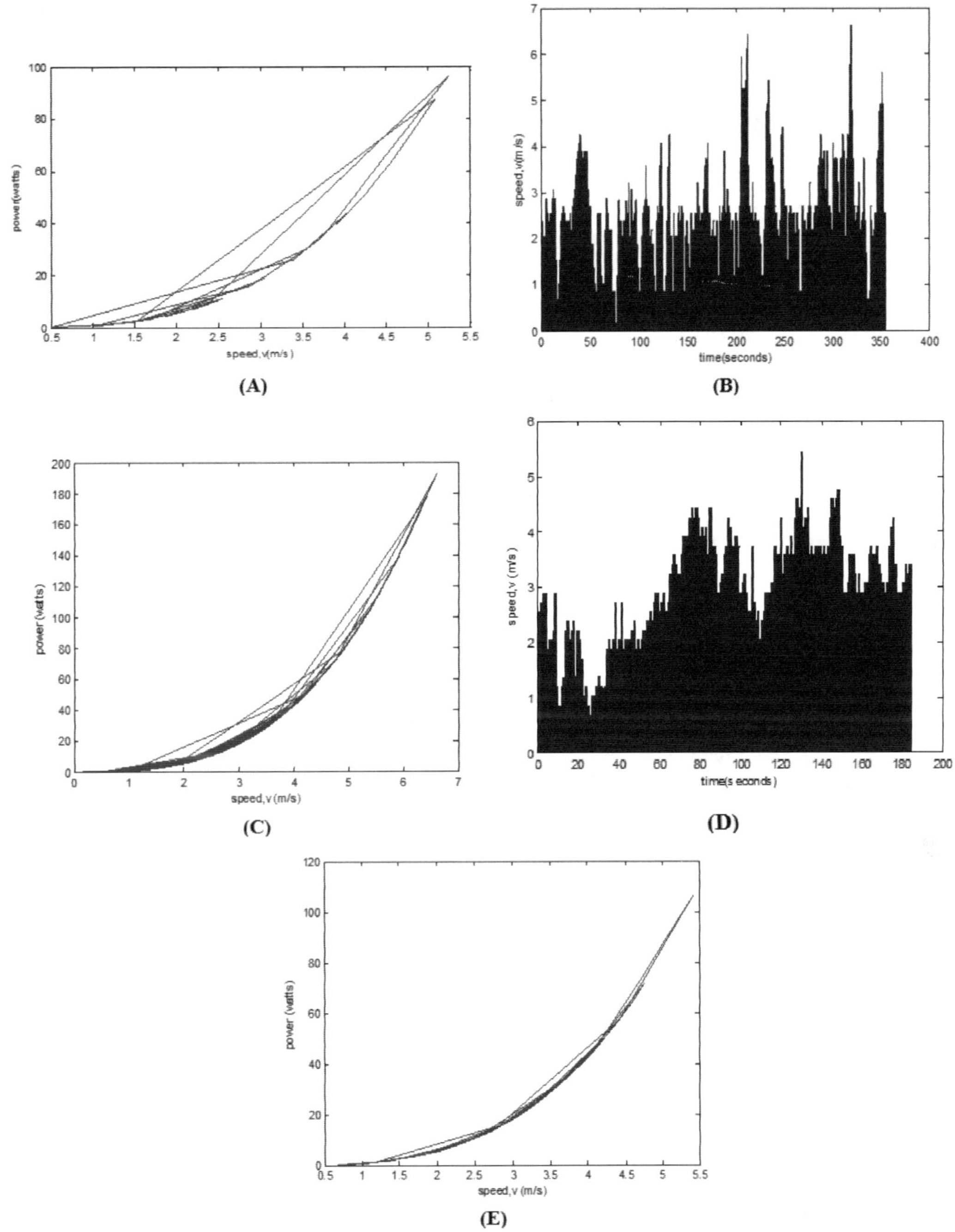

Figure 3. (A) P vs. Speed (m/s) graph, (B) V vs. Time (seconds) graph, (C) Power (watts) vs. Speed (m/s) graph, (D) V vs. Time (seconds) graph, (E) Power (watts) vs. Speed (m/s) graph.

3.5. Parky Beach

The velocity of wind versus time (months) and power (watts) Vs wind speed (m/s) graphs data from one year at 50 m height collected from Pan Asia Power Limited is as shown below respectively in Figures 4(A) and 4(B). Figure 4(A) illustrates that during the months of May, June, July, August and September, the wind speed is high while in other months it is lower with the mean wind speed of 6.73

m/s. Figure 4(B) shows that the power output increases as the wind speed increases. From the mean wind speed, the theoretical power output is about 182.48 W/m^2.

3.6. *Mognamaghat (Cox's Bazar)*

The velocity of wind versus time (months) and power (watts) Vs speed (m/s) graphs of wind speed data from one year at 50 m height collected from Pan Asia Power Limited are shown below respectively as in Figures 4(C) and 4(D) where Figure 4(C) shows that during the months of March, April, May, June, July, August, September and October, the wind speed is high and in other months it is lower. The mean wind speed is 6.90 m/s. Figure 4(D) shows that the power output increases as the wind speed increases and from the mean wind speed, the theoretical power output is about 197.11 W/m^2.

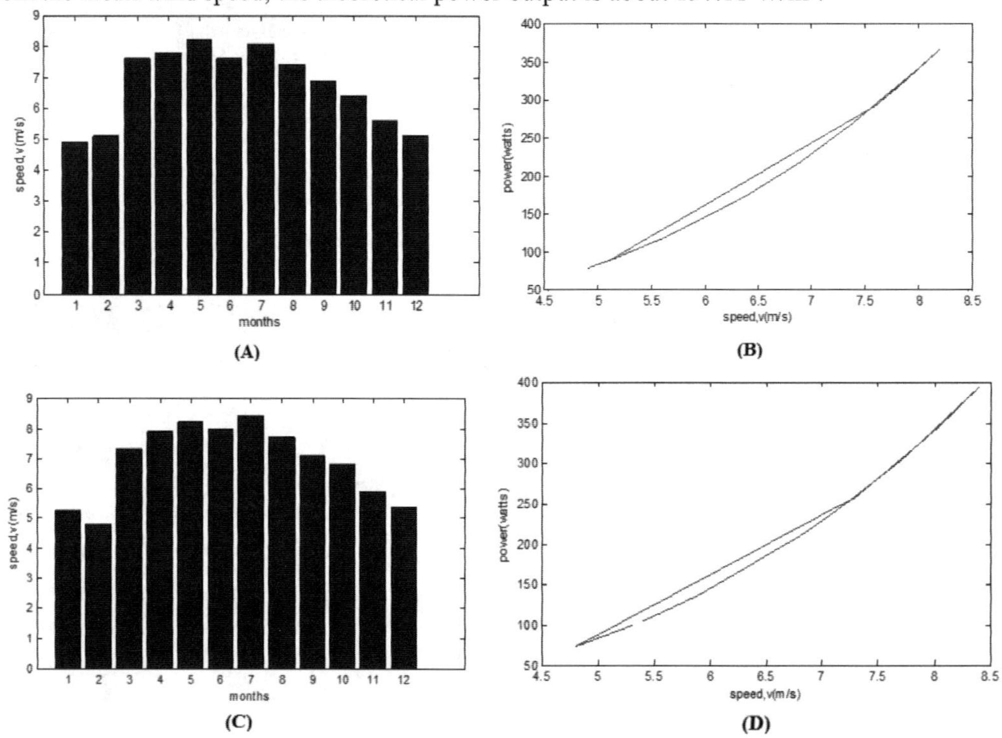

Figure 4. (A) V vs. Time graph, (B) P vs. Speed (m/s) graph, (C) V vs Time graph, (D) P vs. Speed (m/s) graph.

4. Calculation on the number of wind turbines and power generated from the sites

4.1. *Parky beach*

As Parky beach is in the Anwara Upazilla at Chittagong, it has an area of 173.53 sq.km [17]. As it is quite a populated area, 4% of the area is appropriate for building a wind power plant and calculation wise, the total area is 173.53×0.04 = 6.94 = 7 sq.km (approximately) whereas this site has a mean wind speed of 6.73 m/s throughout the year at 50m height. Assuming that 15 wind turbines can be installed in each sq.km with a 2 MW of power that can be produced within an area of 7 sq.km, a total of (15×7) = 105 wind turbines can be installed giving a maximum theoretical power output of (105×2) = 210 MW. If the plant capacity factor of 0.593 (Beltz limit) is considered, the maximum practical power output generated can be about (210×0.593) = 124.53 MW.

4.2. *Mognamaghat*

Mognamaghat at Pekua, Cox's-Bazar has an area of 6.149 sq.km [18] with a low population density. Considering that 20% of the area is appropriate for building wind power plants, the total area for the power plants is = 6.149×0.20 = 1.2298 sq.km. and as this site has a mean wind speed of 6.90m/s throughout the year at 50m height and assuming that 15 wind turbines can be installed in each sq.km which can produce 2 MW power output, the total number of wind turbines that can be erected in an area of 1.2298 sq.km is (15×1.2298) = 19 wind turbines which has the maximum theoretical capacity of (19×2) = 38 wind turbines. If the plant capacity factor of 0.593 (Beltz limit) is considered, the maximum practical power output that can be generated is about (38×0.593) = 22.5 MW.

4.3. Patenga Sea-beach
Patenga, Sea-Beach has an area of 16 sq.km [19] and considering 10% of the area is appropriate for building the wind power plants, the available total area is = 16×0.1 = 1.6 sq.km. As this site has a mean wind speed of about 9.26 m/s throughout the year at 50m height, and assuming that 15 wind turbines can be installed in each sq.km, to produce 3 MW of power output, the total number of wind turbines that can be erected is (15×1.6) = 24 wind turbines which can have the maximum theoretical capacity of (24×3) = 72 MW of power output. Assuming that the plant capacity factor of 0.593 (Beltz limit) is considered, the maximum practical power output that can be generated is about (72×0.593) = 42.7 MW of power output which can be used for centralizing and decentralizing purposes.

4.4. Batali Hill
At a height of 20 m at Batali Hill, a mean wind speed of around 4.036 m/s can be obtained to generate energy which can be used for decentralization programs. Assuming that the plant capacity factor is 0.6 (Beltz limit) and 45% of the area for the power plants is taken into account, the 4 turbines that can produce 3 kW will be $4\times0.6\times(4.036)^3\times0.45 = 0.071$ kW of power output generated which can be used for domestic electrical equipment, like, lights, fans, computers, etc.

4.5. IIUC City Campus
At a height of 20 m at IIUC City Campus, the mean wind speed obtainable is about 3.94 m/s that can generate energy for use in the decentralization programs. Assuming that the plant capacity factor of 0.6 (Beltz limit) and 45% of the area for the power plants with 2 turbines can produce 3kW total power output of which will be $2\times0.6\times(3.94)^3\times0.45 = 0.033$ kW of power output obtainable from this area which can be used to run domestic electrical equipment, like, lights, fans, computers, etc.

4.6. Shah Amanat Bridge
At a height of 30 m at Shah Amanat Bridge, the mean wind speed is about 4.52 m/s in the month of January. Assuming that the plant capacity factor of 0.6 (Beltz limit) and 45% of the area for the power plants together with 5 turbines can produce 3 kW of power output which is $5\times0.6\times(4.52)^3\times0.45 = 0.125$ kW of power output which can be generated for use in lightning the bridge at night.

5. Cost analysis
The total cost analysis, such as investment cost, operation and maintenance cost, cost per kWh and pay-back period of the proposed sites can be done for viability aspect. There is also a comparison between the costs of a wind power plants and the cost of a conventional power plants, such as coal power plants, oil power plants, gas power plant, etc.

The cost of the wind power plant system has two components, namely, the initial installation costs and the operating expenses. The initial installation cost includes the purchase price of the complete system including the tower, wiring, utility interconnection or battery storage equipment, power conditioning unit, etc., plus delivery and installation charges, professional fees and sales tax [20].

The cost per kWh and payback period formulas are as shown below. The calculation for the cost per kWh is a two-step process:

Step 1

$$Annual\ Cost = \left(\frac{Initial\ Cost}{Expected\ Life}\right) + Annual\ Operating\ Costs \quad (1)$$

This formula defines the annual cost over the wind system energy production life span. Assume a 20-year estimate of the wind power plant useful life span.

Step 2

$$Cost \quad Per \quad kWh = \frac{Annual \quad Cost}{Annual \quad Energy \quad Output} \tag{2}$$

5.1. Payback Period

A common and simple way to evaluate the economic merit of an investment is to calculate its payback period which actually is the number of years of energy-cost savings it takes to recover an investment's initial cost. To determine the payback period, the investor first estimates the wind turbine's total initial cost, annual energy cost savings, and annual operating costs. Dividing the total initial cost by the difference between the annual energy cost savings and the annual operating costs gives the payback period:

$$\frac{Total \quad Initial \quad Cost}{(Annual \quad Energy \quad Cost \quad Saving) - Annual \quad Operating \quad Costs} = Payback \quad time, \quad in \quad years \tag{3}$$

A grid-connected residential-scale system generating of 1-10kW power output generally costs between $2,400 and $3,000 per installed kilowatts which come to $24,000-$30,000 for a 10 kW system. A medium-scale of a commercial system generating 10-100kW of power output is more cost-effective, costing between $1,500 and $2,500 per kW. The cost of large-scale systems of greater than l00kW of power output can be in the range of $1,000 to $2,000 per kW, with the lowest costs achieved when multiple units are installed in one location. In general, the cost rates decrease as the machine capacity increases [20]. The other cost components, such as Operation and Maintenance costs are related to a limited number of cost components which are insurance, regular maintenance, repair, spare parts, administration and incidental costs arising therefrom [21]. A rule of thumb estimate for the annual Operation and Maintenance expenses is 2% to 3% of the initial system cost [20].

5.2. Cost of a wind power plant

Generally, 10-15 wind turbines of 2 MW capacity in 1 sq.km area can be installed. If 20% of the total coastal areas are used, about 28,000 MW power output can be obtained from the wind turbines [22].

Investment cost:

Chittagong has a coastal line of about 724 km long, which means an area of (724×10) = 7240 sq.km can be mobilized. If 20% of the area is considered to be used for harnessing the wind energy, 10 wind turbines of 1.8 MW capacity in each sq.km area can be installed giving the total generation of wind energy of about 7240×0.2×10×1.8×0.45 = 11,729 MW and taking into consideration a plant factor of 45%. It is recommended that to gain this generation of wind energy power output from the wind turbine which has been commercially referred to as, V-I 00 1.8 MW, made by the renowned wind turbine manufacturing company, can be made available [8, 23].

At 20 m height at Batali Hill and IIUC City Campus, a mean wind speed around 3.99 m/s was obtainable. The generating energy output from those sites can be used for the decentralization program. If 3 kW rated wind turbines are used in each site, a power output of 0.538 kW generated can be put to use for domestic electrical equipment, such as lights, fans, computers, etc.

At 30 m height at Shah Amanat Bridge, a mean wind speed of about 4.52 m/s is achievable in the month of January where 5 numbers of 3 kW wind turbines can be installed in 5 towers situated on the bridge from which about 1.94 kW of energy output can be generated for use in lightning the bridge at night. The Operation & Maintenance Cost is represented by a pie chart in the operation & maintenance cost for a typical wind power plant as shown in Figure 5.

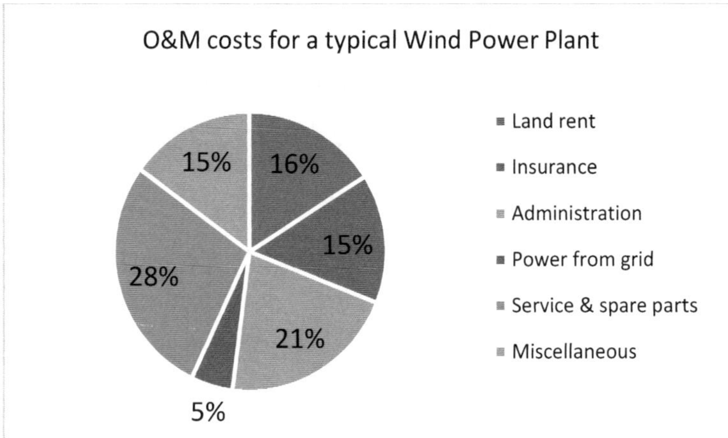

Figure 5. Pie chart of operation & maintenance cost for a typical Wind Power plant.

From the pie chart as given above, it is observed that the total cost of operation & maintenance for a typical wind power plant can be divided into, i) land rental at 16%, ii) insurance protection of 15%, iii) service & spare parts for replacement/repair at 28%, iv) power from grid at 5% and, v) administration cost at 21% [21] where the proportion of percentages can be altered to suit circumstantial requirements when and as necessary.

6. Conclusion

Nowadays, the emission of undesirable and obnoxious carbon dioxide gas is fast growing through the usage of the conventional power plants using fuels, such as, gas, coal, etc. which are creating long term pollutions into the environmental atmosphere. As wind energy is renewable and economical, it can be used as a deciding mitigating factor for a massive reduction of this untenable emission. In this paper, for the sixth sites, the theoretical power output is calculated based on 1.62 m heights which comes to around 298, 389.82, 98, 192 W/m^2 and at 50 m height is about 102 and 197.11 W/m^2. Moreover, a feasibility study is conducted on the wind energy project in Chittagong by selecting suitable sites and analyzing the wind speed data of those sites. The number of wind turbines needed is also calculated for the cost of investment, operation and maintenance, cost per kWh and pay-back period of those power plants. It is assumed that the maximum practical power output generated can be 124.53 MW at Parky beach. It is quite evident that there is a great prospect of wind energy abstraction in Chittagong, whilst proactively, Bangladesh Power Development Board has also taken some beneficial steps to develop the wind power plants in the country at the same time encouraging the private sectors to participate and help to contribute to developing this bright sector.

Acknowledgements

The authors would like to thank the financial and technical assistance provided by AUN-SEED/Net Japan ASEAN Collaborative Education Programme (JACEP). The authors thankful to UM Power Energy Dedicated Advanced Centre (UMPEDAC), the Higher Institution Centre of Excellence (HICoE) Program Research Grant, UMPEDAC - 2018 (MOHE HICOE - UMPEDAC), Ministry of Education Malaysia, RU007-2018 and RU012-2019, University of Malaya.

References

[1] Mominul Islam Mukutl, A.N.M., M. Quamrul Islam, and M.M. Alam, *Anal ysis of wind characteristics in coastal areas of bangladesh.* Journal of Mechanical Engineering, 2008. 39: p. 45-49.

[2] Department, D.E.a.S.A.P., *World Population Prospects, Table A.1.2008 revision. United Nations.* Department of Economic and Social Affairs Population Division 2009.

[3] USAID, *From the American people.* http://www.usaid.gov/bd/programs/energy.html. Internet source, 2013.
[4] NASA.gov, *Non-renewable resource.* http://en.wikipedia.org/wiki/Non-renewable_resource. Internet source, 2013.
[5] Wikimedia., *History of wind power,* http://en.wikipedia.org/wiki/History_of_wind_power. Internet source, 2013.
[6] Shahed Hossain, M., et al., *Strategy for Promotions and Development of Wind energy in Bangladesh.* National Seminar on Renewable Energy, 2011(Paper ID: 2011-11): p. 1-5.
[7] Lancashire, S., J. Kenna, and P. Fraenkel, *Wind pumping Hand-book.* IT Publication, London, 1987: p. 1-4.
[8] MWPC, *Invitation for participation of 100-200 MW offshore grid connected wind power project On a Build Own Operate (BOO) basis - See more at:* http://www.mywindpowersystem.com/2010/05/08/bangladesh-power-development-boardbpdb-announces-100200-mw-project/#sthash.PVglp4hF.dpuf. Database, 2013.
[9] Khadem, K.S., *Feasibility study of Wind Home System in Coastal Region of Bangladesh,* http://homerenergy.com/webcast-downloads/WE58_FeasibilityWHS_Bangladesh.pdf. Renewable Energy Research Centre, 2012: p. 1-7.
[10] BUET, *Fundamentals of Mechanical Engineering,* Department of Mechanical, 2013.
[11] EIS., *Wind Energy Basics,* http://windeis.anl.gov/guide/basics/. Wind energy development programmatic EIS, 2011.
[12] Sterzinger, G. and M. Svrcek, *Wind Turbine Development: Location of Manufacturing Activity.* Renewable Energy Policy Project, 2004.
[13] AL-SHEMMERI, T., *Wind Turbines,* www.bookboon.com. Book, 2010. ISBN: 978-87-7681-692-6.
[14] R & D notebooks., *Low Power Windmills and Wind Turbine Electrical generators for rural homes, amper.ped.muni.cz/~miler/indi/docs/.../Turbinearticleinconstruction.doc.* The group faisca, catembe, Mozambique, 2008.
[15] Turbines info., *Horizontal Axis Wind Turbines – HAWT.* Internet source, 2012.
[16] Ahmmed, S. and M.Q. Islam, *Wind Power for Rural Areas of Bangladesh.* 3rd International Conference on Electrical & Computer Engineering, 2009.
[17] Wikimedia., *Anwara Upazila,* http://en.wikipedia.org/wiki/Anwara_Upazila. Internet source, 2009.
[18] Viklund, A., *Free map tools,* www.freemaptools.com/area-calculator.htm. internet source, 2013.
[19] ConnectBD Ltd., *Bangladesh Navy,* http://www.bangladeshnavy.org/. Internet source, 2013.
[20] WEM., *Wind Energy Manual: Table of contents- Iowa Energy Center Homepage,* http://www.energy.iastate.edu/renewable/wind/wem/wem-13_econ.html. Iowa Energy Center, 2006.
[21] Morthorst, P.E., *Costs & prices.* Wind energy - the facts, 2008. 2.
[22] REIN, *Wind Energy Programme by Bangladesh Power Development Board (BPDB),* http://www.lged-rein.org/database.php?pageid=67. Database, Wind Energy Programme in Bangladesh, 2013.
[23] Bashar, R., *Indo-Bangla Electricity gridline opens, 15.* http://www.energybangla.com/index.php?mod=article&cat=SomethingtoSay&article=193. Energy bangla, 2013.

PV integrated on-demand water electrolysis system for HHO generation and it application as primary fuel in combustion

[1]*Muhammad Shakeel Ahmad, [2]T. Z. Butt, [3]A.K Pandey and [1]Nasrudin Abd Rahim

[1]UM Power Energy Dedicated Advanced Center (UMPEDAC) Jalan Pantai Baharu, Level 20, Wisma R&D University of Malaya,59990 Kuala Lumpur, Malaysia

[2]Depaerment of metallurgy and materials engineering, faculty of engineering and technology, University of the Punjab, Lahore, Pakistan

[3]Research Centre for Nano- Materials and Energy Technology (RCNMET) School of Science and Technology, Sunway University, Malaysia

E-mail: shakeelalpha@gmail.com

Abstract. The brown's gas also known as HHO gas has been produced using dye cell alkaline water electrolyser using 316L stainless plates. The electrolyzer has been integrated with solar panel using battery bank and charge controller to develop on-demand HHO generation system. The produced gas has been employed as primary fuel for combustion and its effectiveness has been investigated. Preliminary observations revealed that the photovoltaic (PV) integrated alkaline water electrolysis system exhibits smooth operation and provides proper control over the production process. On the other hand, the utilization of HHO gas as primary fuel displayed highly satisfying results. The developed system has the capacity to generate HHO on the order of 3L/min at 15Amps and it almost instantly burn paper and acrylic sheet.

1. Introduction

The industrialized modernization and human growth is much reliant on the high calorific value energy sources such as fossil fuels which are already on the brink of extension. Figure 1a shows the forecasted life of fossil fuels, which forecasted life of our fossil fuel reserves due to high excavation rates. Other associated problem is our inappropriate technology to harness useful energy from fossil fuels which cause high emissions of greenhouse gases and increase in CO2 concentration in global environment. Figure 1b is showing CO2 concentration for the past 400,000 years, where CO2 concentrations have shown several cycles of variation from about 180 ppm to 280 ppm. The concentration has increase dramatically from past 200 years due to industrial revolution and the CO2 concentration reached to 400ppm [1].

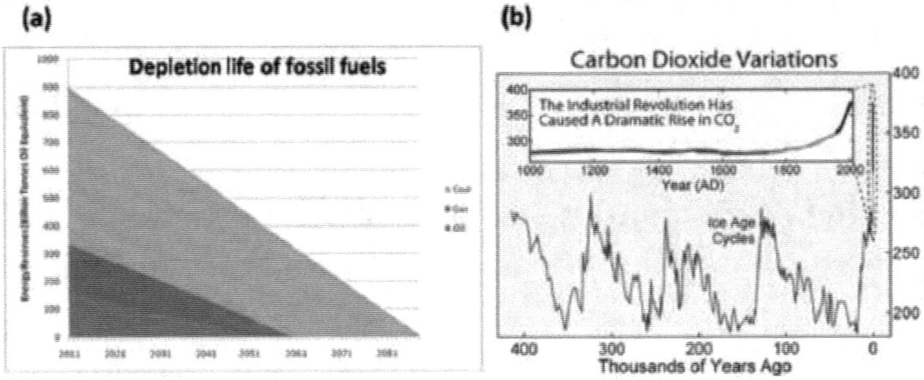

Figure 1. (a) depletion of fossil fuels and (a) golbal CO2 emission history [1].

Much emphasis has been given to renewable energy systems and their effectiveness have been successfully demonstrated [2]. In this regards, chemical storage and on demand use in the form of hydrogen is a lucrative option. Water can be electrolyzed using electricity from solar panels and used to generate electricity or heat energy when required.

Water splitting composed of two half reactions including hydrogen evolution reaction (HER) and oxygen evolution reaction (OER) but efficient water splitting can be achieved by efficient electrocatalyst to facilitate HER at the cathode and OER at the anode. Currently, Pt group metals and Ru/Ir-based compounds are the state-of-theart, noble, efficient HER and OER electrocatalysts respectively with low overpotential and Tafel slope. However, these noble electrocatalyst suffer from high cost, scarcity that limiting its application to produce hydrogen resource economically by water splitting. Moreover, the cell potential of the commercial water electrolyzers (1.8e2.0 V) are about 570 - 770 mV higher than the theoretical minimum value (1.23 V). Since hydrogen production through water splitting is energy intensive process, much of the efforts have been put forwarded to integrate the hydrogen with renewable energy sources. The second most crucial task is hydrogen storage and transportation. As H2 is highly explosive gas, great care is needed to store it with high capital investment and safety infrastructure. These safety issues become more pronounced when talking about HHO gas because this gas itself contains oxidizer in exact stoichiometric ratio. A new approach for efficient utilization of HHO gas is its on-demand production which attracted the attention of various research communities and policy maker circles.

The utilization of HHO in combustion process has been extensively investigated with impressive results. HHO when mixed with other combustion fuels such as natural gas, petrol and liquid petroleum products increase their combustion efficiency [3]. Attempts have been made to substitute some forms of petroleum products altogether with H2 and/or HHO with successful results. HHO has high potential in metal welding and cutting industry and bound to replace unsafe oxyacetylene welding and cutting apparatus. H2 has also been utilized in melting furnace in different proportions. Authors claimed improvement in combustion efficiency and cost effective process [4-6].

In this work, we design dry cell alkaline water electrolyzer and integrated it with solar PV for the production of HHO gas. Another objective of the study was to investigate the utilization of HHO gas as primary fuel for combsution.

Water electrolyzer setup:

A 30 plate dry cell alkaline water electrolyzer kit has been purchased from "greenfuel H2O" (Figure 2).

Figure 2. 30 plate dry cell water electrolyzer

It contains 30 316L stainless steel plates with 5inches active area. No neutral plates have been utilized in cell configuration with 2 volt plate gap based on 12 volt power source. Note that this information has been provided by the supplier. The system has been initially installed in lab to test its performance. Figure 3 shows inside setup of the system. A 64Ah, 12V car battery has been utilize to energize the electrolyzer.

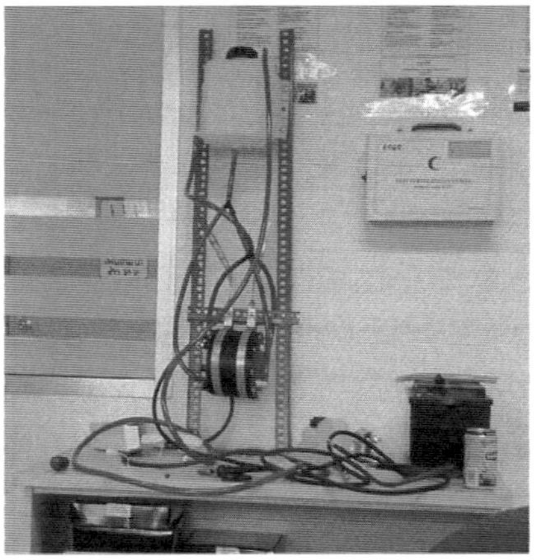

Figure 3. indoor setup of the system

The current has been limited to 15Apms as recommended by the supplier. A maximum of 3L/min HHO gas flow has been measured (using inverted water bottle) at 15Amps after 15min of running.

2. PV integrated water electrolysis system

The system has been integrated with PV panel with rated power of 300W. A charge controller has also been installed to properly charge the battery. The electrolyzer has been directly coupled with the battery and an on/off breaker has been installed between electrolyzer and the battery to function as on/off switch and circuit breaker. Figure 4 shows outdoor setup of the system. Figure 5 shows the electrical wiring and battery connections of the system. Water bubblers and moisture filters have also been installed for safety purpose and removal of water respectively. Flashback arrestor has also been employed to avoid any flash backs.

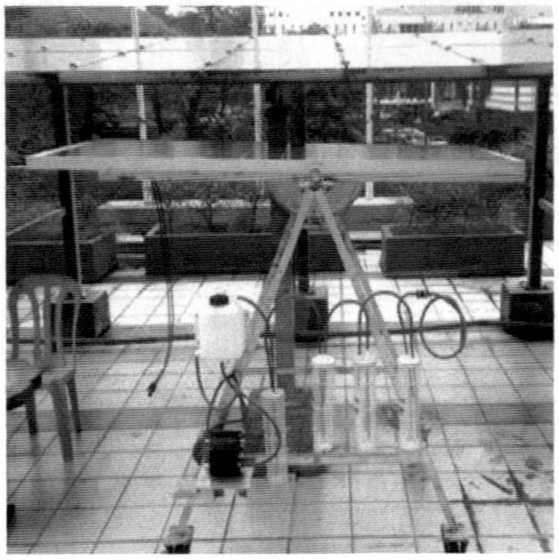

Figure 4. PV integrated water electrolysis system

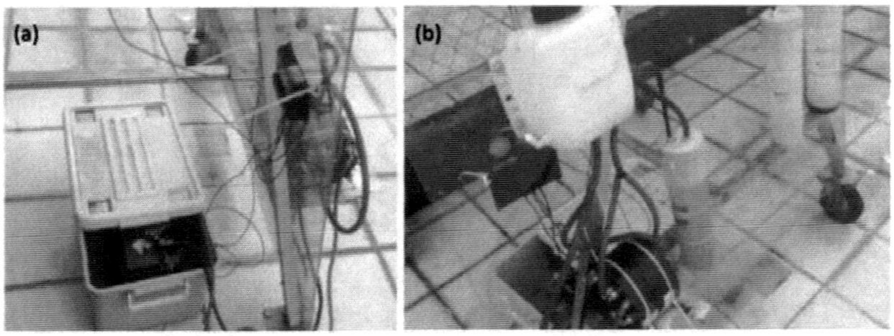

Figure 5. (a.b) different views of the system with electrical wiring

Combustion nozzle:
An appropriate torch with nozzle orifice 0.3mm has been utilized for proper control over the flame and to arrest the flashback. Figure 6 shows the torch employed for this purose

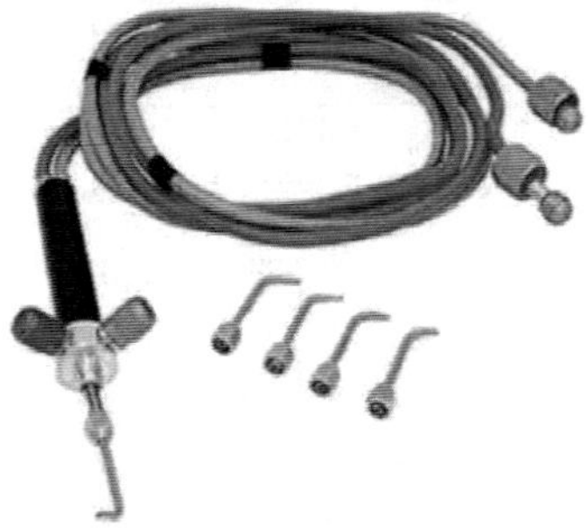

Figure 6. viviroom-Mini Jewelry Gas Welding Micro Torch

Normal paper and acrylic sheets have been tested to study the proper functioning of the flame. The flame temperature according to literature is approximately 2600-2800°C [7]. The flame was almost invisible in broad day light. It take only 3 seconds for the flame to inflame acrylic sheet (1cm thick). And the paper has been instantaneously got flamed when it came in contact with the flame. Figure 7 shows inflamed substrates.

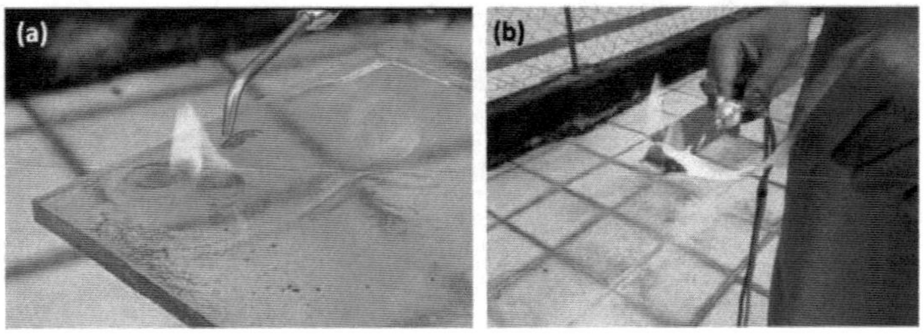

Figure 7. (a) inflamed acrylic sheet and (b) inflamed paper

3. Conclusions

In this study, PV integrated water electrolysis system has been developed to generate HHO/brown's gas. A safe and controlled flame has been achieved by burning the HHO gas employing suitable torch with orifice size 0.3mm. flashback arrestor has also been installed to avoid dangerous flashbacks. The system generated capacity was calculated to be 3L/min. The flame almost instantly burn paper and acrylic sheet then came in contact.

4. Future work

The present will prove to be a starter in the field of on-demand HHO utilization. The system will be extended to metal welding/cutting and aluminum melting furnace operations.

Acknowledgement

The authors thank the technical and financial assistance of UM Power Energy Dedicated Advanced Centre (UMPEDAC) and the Higher Institution Centre of Excellence (HICoE) Program Research Grant, UMPEDAC - 2016 (MOHE HICOE - UMPEDAC).

References

[1] Pandey, A., et al., Natural Sensitizers and Their Applications in Dye-Sensitized Solar Cell, in Environmental Biotechnology: For Sustainable Future. 2019, Springer. p. 375- 401.

[2] Ahmad, M.S., A. Pandey, and N.A. Rahim, Advancements in the development of TiO2 photoanodes and its fabrication methodsfor dye sensitized solar cell (DSSC) applications. A review. Renewable and Sustainable Energy Reviews, 2017. 77: p. 89-108.

[3] Nabil, T. and M.M.K. Dawood, Enabling efficient use of oxy-hydrogen gas (HHO) in selected engineering applications; transportation and sustainable power generation. Journal of Cleaner Production, 2019. 237: p. 117798.

[4] Arjun, T.B., et al., A review on analysis of HHO gas in IC engines. Materials Today: Proceedings, 2019. 11: p. 1117-1129.

[5] Subramanian, B. and S. Ismail, Production and use of HHO gas in IC engines. International Journal of Hydrogen Energy, 2018. 43(14): p. 7140-7154.

[6] Thangaraj, S. and N. Govindan, Evaluating combustion, performance and emission characteristics of diesel engine using karanja oil methyl ester biodiesel blends enriched with HHO gas. International Journal of Hydrogen Energy, 2018. 43(12): p. 6443-6455.

[7] Baltacioglu, M.K., R. Kenanoglu, and K. Aydın, HHO enrichment of bio-diesohol fuel blends in a single cylinder diesel engine. International Journal of Hydrogen Energy, 2019. 44(34): p. 18993-19004.

IOP Conf. Series: Materials Science and Engineering   1127 (2021) 012023   doi:10.1088/1757-899X/1127/1/012023

A Newly-Developed Control Method for Harmonic Mitigation in PV Grid-Connected Inverters Under Electric Arc Furnaces in Industrial Microgrids

Amirreza Naderipour[1], Zulkurnain Abdul-Malek[1,*], Saber Arabi Nowdeh[2], and Josep. M. Guerrero[3]

[*1]Institute of High Voltage & High Current, School of Electrical Engineering, Faculty of Engineering, Universiti Teknologi Malaysia 81310 Johor Bahru, Malaysia

[2]Golestan Technical and Vocational Training center Golestan, Iran

[3]The Villum Center for Research on Microgrids CROM, Department of Energy Technology, Aalborg University, Aalborg East, Denmark, Aalborg, Denmark www.crom.et.aau.dk

E-mail: zulkurnain@utm.my

Abstract. A microgrid (MG) is a discrete energy system consisting of distributed generation (DG) and loads or nonlinear loads that is able to operate at the same time the main grid is operating. The DG units can operate in parallel with the main grid or in an MG mode. The unbalanced and nonlinear characteristics of numerous loads that are in connection with the power system have the potential to bring about some problems, in terms of power quality, which can affect other consumers. For instance, the electric arc furnaces (EAFs) cause considerable problems for power quality. The instantaneous power theory (*pq*) and the coordinate formulation (*dq*) are proposed in this paper for harmonic current compensation for PV grid- connected inverters and EAF in MGs. With the proposed control strategy under EAF conditions, the total harmonic distortions of the system current were decreased from 38.24% to 3.25%, which duly satisfies the IEEE 519-1992 standard.

1. Introduction

The warning increase of CO2 emissions and global warming are two significant signs of climate change, which indicates a pressing need to find out innovative ways to generate energy using renewable sources. A solution to this problem is producing electricity in a place closer to final consumers, which is recognized as a distributed energy generation. This strategy decreases our dependency on long- distance power lines. Making a reliable connection between renewable energy sources and the utility grid, however, can result in new challenges.

New Renewable Energy Sources (RESs), such as Photovoltaic cell (PV), is often intermittent [1]. These energy systems can be combined or connected to a local energy storage system to maintain a continuous power flow between the mains grid and the local network [2].

When smaller producers are connected to grids via distributed electricity generation methods and can produce their own share of electricity, planning for this increase in RESs is one of the unavoidable future problems. Nonetheless, integrating RESs has become a priority in these stand-alone distribution networks. The diffusion of RESs with the MG concept has evolved into clusters of loads and paralleled RESs that can operate as a single power system to provide power to its local area [3].

Fossil energy reserves are getting exhausted due to a substantial increase of demands for electrical energy in current world [4]. On the other hand, the currently-used energy resources must be exploited more efficiently and cautiously because of the high price of energy in today's (and probably future) market. As a result, power electronic converters are getting more and more popular due to their high capacity for converting and controlling electrical power in a broad range from milliwatts to gigawatts using power semiconductor devices [5]. Nowadays, power electronics are processing over 70% of all electricity [6]. For the purpose of minimizing the energy waste, enhancing power quality, and reducing the cost of power production processes and power distribution/transmission, need to power electronics systems with a high degree of efficiency, sustainability, reliability, and quality [7]. With increasing power densities, challenges related to the quality of the power electronic systems have been more significant. The power electronic converters are typically utilized in microgrids aiming at controlling the power flow, and effectively converting power into appropriate DC or AC form [8].

A great deal of research has been conducted into power system for the purpose of attenuating the damping of low-frequency oscillations, whereas literature lacks adequate research into the use of nonlinear loads like EAFs and the way they can affect the network power quality [9].

Load flow calculations are vital for power flows, voltage profile, and losses determination [10]. These calculations are also used to assess voltage regulation issues and the basic capacity that is incorporated into the distributed generator interconnection. These calculations can also support other analyses, such as reliability, contingency, as well as power quality or transients. By finalizing these calculations, the model of the power system can be tuned according to its operational limits [11]. Adjustments may consist of selecting different transformer taps, generator working setpoints, reactive power compensators, and spinning reserve.

The addition of significant levels of renewable generators, such as photovoltaics (PVs), may increase the complexity of these analyses because of the uncertain nature of the energy sources [12]. For example, the time and location dependency of wind generators require extra care when combined with feeder location and load variability. Thus, further studies are required to determine the operating conditions that the new power systems will experience. This situation gets even worse when energy storage devices are employed. In such conditions, the calculations may be done over a longer period.

DG creates several challenges in load flow calculations such as modeling of the transmission or sub-transmission system, simulating the equipment's for voltage-control, embedding single and two-phase lines, single-phase loads, which in the general view cause unbalanced systems in calculations. Therefore, a proper load flow tool for a system containing distributed renewable generation, besides the conventional power system components, must contain a vast number of various generation and energy storage models combined with analysis capabilities.

2. Grid-connected inverter control methods

2.1. Instantaneous Reactive Power Theory

In 1984, Akagi pioneered the control method on the basis of the instantaneous reactive power theory. This method aimed at fixing the instantaneous active power in the feeding network. Instantaneous active and reactive power's theory was originally on the basis of Clarke Transform. The instantaneous three-phase currents and voltages are easily converted into the αβ orthogonal coordinates using Clarke Transform [13, 14]. Where the α and β, axes are the orthogonal coordinates. Presently, $v\alpha$ and $i\alpha$ are on the α axis, $v\beta$ and $i\beta$ are on the axis β. The following relations show this conversion:

$$\begin{bmatrix} v_\alpha \\ v_\beta \end{bmatrix} = \sqrt{\frac{2}{3}} \begin{bmatrix} 1 & -\frac{1}{2} & -\frac{1}{2} \\ 0 & \frac{\sqrt{3}}{2} & \frac{\sqrt{3}}{2} \end{bmatrix} \begin{bmatrix} v_a \\ v_b \\ v_c \end{bmatrix} \quad (1)$$

$$\begin{bmatrix} i_\alpha \\ i_\beta \end{bmatrix} = \sqrt{\frac{2}{3}} \begin{bmatrix} 1 & -\frac{1}{2} & -\frac{1}{2} \\ 0 & \frac{\sqrt{3}}{2} & \frac{\sqrt{3}}{2} \end{bmatrix} \begin{bmatrix} i_a \\ i_b \\ i_c \end{bmatrix} \quad (2)$$

The instantaneous real power (p) and the instantaneous imaginary power (q) is defined as follows:

$$p_{\alpha\beta} = v_a i_a + v_b i_b + v_c i_c \quad (3)$$

In this method, a set of currents (i_a, i_b, i_c) and voltages (v_a, v_b, v_c) from phase coordinates are first transferred. Furthermore, the three-phase instantaneous active power can be obtained through the normal relations among the coordinates abc as follows:

$$p_{\alpha\beta} = v_a i_a + v_b i_b + v_c i_c \quad (4)$$

Instantaneous three-phase reactive power in $\alpha\beta$ reference frame is defined as:

$$q_{\alpha\beta} = v_\alpha i_\alpha + v_\beta i_\beta \quad (5)$$

Active p and reactive q powers are defined as:

$$\begin{bmatrix} p_{\alpha\beta} \\ q_{\alpha\beta} \end{bmatrix} = \begin{bmatrix} v_\alpha & v_\beta \\ -v_\beta & v_\alpha \end{bmatrix} \begin{bmatrix} i_\alpha \\ i_\beta \end{bmatrix} \quad (6)$$

Equation (6) can be rewritten as:

$$\begin{bmatrix} i_\alpha \\ i_\beta \end{bmatrix} = \begin{bmatrix} v_\alpha & v_\beta \\ -v_\beta & v_\alpha \end{bmatrix}^{-1} \begin{bmatrix} p_\alpha \\ q_{\alpha\beta} \end{bmatrix} \quad (7)$$

In (6) $v_\alpha * i_\alpha$ and $i_\beta * v_\beta$ obviously mean instantaneous power since they are defined by the product of the instantaneous voltage in one axis and the instantaneous current in the same axis. As a result, using inverse transformation, current components in $\alpha\beta$ can be calculated using active and reactive instantaneous powers as:

$$\begin{bmatrix} i_\alpha \\ i_\beta \end{bmatrix} = \frac{1}{v_\alpha^2 + v_\beta^2} \begin{bmatrix} v_\alpha & -v_\beta \\ v_\beta & v_\alpha \end{bmatrix} \begin{bmatrix} p_{\alpha\beta} \\ q_{\alpha\beta} \end{bmatrix} \quad (8)$$

The $v_\alpha i_\alpha$ and $v_\beta i_\beta$ components in above are active power elements.

2.2. *DQ Synchronous Reference Frame Control*

Instantaneous voltage and current quantities are transformed to synchronous reference frame with Park's transformation matrix $[L]$ which is as (9).

$$[L] = \sqrt{\frac{2}{3}} \begin{bmatrix} \frac{1}{\sqrt{2}} & \frac{1}{\sqrt{2}} & \frac{1}{\sqrt{2}} \\ cos\theta_a(t) & cos\theta_b(t) & cos\theta_c(t) \\ -sin\theta_a(t) & -sin\theta_b(t) & -sin\theta_c(t) \end{bmatrix} \quad (9)$$

$$\theta_j(t) = \theta\left(t - \frac{2\pi}{3}(j-1)\right),\ j = a, b, c \quad (10)$$

Considering $\Theta j\ (t) = \omega t$, $0\alpha\beta$ components are transformed to $0dq$ components using (11) as:

$$\begin{bmatrix} i_0(t) \\ i_d(t) \\ i_q(t) \end{bmatrix} = \frac{1}{e_{\alpha\beta}} \begin{bmatrix} e_{\alpha\beta} & 0 & 0 \\ 0 & e_\alpha & e_\beta \\ 0 & -e_\beta & e_\alpha \end{bmatrix} \begin{bmatrix} i_0(t) \\ i_\alpha(t) \\ i_\beta(t) \end{bmatrix} \tag{11}$$

So that voltage *d*-component is calculated as:

$$e_d(t) = |E(t)| = e_{\alpha\beta} = \sqrt{e_\alpha^2 + e_\beta^2} \tag{12}$$

Fundamental positive sequence current component in ABC frame is transformed into a DC component in 0dq frame. Remaining harmonic current components including negative and zero sequences are appearing as alternative components in *0dq* frame. It means that there is a frequency shift (ω) in the mentioned components. Consequently, the compensator reference current is I_{cref} is as:

$$I_{cref}(t) = I_L(t) - I_s(t) \tag{13}$$

$$I_s = \overline{i_{Ld}} \frac{1}{e_{\alpha\beta}} \begin{bmatrix} 0 \\ e_\alpha \\ e_\beta \end{bmatrix} \tag{14}$$

Considering (13) and (14), compensator reference $0\alpha\beta$ current is as:

$$\begin{bmatrix} i_{L0}(t) \\ i_{Ld}(t) \\ i_{Lq}(t) \end{bmatrix} - \left(\frac{p_{L\alpha\beta}(t)}{e_{\alpha\beta}} \right)_{dc} \frac{1}{e_{\alpha\beta}} \begin{bmatrix} 0 \\ e_\alpha(t) \\ e_\beta(t) \end{bmatrix} \tag{15}$$

In contrast to other methods, active power is necessary at the compensator side in *0dq* control method which can be accomplished by a correction in compensator reference current (1).

3. Simulation results

Fig. 1 displays the single-line diagram of the system, in which the microgrid system that is considered in the present paper is depicted. There was one DG, while the others were converter interfaced DGs (with PV). The microgrid was able to operate in the grid-connected mode of operation. A basic photovoltaic cell is a low-power electric generator.

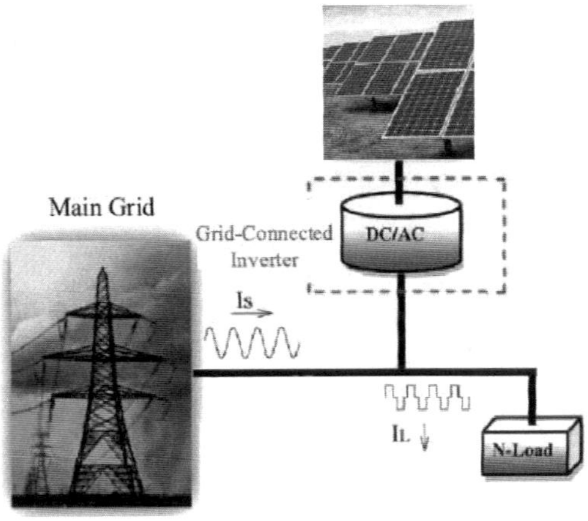

Figure 1. Configuration of the system

The MATLAB Simulink was used to perform the required simulations, where system voltage is assumed sinusoidal. Figure 2 displays the system waveforms with no compensation; The PV source deliver a nonsinusoidal current to the system while nonlinear loads flow harmonic currents from the system. Nonlinear loads including electric arc furnace make a system current nonlinear and unbalanced as shown in Figure 3.

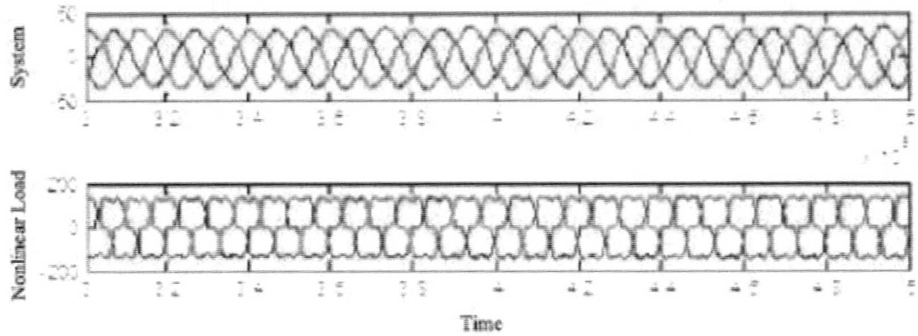

Figure 2. Current waveforms without compensation.

At 4 seconds, proposed control methods are connected to the system; system current harmonics and unbalance is compensated. Remember that *p-q* and *d-q* control methods are used in Figure 3.

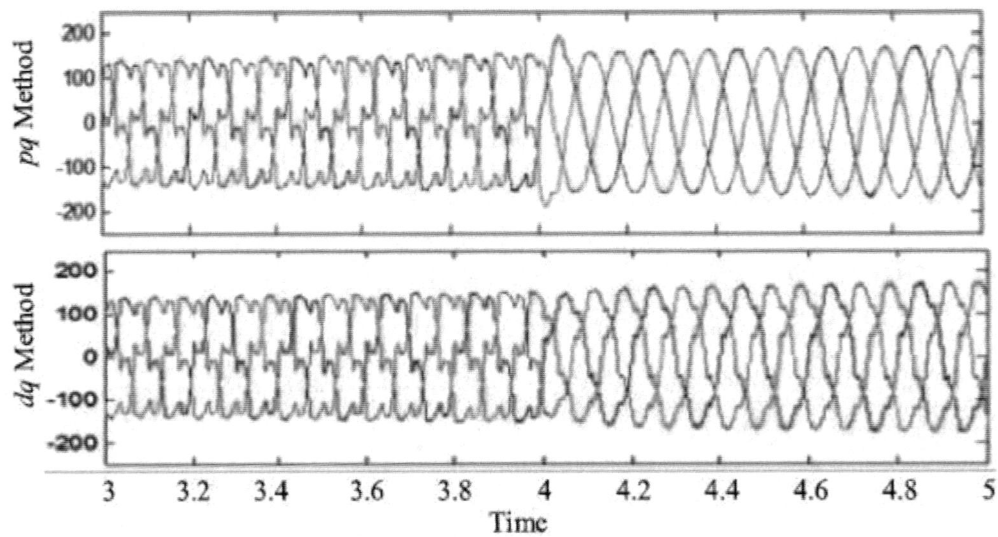

Figure 3. Configuration Current waveforms after using *p-q* and *d-q* control methods

The grid-connected inverter is connected, the source current becomes balanced and sinusoidal. GCI demonstrated in Fig. 3 using *dq* control method since the system voltage is sinusoidal, there is not a major difference between *pq* and *dq* control methods simulation waveforms. For instance, in the *dq* control methods, there is a need for phase detection, which increases the calculations and complex control method while this circuit is not necessary at *pq* method. With no compensation, system current THD was recorded 38.24 %, while with implementing compensation, it showed a reduction down to 3.25 % in the *pq* method. But in *dq* it was 11.4 %.

Table 1. Parameters of the System.

Parameter	*Values*
AC supply voltage	35 V
AC inductor of the diode rectifier: (L_{ac})	5mH
AC resistor of the dioderectifier: (R_{ac})	3 Ω
System resistance: $Rs(\Omega)$	0.3

Load resistance: $RL(\Omega)$	165
Active filter ripple filter inductance Rf	12 Ω
Line Frequency	50Hz
System equivalent inductance:	0.30
Load input inductor: L_{AC}(mH)	3.8
Dc capacitor: C_{dc}(μf)	470

4. Conclusion

With regard to the growth of power electronic devices and non-linear loads, power filters are used as one of the methods to compose harmonics and reactive power due to non-linear loads. Combining active and passive filters (combined filters) is an efficient and economical method. This study proposes the use of the grid-connected inverter control method, including instantaneous power theory (pq) and the coordinate formulation (dq). Studies on a group of nonlinear loads (electric arc furnaces) as harmonic current sources as well as distributed generation source (photovoltaic) show that with the proposed control strategy under EAF conditions, the total harmonic distortions of the system current were decreased from 38.24% to 3.25%.

Acknowledgements

The authors wish to thank the Ministry of Education, Universiti Teknologi Malaysia, and Universiti Sriwijaya (Grant numbers 4F828, 01M44, 18H10, 4B345, 4B379 and 04E54) for the financial support.

References

[1] Naderipour A, Abdul-Malek Z, Vahid MZ, Seifabad ZM, Hajivand M, Arabi-Nowdeh S. Optimal, Reliable and Cost-Effective Framework of Photovoltaic-Wind-Battery Energy System Design Considering Outage Concept Using Grey Wolf Optimizer Algorithm—Case Study for Iran. *IEEE Access*. 2019;**7**:182611-23.

[2] Naderipour A, Abdul-Malek Z, Heidari Gandoman F, Nowdeh SA, Shiran MA, Hadidian Moghaddam MJ, et al. Optimal designing of static var compensator to improve voltage profile of power system using fuzzy logic control. *Energy*. 2020;**192**:116665.

[3] Lasseter B, editor Microgrids [distributed power generation]. 2001 IEEE Power Engineering Society Winter Meeting. Conference Proceedings (Cat. No.01CH37194); 2001 28 Jan.-1 Feb. 2001.

[4] Naderipour AR, editor Voltage and current compensation in dispersed generation systems. 2010 First Power Quality Conferance; 2010 14-15 Sept. 2010.

[5] Naderipour AR, editor A hybrid solution to improve power quality in dispersed generation systems. 2010 First Power Quality Conferance; 2010 14-15 Sept. 2010.

[6] Bahman AS. Multidisciplinary Modelling Tools for Power Electronic Circuits: with Focus on High Power Modules. Department of Energy Technology, Aalborg University: Aalborg University; 2015.

[7] Naderipour A, Abdul-Malek Z, Abohamzeh E, Ramachandaramurthy VK, Miveh MR, editors. Control strategy of Grid-Connected PV Inverters in Microgrid with Nonlinear Operating Conditions. 2018 IEEE 7th International Conference on Power and Energy (PECon); 2018 3-4 Dec. 2018.

[8] Sharkh SM, Abu-Sara MA, Orfanoudakis GI, Hussain B. Power Electronic Converters for Microgrids: Wiley; 2014.

[9] Ozgun O, Abur A, editors. Development of an arc furnace model for power quality studies. 1999 IEEE Power Engineering Society Summer Meeting. Conference Proceedings (Cat.

No.99CH36364); 1999 18-22 July 1999.
[10] Hertem DV, Verboomen J, Purchala K, Belmans R, Kling WL, editors. Usefulness of DC power flow for active power flow analysis with flow controlling devices. The 8th IEE International Conference on AC and DC Power Transmission; 2006 28-31 March 2006.
[11] Kulkarni N, Kamalasadan S, Ghosh S. An Integrated Method for Optimal Placement and Tuning of a Power System Stabilizer Based on Full Controllability Index and Generator Participation. *IEEE Trans Ind Appl*. 2015;**51**(5):4201-11.
[12] Abdel-Nasser M, Mahmoud K. Accurate photovoltaic power forecasting models using deep LSTM-RNN. *Neural Computing and Applications*. 2019;**31**(7):2727-40.
[13] Akagi H, Kanazawa Y, Nabae A. Instantaneous Reactive Power Compensators Comprising Switching Devices without Energy Storage Components. *IEEE Trans Ind Appl*. 1984;**IA-20**(3):625-30.
[14] Fang Zheng P, Ott GW, Adams DJ. Harmonic and reactive power compensation based on the generalized instantaneous reactive power theory for three-phase four-wire systems. *IEEE Trans Pow Electr*. 1998;**13**(6):1174-81.

Simulation Analysis for Induction Motor Drive Fault Detection and Localization Under Variable Load and Speed Operation

Sazali Yaacob*, Amir Rasyadan †, Pranesh Krishnan†

* UniKL Malaysian Spanish Institute, Malaysia

† UniKL Malaysian Spanish Institute, Malaysia

E-mail: sazali.yaacob@unikl.edu.my

Abstract. The induction motors are widely used in the industry for their advantage of being reliable, rugged, and simple in construction. Nonetheless, the use of inverter-based drives has increased the overall system failure possibility mainly because of the power electronics switches used in the inverter itself. For that reason, it is crucial to monitor the inverter switches health during operation to avoid total system breakdown. A method to do so is by using the current vector trajectory analysis on the induction motor stator current signal. However, one disadvantage of using this method is that it is negatively affected by the load and speed variations. This paper is intended to discuss further on the matter and look into the methods that can be used to minimize the effects associated with the load and speed variations using computer simulation analysis.

1. Introduction

Traditionally, the induction motors were driven by fixed speed drives limited for use in applications that only requires simple electrical to mechanical power conversions system. However, this limitation has been lifted off with the introduction of inverter-based drives, the use of inductions motors has now been extended to myriad of new applications that may also involve complex and precise speed control. Nonetheless, the use of the power electronic switches in the inverter have also introduce new failure possibilities. The power inverter is known to be the weakest component in the drive system with around 38% of the drive faults are due to their power switch devices [1-3].

Generally, the faults can be classified into open-circuit and short-circuit fault. A short-circuit switch can generate abnormal overcurrent that could quickly destroy the motor whereas an open-circuit switch fault does not completely halt motor operation, but it stresses the system, noise can be induced causing performance degradation and by time will also lead to total system failure [4-6]. This paper only focuses on the open-circuit fault analysis since the short-circuit faults are mostly detected by means of hardware circuit [7-9].

Based on previous literatures, there are a number of works that focuses on the inverter open-circuit Fault Detection and Localization (FDL). One of the most popular FDL method is the current vector trajectory analysis which uses the Lissajous Curve of the phase current signals to detect the faulty inverter leg and then calculating the average value within one current period to localize the faulty switch. It is simple, does not requires extra sensors, and produce considerably fast detection time [8, 10].

The downside of this method is that it is dependent on the load and the supply frequency variations applied to control the speed of the motor [7, 11]. In separate works, different current signal normalization techniques were discussed, namely; Maximum Current normalization [7], Schmitt Trigger model [11, 12], and Park's Vector Modulus (PVM) Normalization [8, 13]. However, their works are short in terms of comparison between the three methods. Further studies are required to understand why the current normalization is needed and the difference on how the three methods work.

This paper is intended to present an analysis on the induction motor operation under variable load and speed especially on the stator current signal which is the one being monitored for fault detection and localization. The use of the three current normalization methods are then discussed and being compared to see which performs the best.

2. Induction Motor Drive with Fault Detection and Localization Model

To further look into the details of the current vector trajectory based FDL under variable load and speed operation, it is convenient to analyze the system in a computer simulation software.

2.1. Induction motor drive model

The induction motor drive model is built in MATLAB Simulink. The model consists of three main parts; a Sinusoidal Pulse Width Modulation (SPWM) signal generator, a three-phase voltage source inverter, and a three- phase induction motor model. The SPWM generator is built customly by using three sinusoidal reference signals that each set 120-degree phase apart from other; V_a, V_b, V_c are compared with a high frequency triangular signal V_T. V_a, V_b, V_c are generated using a phase accumulator. The reference signals modulate the three-phase PWM duty ratio, hence, their frequency is the desired fundamental frequency f_f of the inverter. This way, the frequency of the SPWM signal can be varied as intended to provide control over the speed of the motor. The fault control block is then used to simulate the open-circuit conditions by overriding any of the six switches PWM. Figure 1 (a) shows the SPWM signal generator with the fault control block and Figure 1 (b) shows the three-phase inverter connected to the induction motor model.

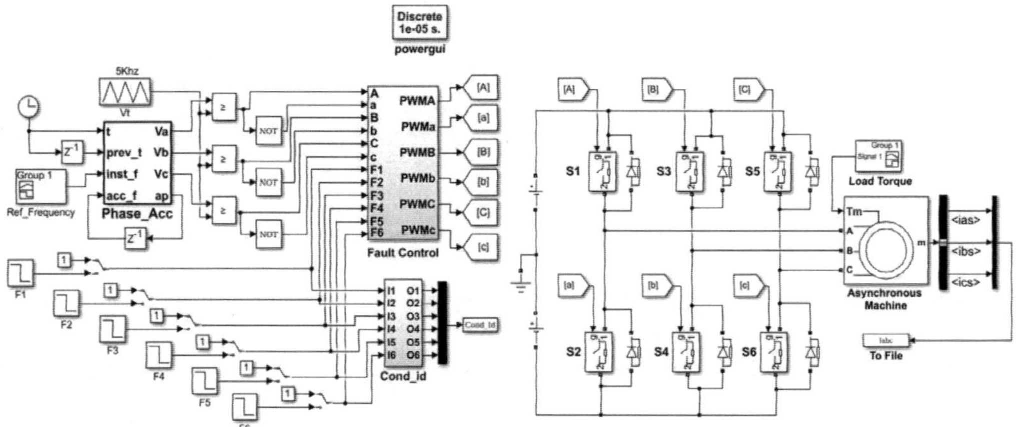

(a) (b)

Figure 1. (a) SPWM generator and fault control. (b) Three-phase inverter connected to Induction motor model.

To simulate the variable load and speed operation, the induction motor DQ-model is supplied with a load torque profile and SPWM reference frequency profile that consists of ramp, step, and sinusoidal inputs built using piecewise function shown in equation (1), (3), and (4) respectively.

$$\text{Ramp:} \quad Y(t) = m \times Y(t-\tau) + A \tag{1}$$

$$m = \frac{Y_2 - Y_1}{t_2 - t_1} \tag{2}$$

$$\text{Step:} \; Y(t) = C \tag{3}$$

$$\text{Sinusoidal:} \; Y(t) = Asin\big(2\pi f(t-\tau)\big) + B \tag{4}$$

Where:
A = Amplitude Scale
B = Amplitude Shift
C = Constant
τ = Time shift
m = Gradient

Based on these equations, the load and SPWM reference frequency profile can be built as intended.

2.2. *Fault Detection and Localization algorithm*

The Fault Detection algorithm requires the monitored three- phase current to be transformed into an equivalent two-phase component by using Park's transformation. The relations are given in (5) and (6):

$$i_d = \sqrt{\frac{2}{3}} i_a - \frac{1}{\sqrt{6}} i_b - \frac{1}{\sqrt{6}} i_c \tag{5}$$

$$i_q = \frac{1}{\sqrt{2}} i_b - \frac{1}{\sqrt{2}} i_c \tag{6}$$

From the two-phase current components i_d and i_q, the Lissajous figure can be plotted to get a good view of the phase current trajectory. Under healthy inverter operation, a circular figure is formed within one fundamental current period, and if faults occur the figure should deform accordingly. To extract valuable information from the Lissajous figure, the slope P between two points of the Lissajous figure is calculated, this value reveals the condition of the inverter. The use of the slope calculation method for inverter fault detection was discussed in [8, 12] with the equation as follows.

$$P = \frac{\Delta i_d}{\Delta i_q} = \frac{i_d(k-h) - i_d(k)}{i_q(k-h) - i_q(k)} \tag{7}$$

Where k is the present sample and h is the previous number of sample.

During healthy inverter operation, the Lissajous figure is circular in shape, hence the slope P values are always varying throughout one current period, but during open switch faults, P can take only one constant value (zero when leg 2 faulty, √3 when leg 1 faulty, or -√3 when leg 3 is faulty) during half a current period depending on which leg is faulty.

The Fault Localization algorithm is then used to localize the faulty switch within the faulty leg. The strategy is based on monitoring of the phase current polarity within one current period as an indicator of either the upper, lower or both the switches is in faulty condition. This can be done by calculating the average values within one current period as shown in the following equation.

$$\langle Mu_N(k) \rangle = \frac{1}{M} \sum_{j=0}^{M-1} i_n(k-j) \tag{8}$$

Where:
Mu_n = Phase a, b, c average current variables.
M = Number of samples per current period

The average values are calculated using a sliding window so that the mean value is updated at every k^{th} sample for faster fault localization.

However, under variable load and speed condition, transient current condition will occur before the motor gets into steady state. Taking these into consideration, some methods are required to handle such variations without causing false alarms.

3. FDL Under Variable Load and Speed Operation

The variable load condition is simulated by using a load profile. In the time span of ten second, a few load variation types are introduced, the inverter is working in healthy condition with the SPWM reference signals frequency f_{ref} being fixed at 60Hz. Fig. 2 shows the load torque profile applied, the phase currents, and the rotor speed.

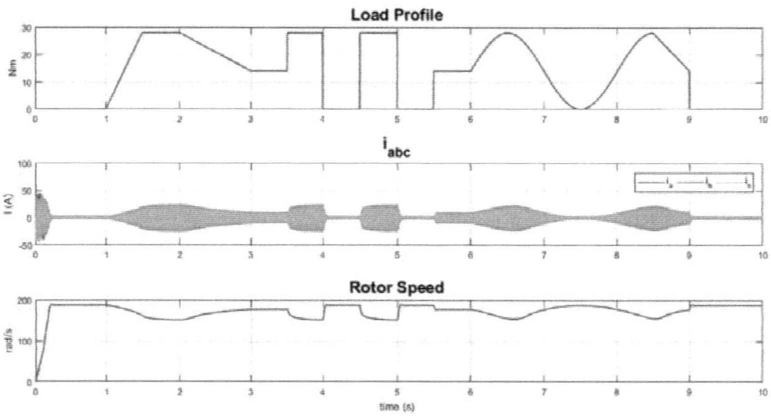

Figure 2. Load torque profile applied, phase currents, and rotor speed.

As observed, load variation directly affects the phase current amplitude, transient region can also be seen when load change occurred before the motor gets into a steady state condition. Yet, the phase currents change smoothly even under a step load increase from 0% to 100% at time 4.5s. It should be noted that under load variation, only the current amplitude changes while the frequency remains constant.

In the case of variable speed operation, the condition is simulated by supplying the SPWM generator model with a custom frequency profile as the SPWM reference signals frequency f_{ref}. In the time span of ten second, a few frequency variations are introduced, the inverter is working in healthy condition with the load being constant at 25%. Figure 3 shows the frequency profile and the effects of the frequency variations to the phase currents and the rotor speed.

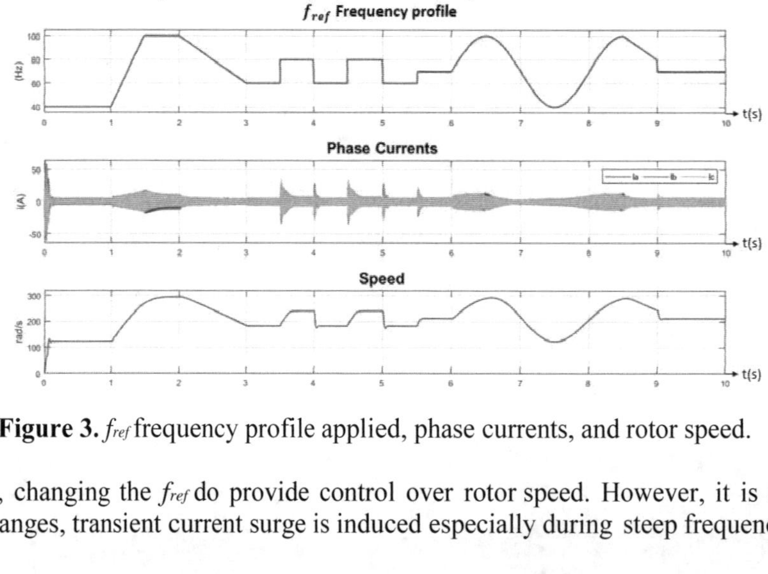

Figure 3. f_{ref} frequency profile applied, phase currents, and rotor speed.

As expected, changing the f_{ref} do provide control over rotor speed. However, it is also observed that when f_{ref} changes, transient current surge is induced especially during steep frequency change.

These transient current condition during load change and frequency change does not affecting the fault detection algorithm that is based on the Lissajous figure slope calculation. As an example, Figure 4 shows the Lissajous figure of the DQ current components at time between time 4.4s to 4.6s where at time 4.5s, a step load increase from 0% load to 100% load is introduced.

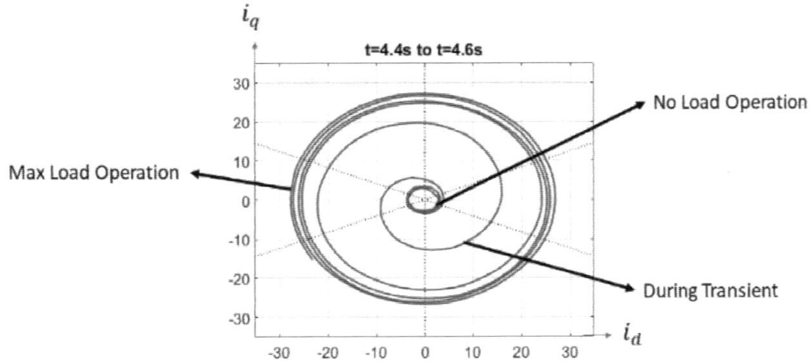

Figure 4. Lissajous figure of DQ current components between t=4.4s to t=4.6s.

The Lissajous figure shows a smooth transition from zero load to max load operation. Fig. 5 shows the zoomed in plot of the phase currents i_{abc} and the slope variable P between time t=4.4s to t=4.6s.

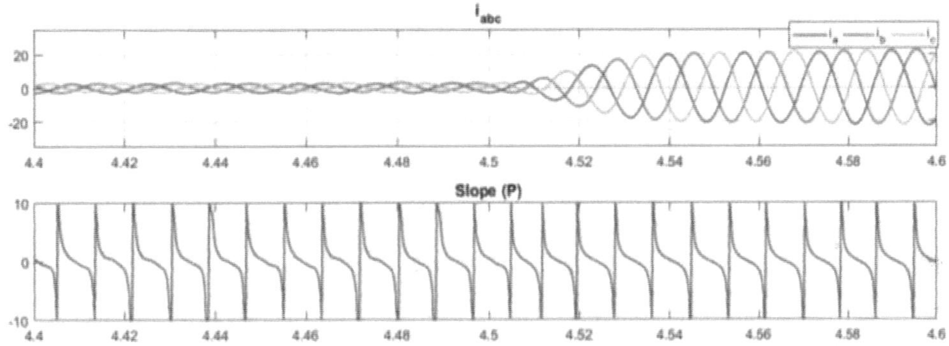

Figure 5. Phase currents and the slope variable.

This shows that even under a step load transition from 0% load to 100%, the current still changes smoothly, hence the slope variable is not affected of this condition and no sign of false alarm occurred.

However, the transient current condition does affect the fault localization variable. Fig. 6 shows the phase-a average current variable Mu_a in the time span of 10 seconds, high spikes are observed especially at fast transient regions which occurs during motor starting, and the regions where step load changes are introduced.

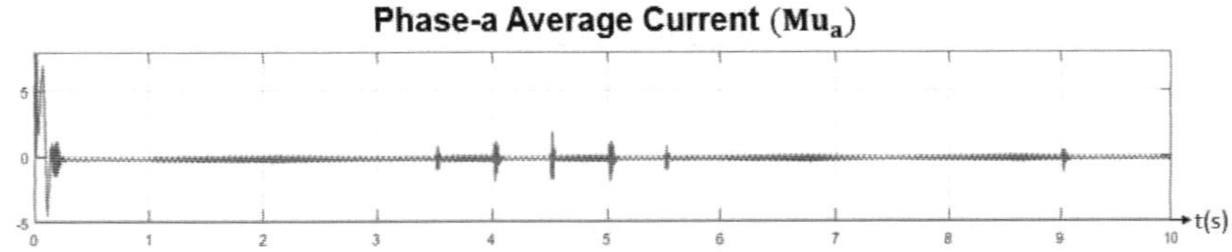

Figure 6. Phase-a average current variable.

This is due to the unbalance of the current flow in negative and positive direction within one current period. For this reason, current signal normalization methods are required to first normalize the current signals before the Fault Localization algorithm is applied.

4. Current Signal Normalization and Phase Current Averaging Compensation

There are three methods being discussed in this work as follows.

4.1. Maximum Current Normalization

One way to normalize the phase current signal is by using the max current value within the sliding window of one current period I_{max}. Dividing each of the instantaneous sample with the obtained I_{max} gives a normalized current signal with the equation as follows.

$$i_{nN} = \frac{i_n}{i_{n\_max}} = \begin{cases} i_{aN} = \dfrac{i_a}{i_{a\_max}} \\ i_{bN} = \dfrac{i_b}{i_{b\_max}} \\ i_{cN} = \dfrac{i_c}{i_{c\_max}} \end{cases} \quad (9)$$

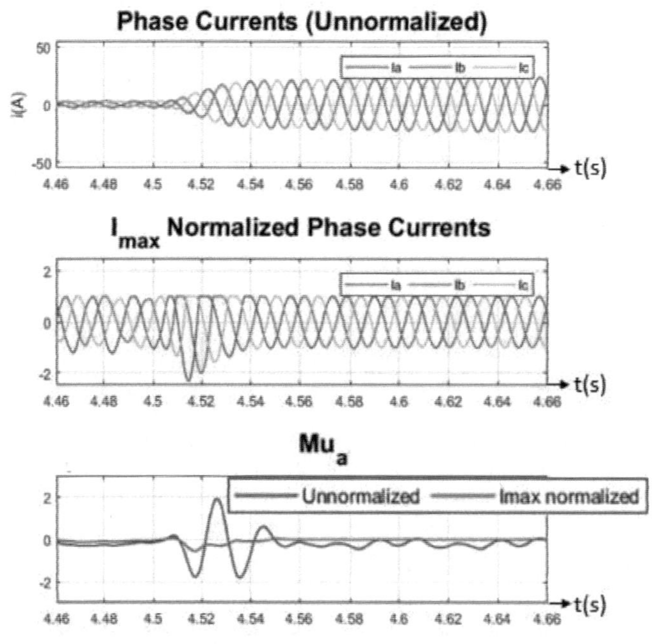

Figure 7. Unnormalized phase current, Maximum current (I_{max}) normalized phase currents, and comparison of phase-a average current variable Mu_a.

Based on the simulation results, the I_{max} normalization method seems to be capable of suppressing the high average current deviation.

4.2. Schmitt Trigger Model

This second method mimics the behavioral model of a Schmitt trigger. Rather than using the maximum current value to normalize the instantaneous phase current amplitude, this method uses two defined threshold values to determine whether the current is in positive or in negative direction as shown in equation (10).

$$i_{nN}(k) = \begin{cases} +1, & i_n > +i_{th} \\ -1, & i_n < -i_{th} \\ i_{nN}(k-1), & otherwise \end{cases} \quad (10)$$

The normalization output is set as high (+1) when the input signal rises above the high-level threshold value ($+i_{th}$) and does not go low (-1) until the input falls below the lower threshold ($-i_{th}$) value. The value of i_{th} used in this simulation is 0.3.

Fig. 8 shows the unnormalized phase current, the Schmitt trigger model outputs, and the comparison of phase-a average current variable Mu_a during the fast-transient region between time 4.4s to 4.6s.

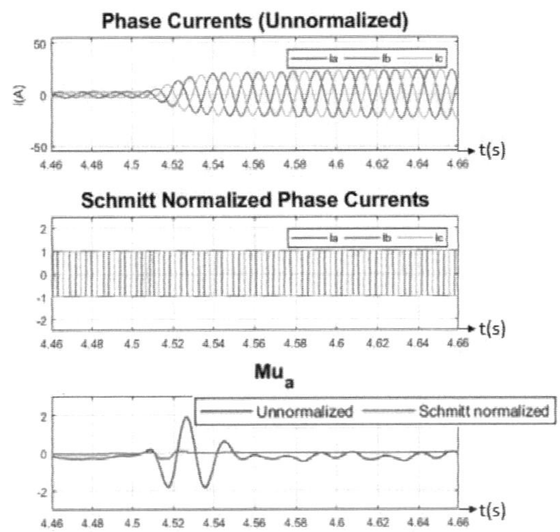

Figure 8. Unnormalized phase current, the Schmitt trigger model outputs, and the comparison of phase-a average current variable Mu_a.

The result shows that this method is also able to suppress the current average value deviations during transient condition.

4.3. Park's Vector Modulus

The other method is known as Park's Vector modulus (PVM) normalization, it is done by using the following equation.

$$|i_{PVM}| = \sqrt{i_d^2 + i_q^2} \quad (11)$$

$$i_{nN} = \frac{i_n}{|i_{PVM}|} = \begin{cases} i_{aN} = \dfrac{i_a}{|i_{PVM}|} \\ i_{bN} = \dfrac{i_b}{|i_{PVM}|} \\ i_{cN} = \dfrac{i_c}{|i_{PVM}|} \end{cases} \quad (12)$$

Where i_d and i_q is obtain from the transformation of the three- phase to two-phase current as describe in equation (11) and (12). Fig. 9 shows the unnormalized phase current, the PVM normalized currents, and the comparison of their average values during the fast-transient region (t=4.4s to 4.6s).

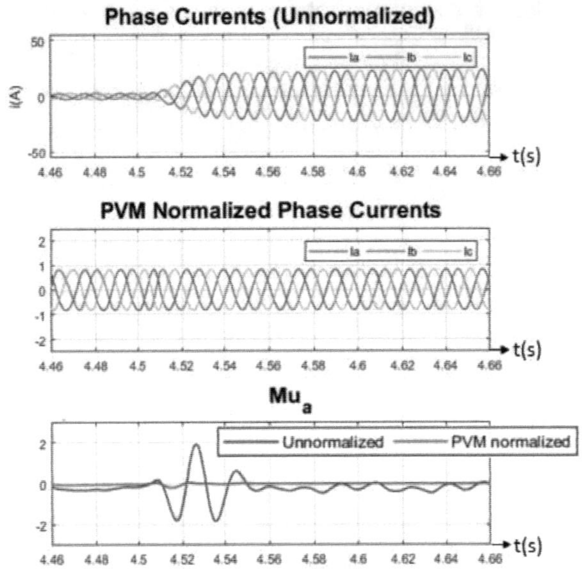

Figure 9. Unnormalized phase current, the PVM normalized currents, and phase-a average current variable Mu_a.

As observed from the results, the PVM normalization method have also shown the capability to suppress the average current value deviations during transient conditions.

4.4. Comparison of Current Signal Normalization Methods
Based on the results, all the three normalization methods presented have shown their ability to suppress the transient current effects on the average current variable (Mu_n). However, in comparison of the three normalization methods, there are some difference between the three. Fig. 10 shows the zoomed in look of the Mu_a variable at the fast-transient region (t=4.4s to 4.6s).

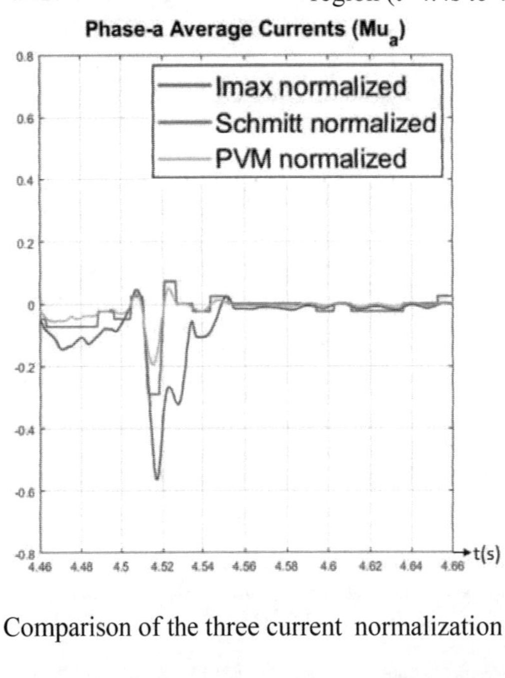

Figure 10. Comparison of the three current normalization methods.

Overall, it is observed that i_{max} normalization gives the poorest results while PVM normalization shows the least deviations during transient condition.

5. Conclusions

This paper presents a simulation analysis on the current vector trajectory fault detection and localization method for inverter open-circuit faults under variable load and speed condition. The simulation shows that during variable load and speed, transient current conditions are induced thus affecting the fault localization process. Three current signal normalization methods that include Maximum Current normalization, Schmitt Trigger model, and the Park's Vector Modulus Normalization (PVM) are discussed to compensate this issue. All the three methods have shown their capability to reduce the effects of transient current condition on the fault localization process, but in comparison of the three, the PVM normalization gives the lowest deviations thus are the most favorable method to be used.

References

[1] Cherif BDE, Bendjebbar M, Benouzza N, Boudinar H, Bendiabdellah A, editors. A comparative study between two open-circuit fault detection and localization techniques in a threephase inverter fed induction motor. 2016 8th International Conference on Modelling, Identification and Control (ICMIC); 2016 15-17 Nov. 2016.

[2] Errabelli RR, Mutschler P. Fault-Tolerant Voltage Source Inverter for Permanent Magnet Drives. *IEEE Trans Pow Electr*. 2012;**27**(2):500-8.

[3] Tabbache B, Benbouzid M, Kheloui A, Bourgeot J, Mamoune A, editors. PWM inverter-fed induction motor-based electrical vehicles fault-tolerant control. IECON 2013 - 39th Annual Conference of the IEEE Industrial Electronics Society; 2013 10-13 Nov. 2013.

[4] Asghar F, Talha M, Kim SH. Neural Network Based Fault Detection and Diagnosis System for Three-Phase Inverter in Variable Speed Drive with Induction Motor. *Journal of Control Science and Engineering*. 2016;**2016**:1286318.

[5] Hang C, Ying L, Shu N. Transistor open-circuit fault diagnosis in two-level three-phase inverter based on similarity measurement. *Microelectronics Reliability*. 2018;**91**:291-7.

[6] Shu C, Ya-Ting C, Tian-Jian Y, Xun W. A Novel Diagnostic Technique for Open-Circuited Faults of Inverters Based on Output Line-to-Line Voltage Model. *IEEE Trans Indust Electr*. 2016;**63**(7):4412-21.

[7] Dhumale RB, Lokhande SD. Neural Network Fault Diagnosis of Voltage Source Inverter under variable load conditions at different frequencies. *Measurement*. 2016;**91**:565-75.

[8] Trabelsi M, Boussak M, Benbouzid M. Multiple criteria for high performance real-time diagnostic of single and multiple open-switch faults in ac-motor drives: Application to IGBT-based voltage source inverter. *Electric Power Systems Research*. 2017;**144**:136-49.

[9] Zhang W, Xu D, Enjeti PN, Li H, Hawke JT, Krishnamoorthy HS. Survey on Fault-Tolerant Techniques for Power Electronic Converters. *IEEE Trans Pow Electr*. 2014;**29**(12):6319-31.

[10] Rasyadan A, Ibrahim Z, Dir TMABT, Yaacob S, editors. Modeling of Time Domain Analysis for Single and Double Open-Circuit Inverter Switch Faults in Three-Phase Induction Motor Drives. 2018 International Conference on Computational Approach in Smart Systems Design and Applications (ICASSDA); 2018 15-17 Aug. 2018.

[11] Yu Y, Hu J, Wang Z, Xu D, editors. IGBT open circuit fault diagnosis in VSI fed induction motor drives based on modified average current method. 2014 9th IEEE Conference on Industrial Electronics and Applications; 2014 9-11 June 2014.

[12] Trabelsi M, Boussak M, Gossa M, editors. Multiple IGBTs open circuit faults diagnosis in voltage source inverter fed induction motor using modified slope method. The XIX International Conference on Electrical Machines - ICEM 2010; 2010 6-8 Sept. 2010.

[13] Estima JO, Freire NMA, Cardoso AJM, editors. Recent advances in fault diagnosis by Park's vector approach. 2013 IEEE Workshop on Electrical Machines Design, Control and Diagnosis (WEMDCD); 2013 11-12 March 2013.

IOP Conf. Series: Materials Science and Engineering 1127 (2021) 012025 doi:10.1088/1757-899X/1127/1/012025

New Series Motor Four Quadrants Drive DC Chopper for Economical EV Part5: Parallel mode

Saharul Arof [a*,b,c], N.H.N.Diyanah[a], Emilia Noorsal[b], Philip Mawby[c], H. Arof[d]

[*a]Universiti Kuala Lumpur, Malaysian Spanish Institute, Kulim Hi-Tech Park, 09000 Kulim, Kedah.Malaysia

, [b]Faculty of Electrical Engineering,Universiti Teknologi MARA ,Pulau Pinang, Malaysia.

[c] School of Engineering, University of Warwick Coventry, CV47AL,United Kingdom.

[d.] Engineering Department,Universiti Malaya, Jalan Universiti, 50603, Kuala Lumpur

E-mail: saharul@unikl.edu.my

Abstract. This is part five of an eight-part series of a new Four quadrant dc chopper. The paper describes the parallel mode of the proposed Four Quadrants DC Chopper (FQDC) for a series motor. The mode is designed to overcome a disadvantage of dc series motors which is the speed reduction under load. First, it is studied via simulation using Matlab/Simulink model. Then the FQDC is experimentally tested on a 0.65kW series motor in a lab setup. Finally, a MATLAB/Simulink model is developed to apply the FQDC to a 35kW series motor that propels an electric car (EV). The results indicate that the proposed technique manages to prevent the speed drop of a dc series motor when loaded. This is evident when the EV is subjected to a hill climb.

1. Introduction

Electric vehicles (EV) are hailed as a means of transportation that is economical and friendly to the environment. However, EVs are still expensive, making them unaffordable to many people. This concern has led to the studies on EVs driven by DC motors that are supposed to be cheaper [1, 2].

1.1. Literature

Producing efficient EVs powered by DC motors is challenging as it requires research and development in mechanical and electrical fields. In the electrical field, efforts have been made in the design of efficient dc motors [2], Four Quadrant DC drive [1, 3], closed-loop control of the DC motor using direct current [4], and closed-loop control of the battery charging and discharging [5]. Some related simulation works that have been reported include testing and implementing the FQDC and EV controllers [6, 7], incorporating a battery charger to the FQDC [8] and testing the FQDC on separately excited dc motor [9]. The FQDC operation controller can also be optimized using Artificial Intelligence to suit different

IOP Conf. Series: Materials Science and Engineering 1127 (2021) 012025 doi:10.1088/1757-899X/1127/1/012025

earth profiles [1 0 - 1 2] . The AI method can also help to improve the performance of the specific chopper operation [13]. This paper describes the operation of the proposed FQDC in parallel mode. This mode is needed to overcome the issue of speed reduction in series DC motors under load. The overall operation of the FQDC in the parallel mode is shown in Figure 1.

The following six equations dictate the voltage and current of the FQDC for all modes.

$$I_a = \frac{V_{batt} - B_{emf}}{R_a + R_f} \tag{1}$$

$$B_{emf} = K_v I_f \omega \tag{2}$$

$$T_d = K_t I_f I_a \tag{3}$$

$$T_d = J\frac{d\omega}{dt} + B_w + T_L \tag{4}$$

where B_{emf} is the back emf of the motor, Va and Vf are the armature and field voltages, K_b is the back emf constant, K_t is the torque motor constant, I_f is the field current, R_a and R_f are the motor coil resistances, I_a is the armature current, ω is the angular speed, and T_d is the motor torque.

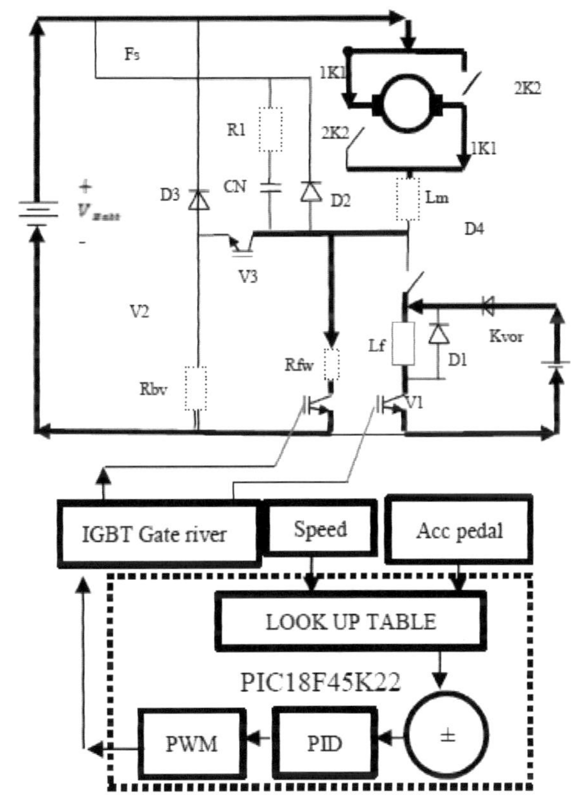

Figure 1. Four quadrants chopper in Parallel mode

The values of the armature and field currents when IGBT V1 and V2 turn ON are

$$\frac{d}{dt}[I_f] = \left[-\left(\frac{R_f}{L_f}\right)\right][I_f]\left[\frac{1}{L_f}\right][V_{ext} \quad K_{drv}] \tag{5}$$

$$\frac{d}{dt}[I_a] = \left[-\left(\frac{R_a}{L_a}\right)\right][I_a]\left[\frac{1}{L_a} \quad -\frac{1}{L_a}\right][V_{ext} \quad K_{drv}] \tag{6}$$

2. Methodologies

2.1. DC series motor in Parallel Mode

The main advantage of a DC series motor is its high starting torque. However, its speed drop drastically when loaded. It is common for an electrical motor to have a slight decrease in speed when loaded but the speed drop shown by a series motor is significant.

To solve the problem, we propose an FQDC which allows the series motor to operate in parallel. In series, the armature and field windings are as shown in simplified Figure 2.

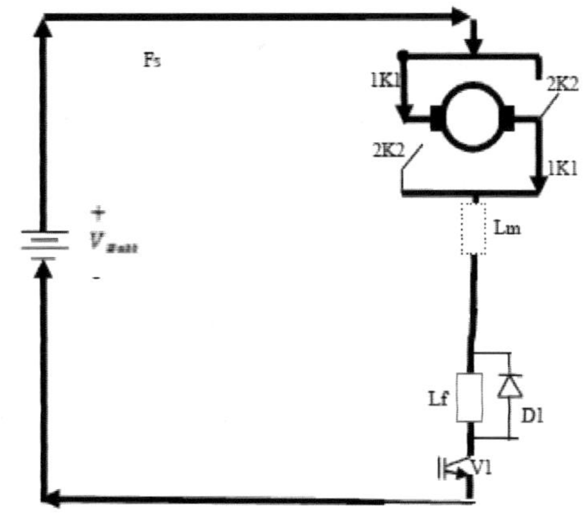

Figure 2. Simplified circuit in series

In series, the armature and field currents are both controlled by varying the PWM value of IGBTV1. But in parallel mode, the field current is separately controlled from the armature current as depicted in Figure 3. While the armature current is controlled by varying the PWM value of IGBT v2, the field current is controlled by varying the PWM value of IGBTV1. Thus, the field current can be changed or maintained independently.

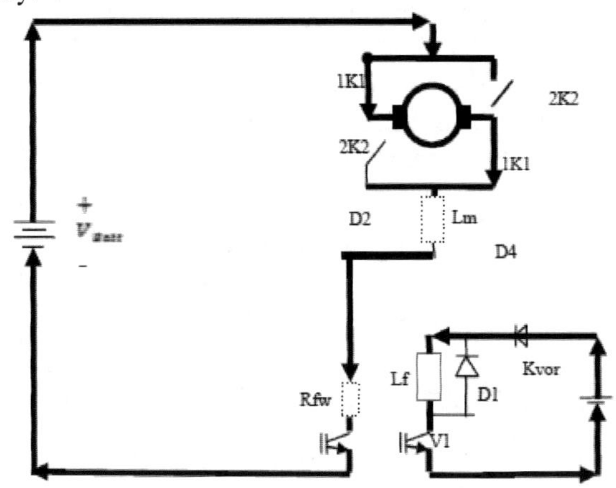

Figure 3. simplified circuit in Parallel mode

When a dc series motor running in series is loaded, its speed decreases which in turn reduces the Bemf voltage. According to Eq. 2. The decrease in Bemf voltage will increase the armature current as stated in Eq. 1. Since the armature and field windings are in series, the field current will also increase. This will increase the Bemf back and offset the earlier decrease caused by the speed drop. However, the Bemf of the motor in parallel connection drops a bit since the field current is separated from the armature current and the field current can be controlled to stay constant. Figure 4 compares the scenarios when the DC motor in series and in parallel when loaded.

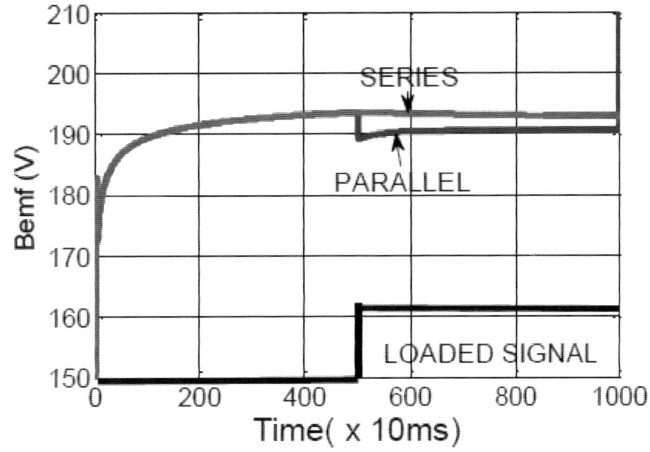

Figure 4. Bemf voltage when motor loaded

As we know the decrease in Bemf voltage will increase the armature current. For the DC motor in series connection, the field current will also increase by the same amount. This increases the Bemf again and reduces the armature current until they settle at new equilibrium values. Thus, the final increase in the armature and field current is not too obvious. However, for the parallel connection, the field current can be maintained separately. Thus, the increase in the armature current is more noticeable as illustrated in Figure 5. Figure 6 shows a slight increase in the field current of the series connection when the load is introduced. On the contrary, the Field current of the parallel connection is maintained. Since the field current of the parallel connection is separately controlled, it can be lowered, raised or maintained.

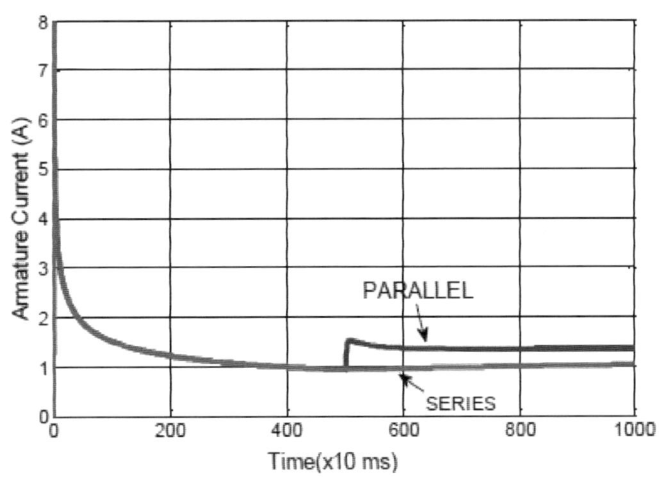

Figure 5. Armature current when the motor is loaded

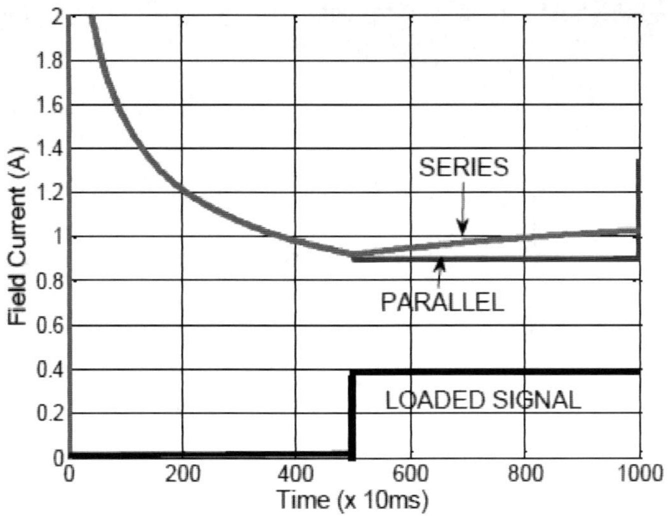

Figure 6. Field current when motor loaded

Since the increase in the armature current is more significant in the parallel connection, torque will also increase following Eq. (3). Figure 7 shows changes in the torque when a load is introduced for the series and parallel connections. For series connection, even though torque increases slightly, the load torque applied is much greater than the generated torque. As a result, the speed drops.

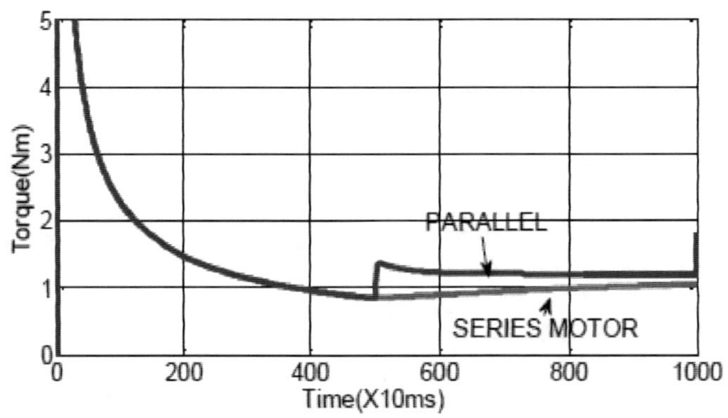

Figure 7. Torque motor when loaded

More specifically, for series connection, due to the applied load, the values of back emf, armature current, field current and torque will increase slightly. But since the counter torque of the load is greater than the slight torque gained, the speed drops. Thus Bemf will continue to drop so that the armature current, field current and torque will increase until the torque is equal to the load torque applied. At this point the speed of the motor will stop decreasing.

For parallel connection, the field current can be controlled separately thus the amount of torque produced can be adjusted to offset the applied torque. When the speed drops due to the load, the Bemf decreases. This increases the armature current while field current is kept constant. Therefore, the torque will increase. This new torque will offset the loading effect and maintain the speed. Figure 8 shows the speed of the motor in series and parallel connections when loaded. Using this approach, a DC EV driven by a series motor can overcome speed drop while climbing a steep hill.

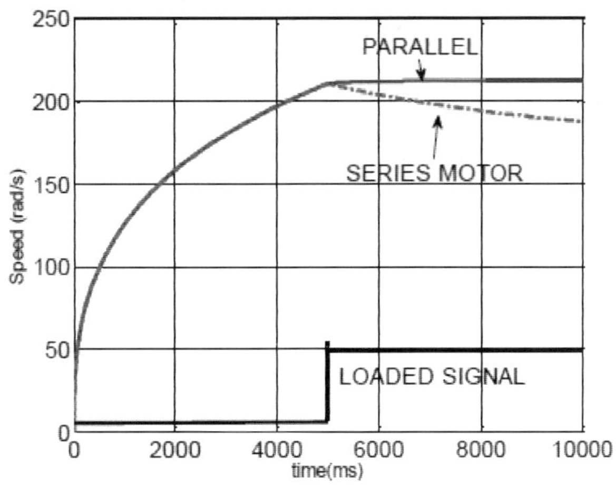

Figure 8. Motor speed when loaded

2.2. Control strategy in Parallel mode

In order to get a high torque, the armature current is set to the maximum. This can be achieved by setting the PWM value of IGBT V2 to the maximum. The field current is adjusted by referring to a Look Up Table so that the Bemf is minimized while the torque is maximized [4, 13]. The field current cannot be too low because it is important in creating magnetic field and flux. These two entities are crucial in producing torque [13].

2.2.1. Control Strategy during parallel mode The process of operating in parallel mode is briefly described in the flowchart of Figure 9. First, a change to the field current path is initiated so that the field current is separated and fed from a different power supply. Next, the armature current is set to the maximum, by firing IGBT V2 at the maximum PWM value. The the field current value is loaded from a look up table and sent to a PID controller.

The Look Up table contains the best field current values with respect to the speed of the motor. The PID controller will determine the PWM signal for IGBTV1. This PWM value is fed to the IGBT in the FQDC. Fig 10 describes the whole process in a block diagram.

2.3. Experimental and Simulation Result in Parallel mode

Simulated mathematical modeling and laboratory experiment are employed to test the effectiveness of the parallel mode on a 0.65 KW series motor. The MATLAB/Simulink mathematical model used for studying the parallel mode operation is shown in Figure. 11 while the hardware experimental setup is shown in Figure 12.

Figure 13 shows the DC EV simulation model developed using a 35kW motor operating at the limited maximum power of 22kW to drive its total weight of 1325kg. It is tested to climb an inclination as shown in Figure 14.

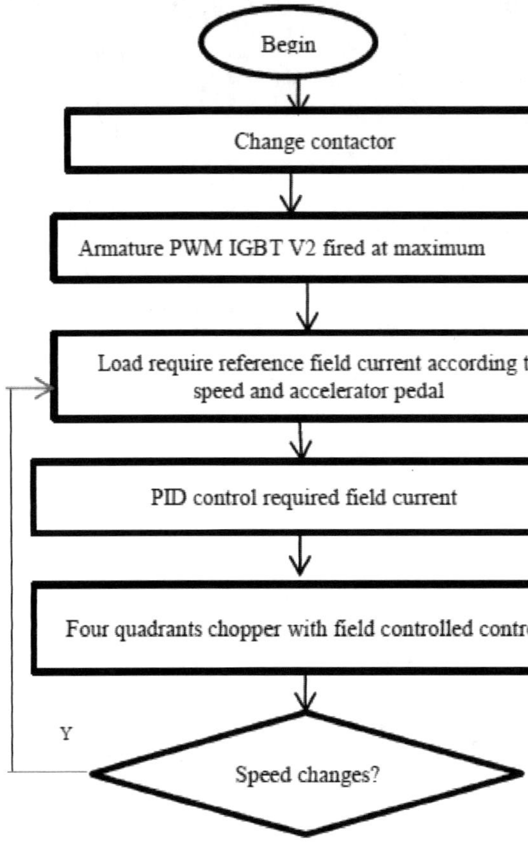

Figure 9. Flow Chart

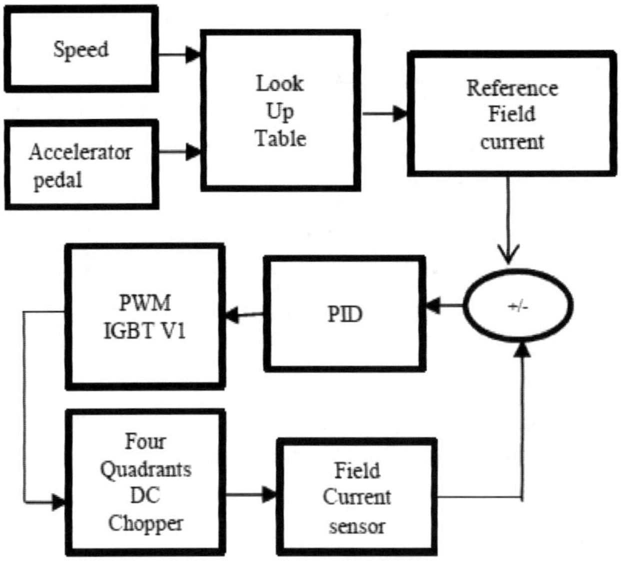

Figure 10. Operation Block diagram

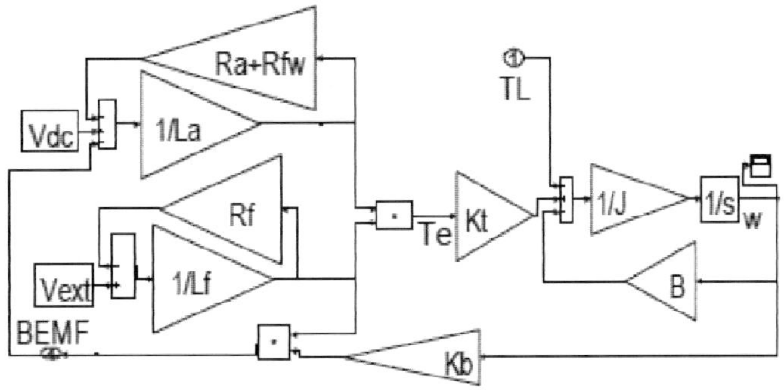

Figure 11. MATLAB/Simulink mathematical model

Figure 12. Experimental setup

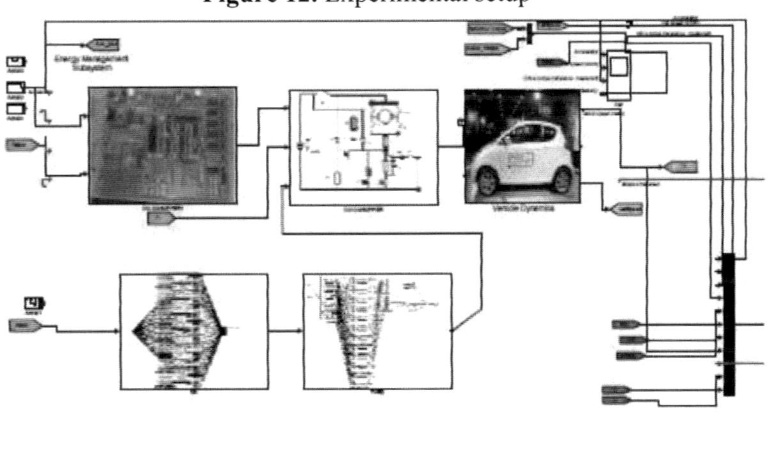

Figure 13. EV simulation model with FQDC and Parallel mode.

Figure 14. DC EV Inclination TEST

3. Results and Discussion

The simulated mathematical modeling and hardware experimental results are shown in Fig. 16. It is observed that when the climbing signal is initiated the speed of the motor decreases. The increase in armature and field current is insignificant. But when the parallel mode is activated, the field current drops while the armature current increases. These result in a higher torque, so the motor speed rises.

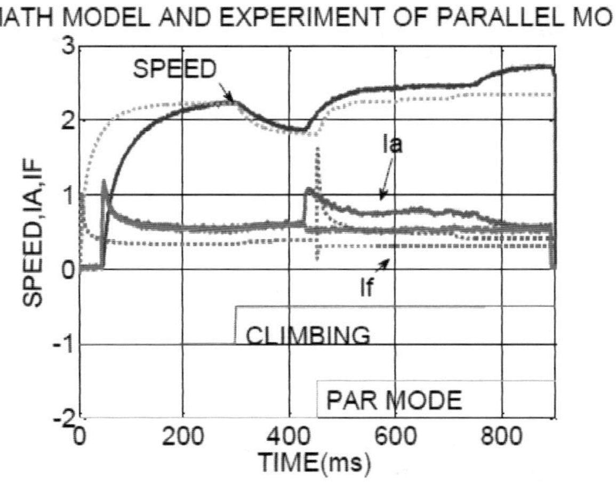

Figure 15. Simulation and experimental results

Fig. 15 simulation (dotted line) and experimental results (straight lines) of the current and speed of 0.65 KWatt for dc series motor in parallel mode of chopper operation is plotted. The MATLAB/simulink results of the DC EV driven by the novel chopper in Parallel mode is shown in Figure 16. At the beginning, the vehicle speed drops due to the climbing effect. When the parallel mode is activated, the speed of the vehicle increases as it overcomes the load effect.

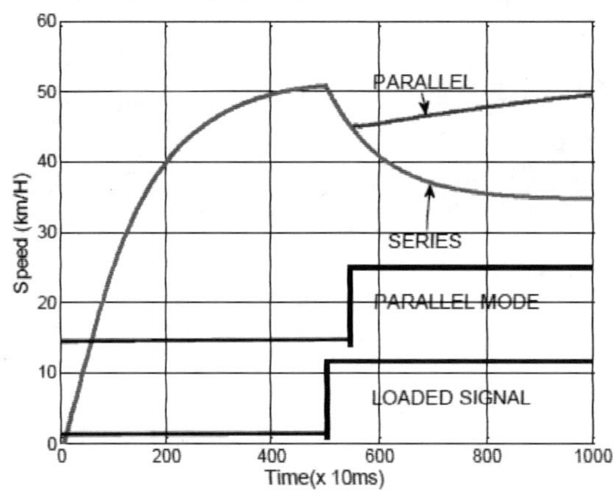

Figure 16. DC EV Test Simulation Results

4. Conclusions

DC drive series motor has a high potential to be utilized in EV. This is due to its simple design, low cost and excellent controllability. In summary, the performance of an optimized DC drive system with the proposed FQDC and Parallel mode is comparable to that of an AC drive and thus is very suitable for application in low cost EVs.

References

[1] Arof S, Yaakop NM, Jalil JA, Mawby PA, Arof H, editors. Series motor four quadrants drive DC

chopper for low cost, electric car: Part 1: Overall. 2014 IEEE International Conference on Power and Energy (PECon); 2014 1-3 Dec. 2014.

[2] Oak Ridge National Laboratory. Advanced Brush Technology for DC Motors 2009 [Available from: http://peemrc.ornl.gov/projects/emdc3.jpg.

[3] Arof S, Noor N, Alias F, Mawby P, Arof H. Investigation of Chopper Operation of New Series Motor Four Quadrants Drive DC Chopper for DC Drive Electric Car. *J Appl Environ Biol Sci.* 2017;**7**(3S):49-56.

[4] Arof S, Jalil J, Kamaruddin N, Yaakop N, Mawby P, Hamzah A, editors. Series motor four quadrants drive DC Chopper part2: Driving and reverse mode with direct current control. 2016 IEEE International Conference on Power and Energy (PECon); 2016: IEEE.

[5] Arof S, Diyanah N, Noor N, Radzi M, Jalil J, Mawby P, et al. Series motor four quadrants drive DC chopper: part 4: generator mode. *Progress in Engineering*. 2019:155-67.

[6] Arof S, Diyanah N, Yaakop N, Mawby P, Arof H. Processor in the Loop for Testing Series Motor Four Quadrants Drive Direct Current Chopper for Series Motor Driven Electric Car. Advanced Engineering for Processes and Technologies: Springer; 2019. p. 59-76.

[7] Arof S, Hassan H, Rosyidi M, Mawby P, Arof H. Implementation of series motor four quadrants drive DC chopper for DC drive electric car and LRT. *J Appl Environ Biol Sci.* 2017;**7**:73-82.

[8] Arof S, Diyanah N, Mawby P, Arof H, Yaakop NM. Low Harmonics Plug-in Home Charging Electric Vehicle Battery Charger Utilizing Multi-level Rectifier, Zero Crossing and Buck Chopper. Progress in Engineering Technology: Springer; 2019. p. 103-18.

[9] Arof S, Diyanah N, Noor N, Jalil J, Mawby P, Arof H. A new four quadrants drive chopper for separately excited DC motor in low cost electric vehicle. Progress in Engineering Technology: Springer; 2019. p. 119-38.

[10] Arof S, Diyanah N, Mawby P, Arof H. Study on Implementation of Neural Network Controlling Four Quadrants Direct Current Chopper. Advanced Engineering for Processes and Technologies: Springer; 2019. p. 37-57.

[11] Arof S, Khairulzaman AM, Jalil J, Arof H, Mawby P, editors. Self tuning fuzzy logic controlling chopper operation of four quadrants drive DC chopper for low cost electric vehicle. 2015 6th International Conference on Intelligent Systems, Modelling and Simulation; 2015: IEEE.

[12] K[1] MKA, Jalil J, Arof H, Mawby P. Artificial intelligence controlling chopper operation of four quadrants drive DC chopper for low cost electric vehicle. *ENERGY.* 2015;**1**:E2.

[13] Arof S, Diyanah N, Noor N, Rosyidi M, Said M, Khairulzaman AM, et al. Genetics Algorithm for Setting Up Look Up Table in Parallel Mode of Series Motor Four Quadrants Drive DC Chopper. Progress in Engineering Technology: Springer; 2019. p. 139-54.

Electromagnetic torque capabilities of permanent magnet flux- switching machine using modular rotor

Irfan Ali Soomro, Erwan Sulaiman, Hassan Ali Soomro, Faisal Amin

Faculty of Electrical and Electronic Engineering University of Tun Hussein Onn Malaysia

E-mail: eng.irfansoomro@gmail.com

Abstract. The use of a segmented rotor electric motor for lightweight electric vehicle applications has gained a lot of attention since the motor must have a high output torque and output power. Segmented rotor motors have performance indices that should be taken into consideration because not every machine is effective. Recently, permanent magnet flux switching machine (PMFSM) in salient rotor has become attractive for provide high torque and power. Undesirably, salient rotor inherited high iron loss and winding loss, Furthermore, PMFSM uses high PM volume and more flux leakage resulting in poor performance. This paper deals with a novel structure of modular rotor PMFSM with non-overlapping winding with minimal iron loss, high torque and high power. 2D finite element analysis (FEA) is used to investigate the performance of proposed motor in JMAG Designer 14.1. Three-phase operation of 12S/10P modular rotor PMFSM was designed and analysed based on the cogging torque, induced back emf, output torque and output power performances. The proposed motor securing the initial torque of 33.30 Nm and power of 12Kw at maximum current density of 30Arms/mm2. From the result it is concluded that modular rotor based PMFSM motor is suitable for light weight electric vehicles.

1. Introduction

Electric motors are essential part of both industrial and domestic appliances. Thus, most of the industrial and domestic applications demand more compact, better efficient, vigorous, light weight and low-priced electric motors. This leads a challenge for researchers to develop an advanced electric motor. Just a decade ago, Flux Switching machine (FSM) has pulled an interest of researchers because of its all excitation sources situated on the stator, come up with simple cooling systems, while the rotor comprises of single piece of iron, which is robust can be used for high speed applications. Flux switching machines are categorized into three kinds as shown in Figure 1, which are (i) field excitation (FE) FSM, (ii) permanent magnet (PM) FSM and (iii) hybrid excitation (HE) FSM. PMFSM and HEFSM both uses permanent magnet as excitation source which generates uniform magnetic flux. While in FEFSM, DC excitation has used which controls the magnet field density [1-3]. The working principle of FSM is conditioned on flux switching. For the better efficiency, easy cooling, robust rotor structure, light weight and low-priced makes FSMs most reliable and effective [4-8]. The flux switching machine (FSM) is known as one of the forms of salient rotor reluctance machine with a novel topology, merging the inductor generator principle [9, 10] and switched reluctance machine (SRM) [9].

In the flux switching machine all the active sources are placed on the stator such as both field and armature winding. Rotor consists of iron core with a simple shape and high strength. The term "flux switching" is to designate machines in which the stator teeth switches flux polarity resulting the motion of a salient pole rotor, and that's basic working principle of operation [11, 12].

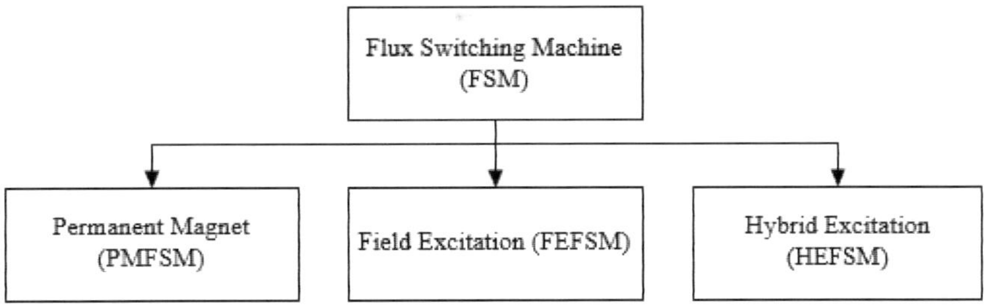

Figure 1. Categorization of FSM

In 1955, Rauch and Johnson had innovated a single-phase alternator [9]. From that time diverse designs of PMSFM has been presented. Figure 2 describes a PMFSM conventional design with the disadvantage of high volume of PM.

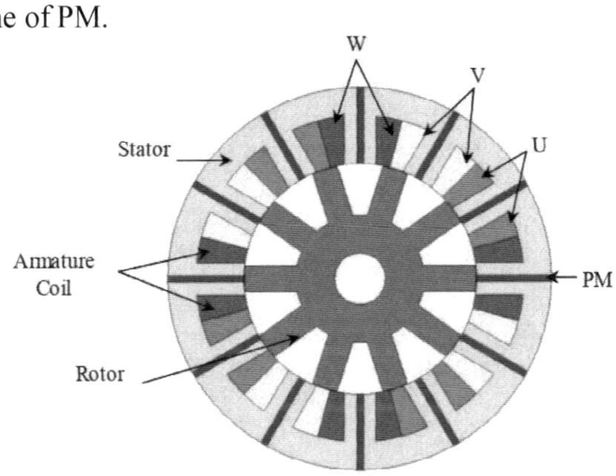

Figure 2. Conventional PMFSM

In order to utilize a smaller number of PM, an E-core topology is set up which replaced the stator poles alternately by stator tooth [13, 14]. Radial and circumferential directions of PM are authorized in E-core configuration. All radial PM are set up alternately while circumferential magnets are remain identical [15, 16]. Furthermore, to increase the armature coil area the middle teeth of E-core has removed which introduces the new C-core topology [17]. Lately another design has introduced in PM with segmented rotor [18, 19]. But it carries a drawback of ambiguous operation of high-speed rotor. Moreover, distribution of flux is not uniform due to the flux leakage at the tip of the PM which decreases the efficiency of motor.

In this paper, a novel topology of PMFSM with 12S-10P modular rotor has been investigated. The main advantage of modular rotor is to reduce the iron loss with marginal reduction in output torque as shown in Figure 3. Modular rotor also known as segmented rotor.it helps to reduce the weight of rotor over salient rotor. Modular rotor firstly was introduced in reluctance machines. Due to the shortest paths produces along the iron core through modular rotor decreases iron losses. The flux flows from one rotor tooth to another rotor tooth [19, 20]. In order to avoid flux leakage through stator core an iron bridge has established. Furthermore, performance of the initial modular PMFSM design has been investigated by 2D FEA (finite element analysis). The performance of the motor is intercept specially at maximum current densities. An optimization technique will be implemented for the better performance of the motor [21].

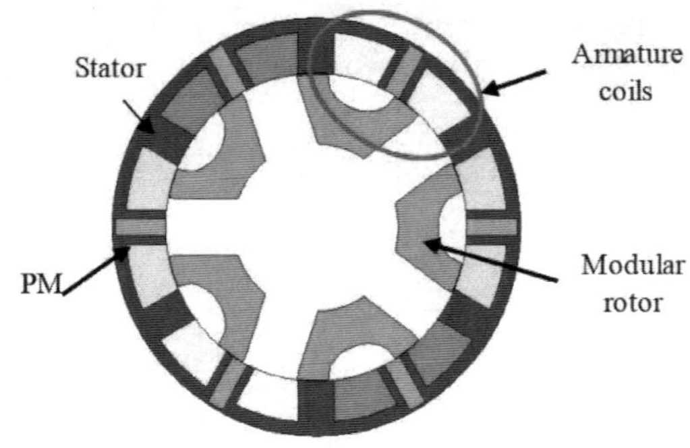

Figure 3. Modular rotor based PMFSM

2. Research method

The novel modular rotor PMFSM design description and limitations are shown in Table 1. Electrical restrictions associated with inverter are set in the same way. Additionally, the current density for the armature coil are varies from 0 A_{rms}/mm^2 to 30 A_{rms}/mm^2. Based on the structure of motor and current density air cooling is feasible for motor. The complete process to accomplish the objective of this paper is shown in Figure 4.

From the arrangement, armature coil is sandwiched in permanent magnets, which are non-overlapping winding. For the purpose of design and scrutiny 2D Finite Element Analysis (FEA) tool is implemented in JMAG Designer 14.1 introduced by Japan Research Institute (JRI). Firstly, the main parts such as stator, rotor, PM and armature coils of the proposed PMFSM modular rotor are built in geometry editor consequently materials, boundary conditions and machine circuitry of machine are set in JMAG designer. The material employed for stator and rotor are steel 35H210, while NEOMAX-35AH is set for the PM.

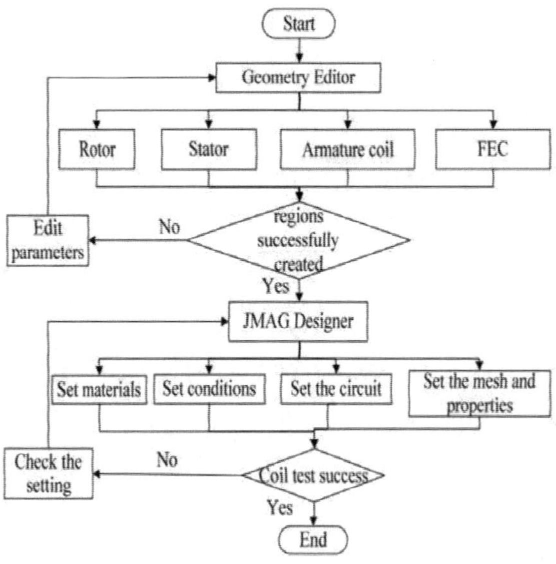

Figure 4. Flow chart for design process

Moreover, to validate the operating principle and the positions of armature coil phase of the proposed design coil arrangement tests are conducted. After the authorisation of the operating principle and coil arrangement test, numerous characteristics of modular rotor PMFSM on no load and at load were examined respectively. In order to inspect the numerous executions of the novel proposed motor at no-load armature current density J_a was set to 0 A_{rms}/mm^2. Furthermore, flux distributions lines, back emf, cogging torque and torque characteristics and efficiency were evaluated.

3. Results and discussions

The outcomes of the proposed 12S-10P modular rotor based PMFSM have been explored in this section. In order to follow through the process results are showed in figures, graphs for better understanding. Table 1 shows the design specifications, parameter requirements and electrical restrictions are provided. There are three methods being discussed in this work as follows.

Table 1. Modular PMFSM design parameters

Items	PMFSM with Modular rotor
Number of rotor poles	10
Stator Outer radius (mm)	75
Stator tooth width (mm)	12.5
Rotor tooth width (mm)	10
Stator back iron depth (mm)	11
Motor stack length (mm)	70
Air gap (mm)	0.3
Radius of rotor (mm)	89.7
Armature slot area (mm^2)	300.8
No. of turns per armature coil slot	44
PM volume (Kg)	0.3

3.1. Coil arrangement test

Coil arrangement test was achieved for each coil individually to validate the armature coil location and operating principle of modular rotor PMFSM. Primarily, Coil arrangement test was achieved for each armature coil 1 to armature coil 12 individually to validate the armature coil location and operating principle of modular rotor PMFSM. The main reason for this process is to categorize the like pattern flux linkage produced in each coil. Furthermore, likewise flux linkage pattern have grouped one another. Consequently, three different phase linkage patterns established namely U, V and W phase as shown in Figure 5. From the figure, it is observed that the flux linkage is not symmetry it is due to the segmental rotor geometry and power loss in relation to stator is unaligned with flux source. Proposed design can further be improved by using optimization method.

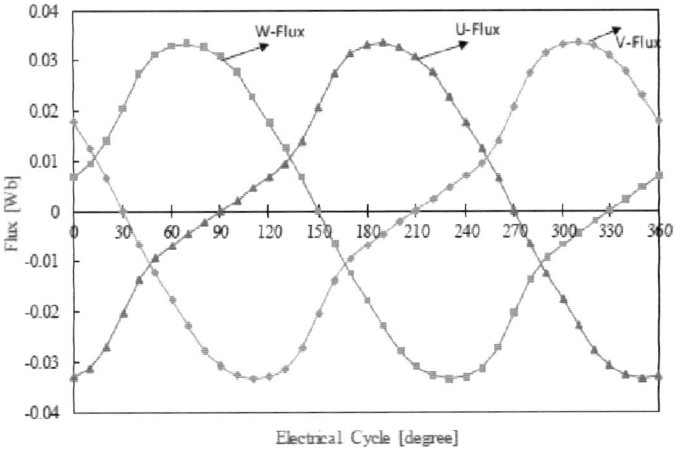

Figure 5. U, V and W phase Flux linkage

IOP Conf. Series: Materials Science and Engineering 1127 (2021) 012026 doi:10.1088/1757-899X/1127/1/012026

3.2. Cogging torque

Cogging torque is collaboration of the armature winding and PM on the stator with rotor teeth. The performance of the motor is affected by the cogging torque if the value is higher than the acceptable value which is 10% of the average torque. The cogging torque analysis for PMFSM modular rotor is shown in Figure 6. From the plot it is obvious that modular rotor PMFSM configuration generates peak to peak cogging torque of 12Nm. It is essential to mention here that the energy variation within the motor was caused by the rotation of rotor results in the cogging torque. The shape of the cogging torque is not uniform throughout because of the odd number of rotor segments regarding the magnetic pull.

3.3. Induced or Back-EMF

The production of voltage in an armature coil at no load condition due to magnetic flux changes through it is known as induced back-emf. As usual, this is represented as a continuous wave that should be exact sine wave without harmonics. However, harmonics seems unavoidable but should present negligible distortions. For the proposed motor, at no load condition and rotating speed of 500 rev/min, 42 V induced voltage generated by the motor is plotted in Figure 7.

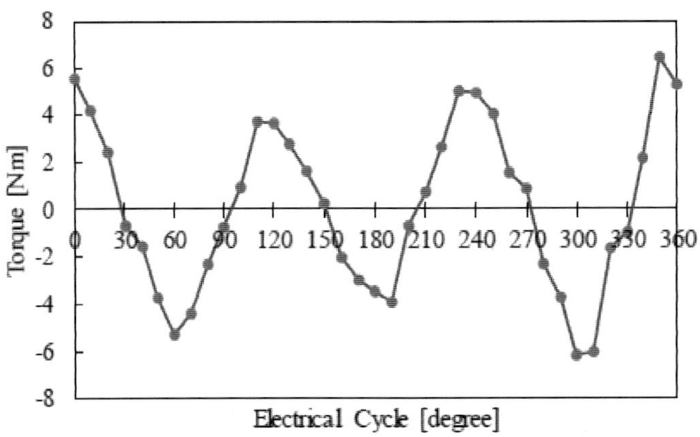

Figure 6. Cogging of Modular rotor PMFSM

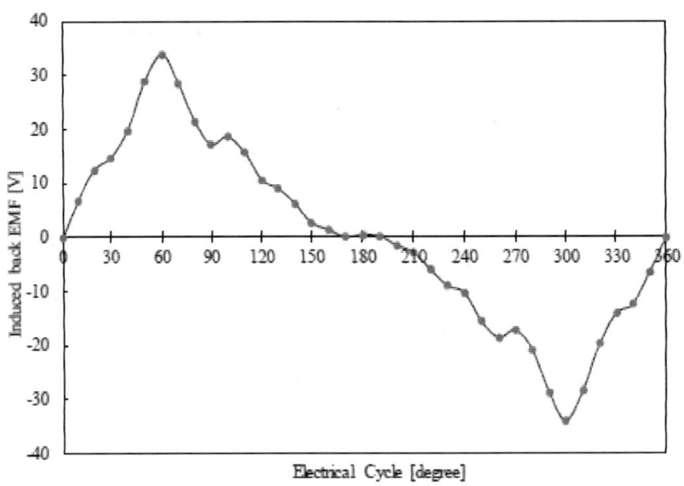

Figure 7. Induced back-emf of Modular rotor PMFSM

3.4. Torque and power at various current densities Ja

The capability of an electric motor is determined by its performance under the given load condition. The plot of the output torque versus armature current density J_a for the proposed motor is shown in Figure 8. In the plot J_a is varies from 0 A_{rms}/mm^2 to 30 A_{rms}/mm^2 and its changes in torque are examined. It is seen from the plot that torque increased with increase in the value of J_a. Thus, maximum output torque achieved is 33.30 Nm at current density of 30 A_{rms}/mm^2.

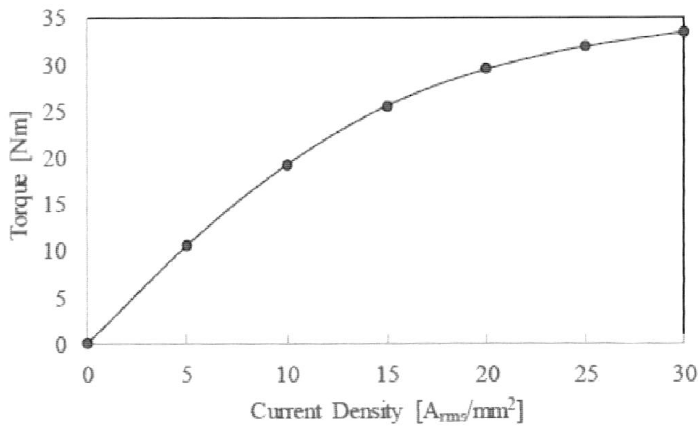

Figure 8. Torque versus armature current density (A/mm^2)

Furthermore, the plot of the output power versus armature current density of the proposed motor is investigated as shown in Figure 9. Form the plot it is clear observed that output power raised as the current density increases. The output power of the proposed motor is observed 10.53Kw. The power of the flux switching machine can be calculated by using Equation (1).

$$P_{out} = \frac{speed*torque}{60}\left(\frac{2\pi}{1000}\right) \tag{1}$$

Where, speed is the maximum speed of the motor after applying the load. Which can be calculated as given in Equation (2).

$$Speed = \frac{V_{max}}{flux_{max}} \tag{2}$$

Where, V_{max} is three phase maximum voltage. Power is directly proportional to the speed, as the speed of motor increases its power increases as shown in figure 10. Form the figure it is observed that after certain speed power start reducing because of the saturation point. Further improvement can be made by optimization of the motor.

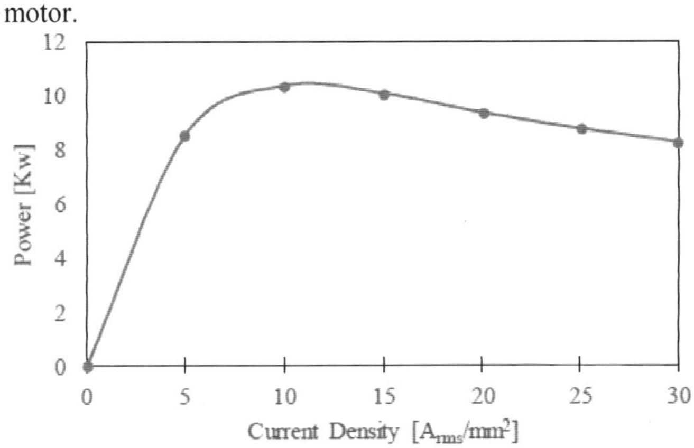

Figure 9. Power versus armature current density (A/mm^2)

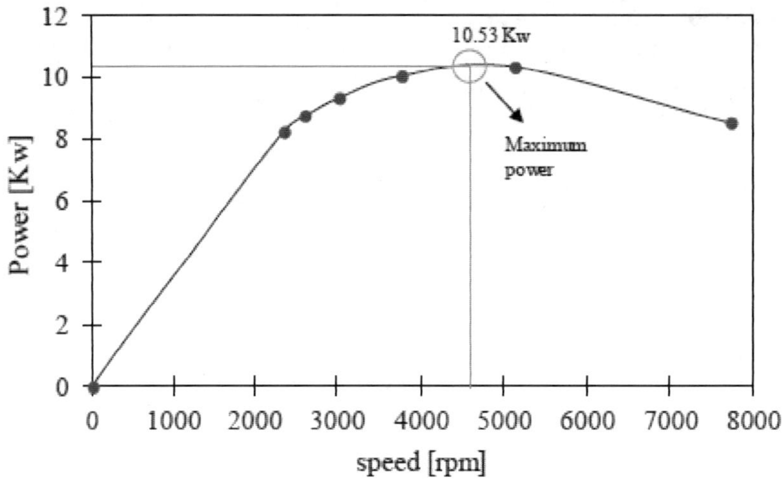

Figure 10. Power versus Speed characteristics

4. Conclusions

Flux switching motor employing modular rotor has been characterised in this paper. The design considered a three- phase arrangement with permanent magnet flux excitation. In this paper a novel modular rotor PMFSM with non-overlapping winding using specification and limitations of the conventional Salient rotor PMFSM. The performance of the modular rotor PMFSM investigated in terms of induced back emf, cogging torque, and average torque and power using JMAG Designer 14.1. thus, the output torque and power of the proposed design is 33.30 Nm and 10.53 Kw respectively. Moreover, this design can be further altered using optimization technique to obtain a optimal torque. Number of PM are reduced in the proposed design which makes it a good option for light weight electric vehicles.

Acknowledgements

"Research and Management Centre, University Tun Hussein Onn Malaysia (UTHM), and FRGS Grants Vote number 1651 under the Malaysian Ministry of Higher Education (MOHE) are completely supporting this research work.

References

[1] Sulaiman E, Kosaka T, Matsui N. High power density design of 6-slot–8-pole hybrid excitation flux switching machine for hybrid electric vehicles. *IEEE Trans Magn*. 2011;**47**(10):4453-6.

[2] Sulaiman E, Kosaka T, Matsui N. Design Optimization and Performance of a Novel 6-Slot 5-Pole PMFSM with Hybrid Excitation for Hybrid Electric Vehicle. *IEEJ Transactions on Industry Applications*. 2012;**132**(2):211-8.

[3] Zhu ZQ, editor Switched flux permanent magnet machines — Innovation continues. 2011 International Conference on Electrical Machines and Systems; 2011 20-23 Aug. 2011.

[4] Pollock C, Brackley M. Comparison of the acoustic noise of a flux-switching and a switched reluctance drive. *IEEE Trans Ind Appl*. 2003;**39**(3):826-34.

[5] Pollock C, Pollock H, Barron R, Sutton R, Coles J, Moule D, et al., editors. Flux switching motors for automotive applications. 38th IAS Annual Meeting on Conference Record of the Industry Applications Conference, 2003.; 2003 12-16 Oct. 2003.

[6] Pollock C, Pollock H, Brackley M, editors. Electronically controlled flux switching motors : a comparison with an induction motor driving an axial fan. IECON'03. 29th Annual Conference of the IEEE Industrial Electronics Society (IEEE Cat. No.03CH37468); 2003 2-6 Nov. 2003.

[7] Pollock C, Wallace M, editors. The flux switching motor, a DC motor without magnets or brushes. Conference Record of the 1999 IEEE Industry Applications Conference. Thirty-

Forth IAS Annual Meeting (Cat. No.99CH36370); 1999 3-7 Oct. 1999.

[8] Pollock H, Pollock C, Walter RT, Gorti BV, editors. Low cost, high power density, flux switching machines and drives for power tools. 38th IAS Annual Meeting on Conference Record of the Industry Applications Conference, 2003.; 2003 12-16 Oct. 2003.

[9] Rauch SE, Johnson LJ. Design Principles of Flux-Switch Alternators [includes discussion]. *Transactions of the American Institute of Electrical Engineers Part III: Power Apparatus and Systems*. 1955;**74**(3):1261-8.

[10] Walker JH. The theory of the inductor alternator. *Journal of the Institution of Electrical Engineers - Part II: Power Engineering*. 1942;**89**(9):227-41.

[11] Husin ZA, Sulaiman E, Kosaka T, editors. Design studies and effect of various rotor pole number of field excitation flux switching motor for hybrid electric vehicle applications. 2014 IEEE 8th International Power Engineering and Optimization Conference (PEOCO2014); 2014 24-25 March 2014.

[12] Kannan S, editor Novel rotor and stator swapped switched reluctance motor. 2012 IEEE International Conference on Power Electronics, Drives and Energy Systems (PEDES); 2012 16-19 Dec. 2012.

[13] Chen JT, Zhu ZQ, Iwasaki S, Deodhar R, editors. A novel E-core flux-switching PM brushless AC machine. 2010 IEEE Energy Conversion Congress and Exposition; 2010 12-16 Sept. 2010.

[14] Hoang E, BEN AHMED H, Lucidarme J, editors. Switching flux permanent magnet polyphased synchronous machines. EPE 97; 1997 1997-09-08; Trondheim, Norwayhttps://hal.archives-ouvertes.fr/hal-00533004/document https://hal.archives-ouvertes.fr/hal-00533004/file/hoang_EPE97.pdf.

[15] Soomro IA, Sulaiman E, Soomro HA, Amin F, editors. Comparative Study on a Modular Rotor and AlCiRaF Permanent Magnet Flux Switching Machine. 2019 IEEE International Conference on Automatic Control and Intelligent Systems (I2CACIS); 2019 29-29 June 2019.

[16] Yang H, Lin H, Zhu ZQ, Fang S, Huang Y. A Winding-Switching Concept for Flux Weakening in Consequent Magnet Pole Switched Flux Memory Machine. *IEEE Trans Magn*. 2015;**51**(11):1-4.

[17] Chen JT, Zhu ZQ, Iwasaki S, Deodhar R, editors. Comparison of losses and efficiency in alternate flux-switching permanent magnet machines. The XIX International Conference on Electrical Machines - ICEM 2010; 2010 6-8 Sept. 2010.

[18] Soomro IA, Sulaiman E, Soomro HA, editors. Modular rotor Based Permanent Magnet Flux Switching Machine for Light Weight EV. 2019 IEEE 15th International Colloquium on Signal Processing & Its Applications (CSPA); 2019 8-9 March 2019.

[19] Zulu A, Mecrow BC, Armstrong M. Permanent-Magnet Flux-Switching Synchronous Motor Employing a Segmental Rotor. *IEEE Trans Ind Appl*. 2012;**48**(6):2259-67.

[20] Jenal M, Sulaiman E. Comparative study on a new permanent magnet flux switching machine configuration over segmental and salient rotor structure. *ARPN Journal of Engineering and Applied Sciences*. 2015;**10**(19):8846-52.

[21] Sulaiman E, Ahmad MZ, Kosaka T, Matsui N. Design Optimization Studies on High Torque and High Power Density Hybrid Excitation Flux Switching Motor for HEV. *Procedia Engineering*. 2013;**53**:312-22.

IOP Conf. Series: Materials Science and Engineering 1127 (2021) 012027 doi:10.1088/1757-899X/1127/1/012027

Influence of Various Tooth Shape Topology for Single-Phase Permanent Magnet Flux Switching Machine

Laili Iwani Jusoh*, Erwan Sulaiman, Md Zarafi Ahmad, Irfan Ali Soomro, Hassan Ali Soomro

Research Center for Applied Electromagnetics,University of Tun Hussein Onn Malaysia, Locked Bag 101, Parit Raja, Batu Pahat, Johor, Malaysia

E-mail: lailiiwani@gmail.com

Abstract. Permanent magnet flux switching machine (PMFSM) has the advantages of solid design structure, high torque performances, and high efficiency. The method to increase performance and decrease the cost of the motor through a selection of numbers of slot and pole is a major issue that needs to be solved. Therefore, to further improve the electromagnetic torque, a novel 4-4S-8P multi-tooth, 4S-8P segmental-tooth and single-tooth PMFSM machine respectively is proposed based on the slot pole analysis performance. Hence, the performance is including the flux linkage, cogging torque, back-electromotive force (back-emf), output torque, and power performance. The results display that the various tooth shape topology have a large impact on the performance design. JMAG Designer version 14.3 under 2D finite element analysis (2D-FEA) are used to complete the analysis. The results show that the right combination of a single-tooth 4S-8P PMFSM structure is the best choice according to the high performance of torque analysis.

1. Introduction

The Flux Switching Machine (FSM) was started in the mid- 1950s and first published [1]. Various new machines have become FSM topologies in a broad range of applications, from commercial vehicles to wind, wind and space, and has become popular lately. A new FSM motor are very easy to construct together by the power of an electronic controller. It requires only two semiconductor switches that drive power and has the potential for high volume, cost-efficient products. Furthermore, the flux switching has an electronically commutated brushless motor that has a much longer life offers highly accurate and more flexible torque and a higher speed at no additional cost [2]. FSM has a variable speed application that has been considered a possible candidate for any potential motor design because of their major improvements [3].

The PMFSM has a brief history and it is an unknown type of PM machines. There were resurrected research interests in PMFSM, certainly due to the number of anticipated benefits. Given that the dynamic parts, for example, the armature windings and the PM configured in the stator directly, but adequate cooling of the machine is easily associated. Moreover, additional advantages, like robust rotor structure, high torque, and flux densities, high performance and improve flux weakening ability are systematically analysed and monitored for different applications. Most recently, the design machines with multi-tooth, segmental and single-tooth design, have been investigated [2, 4, 5].

This analysis, to improve further the performance of motor, a proposed topology 4-4S-8P multi-tooth, 4S-8P for segmental and single-tooth PMFSM design respectively is proposed based on the slot pole analysis performance. Arranging of the paper analysis is being monitored. Formative process of the various tooth shape topology is illustrated, and consequently, operation principle of various tooth shape is compared by 2-D finite-element analysis (2-D-FEA) in the second section. Subsequently, performance comparisons between all three design machines, in conditions of no-load for example, flux linkage and back-EMF performance, and in load performance such as an electromagnetic torque and power are performed in the section 3. Lastly, highlighted conclusions have been illustrated in the section 4.

2. Operation Principle of Various Tooth Shape Topology for Single-Phase Permanent Magnet Flux Switching Machine

Primary machine geometrical dimension is identical in the. Fig. 1. It displays the enlarged motor with the original size of the main machine parts including for example in the space between the rotor and stator, the rotor length and armature coil, and stator yoke diameter.

This model, from a practical point of view, the stator diameter for all the three motors has been fixed at 85 mm. The requirements and characteristics of all three design of PMFSM are presented in the table 1. The cross-section of the multi- tooth, single-tooth and segmental-tooth design has been shown in Fig. 2. Eventually, the design was formed by using a JMAG Software Designer version 14.3.

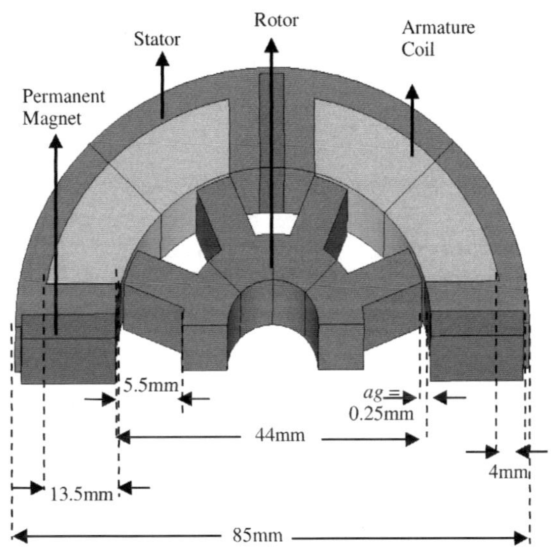

Figure 1. Main machine dimension of the proposed PMFSM

Table 1. Baseline Design Specification of PMFSM

Item	PMFSM
phase	1
Pole/slot	4S-8P
Air gap (mm)	0.25
Current density (mm²)	3.5
Stator Radius (mm)	42.5
Rotor Radius (mm)	22
Stack length (mm)	30
Permanent magnet (g/slot)	14
Stator yoke (mm)	4.5
DC Voltage (V)	36

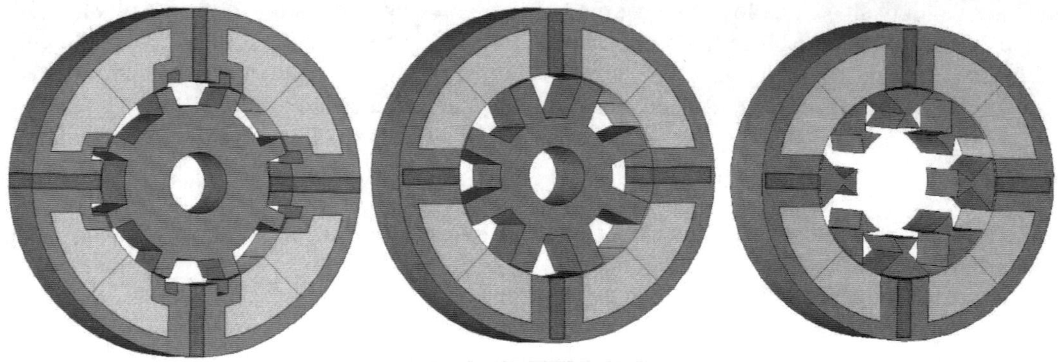
(a) (b) (c)

Figure 2. The topology of the purpose design motor in PMFSM, a) Multi-tooth, b) Single-tooth, c) Segmental.

Design specifications for dimensions of the motors and the workflow for 2D analysis are illustrated in the Fig. 3. The design specification and conditions for all the design motor should be similar. The project implementation has been divided into two phase which are the design and analyse the performance of PMFSM respectively. Geometry editor is applied to design all the motor part separately for example in the rotor, stator and armature coil whereas the simulation and setting of condition of the motor formed by utilizing the JMAG-Designer.

2.1. The Possible Combinations of Stator and Rotor Pole Number for Single-Phase Multi-tooth Design

In order to achieve high performance design, the best possible combination of the number of poles needs to be set [6]. Therefore, some relevant parameters are considered to propose the potential combination of the stator and rotor poles number [7]. This slot and poles combination can be identified using (1) and (2).

$$Ns = km\ (k=2, 4, 6\ldots) \quad (1)$$

$$Nr = Ns\ (2n-1) + 1 \quad (2)$$

The number of stator and rotor poles will be the *Ns* and *Nr*, respectively, while number of phases is *m*. More generally, the number of rotor pole and slot pole especially for single phase should be even in order that the machine can be designed such that a phase winding comprises a coil that has a phase shift of 180 electrical degree among the coils.

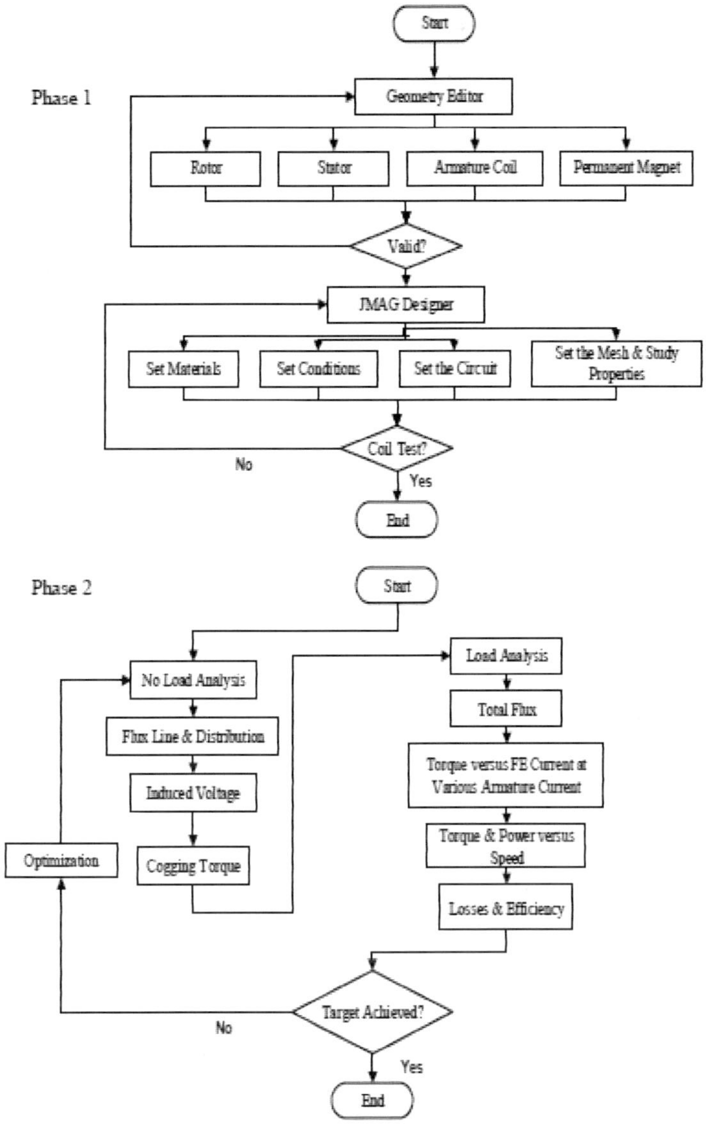

Figure 3. Analyse Performance Based on JMAG Designer with 2D FEA Analysis

2.2. The Possible Combinations of Stator and Rotor Pole Number for Single-Phase Segmental-tooth and Single- tooth Design

Preliminary, according to design of single-tooth and segmental PMFSM, the number of stator slot should be even multiple of phase numbers. In this aspect, the relationship between stator slot number, N_s and rotor pole number, N_r is [8];

$$N_r = N_s(1 \pm \frac{k}{2q}) \quad (3)$$

Therefore, the symbol of (k) and q is a natural entity with its value of 1,2,3 and by the number of phases respectively. Furthermore, the electric frequency (fe), of a proposed motor can be designated as (4).

$$f_e = N_r * f_m \quad (4)$$

Consequently, *fe* and *fm* is the electric frequency and the mechanical rotation frequency. The current of armature coil (I_a) and number of turns (N_a) is determined through the (5) and (6), separately whereas considering diameter specification used for the copper is 0.56mm.

$$I_a = \frac{J_a \alpha S_a}{N_a}\sqrt{2} \quad (5)$$

$$N_a = \frac{\alpha S_a}{\pi\left(\frac{A_{copper}}{2}\right)^2} \quad (6)$$

Subscripts 'a' indicates armature coil, J_a, S_a, and A_{copper} are current density, slot area (mm^2), and area of copper, respectively. Through the analysis, the appropriate electrical limit to the inverter is 240V DC bus voltage and 2.47 Arms for the maximum current of the inverter, while Ja 10 Arms/mm^2 are set to correspond to the current density in the armature coil. Since the design of the multi-tooth structure is different, the slot area of 207.9mm^2 for multi-tooth design and 216.88mm^2 for the single-tooth and segmental-tooth, respectively. In addition, the particle volume is set at 14g maximum for each slot with using an irreversible NEOMAX-35AH material. In the meantime, for the stator and the rotor, an electromagnetic 35A250 type of steel is used. To ensure the flux flows uniformly without any leakage from the stator to rotor, the motor's filling factor (α) are set to 0.5. The power is found using equation (7), and that is proportional to the torque and speed analysis.

$$P = \frac{2\pi}{60} * T * S \quad (7)$$

3. Comparison of the Proposed Design

All the three design performance the comparison is done in the open and close circuit circumstances. The open circuit test includes flux linkage and induced voltages (back-emf) whereas close circuit analysis consists of torque at different armature current densities while torque and power at various speeds respectively.

3.1. Magnetic Flux Characteristics

Before observing the flux linkage performance, the coil arrangement test is analysed on each armature coil separately in order to verify that the principle operation of the PMFSM and to set the position of each phase of the armature coil. By set a rotating rotor at speed of 500 r/min and zero armature current, the magnetic flux linkage is generated by magnitude of PM.

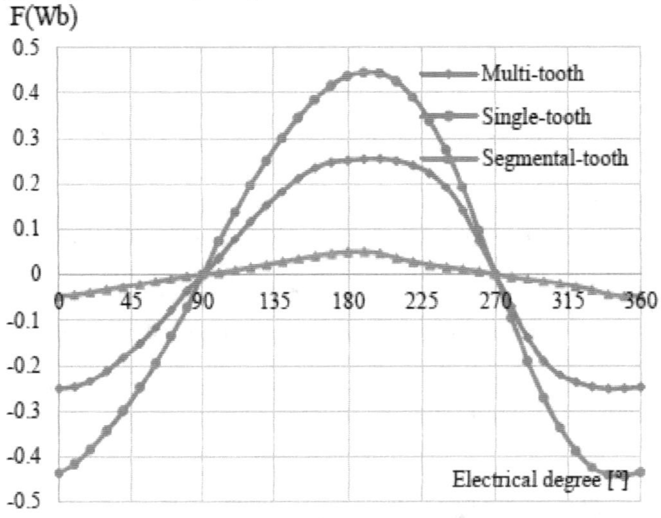

Figure 4. Flux linkage at armature coils.

IOP Conf. Series: Materials Science and Engineering 1127 (2021) 012027 doi:10.1088/1757-899X/1127/1/012027

Consequently, Fig.4 shown the flux linkage for multi-tooth, single-tooth, and segmental-tooth design and findings showed that single-tooth design has the largest flux linkage which it is 0.45Wb supported by multi-tooth design with 0.25Wb. The lowest flux of approximately at 0.05Wb is produced by segmental-tooth design. Initially, flux distortion occurred due to the higher back-emf. The higher back-emf will encourage higher flux distortion. However, the future optimization method analysis can perform a balance magnetic flux.

3.2. Back EMF Characteristics

Results achieved of back-EMF are presented in the Fig.5. From all the three design, waveform of back-EMF is distorted, and that is occurred due to the higher flux linkage performance. The voltages generated by the design of single-tooth are higher in comparison to other design motor roughly at 163V. In general, the back-emf flux is linearly in performance of flux linkage analysis. When the magnitude of the flux linkage increases, the back-emf is will also increase. Furthermore, higher back-emf may be used for regenerative braking systems for charging source. Moreover, voltage from back-emf is even smaller than the voltage applied to the motor that makes it easy to ensure protection due to some faults [9].

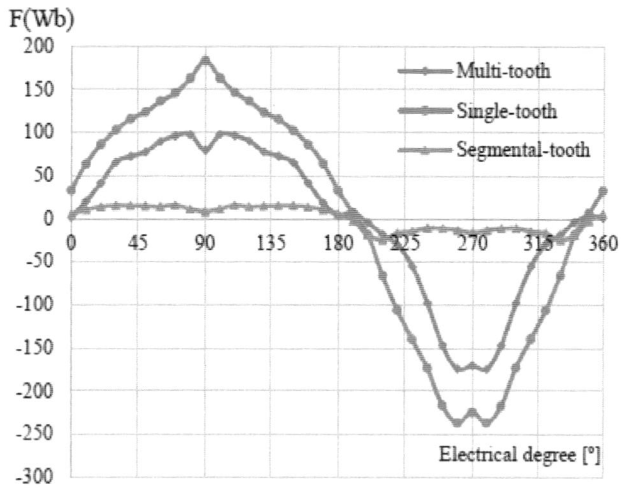

Figure 5. Back-emf at no load performance.

3.3. Load Performance

As a final analysis of load performance, armature current density is set at the maximum current condition. Hence, Fig.6 and Fig.7 proved the result of the torque and power output performance at minimum current density, namely Ja 0 $Arms^2$ to maximum current density, which is Ja 10 $Arms^2$. It proved that the maximum torque and power performance is coming from the single-tooth design motor with approximately 4.1Nm and 166W, accordingly followed by multi-tooth and segmental-tooth design.

IOP Conf. Series: Materials Science and Engineering 1127 (2021) 012027 doi:10.1088/1757-899X/1127/1/012027

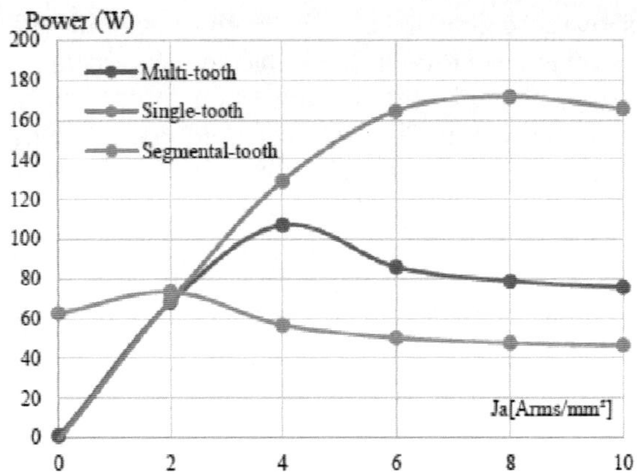

Figure 6. Simulation power at load performance.

However, the increment of power for segmental-tooth design is seem not to be linear may be the cause of the iron and copper losses occur. Meanwhile the design of single-tooth and multi- tooth appears to produce a high torque linear value when zero current density at Ja 0 and slightly increase with high current value at Ja 10. Thus, plot in linear demonstrates that design has ability to achieve very high torque value with increase the current density. A multi-tooth design has been found that 75W of power and 2.1Nm torque performance respectively, compare to the segmental-tooth at 46W and 0.7Nm power and torque performance respectively. This condition revealed that the high value of flux linkage, will develop greater torque performance. It remains, the overall effect on the proposed designs are visualized in Table 2.

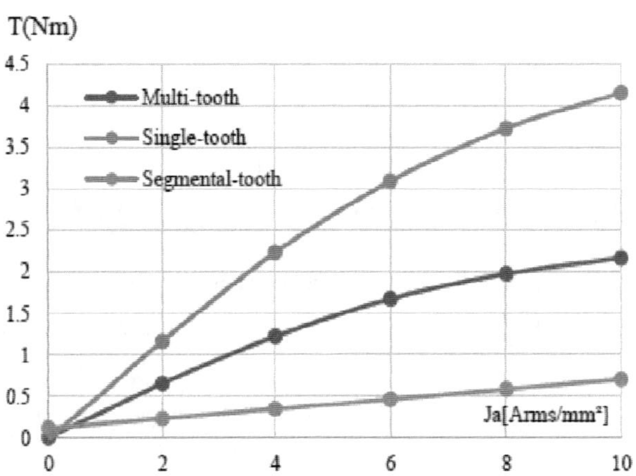

Figure 7. Simulation torque at load performance.

Table 2. Comparison Performance of Various Tooth Shape Topology

Item	Unit	Multi-tooth	Single- tooth	Segmental- tooth
Flux linkage	Wb	0.25	0.45	0.04
Torque	Nm	2.16	4.14	0.70
Power	W	75.64	165.55	46.34
Speed	Nm	333.70	381.00	626.50
Torque ripple	Nm	4.48	2.13	13.8

4. Conclusions
In the present paper, the impact of various tooth shape topology for single-phase permanent magnet flux switching machine has been presented. Three topologies, multi-tooth, single-tooth and segmental-tooth with concentrated windings were examined and compared. To identify the best design among them, especially in terms of flux, torque and power, no load and load test have been carried out. Builds on the 2-D FEA, the single-tooth design has been selected for the future analysis due to the highest torque of 4.1Nm, the largest flux linkage of 0.45Wb, and power of 166W. Furthermore, the voltage of back-emf for this design was considered proportionate and prevalent in motor design due to high performance of flux linkage.

Acknowledgements
These research studies were partly financed in part by the Postgraduate Studies Centre (CGS), the Department of Higher Education of Malaysia (MOHE) and the University of Tun Hussein Onn Malaysia (UTHM).

References
[1] Pollock C, Pollock H, Barron R, Sutton R, Coles J, Moule D, et al., editors. Flux switching motors for automotive applications. 38th IAS Annual Meeting on Conference Record of the Industry Applications Conference, 2003.; 2003 12-16 Oct. 2003.
[2] Khan F, Sulaiman E, Omar MF, Jenal M, editors. Performance comparison of wound field flux switching machines. 2015 IEEE Conference on Energy Conversion (CENCON); 2015: IEEE.
[3] Khan F, Sulaiman E, Ahmad MZ. Review of Switched Flux Wound-Field Machines Technology. *IETE Technical Review*. 2017;**34**(4):343-52.
[4] Chen JT, Zhu ZQ, Howe D. Stator and Rotor Pole Combinations for Multi-Tooth Flux-Switching Permanent-Magnet Brushless AC Machines. *IEEE Trans Magn*. 2008;**44**(12):4659-67.
[5] Omar MF, Sulaiman E, Khan F, Romalan GM, Hassan MK, editors. Performances comparison of various design slot pole of Field Excitation Flux Switching Machines with segmental rotor. 2015 IEEE Conference on Energy Conversion (CENCON); 2015 19-20 Oct. 2015.
[6] Zhu ZQ, Chen JT, Pang Y, Howe D, Iwasaki S, Deodhar R. Analysis of a Novel Multi-Tooth Flux-Switching PM Brushless AC Machine for High Torque Direct-Drive Applications. *IEEE Trans Magn*. 2008;**44**(11):4313-6.
[7] a IJ, a S, a MNSO, editors. Comparative study of single phase FE, PM and HE flux switching motors. 2015 IEEE Student Conference on Research and Development (SCOReD); 2015 13-14 Dec. 2015.
[8] Sulaiman E, Kosaka T, editors. Parameter sensitivity study for optimization of field-excitation flux switching synchronous machine for hybrid electric vehicles. 2012 7th IEEE Conference on Industrial Electronics and Applications (ICIEA); 2012 18-20 July 2012.
[9] Kumar R, Sulaiman E, Ahmad MZ, Othman SMNS, Amin F, editors. Comparative study of initial and optimal outer rotor permanent magnet flux switching machine for downhole application. 2017 First International Conference on Latest trends in Electrical Engineering and Computing Technologies (INTELLECT); 2017 15-16 Nov. 2017.

IOP Conf. Series: Materials Science and Engineering 1127 (2021) 012028 doi:10.1088/1757-899X/1127/1/012028

Series Motor Four Quadrants Drive DC Chopper Part3:Field Weakening mode

Saharul Arof [a]*,N.H.N.Diyanah[a], Emilia Noorsal[b], Philip Mawby[c], H. Arof[d]

[*a]Universiti Kuala Lumpur, Malaysian Spanish Institute, Kulim Hi-Tech Park, 09000 Kulim, Kedah.Malaysia,

, [b]Faculty of Electrical Engineering,Universiti Teknologi MARA ,Pulau Pinang, Malaysia.

[c] School of Engineering, University of Warwick Coventry, CV47AL,United Kingdom.

[d.] Engineering Department,Universiti Malaya, Jalan Universiti, 50603, Kuala Lumpur

E-mail: saharul@unikl.edu.my

Abstract. This paper describes the operation of a field weakening mode of a new four quadrants DC chopper (FQDC) that can be used to extend the speed of an Electric Car (EC). A mathematical model and real experiment are used to verify the effectiveness of the field weakening mode of the proposed FQDC. First, Matlab/Simulink is used to simulate the system under study and the results indicate that the proposed approach is effective. This is followed by a real experiment using a prototype electric vehicle driven by a DC series motor. The results of the experiment show that the proposed FQDC successfully performed the operation as intended.

1. Introduction

Leading car manufacturers in the world have been investing heavily in research and development of efficient Electric Vehicles (EV) and Hybrid Electric Vehicles (HEV) to meet the ever-stringent emission regulation of America and Europe. Using EVs and HEVs have been regarded as one of the ways to curb pollution from the emission of hydrocarbons from conventional internal combustion engines.

1.1. Review Stage

Unfortunately, EVs and HEVs are not widely used as they are still expensive to many people. This shortcoming has led to the study on the possibility of developing cheaper EVs propelled by DC motors. A new breed of Low voltage and high-power DC motors are deemed very suitable for EV and HEV applications [1]. However, high efficiency DC motors alone are not enough for developing complete EV and HEVs. Other research and development areas in electrical engineering domain include power converters/chopper, battery chargers, control, AI, Expert system and electronics [2-13]. The FQDC chopper is designed to perform several modes of operation including driving, field weakening, regenerative braking, resistive braking, generator, reversing and parallel mode driving.

But the scope of discussion of this paper is limited to the Field Weakening mode of the chopper. Field Weakening is a mode of operation of the FQDC and it allows the speed of the DC motor to go beyond its nominal value. The flow of current during this mode is shown in Figure 1.

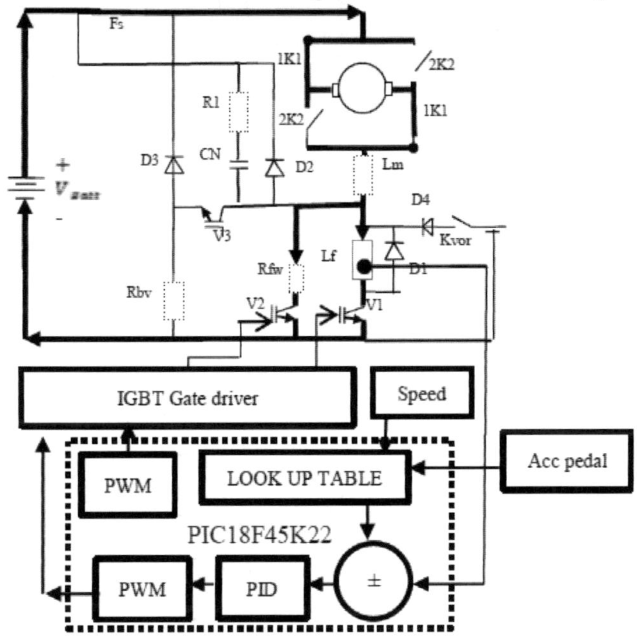

Figure 1. FQDC in field weakening

2. Methodology

2.1. Field Weakening mode of FQDC

The following general equations describe the voltage and current of the chopper. Equations (2) to (4) are general and applicable to all chopper operation modes.

$$I_a = \frac{V_{batt} - I_a(R_a + R_f) - B_{emf}}{L_a + L_f} \quad (1)$$

$$B_{emf} = K_v I_f \omega \quad (2)$$

$$T_d = K_t I_f I_a \quad (3)$$

$$T_d = J\frac{d\omega}{dt} + B_w + T_L \quad (4)$$

Where B_{emf} is the back emf of the motor, K_b is the back emf constant, K_t is the torque constant of the motor, I_f is the field current, R_a and R_f are the motor coil resistances, I_a is the armature current, ω is the angular speed, and T_d is the motor torque.

Field weakening is an operation when the motor operates above base/nominal speed. Field weakening operation is not applicable for every type of driving condition. It is very much dependent on the load [2, 11]. If the load is too heavy due to passengers' weight or the car climbing a steep hill, the field weakening operation should be avoided. In such situations normal driving or parallel mode should be activated instead. Most of the time, the mode selection is made by the chopper operation controller.

The increase in speed is achieved by increasing the torque in field weakening mode. The speed vs torque profile as shown in Figure 2. As seen, the field weakening mode is activated at speed around 50 rad/s. The commencement is accompanied by a jump in torque.

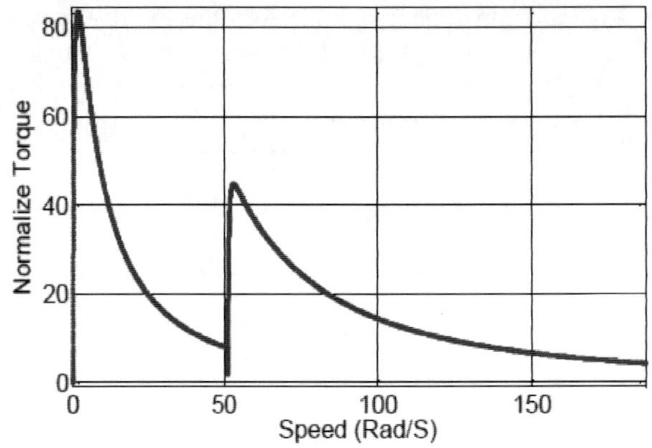

Figure 2. Torque increase at speed 50 during field weakening

The increase in torque is mainly achieved by decreasing the Bemf as stipulated by Equation 1. To reduce the back emf, we need to decrease the field current as stated in equation 2. When BEMF decreases, the armature current increases which leads to the increase in torque as suggested by Equation 3. Figure 3 shows bemf drops when the field weakening mode is activated. Field weakening mode is very much dependent on the correct value of the field current. When the mode is incorrectly executed with a wrong field current value, the chopper would use a lot of energy from the battery to compensate the loss of torque but the generated torque is not enough to counter the load torque.

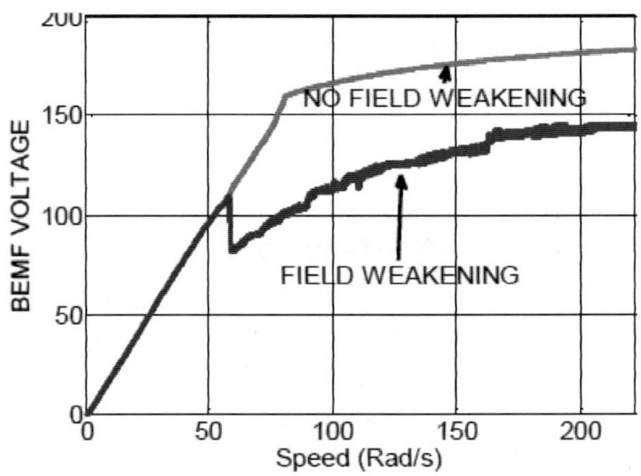

Figure 3. Bemf voltage is reduced during field weakening

As a result, the motor speed drops. Figure 4 shows the speed of the motor during field weakening mode at 2.7s with different values of field current during the test.

IOP Conf. Series: Materials Science and Engineering 1127 (2021) 012028 doi:10.1088/1757-899X/1127/1/012028

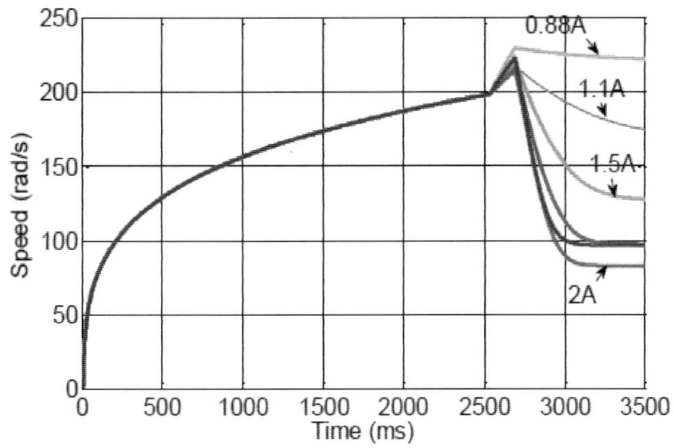

Figure 4. Different field current during field weakening

The simulation results of using different values of Field current in the field weakening mode are also shown in Figure 5. The highest speed is achieved when the field current is set to 0.3A. But when it is further lowered to 0.2A the speed drops. The lowest speed is recorded at 2A field current.

In the figure, the field current that produces the highest motor speed is the one with the medium value, not the highest or the lowest. In short, increasing the Field current increases the Bemf, which in turn decreases the Armature current and consequently the torque. Conversely, reducing the Field current, decreases the Bemf which in turn increases the armature current and consequently the torque.

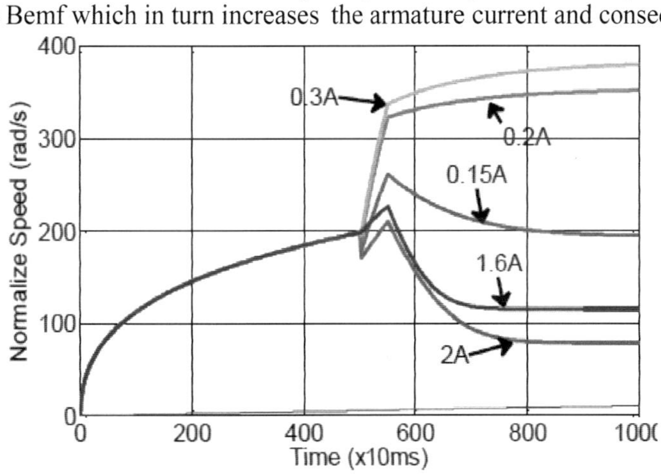

Figure 5. Different field current during field weakening

But if the field current is too low, the torque will also decrease because a low field current generates a low torque according to equation 3.

2.2. Flow of Current during Field Weakening

When the field weakening mode is needed, the drive control unit would control the IGBT V2, now called the field weakening IGBT. The control factor α of the main IGBT V1 is set to almost unity. During this period, the control factor β of IGBT V2 is continuously set to a high value as shown in Figure 6. This induces a current distribution between IGBT V1 and IGBT V2. The value of the field current is inversely proportional to the value of the resistors connected in series with IGBT V2. The control of the different power semiconductors can be seen in Fig. 1. It shows the clock signal, the control α for the main IGBT V1 as well as four different values of control β for the field weakening IGBT V2.

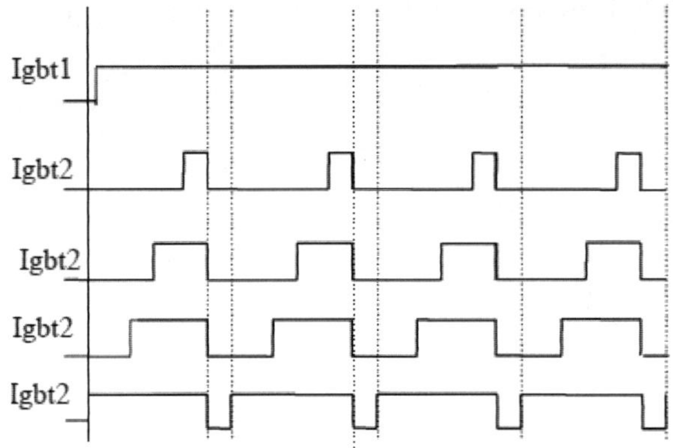

Figure 6. Timing diagram during field weakening

Fig. 1 shows how the motor current distributes through the main IGBT V1 and IGBT V2. When IGBT V2 is fired, some of the current that flows through the field winding detours through the field weakening resistor. In the meantime, the armature current Ia continuously rising when IGBT V2 is fired and its duty ratio is increased proportionally.

While the formula of Back emf and motor torque remains unchanged, the armature and field current when IGBT V1 and V2 are fired and at steady state be classified as below

The condition when Kv1 > Kv2 and Kv2 =0.

$$I_a = \frac{(V_a - B_{emf}) * K_{v1}}{R_a + R_f}$$

$$I_f = \frac{(V_a - B_{emf}) * K_{v1}}{R_a + R_f}$$

The second condition is when Kv1 > 0 and Kv1= Kv2.

$$I_a = \frac{(V_a - B_{emf}) * K_{v2}}{R_a + \frac{R_f * R_{fw}}{R_f + R_{fw}}}$$

$$I_f = \frac{(V_a - B_{emf}) * K_{v2}}{\frac{\frac{R_{fw}}{R_f + R_{fw}}}{R_a + \frac{R_f * R_{fw}}{R_f + R_{fw}}}}$$

The third condition is Kv1 > Kv2 and Kv2 >0

$$I_a = \frac{(V_a - B_{emf}) * K_{v2}}{R_a + \frac{R_f * R_{fw}}{R_f + R_{fw}}} + \frac{(V_a - B_{emf}) * (K_{v1} - K_{v2})}{R_a + R_f}$$

$$I_f = \frac{(V_a - B_{emf}) * K_{v2}}{\frac{\frac{R_{fw}}{R_f + R_{fw}}}{R_a + \frac{R_f * R_{fw}}{R_f + R_{fw}}}} + \frac{(V_a - B_{emf}) * (K_{v1} - K_{v2})}{R_a + R_f}$$

The forth condition is Kv1 and Kv2 > 0 and Kv1 < Kv2

$$I_a = \frac{(V_a - B_{emf}) * K_{v2}}{R_a + \frac{R_f * R_{fw}}{R_f + R_{fw}}} + \frac{(V_a - B_{emf}) * (K_{v2} - K_{v1})}{R_a + R_{fw}}$$

$$I_f = ((V_a - B_{emf}) * K_{v2}) * \frac{\frac{R_{fw}}{R_f + R_{fw}}}{R_a + \frac{R_f * R_{fw}}{R_f + R_{fw}}}$$

The increase in the armature current is caused by two factors. First, according to Kirchoff current law, since the field weakening resistor (RFW) is connected in parallel with the field resistor their total resistance would become smaller. The second factor is the reduction in the field current which decreases the back emf. As the back emf becomes smaller the armature current rises. The influence of the field weakening resistor can be observed in Figure 7. As the field weakening resistor (normalized to field winding resistance) decreases rapidly, the field current decreases slowly but the armature current increases slowly too. So, there exists an optimum value of the RFW which produces the highest torque. Figure 8 shows the influence of the RFW on the values of the back emf (Bemf) and torque in steady state.

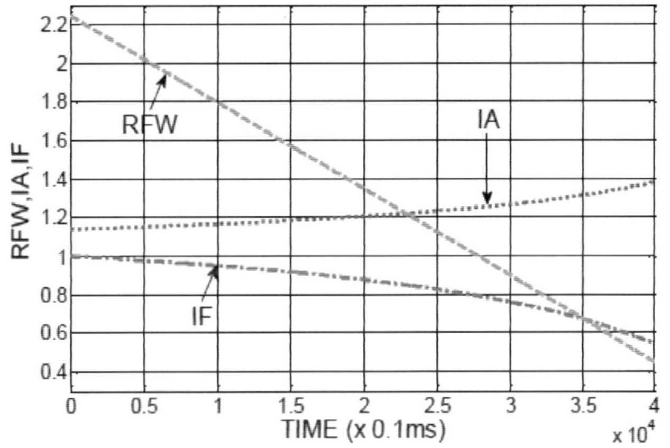

Figure 7. field weakening resistance

As the normalized RFW decreases, the Bemf decreases more slowly but the torque increases gradually until it reaches its maximum value beyond which further reduction in RFW would only reduce its value.

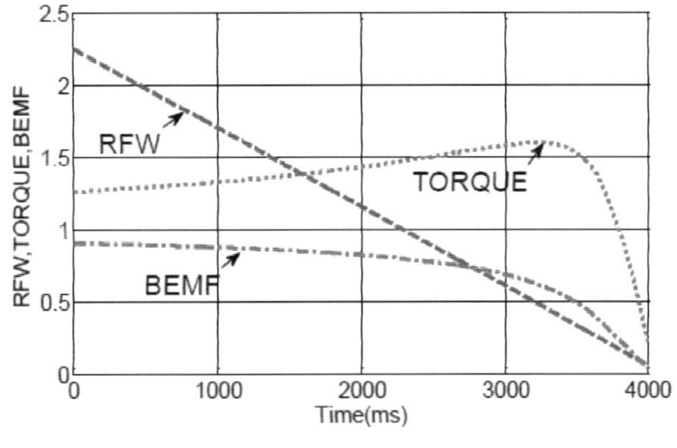

Figure 8. Timing diagram during field weakening

This happens because decreasing the field causes the Bemf to drop which in turn increases the armature current and subsequently the torque. It seems that the effect of the increase in the armature current is bigger than that of the decrease in the field current and thus the torque increases. But further decrease in the field weakening resistor would reduce the field current to the point where the magnetics field and flux start to collapse. As this happens the motor torque drops as shown in Fig. 8. Therefore, the optimum field weakening resistor value to be used is the one where the maximum torque is achieved.

2.3. Control strategy during Field Weakening
First, the system needs to turn on the IGBT V2. This causes some of field current to be diverted from the field winding to the Field weakening resistor. Next, the armature current is set to the maximum by firing IGBT V1 at the maximum PWM value. Then the field current value is loaded from a look up table and sent to the PID controller. The complete process is described in the flowchart of Figure 9. For the field current, a Look Up table like the one in Table 1 will be used so that the best field current is selected for the speed of the motor. The value of the field current is sent to the PID controller that calculates the PWM signal for IGBT V2 to be executed by the four quadrant chopper. The block diagram in Figure 10 describes the whole process.

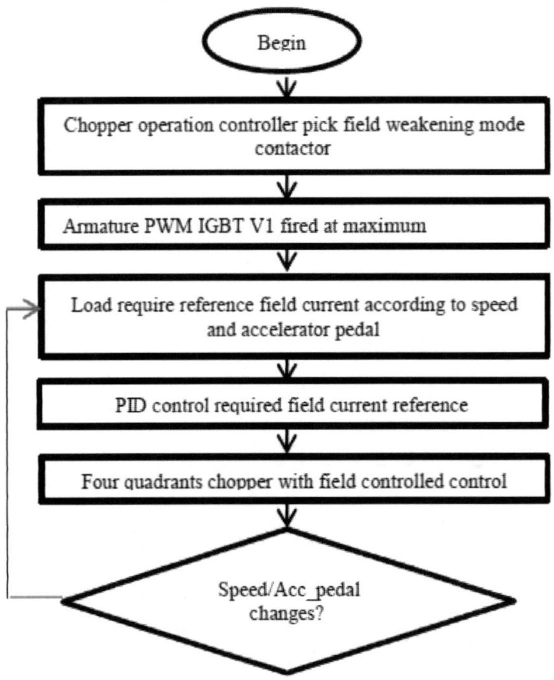

Figure 9. Timing diagram during field weakening

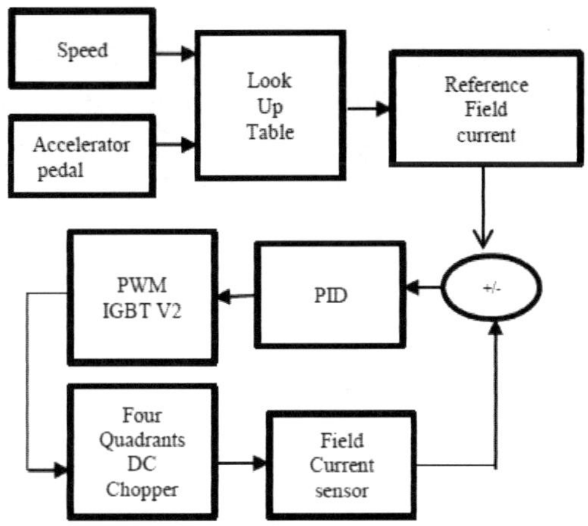

Figure 10. Timing diagram during field weakening

Linear interpolation is necessary to convert the range of the field current in the lookup table to an exact value of the field current for a particular speed as shown in Figure 11.

Table 1. Look Up Table

Speed	Field Current
0-50 rad/s	12-10
50-100	10-8
100-150	8-7

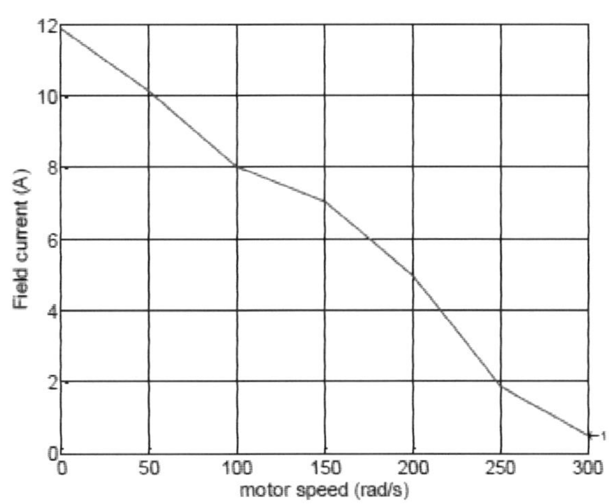

Figure 11. Relationship of speed and field current for Look Up Table.

2.4. Software Development and Simulation Result

Both simulations and experiments were used to test the parallel mode of the Four Quadrants Dc chopper running a 0.65 kW DC series motor. A MATLAB/Simulink mathematical model was developed to study the parallel mode operation. It was constructed by solving linear Differential equations of a DC series motor and is shown in Figure 12. Then the dc series motor model was subjected to the load and controlled by the FQDC based on reference signals from the PID and LUT.

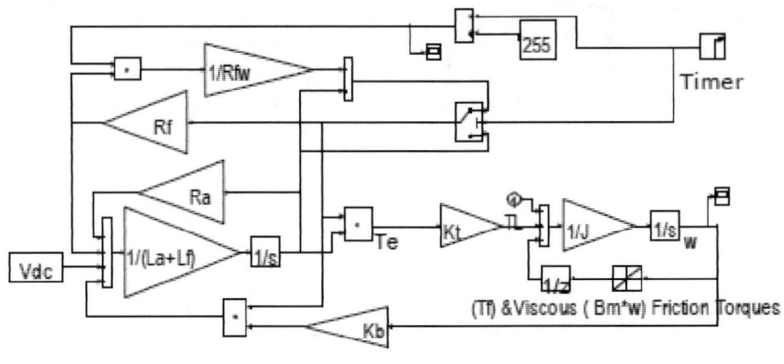

Figure 12. MATLAB/Simulink Mathematical model

Figure 13. Timing diagram during field weakening

Figure 13 shows the hardware experimental setup consisting of a 0.65kW motor attached to a variable load. The model comprises a chopper, vehicle dynamics, igbt firing controller, and lookup table controller .

The complete MATLAB/simulink model is shown in Figure 14.

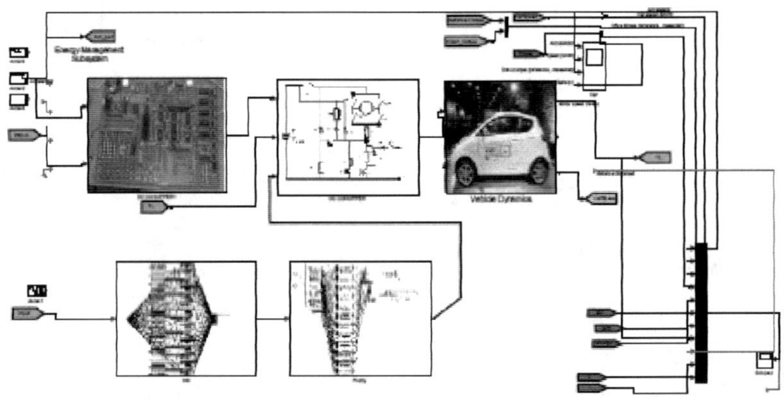

Figure 14. MATLAB/Simulink model

In the vehicle dynamics model, the car experienced hill climbing for load increase test as depicted in Figure 15 below.

Figure 15. field weakening Test with climbing

3. Results and discussion

Fig. 16 shows the simulation result of the speed motor according to the modes of the chopper operation. The straight blue line and the dotted green line represents the experimental result the mathematical model result respectively. The values of the motor speed, armature current and field current in the experiment were captured using Kaptoris software.

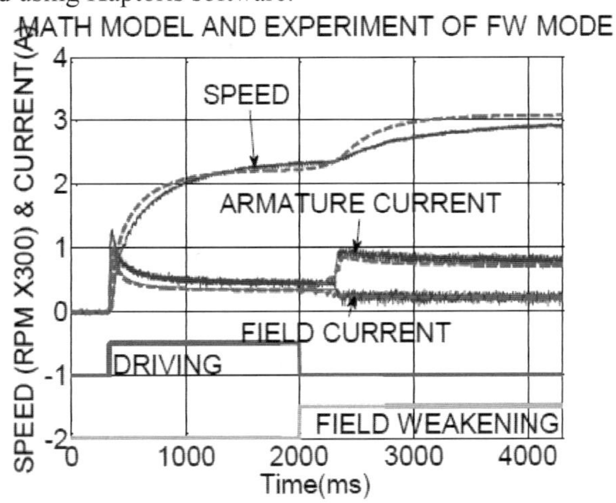

Figure 16. Timing diagram during field weakening.

At the start of the driving mode, the speed of the motor increases until it saturates. Since the armature and field windings are connected in series, they share the same current. The current increases at the beginning and then it decreases due to the increase in Bemf once driving mode is engaged. When the field weakening mode is initiated, the field current drops because of the drop in Bemf as the armature current increases. The reduction in Bemf and increment in the armature current increases the torque and subsequently the motor speed.

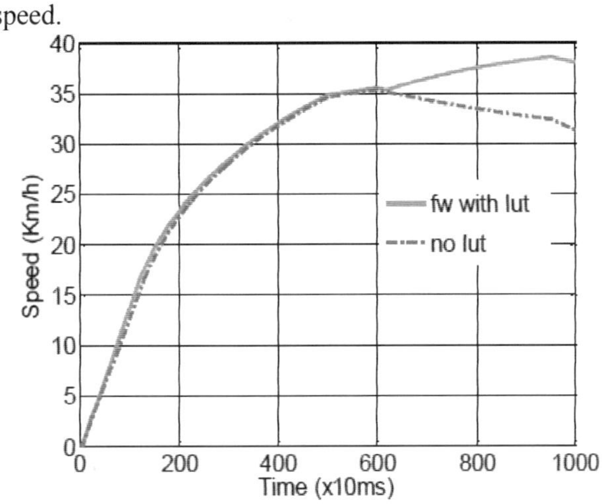

Figure 17. Different Speeds during Field Weakening tested using DC EV Simulation model

Figure 7 shows several signals of FQDC results. Positive torque is for driving, field weakening and parallel mode and negative are for generator, regenerative and resistive brake. Field current is always positive armature current can negative during generator, regenerative and resistive braking. The vehicle speed is always positive. Figure 9 shows the PI index results after defuzzification. The highest PI will be pick as the Chopper operation mode.

The car is tested according to the: accelerator signal (signal 9), brake signal (signal 7) and also with reference to the earth profile shown in Figures 10, 11 and 12 (signal 8).

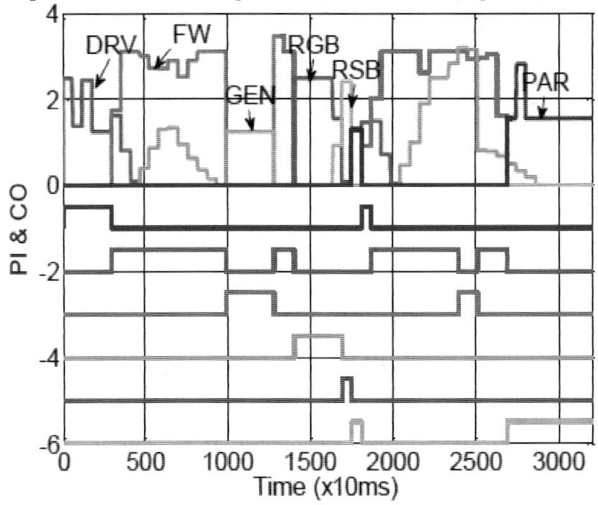

Figure 18. Perfomace Index(PI) and Chopper Operation (CO)

Finally, the FQDC was tested in Field weakening mode on an EV prototype running a 35kW motor operating at the maximum power of 22kW to drive its total weight of 1325kg. In the test, it drove up an inclination of about 7.5 degree. The simulation result is shown in Fig. 17. Two different speeds are shown in the figure and the highest speed is achieved when the system is tested with Field weakening using LUT. The lower speed is when Field Weakening run without referencing LUT.

4. Conclusion

The proposed FQDC successfully performs its field weakening mode which increases the torque and speed of the DC motor. DC drive series motor has a high potential to be utilized in EVs. This is due to its simple design, low cost and excellent controllability.

References

[1] Oak Ridge National Laboratory. Advanced Brush Technology for DC Motors 2009 [Available from: http://peemrc.ornl.gov/projects/emdc3.jpg.

[2] Arof S, Arof H, Khairulzaman AKM, Mawby PA, Jalil JA. Artificial intelligence controlling chopper operation of four quadrants drive DC chopper for low cost electric vehicle. *International Journal of Simulation- Systems, Science and Technology*. 2015;**16**(4):3.1-3.10.

[3] Arof S, Diyanah N, Mawby P, Arof H. Study on Implementation of Neural Network Controlling Four Quadrants Direct Current Chopper. Advanced Engineering for Processes and Technologies: Springer; 2019. p. 37-57.

[4] Arof S, Diyanah N, Mawby P, Arof H, Yaakop NM. Low Harmonics Plug-in Home Charging Electric Vehicle Battery Charger Utilizing Multi-level Rectifier, Zero Crossing and Buck Chopper. Progress in Engineering Technology: Springer; 2019. p. 103-18.

[5] Arof S, Diyanah N, Noor N, Jalil J, Mawby P, Arof H. A new four quadrants drive chopper for separately excited DC motor in low cost electric vehicle. Progress in Engineering Technology: Springer; 2019. p. 119-38.

[6] Arof S, Diyanah N, Noor N, Radzi M, Jalil J, Mawby P, et al. Series motor four quadrants drive DC chopper: part 4: generator mode. *Progress in Engineering*. 2019:155-67.
[7] Arof S, Diyanah N, Noor N, Rosyidi M, Said M, Khairulzaman AM, et al. Genetics Algorithm for Setting Up Look Up Table in Parallel Mode of Series Motor Four Quadrants Drive DC Chopper. Progress in Engineering Technology: Springer; 2019. p. 139-54.
[8] Arof S, Diyanah N, Yaakop N, Mawby P, Arof H. Processor in the Loop for Testing Series Motor Four Quadrants Drive Direct Current Chopper for Series Motor Driven Electric Car. Advanced Engineering for Processes and Technologies: Springer; 2019. p. 59-76.
[9] Arof S, Hassan H, Rosyidi M, Mawby P, Arof H. Implementation of series motor four quadrants drive DC chopper for DC drive electric car and LRT. *J Appl Environ Biol Sci*. 2017;**7**:73-82.
[10] Arof S, Jalil J, Kamaruddin N, Yaakop N, Mawby P, Hamzah A, editors. Series motor four quadrants drive DC Chopper part2: Driving and reverse mode with direct current control. 2016 IEEE International Conference on Power and Energy (PECon); 2016: IEEE.
[11] Arof S, Khairulzaman AM, Jalil J, Arof H, Mawby P, editors. Self tuning fuzzy logic controlling chopper operation of four quadrants drive DC chopper for low cost electric vehicle. 2015 6th International Conference on Intelligent Systems, Modelling and Simulation; 2015: IEEE.
[12] Arof S, Noor N, Alias F, Mawby P, Arof H. Investigation of Chopper Operation of New Series Motor Four Quadrants Drive DC Chopper for DC Drive Electric Car. *J Appl Environ Biol Sci*. 2017;**7**(3S):49-56.
[13] Arof S, Yaakop NM, Jalil JA, Mawby PA, Arof H, editors. Series motor four quadrants drive DC chopper for low cost, electric car: Part 1: Overall. 2014 IEEE International Conference on Power and Energy (PECon); 2014 1-3 Dec. 2014.

Comparison of Fault Tolerant Capability of Three-phase Induction Motor Drives for Automotive Applications

Mahdi Tousizadeh, Che Hang Seng, Nasrudin Abd Rahim*

*Higher Institution Centre of Excellence (HICoE), UM Power Energy Dedicated Advanced Centre (UMPEDAC), Level 4, Wisma R&D, University of Malaya, Jalan Pantai Baharu, 59990 Kuala Lumpur, Malaysia.

E-mail: toosizadeh@gmail.com

Abstract. With the increase proliferation of electric vehicles in the world, the demand for more reliable power electronics drive train is also on the rise. While theoretically the use of multiphase (more than three-phase) motors can increase the reliability of electrical drive train for EV, they are generally not commercially ready and available in the market. Hence, three-phase drive train are still the main stream in current EV technology and the fault-tolerant control of these motor drives remain an interesting and important topic. In this presentation, the existing fault tolerant three-phase motor drive topologies and control methods are reviewed. In particular, the post-fault controls of three-phase motor using 4 legs inverters are discussed and the opportunity for voltage utilization improvement with this configuration is demonstrated.

1. Introduction

In the field of variable speed AC motor drives, the failure of the motors and power electronics are inevitable. In this situation, one solution is to provide the components with higher degree of reliability in order to diminish the chance of failure. The other alternative solution is to utilize some techniques to assure the continuity of the drive in presence of fault, with lower amount of output compared to the healthy drive. Therefore the knowledge of how much power can be delivered by a faulted drive in post-fault mode would be a significant measure since the three phase adjustable speed drives are a candidate to be adopted for automotive applications.

In fact the limits imposed to a motor drive in presence of fault are due to the topology reconfiguration. Once the topology has been reconfigured, anyhow the auxiliary components provide an additional path for the current to ensure balanced operation of the motor in post-fault mode. However, there are a few topology for post-fault operation of fault tolerant three phase drives in which modification on controller side in some cases are mandatory while some other cases can continue operation with topology reconfiguration only. This work is aimed to explore the output limits of a three phase induction motor drive in post-fault mode in order to extend the available output.

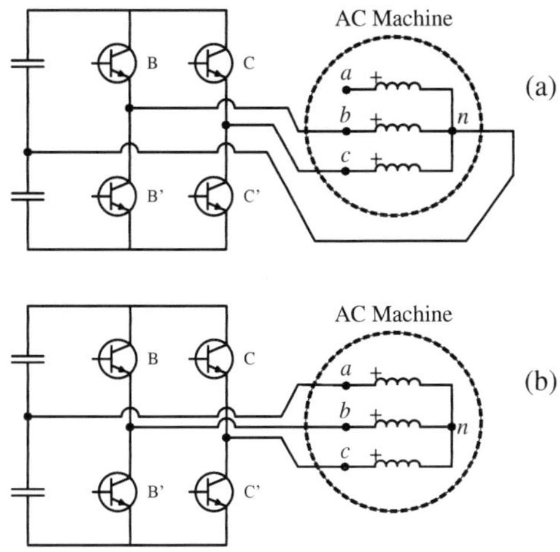

Figure 1. Switch Redundant Post-fault Topology

2. Existing Fault Tolerant Topologies

Numerous researches have been carried out in literature to realize the feasibility of fault tolerant AC motor drives. Depending on where the auxiliary components are placed and how they are going to be utilized in post-fault mode, the fault tolerant topologies can be categorized into four configuration as 1) Switch Redundant Topology, 2) Double Switch-Redundant Topology, 3) Phase Redundant Topology, 4) Parallel Single Phase Topology.

The earliest effort to obtain some degree of fault tolerance on a standard three-phase induction motor drive was proposed in [1] and [2]. This switch-redundant topology incorporates four TRIACs and three fast acting fuses. One of the TRIACs connected to the motor neutral point is in charge of forcing the neutral point of the motor to be connected to the mid-point of DC bus in post-fault mode. The other three TRIACs and three fuses are connected in series with the load phases to protect against the short circuit faults in switch and clearance of connection between the motor and inverter. Anyhow, the post-fault topology of drive with this topology is shown in Fig.1 (a).

The alternative switch-redundant topology to tolerate the switch fault connects the machine phase, corresponding to the faulted inverter leg, to the mid-point of DC bus using a single TRIAC [2, 3]. So, this topology is only able to tolerate the open switch failures on the switch level. If the switch failure is of the short circuit kind, the automatic protection like fuses are necessary to avoid whole drive shut down. The post-fault reconfigured topology is shown in Fig.1 (b) which is in fact same like the structure of four switch driven three phase motor drive.

The Double Switch Redundant Topology, employs additional leg parallel to the inverter legs which is connected to the machine neural point permanently [4-6]. In case of healthy functioning, this leg is inactive and in case of failure the neutral point of the motor would be supplied with this additional leg. Fuses and SCRs are also utilized on each leg to isolate the faulty leg in case the switch is shorted. The post-fault topology of this method is shown in Fig 2.

Introducing a spare inverter leg for improved fault tolerance is referred to Phase Redundant Topology [7, 8]. Here the faulty inverter leg would be isolated by means of SCRs and fuses and instead the spare leg would be replaced by activating of TRIACs. Therefore the drive topology in post-fault mode, shown in Fig. 3, would have three active leg to supply the three phases of the motor which resembles the standard three phase drive topology.

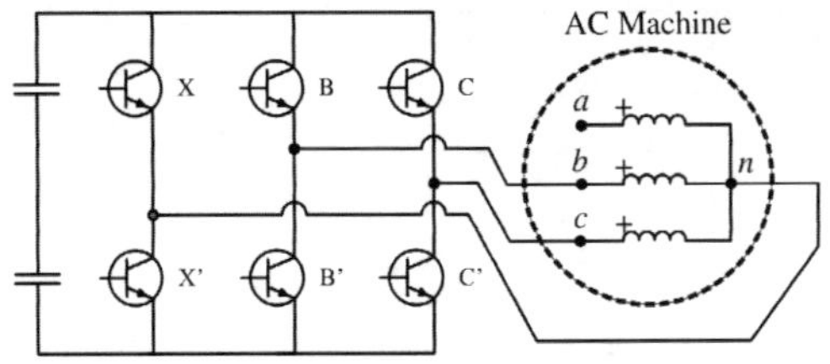

Figure 2. Double Switch Redundant Post-fault Topology

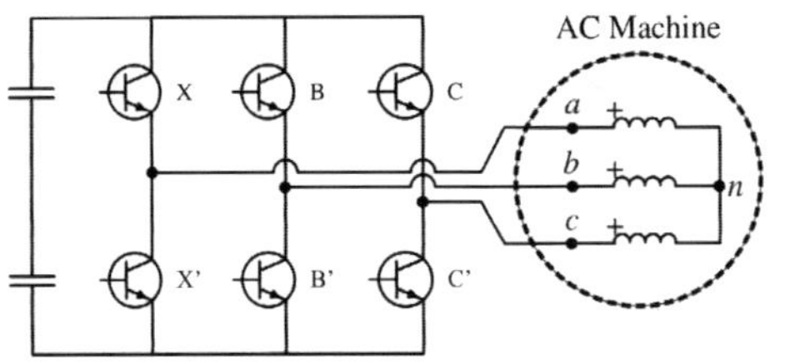

Figure 3. Phase Redundant Post-fault Topology (Standard)

The general theory of parallel single phase inverter for each phase of a multi-phase motor, and particularly for three phase induction motor, was first introduced in order to control each phase of the motor separately [9]. This concept was also utilized in [10-14] and due to having capability of tolerating the fault, it can be considered as one of the fault tolerant topologies for three phase induction motor shown in Fig. 4. Each individual single phase inverter delivers the full DC bus voltage to motor phases, and therefore, in case of failure in either switch or machine phase, the faulted parts could be independently cleared. It means the remaining healthy phases would be capable of operating independently provided that the protecting devices fully open the faulted phase leg from the drive.

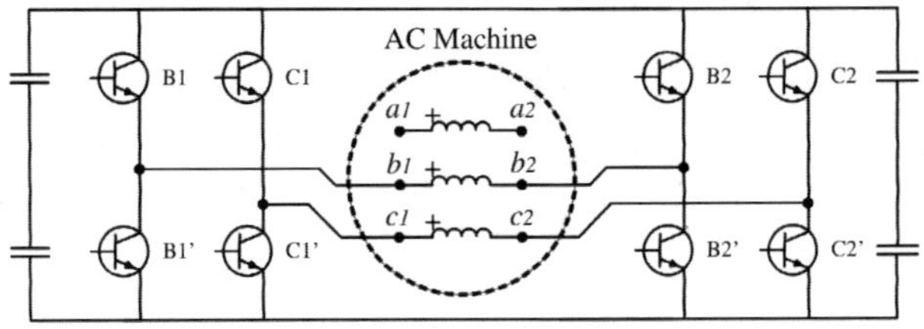

Figure 4. Parallel Single Phase Post-fault Topology

However, the post-fault configuration of fault tolerant three phase drives stated in literature would get obviously one of the configurations depicted in Fig. 1 to Fig. 4, regardless of fault detection algorithm, reconfiguration time and post-fault control strategy. In continue, each topology will be discussed in terms of available input/output power. It should be noted that the auxiliary connecting devices are assumed to be ideal for simplicity.

3. Post-fault Limitations

Generally, for a three phase induction motor under the fault, in order to maintain a circular flux trajectory with fixed magnitude, the phase currents of the remaining phases need to be increased in magnitude by a factor of $\sqrt{3}$ and phase shifted 30° away from the axis of the faulted phase [1]. The phasor diagram of this for an open-circuit fault on phase 'a' is shown in Fig. 5.

This adjustments in phase current grantees the dc magnitude of dq currents in post-fault mode equal to the healthy mode as long as there is path for sine wave zero sequence current to flow. The zero sequence current here would be sine wave with peak magnitude of √3 time bigger the phase current and frequency equal to the fundamental frequency of stator. This concept is applicable to topology of Fig. 1(a) and Fig. 2 and is referred as post-fault balanced flux producing current hereafter with open phase fault in phase 'a'.

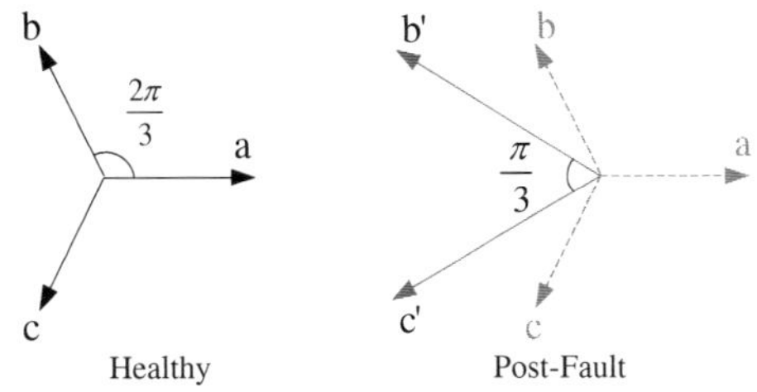

Figure 5. Required Healthy & Post-fault Phase Current

The mathematical model of squirrel cage induction machine in rotating reference frame at steady state is obtained using the following equations.

$$\begin{aligned} V_{ds} &= R_s i_{ds} - \omega \psi_{qs} \\ V_{qs} &= R_s i_{qs} + \omega \psi_{ds} \\ V_{0s} &= R_s i_{0s} + \frac{d}{dt}\psi_{0s} \\ V'_{dr} &= R'_r i'_{dr} - (\omega - \omega_r)\psi'_{qr} \\ V'_{qr} &= R'_r i'_{qr} + (\omega - \omega_r)\psi'_{dr} \end{aligned} \quad (1)$$

where V_{ds} , V_{qs} , V_{0s} are the voltage, i_{ds} , i_{qs} , i_{0s} are current, R is resistance , ω is synchronous frequency, ω_r is rotor frequency and ψ is the flux linkage of the stator with subscript s and of the rotor with subscript r on $dq0$ axis. The flux linkage is determined by

$$\begin{aligned} \psi_{ds} &= L_s i_{ds} + L_m i'_{dr} \\ \psi_{qs} &= L_s i_{qs} + L_m i'_{qr} \\ \psi_{0s} &= L_s i_{0s} \end{aligned} \quad (2)$$

$$\begin{aligned} \psi'_{dr} &= L'_r i'_{dr} + L_m i_{ds} \\ \psi'_{qr} &= L'_r i'_{qr} + L_m i_{qs} \end{aligned} \quad (3)$$

Where L_s and L_r are stator and rotor self-inductances, with mutual inductance of L_m.

The resistive part of the (1) in steady state can be ignored as it is small compared to the back emf term. So, stator voltage near the rated speed would be approximated as follow.

$$\begin{aligned} V_{ds} &= -\omega \psi_{qs} \\ V_{qs} &= \omega \psi_{ds} \\ V_{0s} &= \frac{d}{dt}\psi_{0s} \end{aligned} \quad (4)$$

The zero sequence flux linkage is equal to leakage inductance multiplied by zero sequence current.

$$\frac{d}{dt}\psi_{0s} = l_{ls}\frac{d}{dt}i_0 \quad (5)$$

Let's assume zero sequence current has the form of:

$$i_0 = A\cos(\omega t + \varphi) + B\sin(\omega t + \varphi) \quad (6)$$

Since the zero sequence current in steady state is sine wave with the frequency equal to the stator phase current frequency, the derivative of it would be another sine wave shifted in phase with multiplier of ω.

$$\frac{d}{dt}i_0 = \omega\left[-A\sin(\omega t+\varphi)+B\cos(\omega t+\varphi)\right] \quad (7)$$

Based on this justification, all stator voltage terms in *dq0* axis would be directly a function of speed ω and it can be claimed that maximum speed of the motor is directly proportional to the maximum phase voltage available on machine terminals.

$$V_s^2 = V_{ds}^2 + V_{qs}^2 + V_{0s}^2$$
$$V_s^2 = K.\omega^2 \quad (8)$$

4. Results and Analysis

Referring to post-fault balanced flux producing current, the available torque producing current is reduced by a factor of √3, maintaining a post-fault phase current at rated value. This condition is applied for all post-fault topologies hereafter. For Switch-Redundant Topology in Fig. 1 (a), assuming that the slip speed ω_{sl} is maintained at post-fault mode, the torque producing and flux producing current should be reduced accordingly and therefore maximum available output torque would be reduced by a factor of √3 * √3.

Table 1. Post-fault Capability of Drives

Post-fault Topology	Post-fault Maximum Stator Current p.u.	Post-fault Available Phase Voltage p.u.	Post-fault Maximum Available Torque (%)	Post-fault Maximum Available Speed (%)
Figure 1(a)	$1/\sqrt{3}$	$\sqrt{3}/2$	33.33	86.60
Figure 1(b)	1	1/2	100.00	50.00
Figure 2	$1/\sqrt{3}$	$\sqrt{3}$	33.33	173.21
Figure 2*	1/3	$\sqrt{3}$	11.11	173.21
Figure 3	1	1	100.00	100
Figure 4	$1/\sqrt{3}$	1	33.33	100

Referring to Fig. 1 (a), since the motor neutral is connected to the midpoint of the dc link in the post-fault condition, the inverter still has the capacity to apply 1/2 of the DC bus voltage across each of the remaining phases while it was 1/√3 of DC bus voltage on standard drive. As the voltage drop across the stator resistance is neglected at high speed, the maximum phase voltage will have a direct relation with the maximum available speed of the motor. Therefore it can be concluded that the maximum available speed in this case would be reduced by a factor of √3/2. The maximum available input voltage, maximum stator current, maximum output torque and maximum speed of the motor in post-fault mode for the topologies shown in Fig. 1 to Fig. 4 are stated in Table 1.

For Switch-Redundant Topology in Fig. 1 (b), the motor is able to take full rated current and therefore full output torque. But, the maximum phase voltage would be reduced to 0.5/√3 as the phase '*a*' is connected to the mid-point of DC bus. So the maximum speed would be reduced to half of rated speed accordingly.

The maximum stator current for Double Switch Redundant Topology in post-fault mode, Fig. 2 would be reduced by factor of √3 in the same way as the Switch-Redundant Topology Fig. 1(a). Similarly the maximum output torque would be reduced by the factor of 3 while maintaining the slip speed. It should be noted that the auxiliary leg (X) in this case should be able to carry the current with amplitude of √3 times bigger than rated value. In case the auxiliary leg is aimed to carry up to the rated current only, the phase current should be √3 times smaller than the rated current and therefore the maximum output torque would be 9 times smaller than rated torque. The maximum voltage on the other hand would be increased by the factor of √3 which needs to make sure the motor is able to accommodate this voltage increment. As a result the maximum speed in this case would be √3 bigger.

For Phase Redundant Topology, Fig. 3, as the faulty leg is replaced with the spare leg, the available current rating and output torque of the motor will remain unchanged. Similarly the available phase voltage to the motor would be same as standard drive and consequently rated speed is achievable.

In case of Parallel Single Phase Topology, the available motor current will be same like Double Switch Redundant Topology as one of the motor phases are open. And similarly, the maximum available output torque would be reduced 3 times smaller than the rated torque. In this case the available phase voltage to the motor would be equal to rated voltage as each phase is supplied individually from a single phase inverter. And so, the maximum available speed can go up to rated speed.

5. Conclusions

As a conclusion, the post-fault capability of three phase induction motor drive is significantly affected by the topology of which the drive is working in. The Switch Redundant Topologies does not require many auxiliary devices but it have poor output together with necessity of access to neutral point of DC bus. Additionally, tapping the DC bus mid-point introduces significant DC bus voltage fluctuations in post-fault mode. The Double Switch Redundant Topology shows significant gain in available speed beyond the rated value in post-fault mode and inversely lower output torque. The maximum available torque would be around 11% only if the fourth leg is designed to be able to carry rated current only. Provided that the fourth leg in oversized to take √3 times of rated current, this topology could potentially be used for EV applications since the extended available speed could be converted to torque by means of gearbox. The Phase Redundant Topology shows the rated result for speed and torque at post-fault mode in expense of employing several auxiliary devices. Furthermore, this topology can tolerate the semiconductor faults on inverter only. As this topology increases the complexity of the drive, it is not recommended for EV applications. The Parallel Single Phase Topology is also delivering rated speed in post-fault mode with one third of rated torque. It is also an acceptable option for EV applications. Overall, The Double Switch Redundant Topology is chosen to be further studied for future work.

Acknowledgments
The authors would like to thank the financial and technical assistance provided by AUN-SEED/Net Japan ASEAN Collaborative Education Programme (JACEP).

References

[1] Liu T-H, Fu J-R, Lipo TA. A strategy for improving reliability of field-oriented controlled induction motor drives. *IEEE Trans Ind Appl*. 1993;**29**(5):910-8.

[2] Fu J-R, Lipo TA. A strategy to isolate a switching device fault in a current regulated motor drive. *Electric machines and power systems*. 1996;**24**(8):911-20.

[3] Campos-Delgado D, Espinoza-Trejo D, Palacios E. Fault-tolerant control in variable speed drives: a survey. *IET Electr Pow Appl*. 2008;**2**(2):121-34.

[4] Bolognani S, Zordan M, Zigliotto M. Experimental fault-tolerant control of a PMSM drive. *IEEE Trans Indust Electr*. 2000;**47**(5):1134-41.

[5] de Rossiter Correa MB, Jacobina CB, Da Silva EC, Lima AN. An induction motor drive system with improved fault tolerance. *IEEE Trans Ind Appl*. 2001;**37**(3):873-9.

[6] Bianchi N, Bolognani S, Zigliotto M, Zordan M. Innovative remedial strategies for inverter faults in IPM synchronous motor drives. *IEEE Transactions on energy conversion*. 2003;**18**(2):306-14.

[7] Ribeiro R, Jacobina C, Da Silva E, Lima A, editors. A fault tolerant induction motor drive system by using a compensation strategy on the PWM-VSI topology. 2001 IEEE 32nd Annual Power Electronics Specialists Conference (IEEE Cat. No. 01CH37230); 2001: IEEE.

[8] Rodriguez M, Claudio A, Theilliol D, Vela L, Hernandez L, editors. A strategy to replace the damaged element for fault-tolerant induction motor drive. 2008 5th International Conference on Electrical Engineering, Computing Science and Automatic Control; 2008: IEEE.

[9] Jahns TM. Improved reliability in solid-state AC drives by means of multiple independent phase drive units. *IEEE Trans Ind Appl*. 1980(3):321-31.

[10] Stemmler H, Guggenbach P, editors. Configurations of high-power voltage source inverter drives. 1993 Fifth European Conference on Power Electronics and Applications; 1993: IET.
[11] Corzine K, Sudhoff S, Whitcomb C. Performance characteristics of a cascaded two-level converter. *IEEE Transactions on Energy Conversion*. 1999;**14**(3):433-9.
[12] Mohan N, Undeland TM, Robbins WP. Power electronics: converters, applications, and design: John wiley & sons; 2003.
[13] Qin D, Luo X, Lipo T, editors. Reluctance motor control for fault-tolerant capability. 1997 IEEE International Electric Machines and Drives Conference Record; 1997: IEEE.
[14] Novotny DW, Lipo TA. Vector control and dynamics of AC drives: Oxford university press; 1996.

Using Nonlinear Load DC Variables as Inputs for Harmonic Current Reduction

Yushaizad Yusof *, Nasrudin Abd. Rahim †

*INTEGRA, Fakulti Kejuruteraan & Alam Bina, Universiti Kebangsaan Malaysia, Malaysia,

†UMPEDAC, Wisma R&D,Universiti Malaya, Malaysia

E-mail: yushaizad@ukm.edu.my

Abstract. This paper introduces and discusses new inputs for a single-phase shunt active power filter (SAPF) using nonlinear load equivalent resistance estimation control strategy. The inputs are an average DC voltage and current of a current source nonlinear load. By using a MATLAB Simulink simulation tool, a model of single-phase AC system connected to the SAPF and nonlinear load is designed and its vital voltage and current variables are simulated. Various simulation results are presented and analysed in order to validate the effectiveness of proposed inputs. As a conclusion, the proposed control strategy with new inputs successfully achieved its objective to mitigate harmonic current and to compensate the reactive power with good performance.

1. Introduction

Active power filter (APF) especially in parallel or shunt configuration (SAPF) has been the main tool in mitigating harmonic current distortion recently. The SAPF is connected in between the source and the nonlinear load at point of common coupling (PCC) as depicted in Figure 1. It is the typical configuration for single or three phase system implemented for harmonic current reduction and reactive power compensation. Basically, some advantages of using SAPF reported in literatures are based on its basic configuration, good reactive power and voltage flicker compensations capability etc [1]. In addition, a voltage source inverter (VSI) based SAPF is widely utilized [2-5]. Normally, the choice of topology for SAPF is H-bridge or multilevel VSI which bidirectional switches such as combination of MOSFETs/diodes or IGBTs/diodes are frequently selected as shown in Figure 1. Besides, many of the aforementioned topologies also used an electrolytic DC capacitor as their energy storage element. On the other hand, a ripple or filter inductor is applied to interface the VSI and PCC [6, 7].

Nowadays, many loads are considered nonlinear loads. Typical single-phase nonlinear load choice is a full bridge diode rectifier with inductive or capacitive load [8]. The nonlinear load is known as a source of harmonic current, and it draws reactive power from the line network. As a result, a source current, i_s changes from purely sinusoidal waveform into non-sinusoidal waveform. In addition, the phase angle, θ gap between the i_s and its reference, a source voltage, v_s becomes wider which result in much lower power factor for the system that will greatly affect the system efficiency.

IOP Conf. Series: Materials Science and Engineering 1127 (2021) 012030 doi:10.1088/1757-899X/1127/1/012030

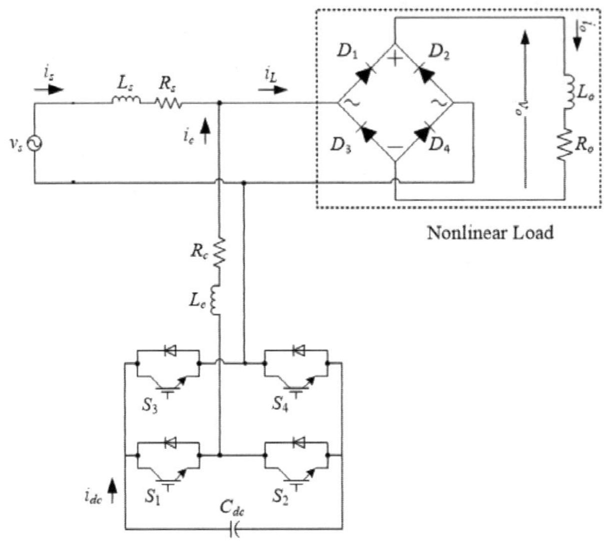

Figure 1. SAPF basic configuration

There are some poor impacts occurred due to harmonic current distortion, e.g. heating of conductor, shorten apparatus life expectancy, higher peak current etc. In order to overcome the problems created by harmonic current, a reliable and efficient control strategy for SAPF is required. Before a final product of control strategy which is a proper gating signal, *g* being fed into the switch, three different control stages are required to be passed through as depicted in Figure 2 [9].

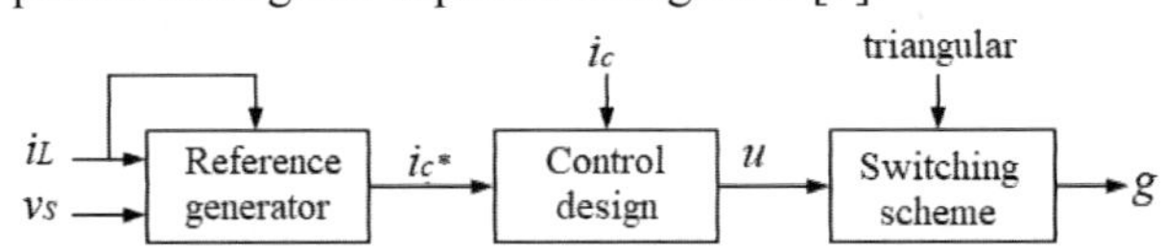

Figure 2. Basic principle of control strategy design [6]

The first stage is called harmonic detection method where the selected variable, load current, i_L which contains rich harmonic component is being filtered using low pass filter (LPF) to develop the fundamental component only. Next, to synchronize the produced signal with the v_s, both are multiplied together to form a sinusoidal reference current, i_s^*. Later, the i_s subtracted by the actual i_L to generate compensated reference current, i_c^*.

The following stage is the control design where the newly generated i_c^* minus its actual signal i_c to yield an error, *e*. This error needs to be compensated using a controller to obtain a control signal, *u*. The desired response of *u* should have a zero steady state error, stable closed loop system, no percentage overshoot as well as fast settling time. In other word, the final value of *u* should be identical to its reference without any error ($u \approx i_c^*$).

The final stage of control strategy adopted by the SAPF is switching scheme technique. Typically, a linear bipolar pulse width modulation (BPWM) is selected. The *u* is compared with the high frequency triangular (carrier) signal to yield the *g*. Sometimes other than the BPWM, hysteresis PWM and space vector PWM were also chosen as the preferred switching scheme technique [8].

The step by step procedure in determining the best option for simplified control strategy has been discussed briefly. Nevertheless, the issue is, almost all the inputs of APF control strategy is a load current, source current and DC-link voltage [1-8]. For many years, in particular for many literatures [1-10], the inputs for SAPF control strategy are the same, without any effort to change it. The exceptional characteristic for the input of SAPF control strategy is that it should have contained a huge amount of harmonic component apart from the fundamental component. Previously, the inputs are dominated by the aforementioned variables which have proven their capabilities for SAPF control strategy, and is categorized as a source of harmonic. Another identical characteristic that is imposed by the previous inputs is prior to be inserted in LPF or other type of filters, its initial AC signal should be transformed into a DC signal first. Particularly, this happened to synchronous reference frame (SRF) and instantaneous reactive power reactive (PQ) theory control strategies for APF [1-5].

Therefore, if the typical source of harmonic is replaced by a DC load voltage and current due to some concrete reasons and as the alternatives for the previous one, which will be explained next. Rapid variation identified on the DC load voltage and current waveforms show that they comprise of high frequency component which in turn will provide harmonic current to the AC system. Besides, since they are already the DC signal, there is no requirement for transformation from AC to DC signal. Moreover, the nonlinear load will also draw reactive power from the supply. Therefore, based on these arguments, the DC voltage and current of nonlinear load has a merit to be implemented as the input of control strategy.

The objective is to implement two new inputs e.g. DC load voltage and current for SAPF control strategy using the nonlinear load equivalent resistance estimation technique proposed in [6]. Next section will provide the methodology and mathematical expression used in suppressing harmonic current and improve the system's power factor. Results and analysis will be given after that, in order to verify the proposed method performance. Finally, the paper ends with conclusion and acknowledgement to the project grant provider.

2. Methodology

Methodology applied in the proposed control strategy can be divided into three sections, where the first section will elaborate about the harmonic detection method, then the control design, and finally the choice of switching scheme.

2.1. Modelling harmonic detection method

By applying the current controlled mode, the compensating current i_c is injected into the line at point of common coupling (PCC) used to force the source current i_s to become pure sinewave and makes in phase with the source voltage v_s, which in turn improves the system power factor to unity. Basically, the currents of SAPF system can be expressed as follows:

$$i_s = i_L + i_c \tag{1}$$

where i_L represents the nonlinear load current. According to [9], since the H-bridge VSI configuration is adopted by the SAPF, the relationship between DC-link voltage v_{dc} and the v_s when the average inductor voltage $v_L = v_{Lc} + v_{Ls} = 0$ for one switching cycle can be summarized as follows:

$$v_{dc} = \frac{v_S - v_R}{1-2D} \tag{2}$$

$$\frac{di_L}{dt} = \frac{(v_S - v_R) + (2D-1)v_{dc}}{L} \tag{3}$$

where v_{Lc}: filter inductor voltage; v_{Ls}: source inductor voltage; v_R: v_L ESR voltage and D: duty ratio. Equation (3) presents the dynamic representation ofthe SAPF.

Since nonlinear load is the main source of harmonic generation, hence the mathematical expression that relates the nonlinear load and harmonic current is discussed first. Nonlinear load such as H-bridge diode rectifier with inductive load is selected as the source of harmonic which has been depicted in Figure 1. Thus, the full-wave rectified sinusoidal voltage can be expressed as a Fourier series [9]:

$$v_o(t) = V_o + \sum_{n=2,4,\ldots}^{\infty} V_n \cos(n\omega t + \pi) \tag{4}$$

where the latter represents an even harmonic series; V_m: peak AC voltage, ω: fundamental frequency (rad/s), and n: harmonic number. The average DC term voltage V_o is represented by:

$$V_o = \frac{2V_m}{\pi}$$

which high frequency component voltage V_n is:

$$V_n = \frac{2V_m}{\pi}\left(\frac{1}{n-1} - \frac{1}{n+1}\right)$$

Using superposition method, the DC load current i_o is calculated by taking each frequency separately before combining the results. Hence, the DC load current is the combination of two components, i.e. the average DC-term current I_o and high frequency component current I_n.

$$I_o = \frac{V_o}{R_o}$$

$$I_n = \frac{V_n}{Z_n} = \frac{V_n}{|R_o + jn\omega L_o|}$$

where Z_n denotes the load's impedance with respect to n. Based on nonlinear load equivalent resistance R_e estimation technique, the i_s^* generation can be easily developed. A simple technique to generate the i_c^* is depicted in Figure 3 since it does not require any tedious mathematical transformation which will save a lot of elapsed time needed for calculation.

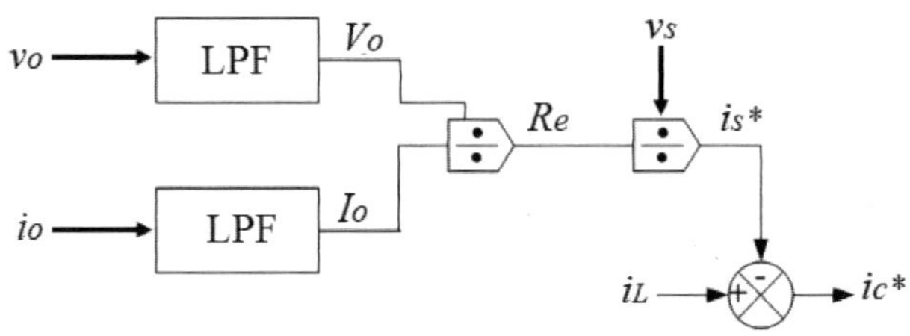

Figure 3. The control strategy with the proposed inputs

Another vital component that has great impact in separating the harmonic component and fundamental component of the load voltage and current is the LPF application. Typical LPF allows a signal below its cut-off frequency f_{co} to pass through, albeit blocks the signal above its cut-off frequency. In this case, the choice of f_{co} is 1 Hz. An expression of a first order Butterworth LPF in Laplace transform is presented as follows:

$$LPF(s) = \frac{1}{\tau s+1} \tag{5}$$

where τ: a time constant, and τ = 0.159 s.

2.2. Controller design and switching scheme

The generated i_c^* signal is used as the reference for the actual compensated current, i_c signal to follow it traits exactly. Therefore, the difference between both signals yield the error, e. After that, the e is controlled through a PI controller such that its product, a control signal u tracks the reference signal exactly. To generate a gating signal g, a bipolar pulse width modulation (BPWM) switching scheme is selected, where the u signal is being compared with the carrier signal in order to form the g. Then, it is fed to the gate of MOSFET switches in two states to operate the SAPF circuit.

Normally, to design the controller for the SAPF, the time domain mathematical representation of SAPF initially is changed to a Laplace transform. Figure 4 illustrates the closed loop control for the compensated current i_c. Hence, the forward transfer function $G(s)$ and the closed loop transfer function $C(s)$ (assuming a unity feedback) for the i_c is expressed as:

$$G(s) = \frac{e(s)}{i_c(s)} = PI(s) \times PWM \times \frac{1}{L_c s+R_c} \tag{6}$$

$$C(s) = \frac{G(s)}{1+G(s)} = \frac{5(K_p s+K_i)}{L_c s^2+(K_p+R_c)s+K_i} \tag{7}$$

$$C(s) = \frac{5K_p}{L_c} \frac{(s + K_i/K_p)}{s^2 + (K_p + R_c)s/L_c + K_i/L_c}$$

where,

$$PI(s) = \frac{K_p s + K_i}{s} = K_p \frac{(s + K_i/K_p)}{s}$$
$$PWM = 5$$

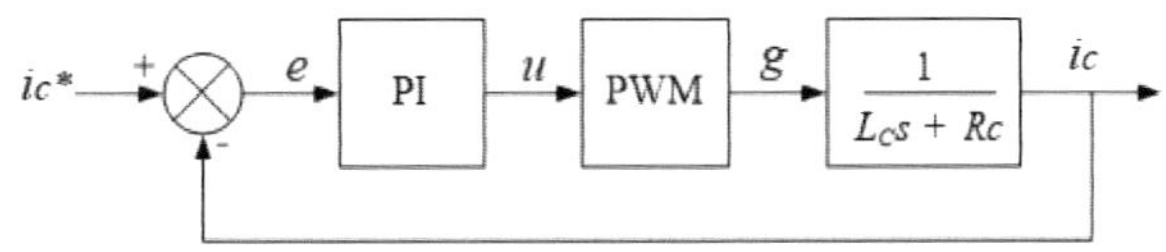

Figure 4. Closed loop control for i_c

With regards to (6), and to obtain a desired response of u, i.e. small overshoot, fast settling time, stable as well as zero steady state error, the respective proportional gain K_p and the integral gain K_i should be tuned accordingly. Using a Bode plot, the K_p and K_i gains are tuned and refined several times before the satisfied results are achieved. Therefore, to figure out the designed closed loop control response, a standard step input has been implemented, and its output response is analysed. Using [9], by equating (6) with (7), the initial value of K_p and K_i gains can be obtained. Hence,

$$C(s) = \frac{\omega_n^2}{s^2 + 2\zeta\omega_n s + \omega_n^2} \tag{8}$$

$$K_p = 2\zeta\omega_n L_c - R_c \tag{9}$$

$$K_i = L_c \omega_n^2 \tag{10}$$

where ω_n: natural frequency (rad/s) and ζ: damping ratio. The calculation of ω_n is given as follows:

$$\omega_n = \frac{1}{\sqrt{(L_s + L_c)C_{dc}}} \tag{11}$$

A linear BPWM has been chosen as the switching scheme, where its carrier frequency is set at 20 kHz. The u signal is being compared with the carrier signal to generate a series of pulse train which also known as the gating signals g, using logic circuit. This signal is then being split into two opposite pulses, when one at ON state, the other will be OFF by inverting the g signal. Both signals are fed to each complement pair of switches' gates in order to turn-on and turn-off the switches according to the provided gating signal state. To measure the quality and total harmonic distortion (THD) of source current as well as the system power factor PF, the following expression is implemented.

$$\text{THD} = \frac{\sum_{n\neq 1}^{\infty} I_{n,rms}}{I_{1,rms}} \tag{12}$$

$$\text{PF} = \cos\theta \cdot \sqrt{\frac{1}{1 + \text{THD}^2}} \tag{13}$$

where $\cos\theta$ is referred to a displacement power factor and the second notation of (13) represents a distortion factor (DF).

3. Simulation results and discussion

The MATLAB Simulink simulation tool is implemented to investigate and verify the proposed inputs of control strategy for harmonic current mitigation and reactive power compensation using SAPF. The model of test system is designed based on Figure 1 where the system parameter is listed in Table 1. By employing fixed step discrete solver and the sampling time was set at 5 μs, the simulation ran about 1 s. A Fast Fourier Transform analysis has been used to analyse the source current THD.

Table 1. System parameter for simulation

Parameter	*Value*
Source voltage	V_s = 110 V_{rms}
System frequency	f_1 = 50 Hz
Line impedance	L_s = 0.5 mH, R_s = 0.05 Ω
Diode rectifier	L_o = 50 mH, R_o = 20 Ω
SAPF	L_s = 3 mH, R_s = 0.1 Ω, C_{dc} = 470 μF
Switching frequency	f_s = 20 kHz
Proportional gain	K_p = 3.2
Integral gain	K_i = 1824

The main objective of SAPF application is to establish pure sinusoidal source current as well as to fix it becoming in phase with the source voltage. In order to show the effectiveness of the proposed control strategy in compensating reactive power and reducing the total harmonic distortion (THD) under 5%, the i_s waveform and THD spectrum are depicted in Figure 5. As a result, after SAPF connection, the i_s becomes exactly a sinewave with its THD around 4% from 28% previously and in phase with the v_s. In addition, the system PF also improved from 0.931 to 0.999. The reduction of i_s THD and the increment of system PF caused the system active power increased from 542 W to 606 W, while the system reactive power reduced from 151 Var to nearly 0 Var. Therefore, the SAPF successfully compensated the system reactive power.

Moreover, the THD spectrum measured until the 40th harmonic order is depicted in Figure 6 shows a significant improvement after SAPF connection. The peak magnitude of fundamental harmonic current has increased significantly to 7.8 A after filtering, from 7.5 A before filtering. However, the rest of harmonic orders indicate significant magnitude reduction especially the dominant harmonics after filtering, which is expected for successful suppression of dominant harmonics after SAPF implementation. Careful observation shows that at the vicinity of highest harmonic order, its particular magnitude is greater when the SAPF is connected than the disconnected one. The reason could be the injection of high frequency i_c that incorporates with overmodulation PWM switching.

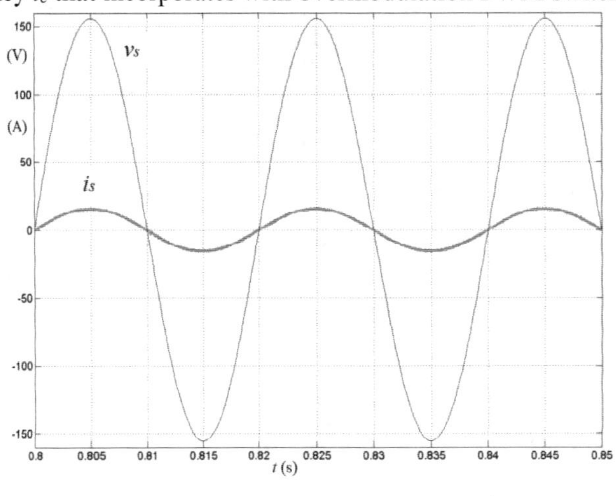

Figure 5. Steady state response of the i_s and v_s

IOP Conf. Series: Materials Science and Engineering 1127 (2021) 012030 doi:10.1088/1757-899X/1127/1/012030

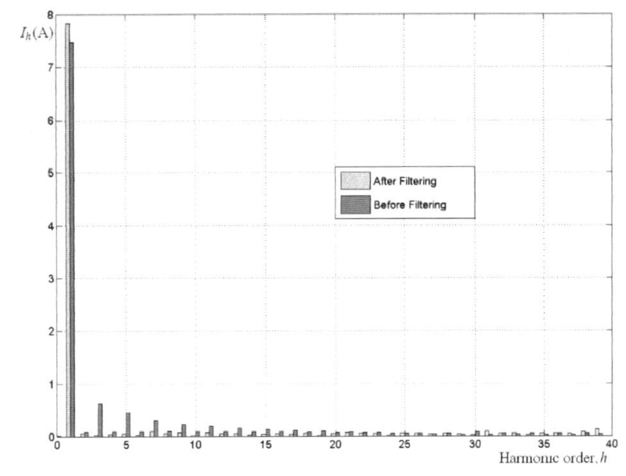

Figure 6. THD spectrum of the i_s before and after filtering

As the SAPF connected to the mains, the waveforms of i_L, i_c, v_o and i_o are illustrated in Figure 7 and 8 respectively. The i_L maintained its feature except the waveform becomes thicker than the previous one. One reason behind this is due to the injection of high frequency component of the i_c. From the observation, the THD of i_c is higher than the i_L, but the magnitude of i_c is lower than the i_L. The i_c waveform is identical to its reference i_c^* and owns rich of harmonic content.

On the other hand, the waveforms of v_o also indicates an identical characteristic like the i_L, e.g. its line variation is thicker than before the connection of SAPF. Again, the reason is because the magnitude of higher harmonic order (40th harmonic order) has been increased a little bit from its origin like depicted by THD spectrum of Figure 6. Moreover, the high switching frequency of 20 kHz injected by the SAPF also contributes to the noisy waveform of v_o, besides the absent of a shunt capacitor across the load which can act as a natural filter for the v_o. Nevertheless, the waveform of i_o is unchanged, since its waveform is being filtered by the L_o all the time.

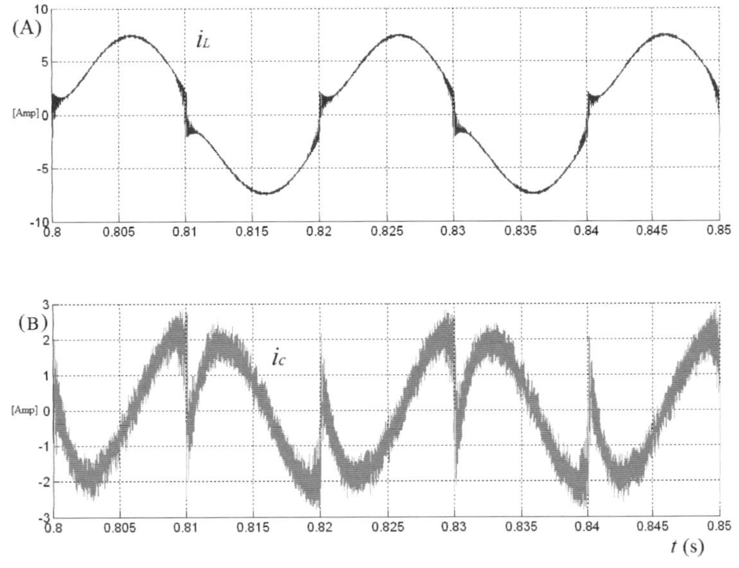

Figure 7. i_L and i_c waveforms after SAPF connection

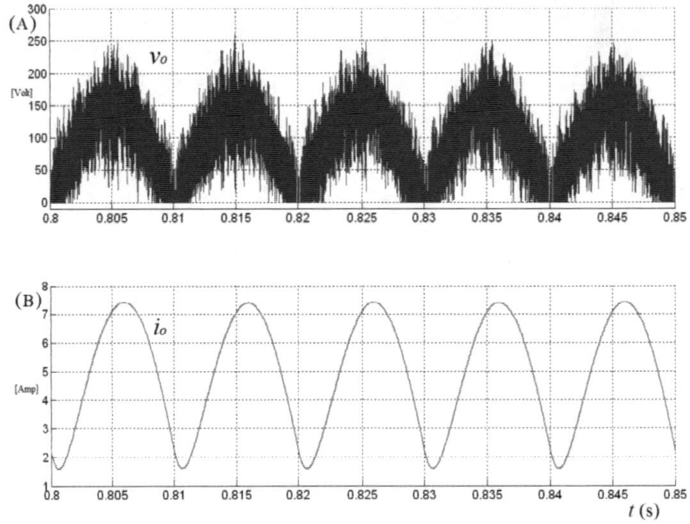

Figure 8. v_o and i_o waveforms after SAPF connection

The output signals responses of LPF and their final product, the R_e signal are depicted in Figure 9. The filtered signals are smoother and slicker compared to the original waveforms of v_o and i_o respectively. Their ripple magnitudes have been reduced significantly which only small variation left on the V_o signal, while the I_o signal features an almost constant DC signal. The settling time for V_o took 0.5 s while the I_o took only 0.2 s. The reason for slow transient response of V_o in particular is due to the time delay introduced by the first order Butterworth LPF. Nevertheless, the R_e signal indicates much faster response by having reached the value of 20 at less than 0.02 s. Since the R_e is a ratio of V_o and I_o, even during transient response, it correctly estimates the value of load resistance R_o, the purpose of estimation.

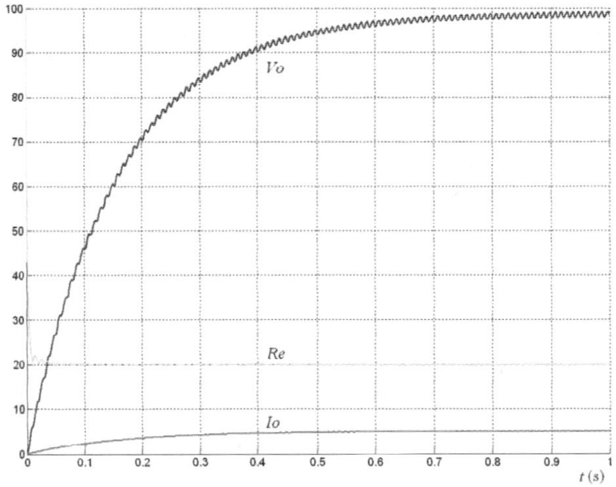

Figure 9. V_o, I_o and R_e signals responses

In order to have good control design, the controller must have a correct and accurate estimation of sinusoidal reference current i_s^* and compensated reference current i_c^*. Using nonlinear load equivalent resistance estimation technique, the harmonic detection method is made simple with short mathematical expression. Therefore, Figure 10 depicts the i_s^* and i_c^* waveforms appearance which are periodical sine wave and highly distorted waveform respectively, same as expected. By comparing the i_s^* with i_s and i_c^* with i_c, clearly shown that the actual waveforms are identical to their reference signals in term of magnitude, phase and shape.

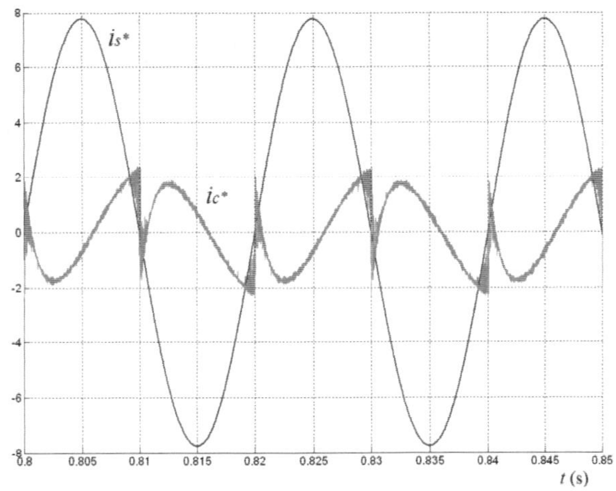

Figure 10. i_s^* and i_c^* signals at steady state

Finally, Figure 11 shows the signal response of u and e respectively during SAPF operation. Both signals exhibit a constant variation within the predetermined boundary which signifies the actual i_c keeps tracking its reference signal accordingly and the control system is stable. As a result, the u signal response indicates that the designed PI controller successfully compensated the e and produced a good control signal for SAPF application. The K_p and K_i tuning definitely give a great impact on the SAPF dynamic response, accuracy and stability. Without accurate K_p and K_i tuning, the SAPF cannot achieve its goal to reduce current harmonic distortion and compensate the reactive power.

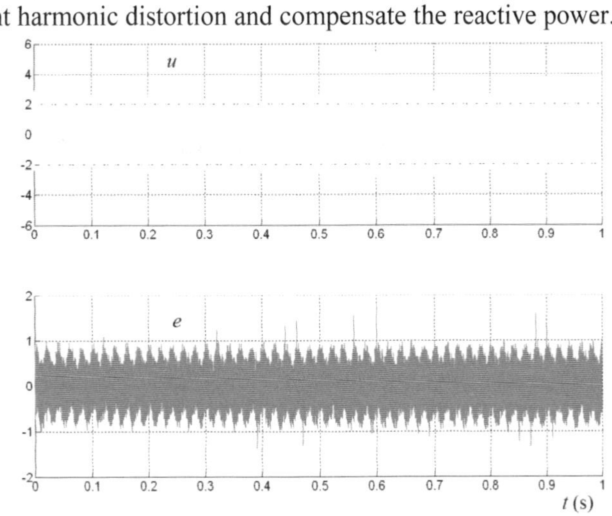

Figure 11. u and e signals responses

4. Conclusion

The new inputs of SAPF control strategy based on nonlinear load equivalent resistance estimation technique has been presented and discussed. The proposed control strategy has a simple design in term of harmonic detection method calculation, introducing the DC load voltage and current as the input variables of harmonic detection method [11], utilizing a PI controller, a bipolar PWM as well as a H-bridge VSI configuration. Besides it used the basic control strategy developed in [6] but with different source of inputs. Thus, the cost to develop the SAPF prototype is low albeit the performance is on par with the other control strategies. It is validated by the simulation result, and has been shown that the equivalent resistance R_e can be estimated accurately which lead to attain a pure sinewave for source current and in phase with the source voltage. As a conclusion, the objective of the proposed control strategy has been achieved after the source current THD successfully obtained below the limit specified by the IEEE-519 standard.

Acknowledgements
This research is supported by the Ministry of Education, Malaysia (KPM) and Universiti Kebangsaan Malaysia (UKM) under FRGS/1/2018/TK04/UKM/03/4 research grant.

References

[1] Devassy S, Singh B. Implementation of solar photovoltaic system with universal active filtering capability. *IEEE Trans Ind Appl*. 2019;**55**(4):3926-34.
[2] Zainuri MAAM, Radzi MAM, Soh AC, Mariun N, Rahim NA. DC-link capacitor voltage control for single-phase shunt active power filter with step size error cancellation in self-charging algorithm. *IET Power Electronics*. 2016;**9**(2):323-35.
[3] Mazari B, Mekri F. Fuzzy hysteresis control and parameter optimization of a shunt active power filter. *Journal of information science and engineering*. 2005;**21**(6):1139-56.
[4] ImaniJajarmi H, Mohamed A, Shareef H. A new method for calculating the reference current of shunt active power filters based on recursive discrete Fourier transform. *International Review on Modelling and Simulations*. 2012;**5**(4).
[5] Ye S, Zhang Y, Xie L, Lu H. Shunt active power filter based on proportional integral and multi vector resonant controllers for compensating nonlinear loads. *Journal of Electrical and Computer Engineering*. 2018;**2018**.
[6] Yusof Y, Umar RZR, Rahim NA. Penuras kuasa aktif pirau satu fasa menggunakan strategi kawalan ringkas. *Jurnal Teknologi*. 2017;**79**(6).
[7] Singh Y, Hussain I, Singh B, Mishra S. Single-phase solar grid-interfaced system with active filtering using adaptive linear combiner filter-based control scheme. *IET Generation, Transmission & Distribution*. 2017;**11**(8):1976-84.
[8] Tuyen ND, Fujita G. PV-active power filter combination supplies power to nonlinear load and compensates utility current. *IEEE Power and Energy Technology Systems Journal*. 2015;**2**(1):32-42.
[9] Rahmani S, Mendalek N, Al-Haddad K. Experimental design of a nonlinear control technique for three-phase shunt active power filter. *IEEE Trans Indust Electr*. 2010;**57**(10):3364-75.
[10] Rohouma W, Balog RS, Peerzada AA, Begovic MM. D-STATCOM for harmonic mitigation in low voltage distribution network with high penetration of nonlinear loads. *Renewable Energy*. 2020;**145**:1449-64.
[11] Yusof Y, Za'im R, Ayob A, Muad AM, Mohamed R, editors. Harmonic Current Detection Method Using Nonlinear Load's DC Output Voltage and Current. 2019 7th International Conference on Smart Computing & Communications (ICSCC); 2019: IEEE.

IOP Conf. Series: Materials Science and Engineering 1127 (2021) 012031 doi:10.1088/1757-899X/1127/1/012031

Control Strategies for Bidirectional DC-DC Converters: An Overview

Muhammad Husnain Ashfaq*, Jeyraj A/L Selvaraj, Nasrudin Abd Rahim

UM Power Energy Dedicated Advanced Center (UMPEDAC) Jalan Pantai Baharu, Level 4, Wisma R&D University of Malaya, 59990 Kuala Lumpur, Malaysia

E-mail: husnainie14@yahoo.com

Abstract. The energy produced by renewable energy sources has variations as these sources are sensitive to weather conditions. To overcome this problem, storage devices like batteries are used for stabilization. Bidirectional DC-DC converter is the main component to interface renewable energy sources with energy storage devices. With the help of proper switching techniques, these converters have the ability to increase or decrease the level of the voltages and can handle the power flow in both directions, from source to storage device (forward direction/charging mode), as well as from storage device to source (backward direction/discharging mode). Hence for an efficient system, the control of power flow must be very effective. This paper presents an overview of various bidirectional DC-DC converter's classical and supervisory control methods that have been proposed in recent years.

1. Introduction

The global energy demand proliferates every year with increasing technology and fast industrial growth rates of high dense regions like South East Asia, Europe, and USA. However, the fossil fuels are in rapid extinction. On the other hand, supplying electric power with quality and reliability has been a challenge due to the overloads in the grid transmission systems. Renewable energy sources provide the ideal answer to the above questions, but the variable nature of these sources does not allow having continuous and reliable source of energy when these resources are used as sole source of power and cause unfavourable effects on the power quality of the connected grid. To overcome this issue, energy storage devices use besides renewable sources. With the use of these devices such as battery packs or super-capacitors, power fluctuations can be eliminated in order to deliver smooth power to the consumer and maximize the energy output of the renewable energy sources.

A bidirectional DC-DC converter is the key element to interface between high voltage DC buses with distributed energy storage devices and low voltage DC buses with distributed energy storage devices in micro grids as shown in figure 1. A bidirectional DC-DC can be used to step down the DC voltage level "buck- converter", step up the DC voltage level "boost- converter" or both step down-step up converter "buck-boost converter", and can transfer power in both directions, forward as well as backward as shown in the figure 2. In classical DC-DC converter a bidirectional characteristic can be achieved by the use of anti-parallel diode with MOSFETs or IGBTs. With the help of controlled switching methods, this configuration allows the flow of current in forward direction as well as in backward direction. Due to its bi-directional ability, it reduces the size, improves the performance and overall conversion efficiency of the system as it does not require two different converters for forward and backward flow of power. Recently it has been getting a lot of attention in academics and vast amount of research studies have been conducted on the controlling techniques for DC-DC converters. The fundamental aim of the controlling method is to provide a fast dynamic response and must have a stable DC output voltage in both power flows (forward and backward) despite the limited amount of sensors. For better and efficient performance of the system, a high efficient bidirectional DC-DC converter is required in order to control the flow of power in both directions by using switching techniques. So an efficient control technique is important for a DC-DC converter as it has two conversion stages (Inverter stage and Rectifier stage). In this paper, an overview of different control methods for bi-directional DC-DC converters is done, and these techniques are critically compared.

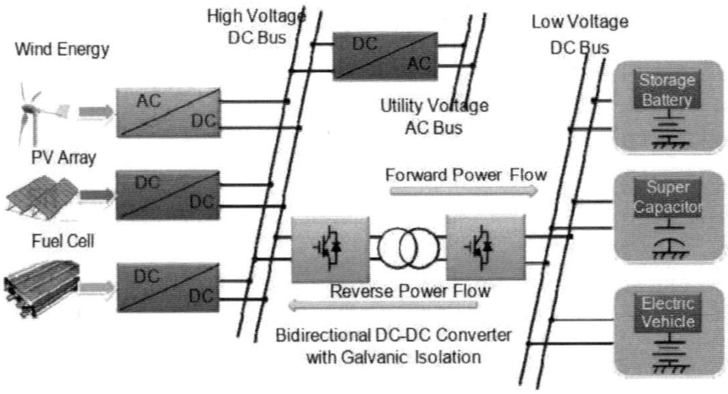

Figure 1. Micro grid with Energy storage devices

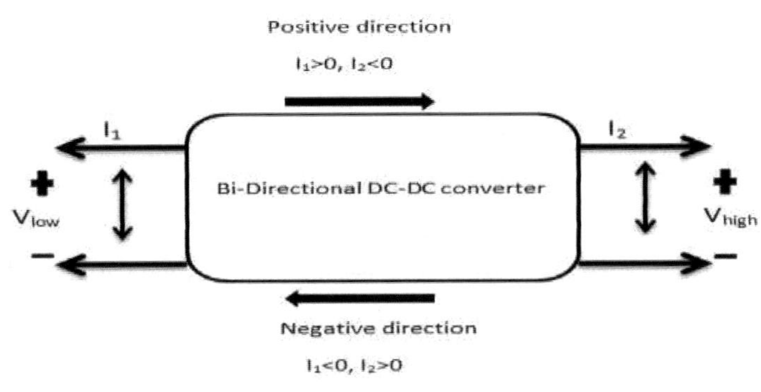

Figure 2. Bidirectional flow of power

2. Proportional-Integral-Derivative Control

A classical PID control is the most simple and easy to implement control method. Due to its excellent control dynamics with zero steady-state error, very swift response, and higher stability, it is the first choice for designing a control algorithm. The generic block diagram of a PID controller is shown in the figure 3. In [1] a pulse width modulation (PWM) is used in conventional bidirectional DC- DC converter for generating the triggering pulses of the switches. As firing angle is fixed, so due to a small variation in input voltage, a high fluctuation in output voltage is observed. To overcome this issue PID controller is proposed. The operating principle of PID controller is discussed in [2, 3]. Despite the change in input supply voltage the output remained constant because the output voltage of the converter is compared with the desired set value, and triggering signals are given to the switches to obtain the fixed desired output voltage.

IOP Conf. Series: Materials Science and Engineering 1127 (2021) 012031 doi:10.1088/1757-899X/1127/1/012031

In [4] a comparison between PID and PI controllers for bidirectional DC-DC converters is discussed in terms of rise time, fall time, steady-state error, and peak overshoot. It is found; PID produced less settling time and steady-state error. During boost mode, PID controller had a settling time of 0.6 sec, and a steady-state error of 0.5 volts, whereas PI controller had a settling time and steady-state error of 0.81 sec and 0.8 volts respectively. Similarly, during buck mode, a steady-state error of 0.71 volts and 0.92 volts is observed with the PID controller and PI controller respectively. As in PI controller, oscillations and overshoot occurred in response to the output of the system. So, an additional derivative gain component with PI is used to eliminate overshoot. Similarly in [5], another comparison in terms of output voltage's stability is made between PID and PI controllers for bidirectional converters used in electric vehicles. In case of PI control method, for different input supply voltage, less variation in output voltage is observed. So while choosing PID or PI, there is always a trade-off between output voltage stability and overshoot. Although PID controllers have high reliability, high dynamics, and are suitable for many control problems, however, they have lack of robustness in the presence of uncertainty and large disturbances. There are many applications where they have very low efficiency especially when non-linearity is present in the system.

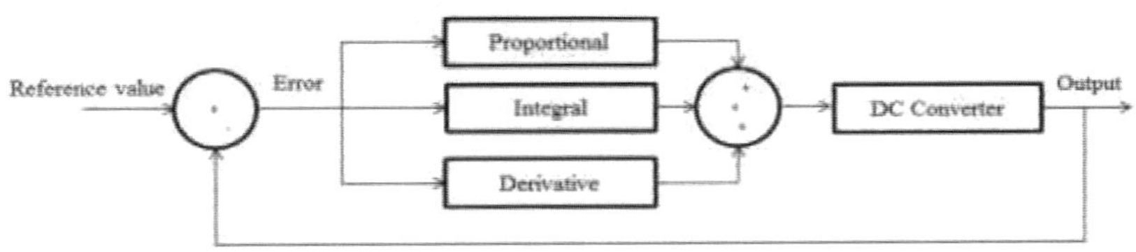

Figure 3. Block diagram of PID control

3. Phase Shift Control

Single-phase shift (SPS) control is a widely used control method for isolated dual active bidirectional DC-DC converters [6, 7]. In single-phase shift control method, a phase shifted square wave with a 50 percent duty ratio is generated by turning on cross-connected switches of both bridges of the dual active bidirectional converter. The voltage of the transformer's leakage inductor (VL-T) can be varied by regulating the phase-shift ratio between V1 (bridge 1 equivalent ac output voltage) and V2 (bridge 2 equivalent output voltage). Through this voltage change, the power flow direction and its magnitude can be controlled. SPS control is very effective because of its various benefits, such as high dynamics, low inertia, and easy implementation. However, the converter cannot work under zero voltage switching (ZVS) because the flow of power and its control depend on the leakage inductance of the transformer. When there is a mismatch between the transformer's primary and secondary voltages magnitude then it causes large circulating power that eventually increases the RMS and peak currents as shown in the figure 4(a). Hence power conversion loss becomes high and the overall efficiency of the converter is reduced.

In power transmission process, there are intervals when there is a backflow of power(negative flow) and as in a single phase shift method, the power flow depends on the leakage inductance of the transformer. So when back power flow increases, then the forward power flow also increases and results in high circulating power. To overcome this issue, an extended phase shift (EPS) is proposed. EPS is an improved method of SPS. The working principle, soft-switching property, and power losses of extended phase shift have been studied in [8-10]. In this method, the cross-connected switch pairs of one bridge are turned on, whereas the cross- connected switch pairs of the second bridge are turned on with an inner phase shift ratio. The ac output voltage of one bridge gives a three level wave and the output of second bridge becomes a two level square wave with a 50% duty ratio as shown in the figure 4(b). During zero- voltage time intervals of the bridge with three-level wave, the backflow of power is zero so the circulating power reduces. Contrast with SPS, EPS has two phase shifts, an outer phase-shift ratio (D1), which is the phase shift between the primary and secondary voltages of high-frequency transformer and is used to control the magnitude and direction of power flow, and the inner phase-shift ratio (D2) which is used to increase the zero voltage operating range and decrease the circulating power. However in this control technique, when the power flow direction is changed between forward and backward and voltage conversions are changed between buck and boost mode, the operating states of both full bridges are needed to be exchanged in order to get the minimized circulating power. The dual-phase shift control is proposed in [11]. The working principle, soft-switching property, and power losses of dual phase shift control have been discussed in [12-15]. In contrast with EPS, the ac output voltages of both full bridges are three-level waves as in this method the cross-connected switch pairs of both full bridges are turned on with an inner phase shift with the same shift ratio as shown in figure 4(c). Because of this, the operating states of the bridges are not needed to be exchanged during power (forward to backward and vice versa) and voltage (buck to boost and vice versa) transitions. Compared to SPS method, DPS has steady-state current, increased zero voltage switching operating range, and decreased current stress. Therefore overall efficiency of the system is increased. Comparing SPS, EPS, and DPS from the view of performance in large practical applications, DPS is a relatively optimal control method because of its fast dynamic response, high efficiency, easy implementation, and ZVS operating range.

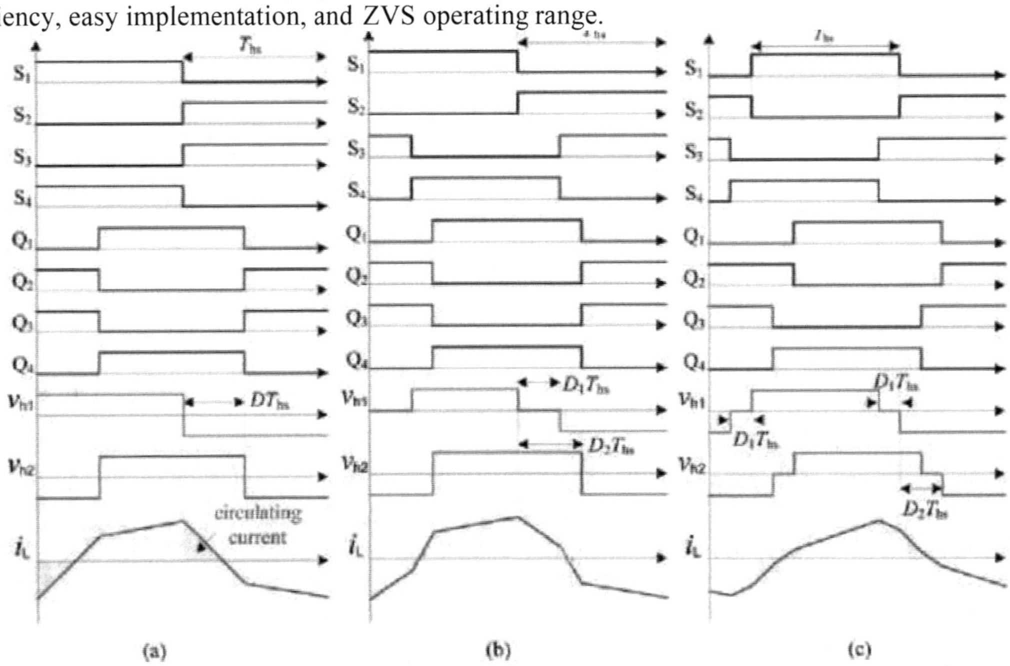

Figure 4. (a) SPS (b) EPS (c) DPS [11]

4. Sliding Mode Control

Due to the presence of non-linear elements in different topologies of bidirectional DC-DC converts, the dynamic equation of the converter becomes non-linear. One approach in designing the control algorithm of that model is to make it linear with the help of linearization methods [16]. But, this method does not represent the exact model as it is based on assumptions and estimations. So, to get a fast dynamic and reliable system that can respond to external disturbances then it is more suitable to use non-linear control method for controlling bidirectional DC-DC converters. Sliding mode is one of the widely used non-linear control method because it has fast dynamic response, robustness against variations in parameters and low sensitivity to external disturbances [17]. The generic block diagram of sliding mode control is shown in figure 5. In [18] for coupled-inductor bidirectional DC-DC converter a sliding mode control is proposed. The proposed sliding mode technique tracked the reference DC voltage of the high side with high performance, which shows its robustness against perturbations. In [19] a sliding mode control of fixed frequency bi-directional DC-DC converter is studied. The comparison is made between PI and sliding mode control in terms of load disturbances and variation in parameters. An overshoot of 5 volts and 11 volts is observed with sliding mode control and PI controller respectively. This shows that sliding mode control is more robust, and has a very low overshoot compared to the PI controller. It is because the sliding mode of control variables is designed according to the need and has nothing to do with the system parameters and external perturbations.

Although this control technique has the robustness and dynamic response. But there are also some limitations, as correct parameters and state information are required. Furthermore, in the sliding mode control technique, there might be unwanted oscillations with finite frequency and amplitude. These undesirable oscillations are known as 'chattering'[20]. To reduce this chattering phenomenon a fuzzy-sliding control is proposed in [20]. With the combination of these two methods, very good robustness can be achieved despite the presence of outer disturbances, and the system will also get stable as the oscillations in the system will be minimized.

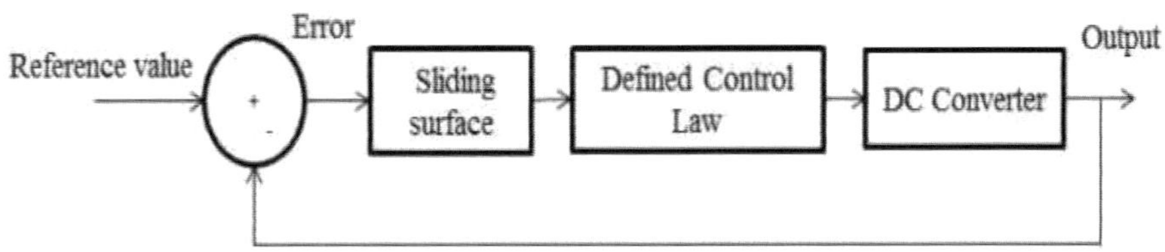

Figure 5. Block diagram of sliding mode control

5. Fuzzy Logic Control

For complex and non-linear system with load disturbance, fluctuation in parameters and uncertainty, a robust response can be achieved with the help of fuzzy logic controllers. These controllers use a very versatile set of if-then rules. Due to their adaptive property these controllers are gaining popularity in controlling bidirectional DC-DC converters. The foremost advantage of using this technique is that no previous information of the system's parameters is needed and contrast with sliding control method, fewer measurements are needed to design the controller. A generic block diagram of fuzzy logic is shown in figure 6. Fuzzy controllers are of two basic types which are FuzzyPI control technique and FuzzyPSOPI (Fuzzy Particle Swarm Optimization Proportional Integral) control technique [21]. In [22] for coupled inductor bi- directional converter a fuzzy logic and PI controller has been designed. The model for both controllers with the voltage control mode is investigated in MATLAB Simulink, steady- state and transient response of the system examined. With the help of simulations it is observed that the fuzzy logic control method had a more stable and dynamic response and had less settling time compared to PI controller. In [21] a comparison between fuzzy logic and dual-phase shift (DPS) is examined for dual active bridge isolated bi-directional DC-DC converters. The study shows high performance results when fuzzy logic is implemented, and had less settling time of 0.32 sec during buck mode, and 0.34 sec during boost mode.

While with DPS, it was 0.53 sec, and 0.35 sec during buck and boost modes respectively. Furthermore, fuzzy logic particle swarm optimization PI control gave more efficient results than the other two logics, because of low proportional (Kp), integral (Ki) gain values with low error of 0.065% and 0.1% during buck and boost modes respectively, and less settling time of 0.3 sec and 0.31 sec during buck and boost modes respectively. However, in order to get minimized errors in fuzzy logic, expert knowledge is required in designing the control algorithm.

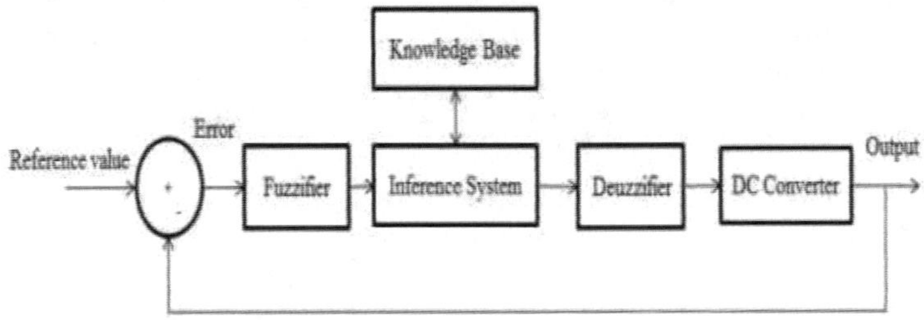

Figure 6. Block diagram of fuzzy logic

6. Model Predictive Control

A model predictive control is a dominant and arising algorithm used in power electronics and energy conversion systems. The above mentioned control techniques perform correction after the error occurs while the main feature of model predictive control is to predict the future behaviour of the control constraints. The block diagram of model predictive control is shown in the figure 7. Because of its very fast dynamic response, and swift reference tracking characteristic this mode has become most attractive control technique comparing with other classical techniques. Although this technique requires a high number of calculations, but with the help of fast, powerful and sophisticated microprocessors easily available today it is very simple to implement. It is used in energy management of fuel cell/battery/ super capacitor hybrid power source [23].

In [24] bidirectional DC-DC converter with model predictive control algorithm is designed, and the efficiency of the system is investigated experimentally with a 2.5kw prototype for a dual phase shift with PWM based control technique, and model predictive control technique. It is found that with Dual- phase shift technique the efficiency of the converter is 89.56% while, with model predictive control method the efficiency of the converter increased to 92.52% because model predictive method has zero phase shift and unity power factor between voltages and currents of high-frequency transformer hence decreasing the overall reactive power of the converter. In [25] a model predictive control method based on cuckoo search algorithm (CS-model predictive method) is proposed for interleaved parallel bidirectional DC- DC converters. The converter is simulated in MATLAB with the PI control method and CS-model predictive method in loading and shedding of the inductor current. It is found that CS-model predictive algorithm had a fast dynamic response with no overshoot, and a rising time of 0.86 milliseconds while PI control method had an overshoot of 6.3% with settling time of 5.02 milliseconds. One of the major problems in this method is, the non-linear behaviour of the model is locally linearized at different operating range, and in order to achieve good performance, linear dynamic design of a bidirectional DC-DC converter is needed inside this technique. To overcome this issue, the multi model predictive control method is proposed.

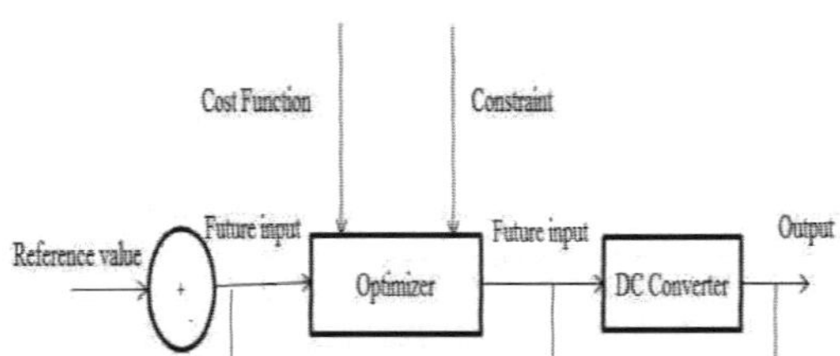

Figure 7. Block diagram of Model Predictive control

7. Artificial Neural Network

An artificial neural network (ANN) is a biological brain representation of a computer model and composed of thousands and millions of interconnected neuron sets bonded with weight connections to process information based on the input signals. It has a layer structure with an input layer, hidden layer, and an output layer as shown in the figure 8. In [26] ANN method is proposed for controlling bidirectional DC-DC converters used in standalone DC micro grids. On the bases of approximate dynamic programming; ANN is trained to have optimum control. From the studies, it is found that very high voltage stability of standalone dc micro gird can be maintained by using artificial neural networks. Whenever there are load changes, the output of the converter can be changed rapidly to maintain the dc bus voltages. This shows that ANN has very fast response time, very insensitive to external disturbances, and can track the reference DC voltage very rapidly.

In [27] an adaptive artificial neural network based PI (AANN-PI) is proposed for bidirectional DC-DC converters. A performance comparison is made between AANN-PI, classical PI, and sliding control in regards of oscillations, rise time and settle time. All three controllers were tested on 150 watt bidirectional DC-DC prototype under various variations in the supply input voltage. From the results it is cleared that AANN-PI controller gives better performance and has high fast dynamic response with insensitivity against external disturbances. As it is discussed, to get minimum errors in fuzzy control logic, precise expert knowledge is very important in designing a control algorithm. To overcome this limitation an Artificial Neuro-Fuzzy interference system is proposed in [28] to control the speed of brushless DC (BLDC) motor with bidirectional DC-DC converters. It is observed that despite the load variations, excellent speed control of BLDC and efficient energy conversation are achieved by combing fuzzy with neural networks. ANN can learn and model no-linear complex systems that is why it is getting very popular in modern technology era but it also has some drawbacks such as time-consuming as they need extensive training, take too much time to process large neural network, and massive amount of data is required for quality prediction.

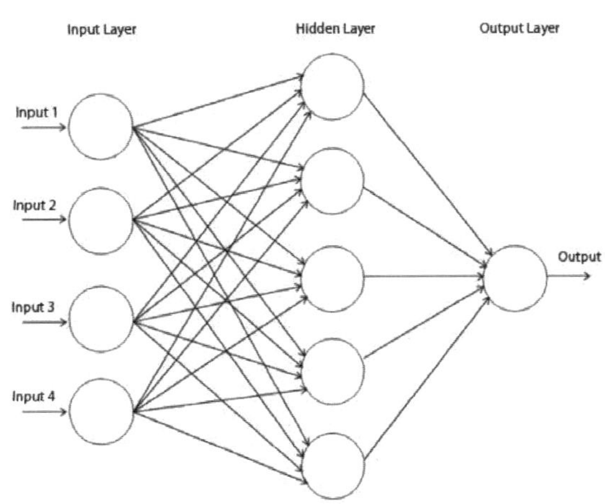

Figure 8. Layer structure of ANN

8. Conclusion

Each control technique has its merits and demerits. Classical control methods are easy to implement and can be used for various applications but they are not effective during the presence of external disturbances. Whereas the supervisory control techniques i.e. fuzzy logic, neural networks are very suitable for non-linear systems and have a very fast dynamic response but they are complex compared to classical methods and also have some limitations. For the very fast dynamics and robustness it is recommended that use classical technique inside the closed loop of supervisory control like Fuzzy-PI or Neural-PI.

Table 1. Brief summary of control methods used in bidirectional DC-DC converters

Control method	Control Problems	Merits	Limitations
PID	-Power flow control -Reduce dead time of switches[2]	-High dynamics -High reliability -Low cost[29]	-Not robust against disturbance and uncertainty - Low efficiency[1]
Single phase shift	-Power flow control - Minimize circulating power[30]	-High dynamics - zero voltage switching and soft switching control[30]	-Backflow of power -High inductor RMS current -Lack of ZVS operating range[31]
Extended Phase shift	-Decrease backflow of power -Minimize back flow of power -reduce circulating current[8]	-High dynamics -Decrease peak current -Increase ZVS operating range[8]	-Exchange of bridge's operating states during power and voltage transitions[11]
Dual Phase shift	-Decrease backflow of power - Eliminate reactive power -reduce circulating power[30, 32]	-ZVS for whole operating range - High efficiency -High dynamics -Improved dead- band effect[32]	-low quality operating modes - Difficult to get high efficiency[21]
Sliding mode control	-Insensitive to external perturbations -Handle severe variations between load and line[31]	-Fast dynamic response -Reference Tracking ability - Robust against external disturbances and uncertainty[31]	To design model accurate parameters and state information are required[31]
Fuzzy control	-Minimize power consumption from grid -minimize control time[21]	-Fast response and robustness - Adaptive property -Applicable to inaccurate and non-linear system[21]	Expert knowledge is required[31]
Model predictiv e	-Power flow control -Regulation of DC voltages and currents[33, 34]	-Fast dynamic response - Reference tracking ability -predict error before it occur[35]	Limitation of using a linear model inside the algorithm[36]
Artificial Neural Network	-Optimal power flow control -model non-linear complex systems[37, 38]	-High stability - Fast dynamic control -Fast reference tracking ability[39]	- Controller need to be trained - Large amount of previous data required - Time Consuming [39]

Acknowledgement
The grant for this research was supported by UMPEDAC- 2018, HICoE Ministry of Education Malaysia.

References

[1] DAS R. PI CONTROLLED BI-DIRECTIONAL DC-DC CONVERTER AND HIGLY EFFICIENT BOOST CONVERTER FOR ELECTRIC VEHICLE: INTERNATIONAL ISLAMIC UNIVERSITY CHITTAGONG; 2014.
[2] Lemus J, Vélez GC, Rodrıguez NJC. Pid controller design for dc motor. *Contemp Eng Sci.* 2018;**11**(99):4913-20.
[3] Khan MR, Khan AA, Ghazali U. Speed Control of DC Motor under Varying Load Using PID Controller. *International Journal of Engineering (IJE)*. 2015;**9**(3):38-48.
[4] Ramya K, Jegathesan V. Comparison of pi and pid controlled bidirectional dc-dc converter systems. *International Journal of Power Electronics and Drive Systems*. 2016;**7**(1):56.
[5] Das R, UddinChowdhury MA, editors. PI controlled Bi-directional DC-DC converter (BDDDC)

and highly efficient boost converter for electric vehicles. 2016 3rd International Conference on Electrical Engineering and Information Communication Technology (ICEEICT); 2016: IEEE.
[6] Monfared KK, Iman-Eini H, Razi R, editors. Control of single-phase bidirectional pev/ev charger based on fcs-mpc method for v2g reactive power operation. 2019 10th International Power Electronics, Drive Systems and Technologies Conference (PEDSTC); 2019: IEEE.
[7] Zhao B, Song Q, Liu W, Liu G, Zhao Y. Universal high-frequency-link characterization and practical fundamental-optimal strategy for dual-active-bridge DC-DC converter under PWM plus phase-shift control. *IEEE Trans Pow Electr*. 2015;**30**(12):6488-94.
[8] Zhao B, Yu Q, Sun W. Extended-phase-shift control of isolated bidirectional DC–DC converter for power distribution in microgrid. *IEEE Trans Pow Electr*. 2011;**27**(11):4667-80.
[9] Zheng M, Wen H, Shi H, Hu Y, Yang Y, Wang Y. Open-circuit fault diagnosis of dual active bridge DC-DC converter with extended-phase-shift control. *IEEE Access*. 2019;**7**:23752-65.
[10] Naayagi R, Forsyth A, Shuttleworth R, editors. Performance analysis of extended phase-shift control of DAB DC-DC converter for aerospace energy storage system. 2015 IEEE 11th International Conference on Power Electronics and Drive Systems; 2015: IEEE.
[11] Gao S, Lu X, Liu X, editors. Minimum reflux power control of bidirectional DC-DC converter based on dual phase shifting. 2018 IEEE 2nd International Electrical and Energy Conference (CIEEC); 2018: IEEE.
[12] SAVRUN MM, KOROGLU T, Adnan T, CUMA MU, BAYİNDİR KC, TUMAY M. 1P3S HIGH FREQUENCY TRANSFORMER BASED DUAL-ACTIVE BRIDGE DC-DC CONVERTER. *International Journal of Energy Applications and Technologies*.**4**(4):174-81.
[13] Costinett D, Maksimovic D, Zane R. Design and control for high efficiency in high step-down dual active bridge converters operating at high switching frequency. *IEEE Trans Pow Electr*. 2012;**28**(8):3931-40.
[14] Kim M, Rosekeit M, Sul S-K, De Doncker RW, editors. A dual-phase-shift control strategy for dual-active-bridge DC-DC converter in wide voltage range. 8th International Conference on Power Electronics-ECCE Asia; 2011: IEEE.
[15] Hong M, Xuanjie G, Chengbi Z, Shujiang D, editors. An improved dual phase shift control strategy for dual active bridge DC-DC converter with soft switching. 2018 International Power Electronics Conference (IPEC-Niigata 2018-ECCE Asia); 2018: IEEE.
[16] Filsoof K, Lehn PW. A bidirectional multiple-input multiple-output modular multilevel DC–DC converter and its control design. *IEEE Trans Pow Electr*. 2015;**31**(4):2767-79.
[17] Cao J, Chen Q, Zhang L, Quan S, editors. Sliding mode control of bidirectional DC/DC converter. 2018 33rd Youth Academic Annual Conference of Chinese Association of Automation (YAC); 2018: IEEE.
[18] Ciccarelli F, Lauria D, editors. Sliding-mode control of bidirectional dc-dc converter for supercapacitor energy storage applications. SPEEDAM 2010; 2010: IEEE.
[19] [19] Jeung Y-C, Choi I-C, Lee D-C, editors. Robust voltage control of dual active bridge DC-DC converters using sliding mode control. 2016 IEEE 8th International Power Electronics and Motion Control Conference (IPEMC-ECCE Asia); 2016: IEEE.
[20] Utkin V. Chattering problem. *IFAC Proceedings Volumes*. 2011;**44**(1):13374-9.
[21] Kayaalp Rİ, Demirdelen T, Tümay M, editors. A novel fuzzy logic control for bidirectional DC-DC converter and comparison with dual phase-shift control method in medium voltage applications. 2016 IEEE International Conference on Computational Intelligence and Virtual Environments for Measurement Systems and Applications (CIVEMSA); 2016: IEEE.
[22] Narasimharaju B, Dubey SP, Singh S, editors. Intelligent technique for improved transient and dynamic response of bidirectional DC-DC converter. 2010 International Conference on Power, Control and Embedded Systems; 2010: IEEE.
[23] Bambang RT, Rohman AS, Dronkers CJ, Ortega R, Sasongko A. Energy management of fuel cell/battery/supercapacitor hybrid power sources using model predictive control. *IEEE Transactions on Industrial Informatics*. 2014;**10**(4):1992-2002.

[24] Akter P, Uddin M, Mekhilef S, Tan NML, Akagi H. Model predictive control of bidirectional isolated DC–DC converter for energy conversion system. *International Journal of Electronics*. 2015;**102**(8):1407-27.
[25] Sun W, Chen Q, Zhang L, editors. Model Predictive Control Based On Cuckoo Search Algorithm of Interleaved Parallel Bi-directional DC-DC Converter. 2019 34rd Youth Academic Annual Conference of Chinese Association of Automation (YAC); 2019: IEEE.
[26] Li S, Fu X, Weizhen D. Systems, methods and devices for control of dc/dc converters and a standalone dc microgrid using artificial neural networks. Google Patents; 2019.
[27] Ibeas A. Artificial Neural Network Based Adaptive Control of Single Phase Dual Active Bridge with Finite Time Disturbance Compensation.
[28] Kavathe R, Chandle JO, Patil N, Kokare M. ANFIS Based Speed Control of BLDC Motor with Bidirectional DC-DC Converter. *International Journal of Research and Scientific Innovation (IJRSI)*. 2018;**5**(6):153-8.
[29] Knospe C. PID control. *IEEE Control Systems Magazine*. 2006;**26**(1):30-1.
[30] Kumar BM, Kumar A, Bhat A, Agarwal P, editors. Comparative study of dual active bridge isolated DC to DC converter with single phase shift and dual phase shift control techniques. 2017 Recent Developments in Control, Automation & Power Engineering (RDCAPE); 2017: IEEE.
[31] Gorji SA, Sahebi HG, Ektesabi M, Rad AB. Topologies and control schemes of bidirectional DC–DC power converters: an Overview. *IEEE Access*. 2019;**7**:117997-8019.
[32] Hou N, Song W, Wu M. Minimum-current-stress scheme of dual active bridge DC–DC converter with unified phase-shift control. *IEEE Trans Pow Electr*. 2016;**31**(12):8552-61.
[33] Mayne DQ. Model predictive control: Recent developments and future promise. *Automatica*. 2014;**50**(12):2967-86.
[34] Grüne L, Pannek J. Nonlinear model predictive control. Nonlinear Model Predictive Control: Springer; 2017. p. 45-69.
[35] Parisio A, Rikos E, Glielmo L. A model predictive control approach to microgrid operation optimization. *IEEE Transactions on Control Systems Technology*. 2014;**22**(5):1813-27.
[36] Batliner M, König O, Jakubek S, Prochart G. Method and controller for model predictive control of a multi-phase DC/DC converter. Google Patents; 2019.
[37] Qasim M, Khadkikar V. Application of artificial neural networks for shunt active power filter control. *IEEE Transactions on industrial informatics*. 2014;**10**(3):1765-74.
[38] Van Gerven M, Bohte S. Artificial neural networks as models of neural information processing. *Front Comput Neurosci*. 2017;**11**:114.
[39] Zulu A, John S. A review of control algorithms for autonomous quadrotors. *arXiv preprint arXiv:160202622*. 2016.

Design and Analysis of Multilevel Inverter based on Cascaded Half-Bridge and Three-Phase Unfolder Circuit

Farhana N. Rosli, M. F. M. Elias, N. A. Rahim

Higher Institution Centre of Excellence (HICoE), UM Power Energy Dedicated Advanced Centre (UMPEDAC), Level 4, Wisma R & D, University of Malaya, Jalan Pantai Baharu, 59990 Kuala Lumpur, Malaysia

E-mail: nurfarhanarosli@gmail.com

Abstract. In this paper, a multilevel inverter based on cascaded half- bridge and three-phase unfolder is proposed. The proposed inverter design is simulated by using MATLAB Simulink. The converter is designed with four series-connected half-bridge cells and a three-level neutral-point-clamped inverter. The proposed converter is able to produce an output voltage of 230V and a frequency of 50 Hz. There are two modulation techniques implemented for the proposed inverter topology namely; phase-shifted modulation and phase opposition disposition modulation. The basic principle of operations using both modulation techniques including the performance analysis are presented.

1. Introduction

Multilevel inverters are currently a promising power converter especially for medium and high-power applications [1]. The most commonly used MLI configurations are cascaded H- Bridge (CHB), neutral-point-clamped (NPC) and flying capacitor (FC). Generally, MLI has better quality of harmonic profile, reduced dv/dt stress on load voltage and enables the use of power switches with lower voltage rating. MLI also can be implemented in low voltage applications such as integration of renewable energy into power grids due to the advantages of decreased output harmonics and switching losses [2, 3].

MLI can be arranged in symmetric, asymmetric, unipolar and bipolar configurations as listed in [4-12]. Recently, the MLI topologies categorized as reduced device count topologies have been introduced [13]. The features and analysis of several MLI topologies were summarized and discussed in [13, 14]. The utilization of multiple dc sources and switches is a known concept to generate multilevel output voltage in MLI especially using CHB inverters. Owing to many advantages it has, MLI is particularly suited for renewable energy systems. However, if a CHB is extended to three-phase application, it will have a significant increase in the number of devices count such as power switches, dc sources or capacitors. In [9], half-bridge (HB) modules are connected to each dc source and a full-bridge inverter is used to unfold the rectified dc-link into AC output voltage. This configuration is called bipolar as the circuit composed of level and polarity circuits.

To generate a 3-phase 5-level output line voltage for the topology in [9], it requires 3 dc sources and 18 power switches. For 3- phase CHB configuration, 12 power switches and 3 dc sources are required to generate 5-level output line voltage. A combination of FC and CHB topology was implemented in [15]. However, the topology has shortcomings in terms of the utilization of large electrolytic capacitor and it requires additional control strategy to balance the capacitors in FC.

Recently, a series resonant converter based on a dual-active bridge with a three-phase NPC unfolder circuit has been introduced in [16] in which the converter is able to reduce dc- link capacitance requirement and switching losses of the inverter by using an unfolder circuit. Thus, with the advantages of NPC unfolder and half-bridge circuits, a new three-phase multilevel inverter topology based on half-bridge cell and NPC unfolder circuit is proposed in this paper. Since the half-bridge inverter is a multilevel configuration circuit, the harmonic distortion of the output voltage can be further reduced. Moreover, the converter size and control parameters are further reduced as the usage of capacitor is eliminated from the proposed topology. The advantage of the proposed topology based on half-bridge cell and NPC unfolder circuit is that the maximum level of output voltage that can be generated is seven at the maximum modulation index. Thus, the total dc-link voltage of the inverter is higher than the total input voltage. Another advantage of the proposed topology based on HB and NPC is the utilization of power switches are reduced from 36 to 32 as compared to 3-phase configuration producing 13-level output line voltage in CHB.

In Section 2, the circuit configuration and operation of half- bridge and NPC inverter are presented. In Section 3, phase- shifted modulation and phase opposition disposition modulation are explained and compared. In Section 4, the simulation results of both modulation techniques are shown and compared in terms of output voltage levels, harmonic distortion, modulation index, and filterrequirement.

2. Design and Operation Principles

Circuit configuration of the proposed three-phase unfolder topology based on multilevel inverter is illustrated in Fig. 1.

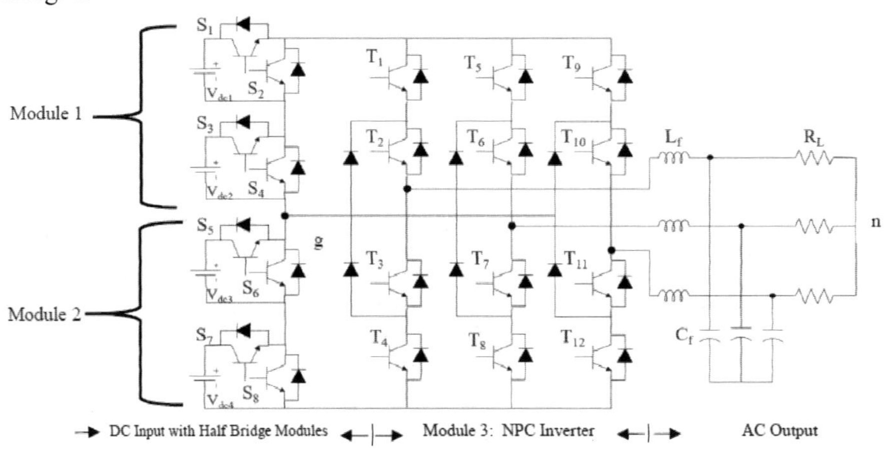

Figure 1. Circuit Configuration of Half Bridge Modules with Three Phase NPC Unfolder

The proposed topology consists of three modules. Module 1 and Module 2 form an arrangement of series connection of half-bridge for top and bottom dc-link of the inverter. The combination of the two modules acts as level generation part of the unfolder circuit. Module 3 consists of three-phase neutral-point-clamped inverter (NPC) and act as a polarity circuit. The half-bridge inverter is chosen since it is modular in nature and it can generate multilevel dc-link easily by connecting them in series [5, 9]. Whereas, the NPC inverter is selected since it can provide additional voltage level at higher modulation index compared to the two-level conventional inverter, thus producing higher output quality.

Level generation modules consist of 4 half-bridge cells in series with each cell has one direct current (dc) source and two switches operate in a toggle mode. Therefore, 8 switches and 4 dc sources are used to design the level generation modules as depicted in Fig.1. The first four switches (S1 to S4) are combined to form top module of dc-link while the later four switches (S5 to S8) are joined to form bottom module of the dc- link. There are 4 possible switching states of the half-bridge cell as listed in Table 1.

Table 1. Switching States of Half-Bridge Inverter

State	Voltage level (Vtop/Vbot)	S1/S5	S2/S6	S3/S7	S4/S8
1	0Vdc	0	1	0	1
2	Vdc	1	0	0	1
3	Vdc	0	1	1	0
4	2Vdc	1	0	1	0

Note: S1-S2, S3-S4, S5-S6, S7-S8 are a complement to each other, 1 and 0 indicate the ON and OFF states of the switches respectively

When S2 and S4 are turned ON, Vtop or voltage across between the switches in the top module is equal to 0. When only S1 or S3 is turned ON, the output voltage of the top module dc-link is equal to Vdc. When both S1 and S3 are turned ON, Vtop is equal to 2Vdc. A similar concept of switching state can also be applied in generating Vbot or voltage across between the switches in bottom module. It can be observed that, to obtain a given level at total dc link, two switches are required to conduct simultaneously in top and bottom modules of the dc-link. It also can be observed that since voltage level is synthesized by combining all the input sources, equal load sharing is possible. These redundancies can provide flexibility if the capacitor is used for voltage balancing method. Based on the switching states shown in Table 1, the output voltage of dc-link can be presented as given in (1) for top dc-link, (2) for bottom dc-link and (3) for total dc-link voltage.

$$V_{top} = V_{dc1}(S1) + V_{dc2}(S3) \quad (1)$$

$$V_{bot} = V_{dc3}(S5) + V_{dc4}(S7) \quad (2)$$

$$V_{tot} = V_{top} + V_{bot} \quad (3)$$

The inverter shown in Fig. 1 is a three-phase or a three-leg of NPC inverter. The leg has a dc link voltage of Vtot which is generally halved into Vtop and Vbot. The leg consists of four bidirectional power switches with the additional two clamping diodes to produce common point 'g'[17]. The three possible states of one leg NPC inverter are listed in Table 2. For analysis, the point 'g' is used as the reference node with a potential of 0V. The potential difference between the point 'a' which represents a leg for phase A and the common point which represents as 'g' is inverter output voltage at common point, Vag. It means that Vag can be in form of -1, 0 and 1.

To produce Vag = 1, switches T1 and T2 are simultaneously turned ON which resulting in Vag = Vtop = 2Vdc in the proposed topology in Fig. 1. When switches T2 and T3 are switched ON, the Vag resulting in 0 which equivalent to Vtop = 0Vdc. Currently, if the load current is positive, the current completes its path through T2 and D1 and if the load current is negative, the current flow through T3 and D2. Next is when Vag = -1 = Vbot = -2Vdc, switches T3 and T4 are turned ON at the same time.

Table 2. Switching States of One Leg NPC Inverter

State	Vag	T1	T2	T3	T4
1	1	1	1	0	0
2	0	0	1	1	0
3	-1	0	0	1	1

The modulation technique for the proposed topology will be discussed later in the next section. Before that, the reference signals of the pulse-width modulation (PWM) technique and switching states of three-phase unfolder are derived based on a three-phase line-to-line voltage equation in the balanced system as listed in (4). Since the NPC has a common point and the dc-link voltage of NPC is halved into two, there are two reference signals at level generation part in which each of the reference signal is represented for top (Vref1) and bottom module (Vref2). The two reference signals are generated after implementing equation (5) and its corresponding switching states are presented in Table 3.

Table 4 lists the switching state of the reference signals at inverter output voltage between each phase to a common point (Vxg) where x represents phase a, b, and c. Equation (6) and (7) show the relationship between inverter output voltage (Vxg), load voltage (Vxn) and line voltage (Vxx). Fig. 2 shows the illustrations of step-by-step procedures in generating two reference signals for half-bridge modules. The switching states of three-phase unfolder can be derived from the Vxg and its corresponding switching states as depicted in Fig 3.

$$\begin{aligned} Vab(\theta) &= Vm \times \sin(\theta) \\ Vbc(\theta) &= Vm \times \sin(\theta-240°) \\ Vca(\theta) &= Vm \times \sin(\theta-120°) \end{aligned} \tag{4}$$

where $\theta = 2\pi ft$, f = frequency, Vm = Peak Amplitude of reference signal

$$\begin{aligned} Vab(\theta) &= | Vm \times \sin(\theta) | \\ Vbc(\theta) &= | Vm \times \sin(\theta-240°) | \\ Vca(\theta) &= | Vm \times \sin(\theta-120°) | \end{aligned} \tag{5}$$

$$\begin{aligned} Van &= (2/3)\, Vag - (1/3)\, Vbg - (1/3)\, Vcg \\ Vbn &= -(1/3)\, Vag + (2/3)\, Vbg - (1/3)\, Vcg \\ Vcn &= -(1/3)\, Vag - (1/3)\, Vbg + (2/3)\, Vcg \end{aligned} \tag{6}$$

$$\begin{aligned} Vab &= Van - Vbn \\ Vbc &= Vbn - Vcn \\ Vca &= Vcn - Van \end{aligned} \tag{7}$$

Table 3. Switching States of Two Reference Signals

Sector	Angle (θ)	Vref1	Vref2
1	$0 < \theta \leq 60$	Vca	Vab
2	$60 < \theta \leq 120$	-Vca	-Vbc
3	$120 < \theta \leq 180$	Vab	Vbc
4	$180 < \theta \leq 240$	-Vab	-Vca
5	$240 < \theta \leq 300$	Vbc	Vca
6	$300 < \theta \leq 360$	-Vbc	-Vab

Table 4. Switching States for Inverter Output Voltage

Sector	**1**	**2**	**3**	**4**	**5**	**6**
Vag	0	1	1	0	-1	-1
Vbg	-1	-1	0	1	1	0
Vcg	1	0	-1	-1	0	1

IOP Conf. Series: Materials Science and Engineering 1127 (2021) 012032 doi:10.1088/1757-899X/1127/1/012032

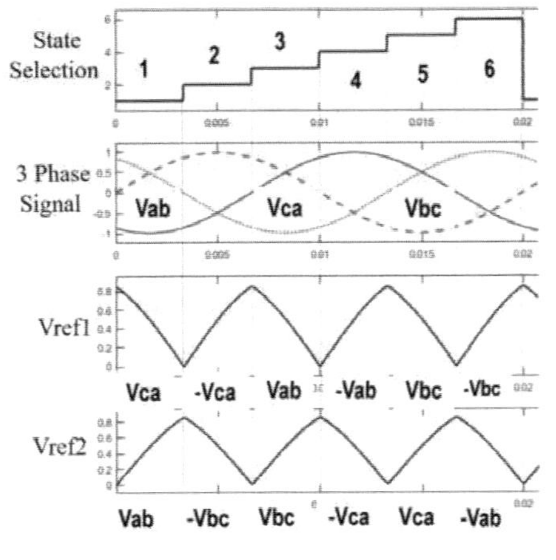

Figure 2. Two Reference Signals

3. Phase-Shifted (PS) and Phase-Opposition Disposition (POD) Modulation Techniques

The structure of converters can directly relate to its modulation techniques. Depending on the converter structure, each topology has its own best modulation technique. One carrier signal modulation method is typically adopted in two-level inverters whereas multicarrier signal method is normally implemented in multilevel inverters to accommodate more switches as compared to the conventional inverters [18, 19]. PWM signals can be produced by using scalar and vector modulation techniques. For scalar PWM modulation technique or sinusoidal PWM (SPWM), the switching signals are generated by comparing a sinusoidal reference signal with a triangular carrier signal.

SPWM method typically uses phase-shifted (PS) and level- shifted (LS) modulation techniques [20, 21].

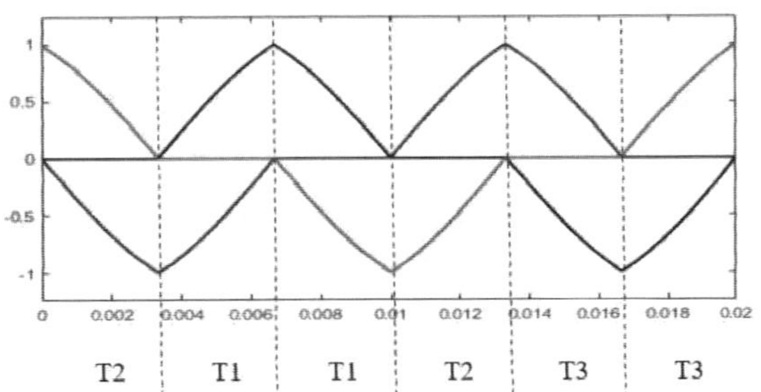

T2	T1	T1	T2	T3	T3
T3	T2	T2	T3	T4	T4
T7	T7	T6	T5	T5	T6
T8	T8	T7	T6	T6	T7
T9	T10	T11	T11	T10	T9
T10	T11	T12	T12	T11	T10

Figure 3. Switching States of Three-Phase NPC Unfolder

The LS modulation technique can be further categorized into phase disposition (PD), phase opposition disposition (POD) and alternate phase opposition disposition (APOD). POD and PS techniques have been identified as a suitable modulation technique for NPC and cascaded inverters. Therefore, PODPWM and PSPWM are adopted to modulate top and bottom modules of dc-link in the proposed multilevel inverter. A simple comparison between both modulation techniques is made for a circuit configuration utilizing, two half-bridge cells for top and bottom modules of the proposed topology.

For PS modulation, N number of the cells of half-bridge inverter require 360°/N number of phase-shifted carrier signals. Each module has two half-bridge cells, thus 2 carrier signals with 0 and 180 phase shift are needed for PS modulation as demonstrated in Fig 4. Each module is modulated independently using the corresponding reference signals as listed in Table 3 and depicted in Fig 2.

The POD modulation technique is arranged in a vertical shift instead of a phase shift as used in PS. A similar concept of PS modulation can be applied by calculating the number of carrier signals required. The amount of carrier signal is directly proportional to the number of cells. Therefore, 2 carrier signals are required, and each carrier is set between two voltage levels since there are 2 cells of half-bridge. The difference between POD and other LS modulation is that the carrier signal for the top module is in phase with each other whereas the carrier signals for the bottom module which represent a negative carrier is in opposite phase of the carrier signal for the top module. Fig.5 shows the arrangement of LS modulation technique with the corresponding reference signals. A stepped multilevel waveform of PS and POD are demonstrated in Fig.6 whereby Equation (8) - (9) show the relationship between the carrier signal (Vc1, Vc2) reference signals (Vm = Vref1, Vref2) and inverter output line voltage (Vout)

$$M = Vout/Vin = Vm / Vc \qquad (8)$$

where M = Modulation Index, Vin = Total dc link, Vo = Output Line Voltage, Vc = Peak Amplitude of carrier signal.

$$(Vref1 \text{ OR } Vref2 >= Vc1 \text{ OR } Vc2) = S1 \text{ or } S3 \text{ or } S5 \text{ or } S7 \qquad (9)$$

where Vc1 = Carrier Signal 1, Vc2 = Carrier Signal 2

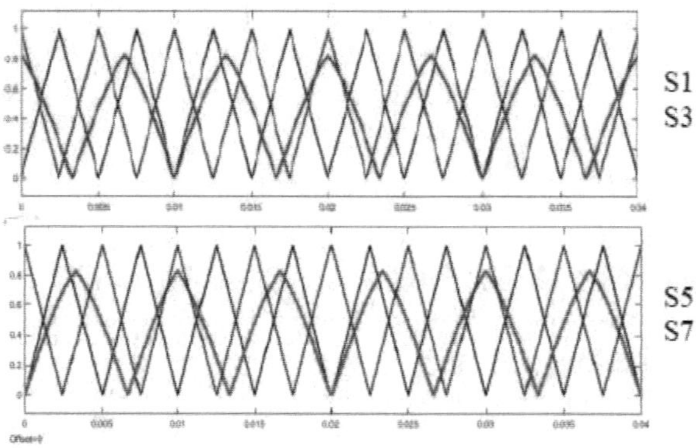

Figure 4. Configuration of Carrier and Reference Signals for PSPWM

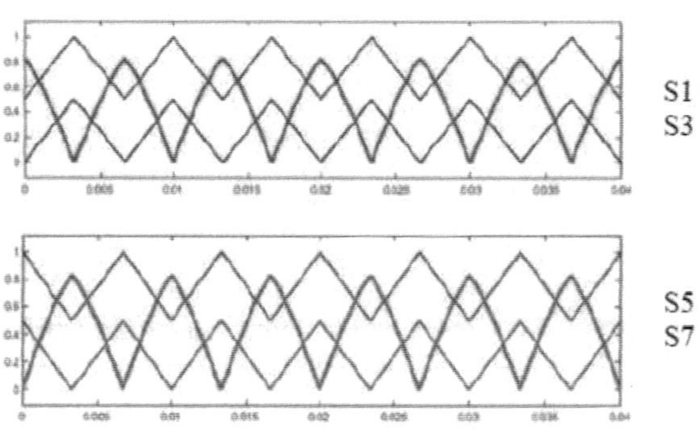

Figure 5. Configuration of Carrier and Reference Signals for PODPWM

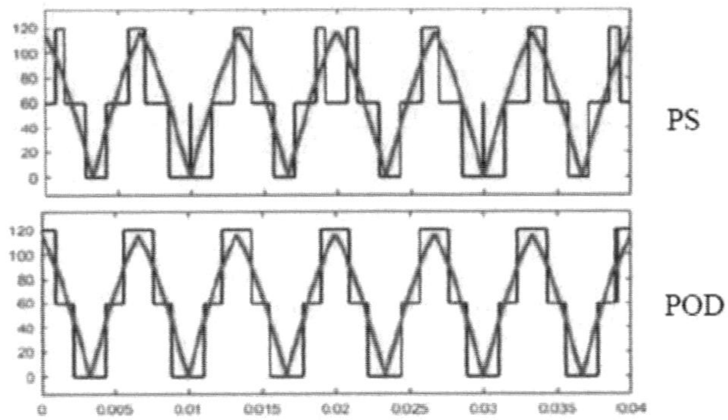

Figure 6. Stepped Waveforms of PS-Vtop (scaled at 200Hz) and POD-Vtop (scaled at 150Hz) with each Vdc = 60V

4. Results and Discussion

The performance of the proposed multilevel converter based on half-bridge cells and NPC inverter is simulated by using MATLAB Simulink.

The converter is designed to provide an output of 260V line voltage. This means that the minimum total dc-link voltage required for top and bottom modules is 320V.This will be the dc input voltage for NPC inverter to produce the desired AC output voltage. In this case, an off-grid system with a resistive load is considered. A low pass LC filter is used to smooth the inverter output [22]. Equation (10) shows the relationship of the peak amplitude of reference signal (Vref1, Vref2) with modulation index (M) whereby the simulation parameters of the proposed topology are given in Table 5.

$$\text{Vref (peak)} = M \times \sin(60°) \tag{10}$$

Table 5. Simulation Parameters

Parameters	Value
DC input voltage, Vdc	160 V x 4
Filter inductor, Lf	3-20 mH
Filter capacitor, Cf	75-450 nF
Load resistor, R	150 Ω
Carrier frequency, fc	10-60 kHz
AC Voltage, Vrms	260 V Power
Frequency, Hz	50 Hz

Fig. 7 shows the output voltage of the top and bottom modules as well as the total dc-link voltage for both PS and POD modulations. The output voltage produced at each module is 320V and this will serve as dc input for the inverter. For a fair comparison, both modulation methods are evaluated at the same operating conditions whereby in this case, the carrier frequency for PS is set to half of the carrier frequency under POD modulation. It means that as fc = 30kHz at POD modulation, the fc for PS modulation is 30kHz/2 = 15kHz. The carrier frequency of PS is divided by half due to the presence of 2 half-bridge cells in one module. Thus, the effective switching frequency for PS will be the same as the POD.

The configuration effect of the carrier frequency (fc) of PS and POD can be seen in the output voltage of top module as shown in the first row of Fig. 8 and Fig.9. Notice that at a certain instant of time, when fc is 30kHz (POD) and 15kHz (PS), there are 3 triggered pulses for output voltage at top module (Vtop). Switches S1 and S3 are turned ON at 1.5 pulses for PS modulation and for POD modulation, S3 continuously switched ON while S1 is switched ON at 3 pulses. This shows that PS modulation has an advantage over POD as it can generate 3 pulses at Vtop with reduced carrier frequency and it has even power distribution among switches as the switches is turned ON at toggle mode continuously. The switching patterns of switches T5 to T8 of the three-phase NPC unfolder are illustrated in Fig. 10. The switches of NPC unfolder are operated at the fundamental frequency of 50 Hz and both modulation methods use the same switching pulses for NPC unfolder. The inverter output line voltage (Vo) produce 3-level, 5-level and 7-level at 0.4, 0.96 and 1.15 modulation index.

Fig.11 shows the three-phase five-level of unfiltered and filtered line voltage with its current waveforms at M = 0.96. Output voltage at M = 0.4 and 1.15 are shown in Fig.12 and Fig.13 respectively. Fig.13 shows that the produced line voltage is 368V and it can generate one extra level at unfiltered output line voltage as compared to when M = 0.96. This is due to the fact that when M is equal to 1.15, the peak amplitude of the reference signals is 1. Thus, the total dc-link voltage can utilize additional 0.15% from the total dc input voltage.

The Fast Fourier Transform (FFT) of the Vab and Iab at M = 0.96 with carrier frequency at 30kHz (POD) and 15kHz (PS) are shown in Fig. 14 and Fig.15 respectively where the total harmonic distortion (THD) for voltage and current are below 5 percent which is still within the acceptable range of THD recommended by IEEE Std. 519, which are around 0.90% to 0.96%. From both figures, the magnitudes of the low order harmonics are very low as compared to the fundamental component.

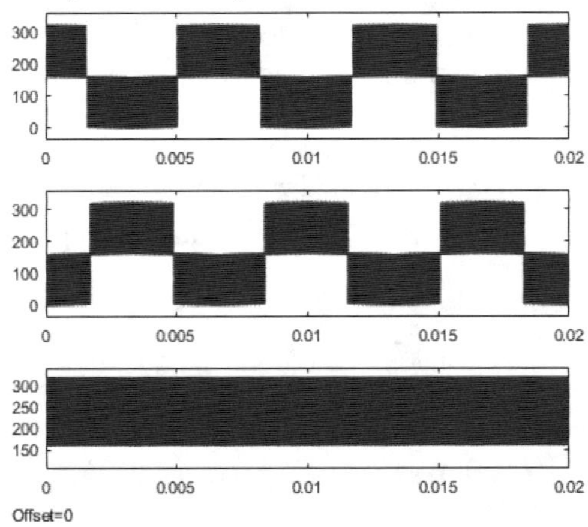

Figure 7. Output Voltage at Vtop, Vbot and Vtot

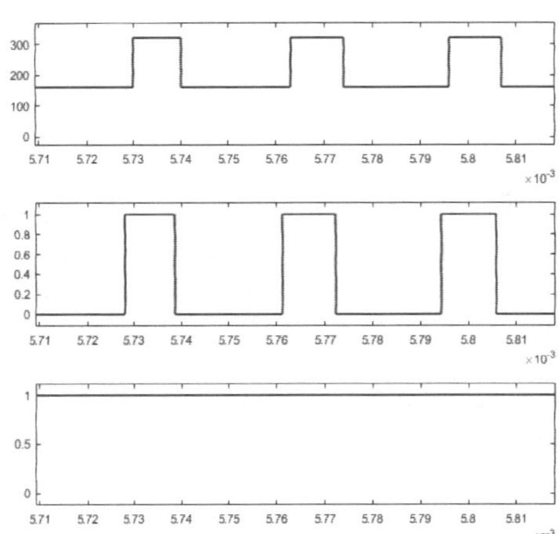

Figure 8. POD Vtop and Switches S1 and S3

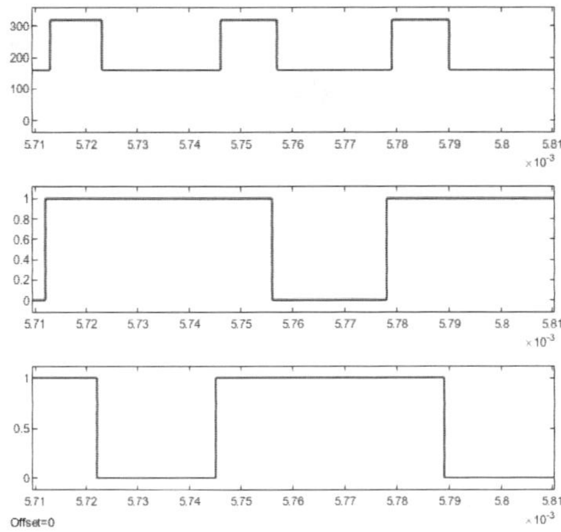

Figure 9. Phase Shifted Vtop and Switches S1 and S3

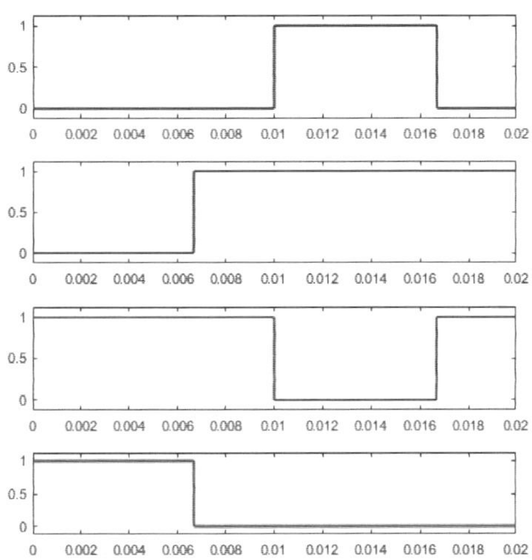

Figure 10. Switching Pulses of Unfolder Switches

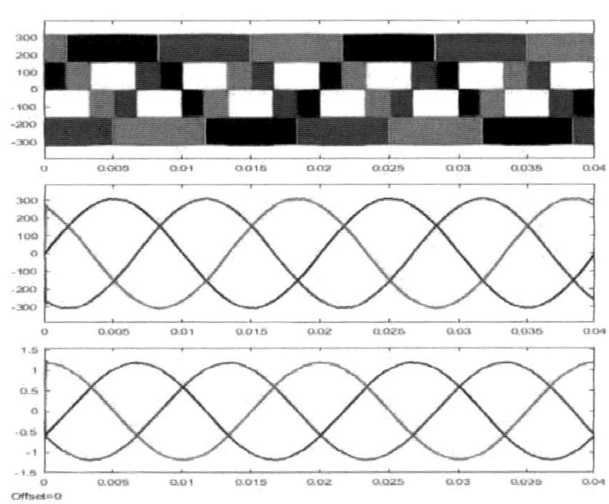

Figure 11. Output Waveform of Three-Phase (Unfiltered and Filtered Voltage, Line current)

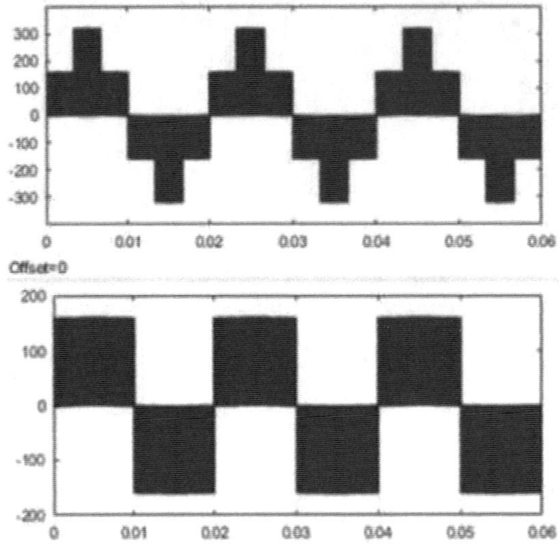

Figure 12. Unfiltered Vab of PS and POD at M = 0.4

THD can be further reduced at M = 1.15 where the range of THD for voltage and current is around 0.80% to 0.82%. At M = 0.4 and 1.15, the difference between PS and POD can be seen clearly in terms of waveform pattern and peak amplitude of line voltage as illustrated in Fig. 12 and Fig. 13.

The value of line voltage at M = 0.4 is 128V in which falls under range that can be supplied using a single dc source = 160V. Therefore, only 1 dc power supply and 1 half-bridge cell are utilized to operate at M = 0.4 as shown in POD modulation. However, this is opposite for PS modulation in which both half-bridge cells are utilized and produce 320V line voltage at M = 0.4. Notice that during PS modulation, there is a small spike on the output voltage at M = 1.15 compared to no spike at POD modulation.

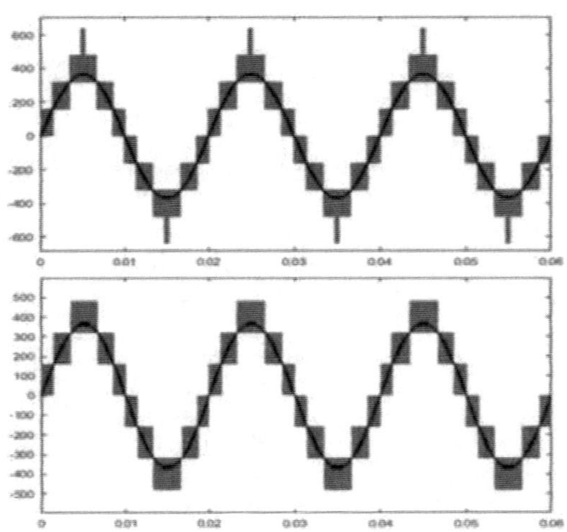

Figure 13. Unfiltered Vab of PS and POD at M = 1.15

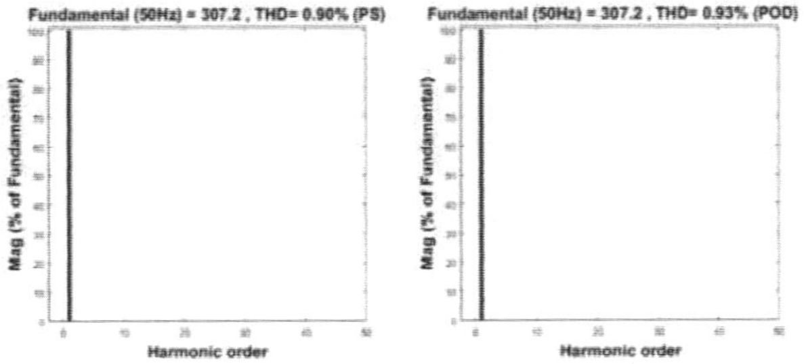

Figure 14. FFT Analysis of Load Voltage (PS and POD)

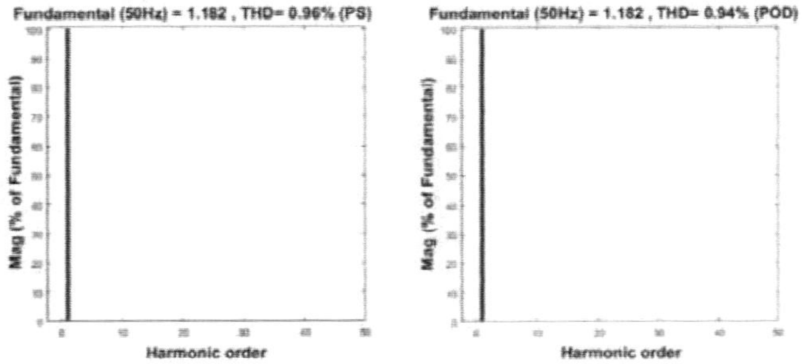

Figure 15. FFT Analysis of Load Current (PS and POD)

By considering Fig. 14 and Fig.15, the load voltage of PS modulation has better THD as compared to POD modulation by 0.03% whereas the THD of load current at POD modulation is slightly lesser by 0.02% over PS modulation. Figure 16 shows THD range of unfiltered and filtered line voltage in terms of carrier frequency.

The results are obtained when the carrier frequency is set at 30 kHz and M = 0.96 as the THD of unfiltered line voltage is lesser than other carrier frequencies. It is also notice that as the carrier frequency increased, the THD is also increased. However, as the carrier frequency increased, the filter requirement on inductor and capacitor values is reduced as shown in Fig. 17. Fig.18 shows the relationship between output voltage and THD for both modulation methods. It shows that as the modulation index increases which contributes to the greater output voltage, the THD is reduced. Comparing both techniques, it shows that POD modulation has better THD and utilize dc-link efficiently over PS modulation as M is increased.

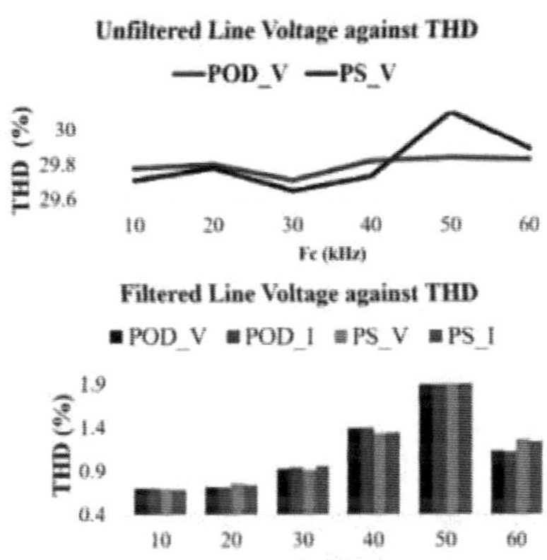

Figure 16. THD Unfiltered and Filtered Line Voltage against Carrier Frequency

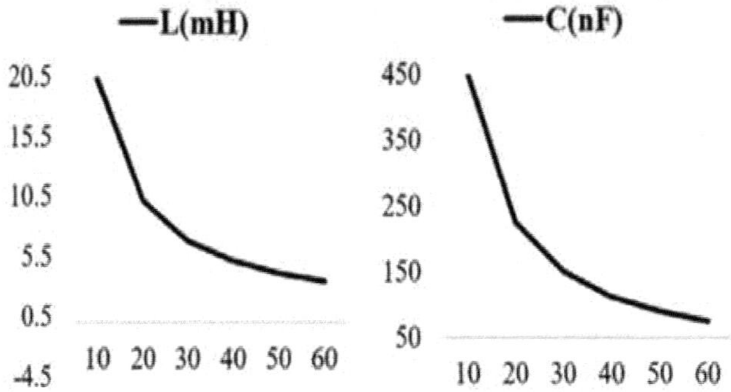

Figure 17. Inductor and Capacitor values against Carrier Frequency

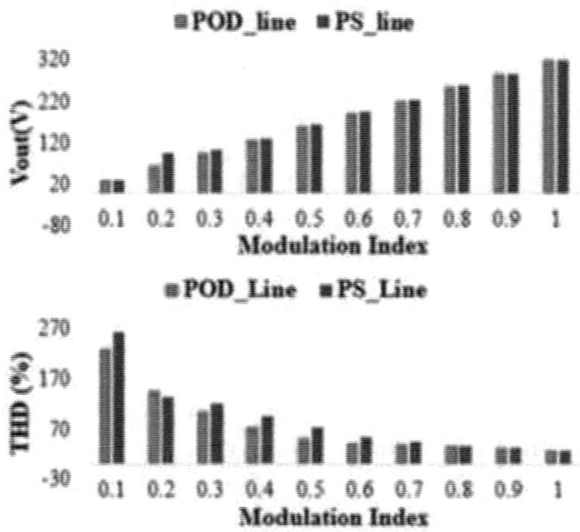

Figure 18. THD and Amplitude of Filtered Line Voltage against M

5. Conclusion

In this paper, the design and analysis of a three-phase multilevel inverter based on cascaded half-bridge cells and three-phase NPC inverter have been presented. It consists of two modules of cascaded half-bridge to produce the dc-link for the NPC inverter. The half-bridge cell can generate 3-level PWM output voltage whereas, the NPC inverter is switched at the fundamental frequency of 50 Hz. From the results, it showed that half-bridge modules modulated using PS has higher switching loss but even power distribution among switches as compared to POD that has lower switching loss but possess uneven power distribution among switches. The simulation results obtained verified the functionality of the proposed inverter employing both modulation technique, POD and PS. Closed-loop control is recommended as further works as to maintain the inverter output voltage with input voltage derived from renewable energy sources such as solar energy. In addition, a prototype of the proposed multilevel inverter topology can be built to verify its performance experimentally.

Acknowledgement

The authors thank the technical and financial assistance of UM Power Energy Dedicated Advanced Centre (UMPEDAC) and the Higher Institution Centre of Excellence (HICoE) Program Research Grant, UMPEDAC - 2018 (MOHE HICOE-UMPEDAC), Ministry of Education Malaysia, RU007-2018 and RU012-2019,University of Malaya.

References

[1] Akagi H. Multilevel converters: Fundamental circuits and systems. *Proc IEEE*. 2017;**105**(11):2048-65.
[2] Yuan X. Derivation of voltage source multilevel converter topologies. *IEEE Trans Indust Electr*. 2016;**64**(2):966-76.
[3] Cecati C, Ciancetta F, Siano P. A multilevel inverter for photovoltaic systems with fuzzy logic control. *IEEE Trans Indust Electr*. 2010;**57**(12):4115-25.
[4] Babaei E, Kangarlu MF, Hosseinzadeh MA. Asymmetrical multilevel converter topology with reduced number of components. *IET Power Electronics*. 2013;**6**(6):1188-96.
[5] Babaei E. A cascade multilevel converter topology with reduced number of switches. *IEEE Trans Pow Electr*. 2008;**23**(6):2657-64.
[6] Najafi E, Yatim AHM. Design and implementation of a new multilevel inverter topology. *IEEE Trans Indust Electr*. 2011;**59**(11):4148-54.
[7] Babaei E, Kangarlu MF, Sabahi M. Extended multilevel converters: an attempt to reduce the number of independent DC voltage sources in cascaded multilevel converters. *IET Power Electronics*. 2014;**7**(1):157-66.
[8] Xiao B, Hang L, Mei J, Riley C, Tolbert LM, Ozpineci B. Modular cascaded H-bridge multilevel PV inverter with distributed MPPT for grid-connected applications. *IEEE Trans Ind Appl*. 2014;**51**(2):1722-31.
[9] Su G-J. Multilevel DC-link inverter. *IEEE Trans Ind Appl*. 2005;**41**(3):848-54.
[10] Saeidabadi S, Gandomi AA, Hosseini SH, Sabahi M, Gandomi YA. New improved three-phase hybrid multilevel inverter with reduced number of components. *IET Power Electronics*. 2017;**10**(12):1403-12.
[11] Rahim NA, Chaniago K, Selvaraj J. Single-phase seven-level grid-connected inverter for photovoltaic system. *IEEE Trans Indust Electr*. 2010;**58**(6):2435-43.
[12] Hasan MM, Abu-Siada A, Dahidah MS. A three-phase symmetrical DC-link multilevel inverter with reduced number of DC sources. *IEEE Trans Pow Electr*. 2017;**33**(10):8331-40.
[13] Gupta KK, Ranjan A, Bhatnagar P, Sahu LK, Jain S. Multilevel inverter topologies with reduced device count: A review. *IEEE Trans Pow Electr*. 2015;**31**(1):135-51.
[14] Vijeh M, Rezanejad M, Samadaei E, Bertilsson K. A general review of multilevel inverters based on main submodules: Structural point of view. *IEEE Trans Pow Electr*. 2019;**34**(10):9479-502.
[15] Lezana P, Aceitón R. Hybrid multicell converter: Topology and modulation. *IEEE Trans Indust Electr*. 2010;**58**(9):3938-45.
[16] Chen WW, Zane R, Corradini L. Isolated bidirectional grid-tied three-phase ac–dc power conversion using series-resonant converter modules and a three-phase unfolder. *IEEE Trans Pow Electr*. 2017;**32**(12):9001-12.
[17] Gupta KK, Bhatnagar P. Multilevel inverters: conventional and emerging topologies and their control: Academic Press; 2017.
[18] Arslan AO, Kurtoglu M, Eroglu F, Vural AM, editors. Comparison of phase and level shifted switching methods for a three-phase modular multilevel converter. 2018 5th International Conference on Electrical and Electronic Engineering (ICEEE); 2018: IEEE.
[19] McGrath BP, Holmes DG. Multicarrier PWM strategies for multilevel inverters. *IEEE Trans Indust Electr*. 2002;**49**(4):858-67.
[20] Al Hadi A, Fu X, Waithaka W, Challoo R, Li S, editors. Comparison and Simulation of the Level-Shifted and Phase-Shifted Modulation for a Five-Level Converter for Integration of Renewable Sources. 2018 Clemson University Power Systems Conference (PSC); 2018: IEEE.
[21] Arazm S, Vahedi H, Al-Haddad K, editors. Phase-shift modulation technique for 5-level packed U-cell (PUC5) inverter. 2018 IEEE 12th International Conference on Compatibility, Power Electronics and Power Engineering (CPE-POWERENG 2018); 2018: IEEE.
[22] Texas Instruments. LC Filter Design. 2016.

IOP Conf. Series: Materials Science and Engineering 1127 (2021) 012033 doi:10.1088/1757-899X/1127/1/012033

NEPLAN-Based Analysis of Impacts of Electric Vehicle Charging Strategies on Power Distribution System

Aslam A Ahmed[1], Majid A Abdullah[1,2], Muhamad Mansor[1], Marayati Bte Marsadek[1], Yong Jia Ying[1], Muhamad Safwan Abd Rahman[1], Nur Ashida Salim[3]

[1]Department of Electrical and Electronics Engineering, College of Engineering, Universiti Tenaga Nasional, 43000 Kajang, Selangor, Malaysia

[2]Hafr Al Batin Community College, The University of Hafr Al Batin, Al Jamiah, Hafar Al Batin 39524, Saudi Arabia

[3]Faculty of Electrical Engineering, Universiti Teknologi MARA, 40450 Shah Alam, Selangor, Malaysia

E-mail: majid.abdullateef@uniten.edu.my

Abstract. This paper aims to analyze the impacts associated with the incorporation of EV charging stations into the distribution network. Using NEPLAN software, a typical distribution network was simulated and analyzed under normal conditions and with integrated coordinated and uncoordinated EV charging stations. A comparative study was then conducted on the main parameters of the distribution system: voltage profiles, transformer loading, as well as power demand and losses. Based on the simulation results, it can be argued that the worst adverse effects are associated with an uncoordinated charging strategy, while the use of a coordinated charging strategy demonstrates significant improvements. Although the distribution network components or requirements of either strategy have not been changed, it is evident, however, that shifting and splitting EV loads has resulted in a coordinated approach that eliminates all adverse effects. Implementing these strategies helps to overcome the challenges of incorporating charging stations into the distribution network, such as equipment overheating, rapid aging issues, as well as the need for additional investment in distribution infrastructure and capacity expansion.

1. Introduction

The world is experiencing increasing environmental problems, so there is an urgent need to replace conventional energy generated by burning fossil fuels with renewable energy generated from clean and sustainable sources. One of the primary sources of environmental pollution is the traditional transportation used in current societies. As it is not possible wholly dispensed transportation, therefore, it must be replaced by relying on electric vehicles (EVs) [1]. However, EV deployment raises a concern about the effect of EV charging on the distribution system. EVs are expected to pose a significant burden on the distribution system in the future, because charging a large number of vehicles at the same time causes a significant load on networks, primarily when this load is concentrated in specific areas [2, 3].

The identified challenges due to EV charging on the distribution system are voltage drop, overloading of transformer and lines, system instability, harmonics and system losses [4-6]. A review in [7] concluded that the main impact of EV integration into the distribution system was the overloading issue, which further led to equipment overheating and rapid aging issues. Moreover, it was concluded in [8] that the severity of impact due to EV charging is highly dependent on the EV connection point. For instance, if EVs are connected to a weak bus and receive charges, the impact of EV charging will be more prominent than the case where EVs are connected to a stronger bus.

The impact of EV charging on the distribution system voltage deviation was extensively investigated in literature [3, 5, 7, 9, 10]. The results of these studies revealed that a large number of fast-charging EVs would introduce serious voltage deviations and voltage instability problems. The voltage deviation does harm to people's life and work, shorten the service life of electric power devices, and make a negative impact on the power system [11]. The power system is complex and fluctuating, so the voltage deviation cannot be avoided, but the voltage deviation should be controlled in a suitable range. According to studies [10, 12, 13], the 11 kV distribution network should be operated with a voltage tolerance of +/-5%, while 400 V low voltage distribution system should be operated with a voltage tolerance of +10% and -6%.

In terms of transformers, it was noticed in literature that as the number of connected EV increases the service transformers are highly susceptible to be overloaded. Previous studies concluded that a higher penetration rate of EV would exert a more negative influence on transformer lifespan, and it could cause transformer failure due to excessive operating temperature [14]. In terms of load profile, the work reported in [15-18] describes the effect of charging strategies on the load profile. Based on the studies, most of the consumers need their plug-in electric vehicles (PEVs) batteries to be charged as soon as they get home after their working hours. Assuming that they are at fully discharged state, all the batteries start charging at the same time. By doing so, the peak demand for the electrical grid increases dramatically, and a new peak load exceeding the distribution system capacity of the system might be introduced. In both cases, the result is the need for additional investments in distribution infrastructure and generation capacity expansion.

In terms of power losses, different PEV penetration levels were considered in [19] to estimate the PEV impact on system power losses. It was found that the system energy losses increase dramatically with increasing EV penetration. A recent study about the impact of the placement of fast charging stations in distribution systems showed a power loss increase of 85% compared to a base layer with no EV integration [20]. Power losses refer to the difference between the amount of energy delivered to the distribution system and the amount of energy clients are charged. Increasing the distribution system power losses reduces the efficiency of transmitting energy to customers. These energy losses cost millions of dollars every year, which means an economic loss to the utility companies. Therefore, energy losses minimization is one of the important objectives in operating the distribution system.

The objective of this paper is to investigate the impacts of integrating EVs into the power distribution network. The investigation is performed by simulating a typical distribution system with different charging strategies and then analyzing the main operational parameters of the distribution system, such as the voltage profile, transformer loading, peak demand, and system losses. The rest of the paper is arranged into sections. Section 2 reviews the concepts of EVs, charging stations, and charging strategies. Then, Section 3 describes the methodology that was followed for achieving the objectives of this paper. Subsequently, Section 4 presents the simulated cases with the corresponding results and discussion. Finally, Section 5 summarises the paper conclusion.

2. Overview of Electric Vehicles and Charging Stations

2.1. Electric Vehicles

An electric vehicle is generally defined as any vehicle in which some or all of the driving energy is supplied through electricity from a battery [21]. There are several types of electric cars currently available. These types are either pure electric vehicles or hybrid electric vehicles. The construction and main components of the different types of EVs are compared in Figure 1.

IOP Conf. Series: Materials Science and Engineering 1127 (2021) 012033 doi:10.1088/1757-899X/1127/1/012033

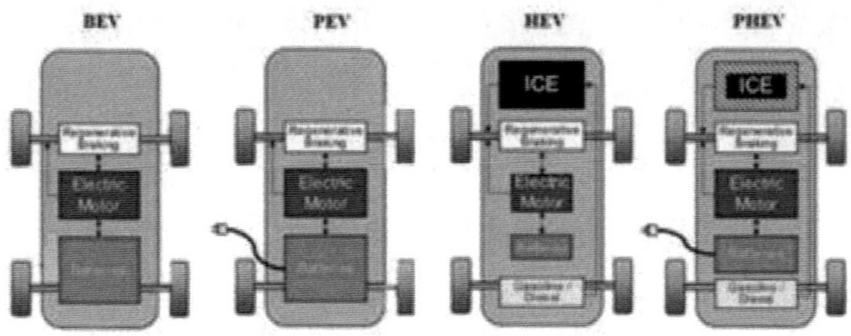

Figure 1. Different types of EVs

Pure electric vehicle depends solely on an electric motor that solely uses a battery as an energy source. However, the hybrid electric vehicle (HEV) combines the features of an electric car and an internal combustion engine [22]. Pure electric vehicles are also classified into two types, battery electric vehicles (BEVs) and plug-in electric vehicles (PEVs). In BEVs, the battery can be recharged with power from only regenerative braking. However, a PEV has a larger battery, which can also be recharged from a grid connection [23].

2.2. Charging Stations

Charging EVs can be done in individual charging points, which are homes and small businesses, or in public stations. The public stations are fed from a three-phase system, making them faster in charging. However, individual charging points are fed from a single-phase system, taking a longer time for charging [12]. Slow charging takes a long time with low charging current and thus low charging power. This type of charging has little effect on battery life. In contrast, fast charging takes a short time, but with substantial charging current or higher voltage. This type of charging may cause some damage to the battery, affect the power quality of the distribution system, and requires high requirements for the technology and safety of the charging equipment [24, 25].

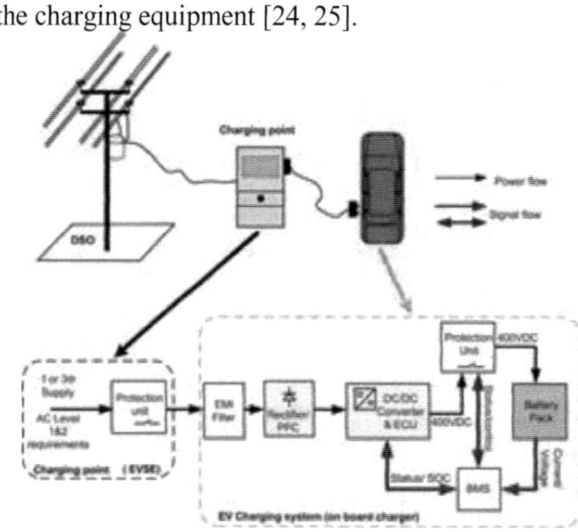

Figure 2. EV Charging configuration at AC level 1and 2 setups (i.e., onboard charger) [7].

IOP Conf. Series: Materials Science and Engineering 1127 (2021) 012033 doi:10.1088/1757-899X/1127/1/012033

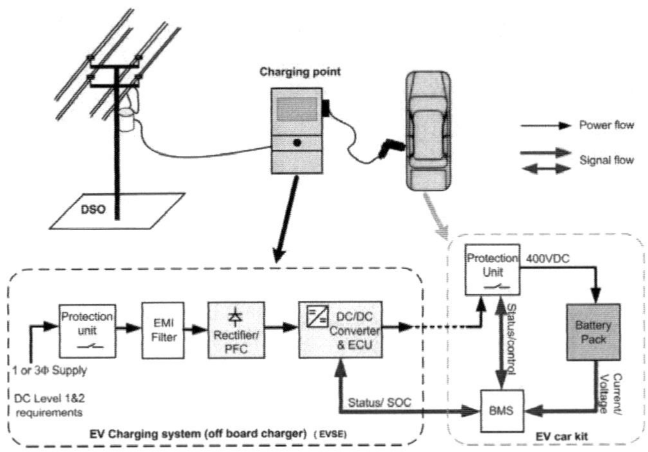

Figure 3. EV charging configuration at DC level 1 and 2 frameworks (i.e., off-board charger) [7].

The diagrams in Figure 2 and Figure 3 illustrate the main functions of different level chargers implementation. As shown in the figures, all charging systems are essentially AC- type, but then the AC is converted to a DC for charging batteries. The process of converting from AC to DC may be performed by converters merged within the EVs or separated converters. The first type is referred to as an on-board charger, while the second type is referred to as an off-board charger [26].

2.3. Charging Strategies

PHEV charging schemes can be classified into uncoordinated and coordinated charging. The coordination refers to the alignment of charging instances and parameters among PHEVs. In the literature, uncoordinated and coordinated charging strategies are also called uncontrolled and controlled charging, respectively. In uncoordinated charging strategy, EVs are automatically charged when connected to the network without any regulation or control from the network operator. Uncoordinated charging strategy is simple, but because of its dependence on drivers' behavior, it makes electrical loads random and highly variable. For example, most EVs are connected at the same time once employees arrive at their companies or once the residents arrive at their homes, leading to a large electrical load on the network. In addition, the occurrence of this large load may coincide with the daily peak hours of regular loads. As a result, the network may experience significant problems, such as voltage drops, power cuts, etc. [13].

The coordinated charging strategy aims at minimizing the harmful effects of integrating EVs charging through the management of the charging process in order to reduce stress on the network. The operator manages the charging by coordinating the EVs connection appropriately, so that does not get large loads at the same time while satisfying the demands of drivers. This coordination prevents network instability and maintains high operational performance [13]. One method for implementing the coordinated charging strategy is by shifting the charging to off-peak hours. By using the peak load shifting method, there is no reduction in the total energy consumed; instead, there is only redistributing of consumption time. Besides, shifting peak load limits the harmful effects of load changes while reducing the cost of energy consumption [27]. During the off-peak period, the control center transmits the information of charging energy to end-users to proceed the operation of charging the EVs.

3. Research Methodology

3.1. Test Distribution System Selection

An existing test system has been selected and modified to include the integration of EV charging stations. The test system is a typical distribution system that is contained in the examples provided in NEPLAN software. Figure 4 shows the single line diagram of the distribution system under consideration in this study. It comprises 29 nodes, 19 lines, and 2 distributed generation units. The total installed capacity of the distributed generation units is 1.1 MW, and the total peak load is 9.218 MW. The distribution system is also supplied by a network feeder through a 110/20 kV, 16 MVA transformer. The system has a total of 15 loads, both high voltage and low voltage customers. The 0.4 kV low voltage (LV) customers are supplied via 20/0.4 kV transformers, and the 20 kV customers are supplied directly.

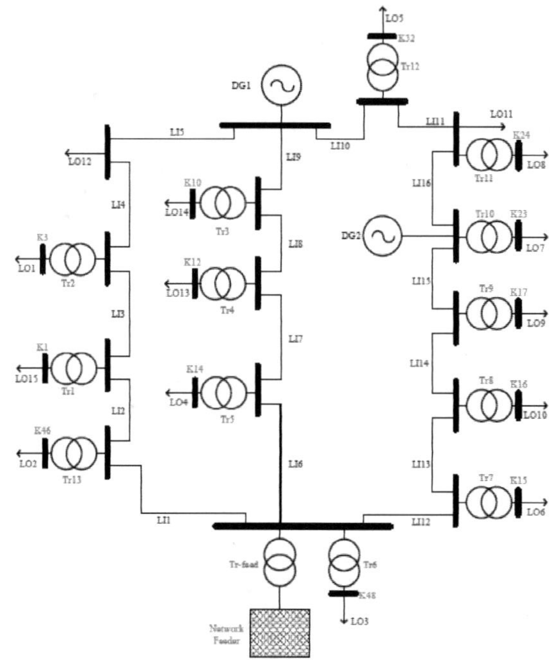

Figure 4. Distribution system single line diagram.

3.2. Modelling Methods

Two distribution generation units are included in the distribution system: hydro and solar power plants. Their generation profiles are as shown in Figure 5.

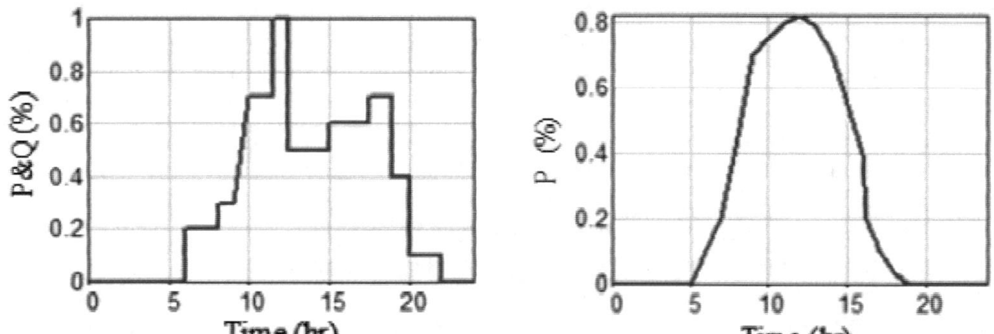

Figure 5. Distributed generation models (left) Hydropower station (right) Solar power station.

Similarly, different types of load profiles are represented in the distribution system. The 15 system loads are assigned as follows: 3 loads are household type, 2 loads are a commercial enterprise, 3 loads are industry type, 1 load is big industry type, 1 load is heating type, and 5 loads are a combination of two or three types. This classification is shown in the appendix, with load profile of each type is shown in Figure **6**.

PHEVs are treated as different types of loads, depending on the charging schemes. They are treated either as a rigid load, where each PHEV is an inflexible load in uncontrolled charging, or a flexible load, where each PHEV is a controllable flexible load in indirectly controlled charging and smart charging. For this study, it was assumed that EVs would be connected to the network at the same point of connection as the conventional household and other loads [28]. The EV charging station was modeled as a constant active power load with unity power factor, following the studies [29-31]. The capacity of each EV charging station was assumed to be proportional to the peak active power of the original load connected to the same node. The charging load was varied based on the case study under consideration.

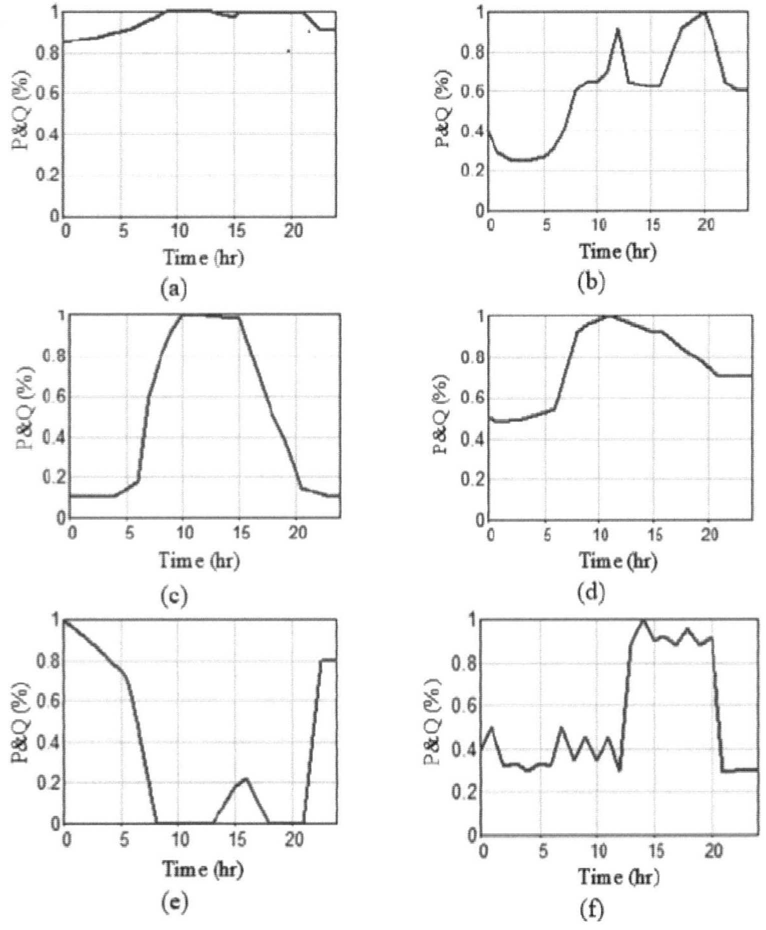

Figure 6. Typical load models (a) Big industry (b) Household Commercial enterprise (d) Industry (e) Heating (f) Motorway service.

3.3. Analysis Method

NEPLAN software was used for the investigation analysis in this study. NEPLAN Power Systems is the most user-friendly and one of the complete planning, optimization and simulation tool for transmission, distribution, generation, and industrial networks. NEPLAN covers all aspects of modern power system planning and analysis. It has been used to take the real load flow for each test. The load flow consists of node voltage magnitudes and node voltage phase angles [32]. In this study, to evaluate the impact of EV loads on the distribution system, load flow analysis was performed for one day at every 60-minutes. The analysis considered the load shape profiles of both EV and conventional loads.

3.4. Study Cases

This paper studies the performance of a distribution system in three different cases. The first case does not contemplate any modification in the system (actual conditions), and the other two cases consider EV charging with uncoordinated and coordinated charging strategies. For the uncoordinated case, it was considered that all EVs need to be entirely charged at the corresponding peak hours of each load in order to study the network's worst conditions. In addition, this case assumed that there is no control in place to modify the load schedule. Thus, domestic users will tend to plug their vehicles into the charging outlets, as soon as they get home from work. Similarly, commercial and industrial users will charge their vehicles during peak hours while they are working or shopping. Figure 7 and Table 1 show the methodology that has been followed for implementing the uncoordinated EV charging stations in this paper. As indicated, the following assumptions were made for the implementation:

- For placing the EV charging stations:
    - The node voltage rating must be 400V.
    - The transformer loading at the respective node must be less than 80%.
- For sizing the EV charging stations:
    - Residential customers: The capacity is equal to the magnitude of demanded active power at the respective node.
    - Commercial customers: The capacity is equal to 50% of the magnitude of demanded active power at the respective node.
    - Industrial customers: The capacity is equal to 20% of the magnitude of demanded active power at the respective node.
    - Big industry customers: The capacity is equal to 20% of the magnitude of demanded active power at the respective node.
    - Heating customers: There is no EV charging station.
- For determining charging time:
    - The worst case is chosen where charging is during peak demand hours.
    - Six hours are needed in slow charging to reach full capacity.
    - Two hours are needed in fast charging to reach full capacity.
- Other considerations:
    - All transformers do not have tap-changing capabilities.

Table 1. Parameters of the uncoordinated charging case.

Load type	Charging type	Penetration level (%)	Charging time (hrs)	Charging period
Residential	slow	100	6	16:00-22:00
Commercial	fast	50	2	10:00-12:00
Industrial		20	2	10:00-12:00

For a coordinated charging case, scheduling of EV charging in order to reduce daily variations in demand was considered. It was assumed that a simple timed controller is added to the households charging circuit to schedule charging to be done during the off-peak period, starting at 01:00 a.m. and remains on until 6:00 a.m. In addition, it was assumed that EV charging load at commercial and industrial nodes is spilled into four schedules– each 1/4 of the total charging load. In other words, instead of charging all vehicles simultaneously for two hours during peak time, these vehicles are distributed to be charged within the work period, between 08:00 a.m. and 4:00 p.m. It is expected that this direct demand management method will prevent overloading of distribution transformers at peak load periods; thus, indirectly improves the voltage profile. This strategy is summarised in Table 2.

Table 2. Parameters of the coordinated charging case.

Load type	Charging type	Penetration level (%)	Charging time (hrs)	Charging period
Residential	slow	100	6	01:00-06:00
Commercial	fast	12.5	8	08:00-16:00
Industrial		5	8	08:00-16:00

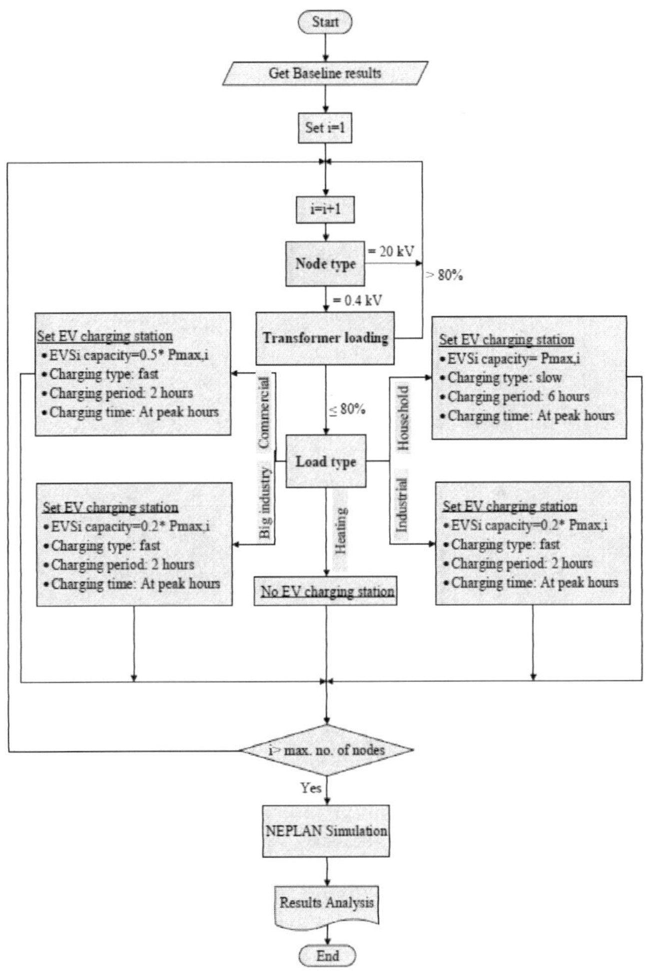

Figure 7. Methodology of implementing uncoordinated EV charging stations.

4. Results and Discussion

This section discusses the simulation results on the impacts and mitigation of EV charging on the distribution system. The discussed impacts include voltage profiles, transformers loading, power demand, and losses increment. All impacts are assessed at the low voltage (LV) nodes.

The percentage voltage band at all LV nodes are plotted in Figure 8. As aforementioned, the allowable voltage drop should not exceed 6% at LV nodes. This limit was not exceeded in all simulation cases; however, it was very close at node K52, down to 94.5% in the base case and with coordinated charging case. This drop was further increased in the case of uncoordinated charging, down to approximately 94.2%. In addition, the voltage dropped further at K24, reaching approximately 94.5% in comparison with 94.8% in cases of base operation and coordinated charging strategy. Note that the voltage drop at K52 under regular operation and coordinated charging is because of overloading caused by the other loads.

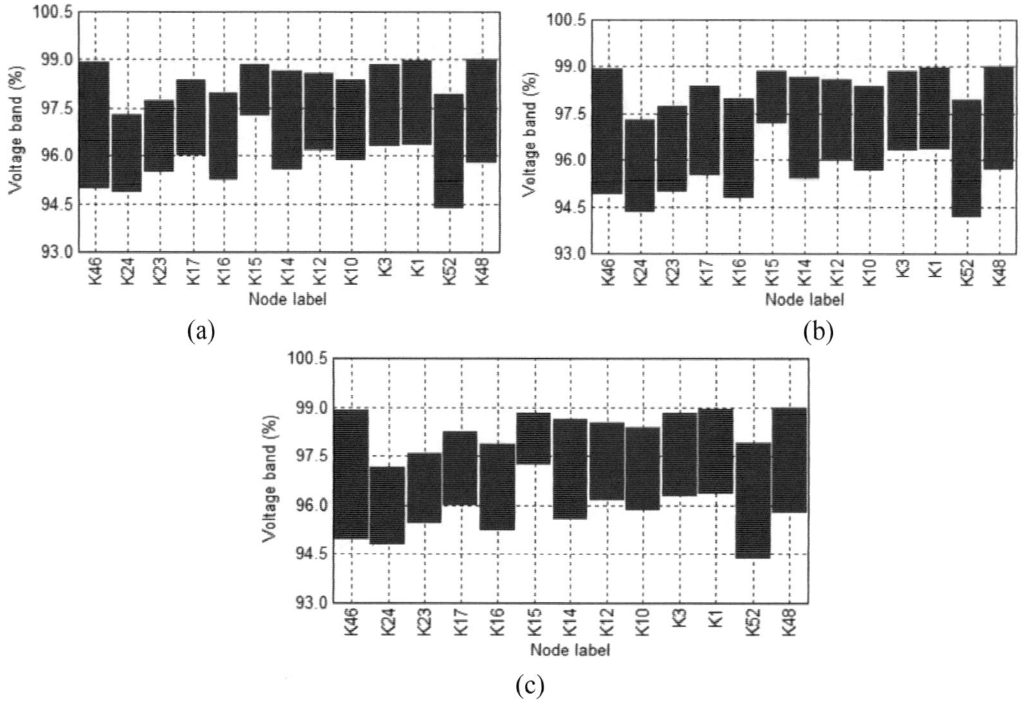

Figure 8. Voltage band of LV nodes for (a) base case, (b) uncoordinated EVs case, and (c) coordinated EVS case.

The transformer's loadings of the three simulated cases are shown in Figure 9. It is demonstrated that the feeder transformer (TR_FEED) maximum loading is approximately 35% in both the base case and the coordinated charging case. Nonetheless, it increased to 40% when uncoordinated charging stations were integrated into the distribution network. The same happened for the transformer Tr9, where its loading percentage increased from 60% in the base case to 98% in the case of uncoordinated charging. The loading of transformers Tr12 and Tr13 increased from 80% and 87% in the base case to 87% and 102% in the case of uncoordinated charging, respectively. For coordinated charging, the simulation results in the figure indicate the positive effect of coordinating EV charging on transformers loading, whereas the maximum loading of the main feeder transformer dropped from 40% in the previous case to 35% in this case. In addition, the maximum loading of transformers 9, 12, and 13 (Tr9, Tr12, and Tr13) dropped from 98%, 88%, and 102% to 63%, 87%, and 81%, respectively. It is also clear that there is no overloading in any transformer, except Tr5 which is originally overloaded even in the base case. Moreover, by comparing the results of this case with the results of the base case, it can be noted that the maximum loading of the transformers is almost identical in both cases. This is because the additional loads resulting from the charging of EV were shifted from the peak hours.

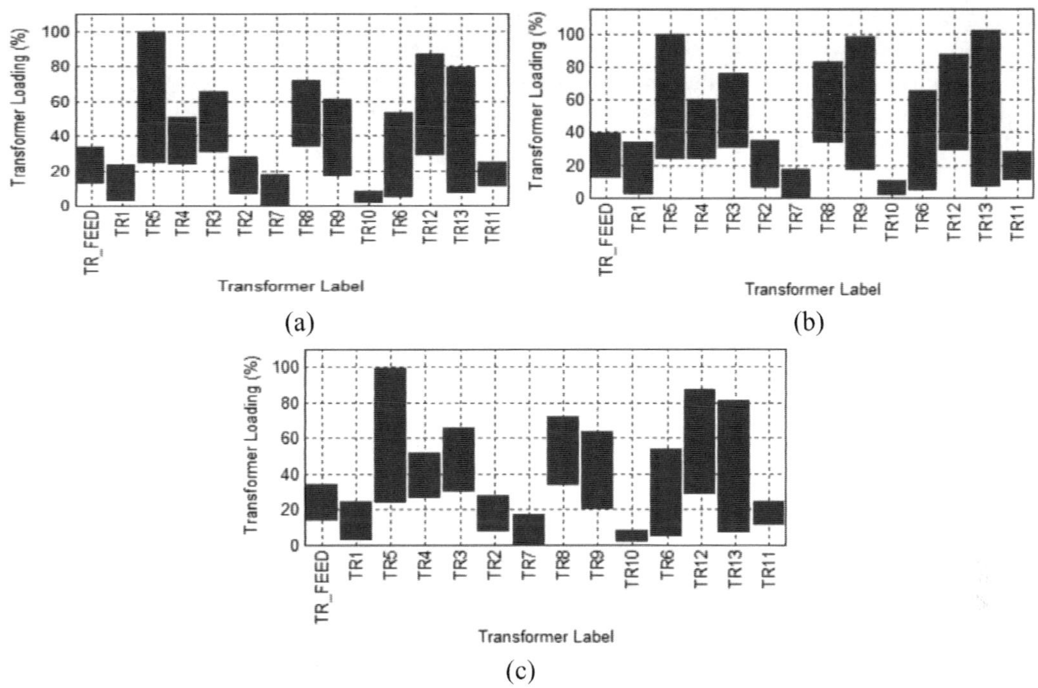

Figure 9. Transformers loading for (a) base case, (b) uncoordinated EVs case, and (c) coordinated EVS case.

Figure 10 depicts the comparison of total load power for the base case, uncoordinated charging, and coordinated charging. In the figure, the blue line is the total load power without EVs, the red and black lines are the load power profile with uncoordinated and coordinated charging strategies, respectively. In the uncoordinated charging strategy, the energy consumption increased in peak hours, reaching a peak demand of 10.87 MW at 11:00 am. While in coordinated charging strategy, the peak value occurred at 2 pm (8.89 MW), which is almost equal to the peak demand of the base case (8.75 MW).

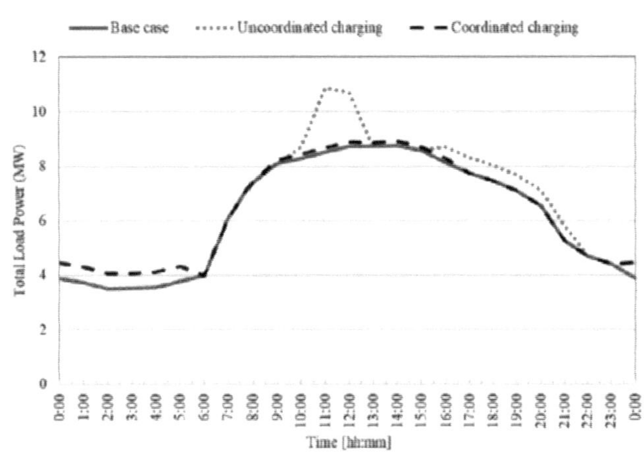

Figure 10. Total load power of the different cases.

Figure 11 depicts the comparison of total power losses for the base case, uncoordinated charging, and coordinated charging. As can be observed, the blue line shows the total power losses without EV penetration, whereas the red and black lines show the changes in the power losses profile due to uncoordinated and coordinated charging strategies, respectively. Uncoordinated charging strategy increased energy losses in peak hours, reaching a peak demand of 103.1 kW at 11:00 am. On the other hand, the peak value for a coordinated charging strategy occurred at 2 pm (78 kW), which is nearby to the peak demand without EV penetration (76 kW). A summary of the comparison results is provided in Table 3.

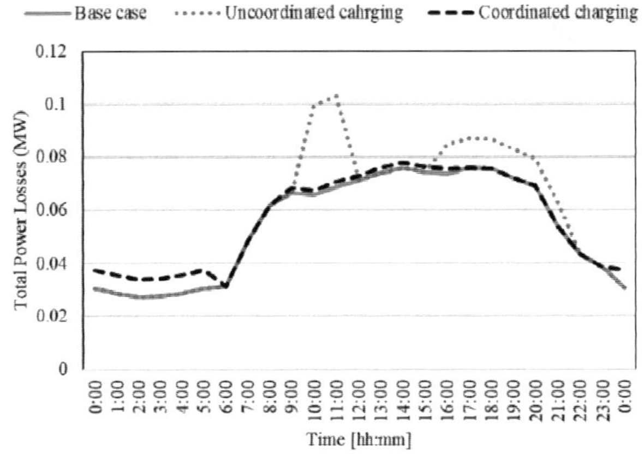

Figure 11. Total power losses of the different cases.

Table 3. Performance comparison of the studied cases.

Strategy	Load energy (MWh)	Max. load (MW)	Energy losses (MWh)	Max. losses (MW)
Base case	150.5	8.75	1.31	0.076
Uncoordinated charging	158.6	10.87	1.44	0.103
Coordinated charging	155.1	8.89	1.37	0.078

5. Conclusion

The objective of this paper is to investigate the impacts of EV charging on the power distribution network. Using NEPLAN software, a typical distribution network was simulated and analyzed under normal conditions and with integrated coordinated and uncoordinated EV charging stations. A comparative study was then made for the distribution system power demand and power losses of all simulated cases. Based on the simulation results, it can be concluded that the uncoordinated EV charging exhibits worst results, while some betterment (less negative impact on the grid) is observed under coordinated EV charging strategy. Charging EVs during peak times increases peak-demand significantly, thereby, requires expansion of generating capacity. By comparing the simulation results of the coordinated charging with the uncoordinated charging, it was found that although there were no changes were carried out in the distribution system components or specifications in either strategy, however, shifting and splitting the EV charging loads made the coordinated strategy able to eliminate all the resulted adverse impacts. The simulation results of the coordinated charging case were almost identical with the simulation results of the base case without any EV charging stations. It is evident that the coordinated charging strategy is the appropriate choice due to its ability to smooth load demand and reduce power losses.

Acknowledgments

The authors acknowledge the Ministry of Education, Malaysia, for supporting this work under grant no. 20190102LRGS. Special thanks to those who contributed to this project directly or indirectly.

Appendix

Table 4. Assigned load profiles

Load label	Load Node	Load Percentage (%)	Load Type
LO1	K3	100	Household
LO2	K46	100	Commercial enterprise
LO3	K48	100	Commercial enterprise
LO4	K14	100	Household
LO5	K52	90	Household
		10	Heating
LO6	K15	100	Heating
LO7	K23	100	Household
LO8	K24	100	Industry
LO9	K17	30	Commercial enterprise
		40	Household
		30	Industry
LO10	K16	100	Industry
LO11	---	100	Big industry
LO12	---	20	Commercial enterprise
		80	Motorway service
LO13	K12	10	Household
		90	Industry
LO14	K10	100	Industry
LO15	K1	90	Commercial enterprise
		10	Heating

References

[1] Zhang F, Zhang X, Zhang M, Edmonds AS, editors. Literature review of electric vehicle technology and its applications. 2016 5th International Conference on Computer Science and Network Technology (ICCSNT); 2016: IEEE.

[2] Dharmakeerthi C, Mithulananthan N, Saha T. A comprehensive planning framework for electric vehicle charging infrastructure deployment in the power grid with enhanced voltage stability. *International Transactions on Electrical Energy Systems*. 2015;**25**(6):1022-40.

[3] Gray MK, Morsi WG. Power quality assessment in distribution systems embedded with plug-in hybrid and battery electric vehicles. *IEEE Transactions on Power Systems*. 2014;**30**(2):663-71.

[4] Galiveeti HR, Goswami AK, Choudhury NBD. Impact of plug-in electric vehicles and distributed generation on reliability of distribution systems. *Engineering science and technology, an international journal*. 2018;**21**(1):50-9.

[5] Tan KM, Ramachandaramurthy VK, Yong JY, Padmanaban S, Mihet-Popa L, Blaabjerg F. Minimization of load variance in power grids—investigation on optimal vehicle-to-grid scheduling. *Energies*. 2017;**10**(11):1880.

[6] Tan KM, Ramachandaramurthy VK, Yong JY, editors. Three-phase bidirectional electric vehicle charger for vehicle to grid operation and grid voltage regulation. 2016 IEEE Transportation Electrification Conference and Expo, Asia-Pacific (ITEC Asia-Pacific); 2016: IEEE.

[7] Mwasilu F, Justo JJ, Kim E-K, Do TD, Jung J-W. Electric vehicles and smart grid interaction: A review on vehicle to grid and renewable energy sources integration. *Renewable and sustainable energy reviews*. 2014;**34**:501-16.

[8] Yong JY, Ramachandaramurthy VK, Tan KM, Mithulananthan N. Bi-directional electric vehicle

fast charging station with novel reactive power compensation for voltage regulation. *International Journal of Electrical Power & Energy Systems*. 2015;**64**:300-10.

[9] Dharmakeerthi C, Mithulananthan N, Saha T. Impact of electric vehicle fast charging on power system voltage stability. *International Journal of Electrical Power & Energy Systems*. 2014;**57**:241-9.

[10] Yong JY, Ramachandaramurthy VK, Tan KM, Arulampalam A, Selvaraj J, editors. Modeling of electric vehicle fast charging station and impact on network voltage. 2013 IEEE Conference on Clean Energy and Technology (CEAT); 2013: IEEE.

[11] Ma G, Jiang L, Chen Y, Dai C, Ju R. Study on the impact of electric vehicle charging load on nodal voltage deviation. *Archives of Electrical Engineering*. 2017;**66**(3).

[12] Shahnia F. Analysis and correction of voltage profile in low voltage distribution networks containing photovoltaic cells and electric vehicles: Queensland University of Technology; 2011.

[13] Ul-Haq A, Cecati C, Ehsan A, Strunz K, editors. Impact of electric vehicles on voltage profile and harmonics in a distribution network. 2015 First Workshop on Smart Grid and Renewable Energy (SGRE); 2015: IEEE.

[14] Yong JY, Ramachandaramurthy VK, Tan KM, Mithulananthan N. A review on the state-of-the-art technologies of electric vehicle, its impacts and prospects. *Renewable and Sustainable Energy Reviews*. 2015;**49**:365-85.

[15] Dubey A, Santoso S. Electric vehicle charging on residential distribution systems: Impacts and mitigations. *IEEE Access*. 2015;**3**:1871-93.

[16] Bunga SK, Eltom A, Sisworahardjo N, editors. Impact of plug-in electric vehicle battery charging on a distribution system. 2014 IEEE Industry Application Society Annual Meeting; 2014: IEEE.

[17] Darabi Z, Ferdowsi M. Impact of plug-in hybrid electric vehicles on electricity demand profile. Smart Power Grids 2011: Springer; 2012. p. 319-49.

[18] Qian K, Zhou C, Allan M, Yuan Y. Modeling of load demand due to EV battery charging in distribution systems. *IEEE transactions on power systems*. 2010;**26**(2):802-10.

[19] Shafiee S, Fotuhi-Firuzabad M, Rastegar M. Investigating the impacts of plug-in hybrid electric vehicles on power distribution systems. *IEEE Transactions on Smart Grid*. 2013;**4**(3):1351-60.

[20] Oliveira-De Jesus D, Paulo M, Rios MA, Ramos GA. Energy Loss Allocation in Smart Distribution Systems with Electric Vehicle Integration. *Energies*. 2018;**11**(8):1962.

[21] Richardson DB. Electric vehicles and the electric grid: A review of modeling approaches, Impacts, and renewable energy integration. *Renewable and Sustainable Energy Reviews*. 2013;**19**:247-54.

[22] Valenzuela Santillan AD. Impacts of Electric Vehicles on distribution networks: Case study Ecuador: Reino Unido/Newcastle Univesity l/2016; 2016.

[23] Bangalore P. Development of test system for distribution system reliability analysis, integration of electric vehicles into the distribution system 2011.

[24] Shi R. The dynamic impacts of electric vehicle integration on the electricity distribution grid: University of Birmingham; 2013.

[25] Su C, Zhang J, editors. Research on the impacts of large-scale electric vehicles integration into power grid. AIP Conference Proceedings; 2018: AIP Publishing LLC.

[26] Electropaedia. Battery and Energy Technologies [Available from: https://www.mpoweruk.com/infrastructure.htm.]

[27] Lin C-C, Deng D-J, Liu W-Y, Chen L. Peak load shifting in the internet of energy with energy trading among end-users. *IEEE Access*. 2017;**5**:1967-76.

[28] Richardson P, Flynn D, Keane A, editors. Impact assessment of varying penetrations of electric vehicles on low voltage distribution systems. IEEE PES general meeting; 2010: IEEE.

[29] Xu DQ. Coordination of Plug-in Electric Vehicles Over a Communications Infrastructure: McGill University Libraries; 2013.

[30] Clement-Nyns K, Haesen E, Driesen J. The impact of charging plug-in hybrid electric vehicles on a residential distribution grid. *IEEE Transactions on power systems*. 2009;**25**(1):371-80.
[31] Deilami S, Masoum AS, Moses PS, Masoum MA. Real-time coordination of plug-in electric vehicle charging in smart grids to minimize power losses and improve voltage profile. *IEEE Transactions on Smart Grid*. 2011;**2**(3):456-67.
[32] NEPLAN. NEPLAN Electricity [Available from: https://www.neplan.ch/neplanproduct/en-electricity/.]

Effects of Generator Ratings on Inertia and Frequency Response in Power Systems

Khaldon Ahmed Qaid, Chin Kim Gan*, Norhafiz Bin Salim, and Khairul Anwar Ibrahim

Faculty of Electrical Engineering, Universiti Teknikal Malaysia Melaka, 76100 Durian Tunggal, Malaysia

E-mail: ckgan@utem.edu.my

Abstract. The increasing share of renewable generation integrated in the traditional power systems network has brought new challenges to the utility. More specifically, the high penetration of solar energy in the network will reduce the total system inertia which could jeopardize the system's stability during contingency. The lack of inertia in the power system will increase the Rate of Change of Frequency (RoCoF), which is harmful in the event of sudden load change. In this regard, this paper analyses the effects of generator ratings on system's inertia and frequency response. The IEEE 9-BUS test system has been utilized in this paper to model the system's inertia response under various contingency scenarios. The results show that small rating of generators can achieve 20% higher inertia response as compared to the case with larger generator ratings. This implies that system with smaller generators can better recover the system's frequency than the large-scale generators of similar total rating.

1. Introduction

Fossil fuel power stations release about 2.2 billion tons of carbon dioxide CO_2 to the air annually in the United States [1]. The major causes of environmental pollution in the world are acid rain and global warming [2]. These problems have prompted governments and agencies around the world to set new goals in increasing the renewable energy sources (RES) share in their electrical generation mix [3]. For example, China has set a goal of generating over 15% of its total power from renewable energy sources by 2020 with 30 GW of biomass, 200 GW of wind, 50 GW solar and 420 GW of hydro. These strategies are critical to address the large rise in world energy demand while at the same time decreasing the quantity of pollutions. However, the reliability of the whole electrical system may decrease due to the high integration of intermittent RESs which may influence the stability of the system's frequency [4]. The inertial response in RESs is naturally low for wind turbines or nonexistent for PV solar systems with no rotational parts [5].

For instance, the unstable speed wind turbines are connected to the grid across a converter. This has efficiently decoupled the wind turbine inertia from relieving system transients. In addition, photovoltaic systems do not have any inertia response contribution to the power system. Consequently, the total inertia of the entire power system will decrease when the share of renewable energy increases. In [6], the authors have reported that the growing number of RESs in the United Kingdom could decrease the inertia constant by up to 70% from 2013 to 2033. The Rate of Change of Frequency (RoCoF) of the power system will be sufficiently high to trip the frequency relay during load-shedding due to the low system's inertia. Therefore, the decreasing of the inertia constant may lead to network blackout during contingency. Study in [7] shows that whenever the percentage of RES installed capacity increases, the inertial constant will be decreased. This will increase the RoCoF of the power system. Moreover, the number of generation units that provide reserve capacity for primary and secondary frequency response will decrease when the penetration level of the RESs increases. This will lead to the increase in frequency deviation, which is undesirable.

This paper presents the impact of generator ratings on system's inertia constant and the associated frequency response. The IEEE 9-BUS system has been utilized as the test system and simulated in DIgSILENT PowerFactory software. This paper is organized in the following order: Section 2 presents the concept of the swing power equation with inertia constant and the rate of change of frequency. Section 3 shows the IEEE 9-BUS test system. Moreover, the methodology of the case study will be presented in this section. Section 4 discusses and analyzes the results of the case study. The findings and summary of this paper are summarized in Section 5.

2. Swing equation

Assume a three-phase synchronous alternator is motivated by a prime mover, the machine rotor motion equation can be written as follows [8]:

$$J\frac{d^2\theta}{dt^2} = T_m - T_e = T_a \tag{1}$$

Where J is total inertia of the rotor mass in kgm_2, T_m is mechanical torque of the prime mover in N-m, T_e is electrical torque output of alternator in N-m. and θ is the angular position of the rotor in rad.

By neglecting the losses, the net accelerating torque, T_a, is generated by the difference between the mechanical and electrical torque. The electrical torque is equivalent to the mechanical torque in the steady-state; hence the acceleration of power will be equal to zero. In this time the rotor will rotate at synchronous speed ω_s in rad/s. The angular position θ is measured with a stationary reference frame. The θ in term of the synchronous rotating frame is given as follows:

$$\theta = \omega_s t + \delta \tag{2}$$

where, δ is angular position in radians. Taking the time derivative of the above equation with respect to the synchronous rotation reference frame can be written as follows:

$$\frac{d\theta}{dt} = \omega_s t + \frac{d\delta}{dt} \tag{3}$$

Defining the angular speed of the rotor as

$$\omega_r = \frac{d\theta}{dt} \tag{4}$$

It can be elaborate (3) as the following equation:

$$\omega_r - \omega_s = \frac{d\delta}{dt} \tag{5}$$

Therefore, the rotor angular speed is equal to the synchronous speed only when dδ/dt is equal to zero. Taking the derivative of (3) and substitute into (1), the torque equation can be written as follows:

$$J\frac{d^2\delta}{dt^2} = T_m - T_e = T_a \tag{6}$$

Multiplying both sides by ω_r,

$$J\omega_r\frac{d^2\delta}{dt^2} = P_m - P_e = P_a \tag{7}$$

Where P_m is mechanical power in MW, P_e is electrical power in MW and P_a is accelerating power in MW.

2.1. *Inertia constant*

The kinetic energy stored in a single generator shaft is often expressed as proportional to its power rating. It is called the inertia constant H. Inertia can be known as the resistant to change and it prevents the frequency from changing rapidly. The inertia constant is measured in seconds and it falls typically in the range of 2-9s. For large power plants, the inertia constant is inversely proportional to the power rating. That means when the generator rating increases the inertia constant will decrease as reported in [9] and as shown in Figure 1.

Two factors mainly affect the system inertia, *i.e.*, the inertia of each generator and the number of operating generators in the system. In the conventional power plants, the synchronous generators can provide the inertia to the system given the strong coupling between electrical frequency and the rotational speed [10].

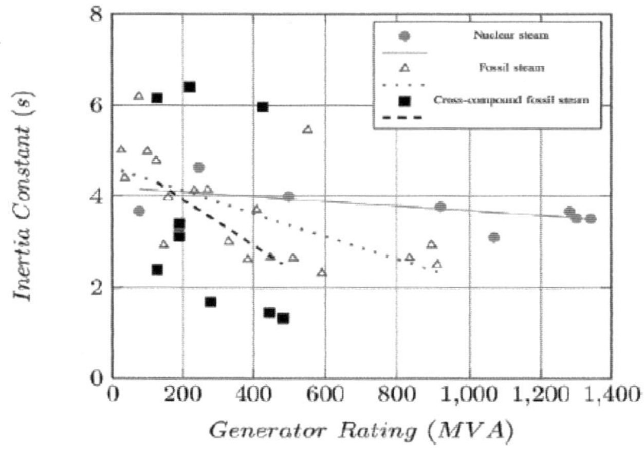

Figure 1. Inertia constant for conventional generators [9].

The equation of motion of the machine rotor is given as below, that consider inertial constant of a machine as [11],

$$H = \frac{Stored\ Energy\ at\ rated\ speed\ in\ MWS}{Generator\ MVA\ rating} \tag{8}$$

$$K.E. = \frac{1}{2} J\omega^2 \tag{9}$$

From Equation (7) and (8)

$$H = \frac{\frac{1}{2} J\omega^2}{MVA} \tag{10}$$

Where H is inertia constant in MWs/MVA, J is the moment of inertia of the rotating mass in kgm^2, ω is the nominal speed of the rotation in rad/s and MVA is nominal generator rating. The total system inertia is calculating from the equation below:

$$H_{sys} = \sum_{i=1}^{n} H_i S_i \tag{11}$$

Where H_{sys} is total system inertia, H_i is generator *i* inertia in second and S_i is generator *i* rating in MVA.

2.2. Rate of change of frequency (RoCoF)

The time derivative of the power system frequency (df/dt) is known as RoCoF. In a traditional power system, RoCoF was not generally taken into consideration because of the nature of the network without any integrated RESs and small distributed generators. The RoCoF value is not necessary for the conventional power system, but by increasing the penetration of RES into the power system, the RoCoF can play an important role to control the frequency of the network. Moreover, RoCoF can be used for knowing the mismatch power generation and load consumption. The mismatch can be caused by the disconnection of either large loads or generators. The high amount of RoCoF might be observed because of the low system inertia caused by the integration of a high penetration level of uncertain generation such as PV system or Wind turbines. The rate of change of frequency is calculated from Equation (12) [12, 13].

$$\frac{Df}{dt} = \frac{\Delta p \times f_0}{2H \times S} \quad (12)$$

Where $\frac{df}{dt}$ is the rate of change of frequency in Hz/s, H is total system inertia in MWs/MVA, Δp is generator loss in MW, f_0 is generator rated frequency in Hz and S is total generators rating in MVA.

3. Methodology

The IEEE 9-BUS test system is used in this paper, to highlight the effects of inertia constant and the generator rating MVA on the frequency response after contingencies. The system has 9 buses, 3 generators, 3 transformers and 6 transmission lines. It consists of 3 general loads which total up to 315 MW and 115 MVar. A turbine governor and AVR exciters are equipped to all the generators. Figure 2 shows the IEEE 9-BUS system single line diagram in DIgSILENT PowerFactory software. The generators data used in this simulation are shown in Table 1 [14]. The inertia constant H which is shown in Table 3 based on the system rated power MVA that has been calculated from the stored energy at nominal speed from Equation (10).

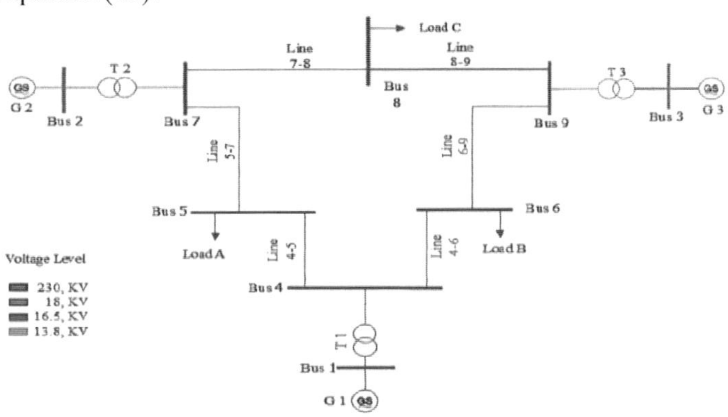

Figure 2. IEEE 9-BUS system single line diagram.

Table 1. Generators data.

Gen. No.	S (MVA)	P (MW)	J (kgm2)	H (s)	Q (MVar)	V (pu)
G1	247.5	71.6	9160.1	2.63	27	1.04
G2	192	163	1115.8	4.13	6.7	1.03
G3	128	85	8592	4.77	-10.9	1.02

Three scenarios have been considered in this study. Each scenario considers a generator to be disconnected from the system and the other two generators remain in operation. The details are as shown in Table 2.

Table 2. Generators operation scenarios.

Scenario No.	G1	G2	G3
1	Off	On	On
2	On	Off	On
3	On	On	Off

Each scenario consists of three cases. The data in Case 2 is the IEEE standard power rating. In Case 1 the power rating is decreased for all the generators by -20% from Case 2 to see the change in Inertia. Opposite of Case 1, the power rating in Case 3 is increasing by +20% to compare with Case 2. Table 3 shows the data used in each scenario for all cases. Figure 3 shows different values of generators power rated with inertia constant calculated by Equation (10).

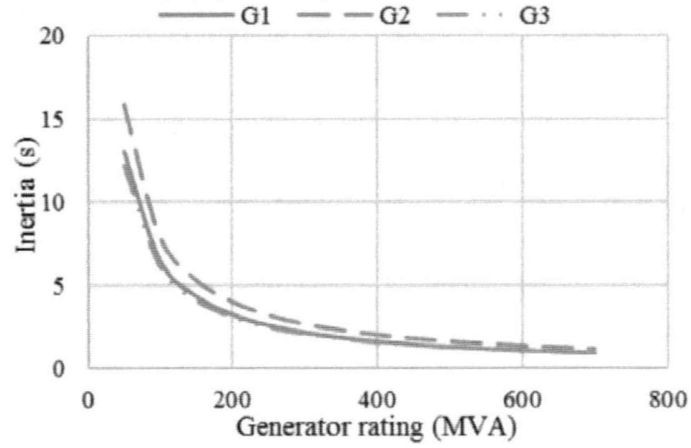

Figure 3. Inertia constant in different power rated MVA.

Table 3. Power and inertia data used in scenarios for each generator.

Case No.	Scenario 1		Scenario 2		Scenario 3	
	G 1		G2		G 3	
	P (MVA)	H (s)	P (MVA)	H (s)	P (MVA)	H (s)
1	200	3.25	150	5.29	100	6.10
2	247.5	2.63	192	4.13	128	4.77
3	300	2.6	250	3.18	200	3.05

4. Result and discussion

Numerical results from DIgSILENT Power Factory simulation shows that, when the power rating of the generator is increased, the inertia constant H of the generator is decreased. This scenario has led to a high change in the RoCoF of the whole system and the system will be in critical status. The RoCoF in all the scenarios is calculated from Equation (12). The total system inertia before and after disconnecting the generators are calculated from Equation (11). The data and results for each scenario are shown in the following sub-sections.

4.1. Scenario 1

Generator 1 (G1) with the capacity of 71.6 MW is equivalent of approximately 22% of total system generation. This generator disconnected from the network after 1 second. The other generators will keep on continuing the operation, but the power rating and inertia constant data of all generators are changed based on Case 1, 2 and 3 respectively as shown in Table 4.

Table 4. Result of Scenario 1.

Case No	Total inertia before disconnecting G1(MWs)	Total inertia after disconnecting G1 (MWs)	RoCoF (Hz/s)	Freq nadir (Hz)	Steady state (Hz)
1	2053.5	1403.50	0.479	52.54	52.96
2	2054.4	1403.52	0.564	59.19	59.73
3	2185.0	1405.0	0.628	59.89	59.95

Figure 4 shows that the frequency is unstable before tripping G1 in Case 1 due to the mismatch between generation and load. In other words, if the demand is higher than the rated generation, the frequency dropped. However, after contingency, the frequency is dropped to 52.546 Hz in Case 1, which result in a drastic change in RoCoF by 0.479 Hz/s just after the contingency. It can be observed from Figure 4 that the system collapse at approximately 0.5 s. That is because of the shortage of active power MW in the system due to the small generator rating MVA and the large load.

In Case 2 which is IEEE standard base case system, the system is recovered (59.379 Hz) in the steady-state after contingency which the frequency nadir is 59.199 Hz. The system is stable because of the balance between the supply and the demand.

In Case 3 the total generator's rating is 450 MVA as shown in Table 4. In this case, G1 is disconnected and the system inertia is low which led to high RoCoF. As a result, the frequency in steady-state is 59.954 Hz and frequency nadir is 59.893 HZ so the system is stable after contingency.

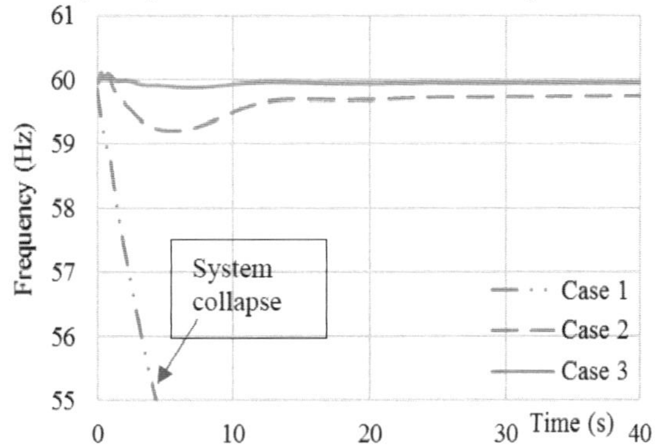

Figure 4. Result of inertia and frequency for Scenario 1.

4.2. Scenario 2

In this scenario, G2 is disconnected after 1 second from the rest of the network. The loading of G2 is 163 MW which is about 51% of total system generation. Like Scenario 1, due to mismatch between generation and demand, the frequency dropped to 53.829 Hz in Case 1 and it tried to recover the frequency during the secondary frequency response. However, the RoCoF experiences the extreme change of 1.210 Hz/s just after the occurrence of the contingency. This is because of the shortage of active power MW in the system due to the small generator rating MVA and the large load. Figure 5 shows that the system failed to recover the network frequency to the nominal range, and the network has collapsed.

In Case 2 which is IEEE standard base case, the system frequency is recovered to 59.816 Hz in the steady-state after contingency which the frequency nadir is 58.702 Hz. The system is stable because of the balance between the load and the generation.

In Case 3 the total generator's rating is 550 MVA as shown in Table 5. In this case, G2 is disconnected and the system inertia is low which led to high RoCoF. As a result, the frequency in steady-state is 59.841 Hz and frequency nadir is 59.049 Hz so the system is stable after contingency as shown in Figure 5.

Table 5. Result of scenario 2.

Case No	Total inertia before disconnecting G2 (MWs)	Total inertia after disconnecting G2 (MWs)	RoCoF (Hz/s)	Freq. nadir (Hz)	Steady state (Hz)
1	2053.5	1260.0	1.210	53.829	57.421
2	2054.45	1261.49	1.284	58.702	59.816
3	2185.0	1390.0	1.336	59.049	59.841

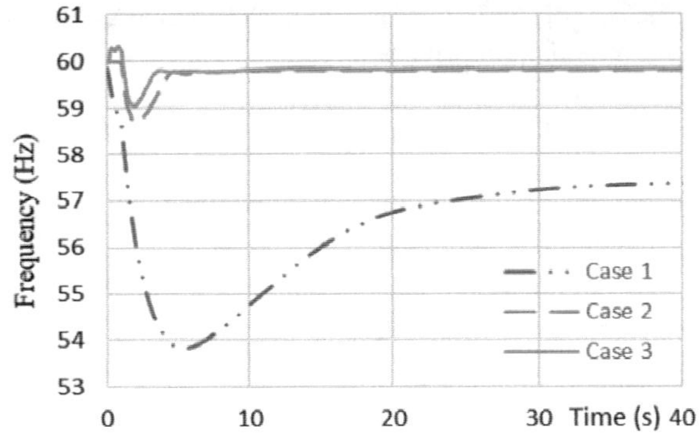

Figure 5. Result of inertia and frequency for Scenario 2.

4.3. Scenario 3

This scenario is almost similar to Scenario 2 as shown in Figure 6, except it operates with higher generator rating. In this scenario, G3 is disconnected after 1 second. Therefore, the network will face the lack of 85 MW (approximately 26% of total generation) that must be compensated by G1 and G2. Thus, as shown in Figure 6 and Table 6, the increasing generator rating value has a significant negative impact on the total inertia constant H of the system. For instance, the increasing of inertia constant from Case 1 with a total of 300 MVA of generation is 8.9768 s in Scenario 2 that compare to 350 MVA with 9.9613 s inertia in Scenario 3.

In Case 1, the frequency nadir is 55.022 Hz and the frequency reach to steady-state with 58.197 Hz. The RoCoF is 0.568 Hz/s, in this case, because of the high inertia of the system. Case 2 is the best in term of inertia because the increased in MVA rating of all generators with total 439.5 MVA. The steady-state frequency, in this case, is 59.917 Hz and the nadir is 59.589 with RoCoF 0.669 Hz/s. Case 3 the RoCoF is high due to the decrease of inertia. The system frequency nadir is 0.775 Hz/s and the system became stable at 59.882 Hz.

Table 6. Result of Scenario 3

Case No	Total inertia before disconnecting G3 (MWs)	Total inertia after disconnecting G3 (MWs)	RoCoF (Hz/s)	Freq. nadir (Hz)	Steady state (Hz)
1	2053.5	1443.5	0.568	55.022	58.197
2	2054.45	1443.89	0.669	59.589	59.917
3	2185.0	1575.0	0.775	59.177	59.882

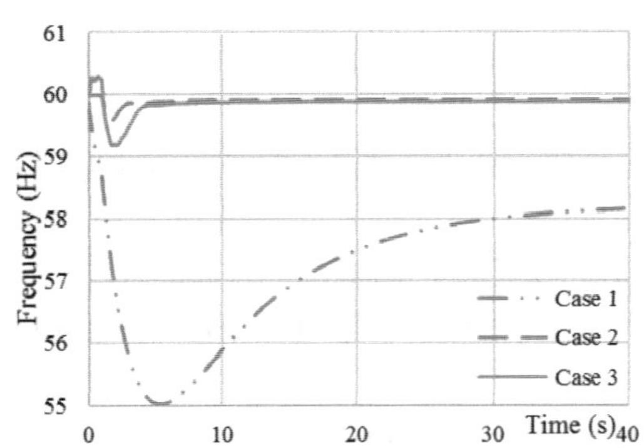

Figure 6. Result of inertia and frequency for Scenario 3.

5. Conclusion

This paper presents the analyses on the rotational inertia of conventional power system, specifically on the impact of generator ratings on the system's inertia constant and the associated frequency response. In general, the analysis shows that the lack of inertia in the power system increases the Rate of Change of Frequency (RoCoF).

The IEEE 9-BUS test system utilized in this paper as a case study to module the inertia response. Three different scenarios have been considered in this study that each scenario consists of three cases. The results show that the small rating of generators produces the highest inertia in the power system, which helps to recover the system frequency better than the large generators. As the findings of this paper, the integration of small rotational generators that can deliver high inertia rate could help to recover the system during the contingency within the primary frequency response. Based on the three scenarios presented, the range of the generator power rating required to get optimal inertia response is between 100 to 400 MVA.

Acknowledgments

The authors would like to gratefully acknowledge the funding provided by the Universiti Teknikal Malaysia Melaka under grant JURNAL/2018/FKE/Q00012.

References

[1] Lashof D, Yeh S, Carter S, Doniger D, Murrow D, Johnson L. Cleaner and cheaper: Using the clean air act to sharply reduce carbon pollution from existing power plants, delivering health, environmental, and economic benefits. *Natural Resources Defence Council (NRDC) Retrieved from http://assets fiercemarkets net/public/sites/energy/reports/nrdcreport1 pdf.* 2014.

[2] Nations U. Kyoto protocol to the United Nations framework kyoto protocol to the United Nations framework. *Rev Eur Community Int Environ Law*. 1998;**7**:214-7.

[3] Dreidy M, Mokhlis H, Mekhilef S. Inertia response and frequency control techniques for renewable energy sources: A review. *Renewable and sustainable energy reviews*. 2017;**69**:144-55.

[4] Secretariat R. Renewables 2012 global status report. *Rep Paris: REN12*. 2012.

[5] Dehghanpour K, Afsharnia S. Electrical demand side contribution to frequency control in power systems: a review on technical aspects. *Renewable and Sustainable Energy Reviews*. 2015;**41**:1267-76.

[6] Smith R. Electricity Ten Year Statement 2015. *United Kingdom National Grid*. 2015:2-13.

[7] Jayawardena A, Meegahapola L, Perera S, Robinson D, editors. Dynamic characteristics of a hybrid microgrid with inverter and non-inverter interfaced renewable energy sources: A case study. 2012 IEEE International Conference on Power System Technology (POWERCON); 2012: IEEE.

[8] Ulbig A, Borsche TS, Andersson G. Impact of low rotational inertia on power system stability and operation. *IFAC Proceedings Volumes*. 2014;**47**(3):7290-7.

[9] Tielens P, Van Hertem D. The relevance of inertia in power systems. *Renewable and Sustainable Energy Reviews*. 2016;**55**:999-1009.

[10] Mancarella P, Puschel S, Zhang L, Wang H, Brear M, Jones T, et al. Power system security assessment of the future national electricity market. *University of Melbourne*. 2017.

[11] Anderson PM, Fouad AA. Power system control and stability: John Wiley & Sons; 2008.

[12] Ahmadyar AS, Riaz S, Verbič G, Riesz J, Chapman A, editors. Assessment of minimum inertia requirement for system frequency stability. 2016 IEEE International Conference on Power System Technology (POWERCON); 2016: IEEE.

[13] Kundur P. Power System Stability and Control. McGraw-Hill; 1994.

[14] DIgSILENT GmbH. DIgSILENT PowerFactory Manual Version 15.2. DIgSILENT GmbH, Germany; 2014.

IOP Conf. Series: Materials Science and Engineering 1127 (2021) 012035 doi:10.1088/1757-899X/1127/1/012035

A review of optimization methods for economic and emission dispatch considering PV and wind energy

M Y Nassar *, M N Abdullah *, A Ahmed *

*Green and Sustainable Energy (GSEnergy) Focus Group, Faculty of Electrical and Electronic Engineering, Universiti Tun Hussein Onn Malaysia, 86400 Parit Raja, Batu Pahat, Malaysia

E-mail: mnoor@uthm.edu.my

Abstract. The power generation system has fossil fuels as their major electricity generation source. This source is not reliable and environmentally friendly so, the alternative source such as renewable energy (RE) is getting much attention in the development of a power generation system. To achieve optimal power output for the power system there is a necessity to solve the economic and emission dispatch (EED) problem considering the operational system constraints. Numerous types of research on optimization of fossil fuel cost and emission roles have been focused to solve the concerned power dispatch issue. This paper represents a comprehensive study on combined economic and emission dispatch (CEED) problem with respect to conventional and unconventional energy sources. To optimize the power dispatch problem, the existing algorithm consisting of optimization methods and objective functions in integrated renewable energy sources (RES) system is tested and reviewed. This paper presented various optimization methods applied to different test systems to minimize emission and cost for power dispatch with PV and wind. Future trends and necessary steps next in the field of power dispatch are also been deliberated.

1. Introduction

Nowadays, the power system is facing more challenges in generation variability as a multitude of RES are accommodated in the operation of the power system. Therefore, the operators of the power system are optimizing the dispatch along with the commitment of generators to supply the demand for electricity while respecting all operational requirements [1].

The incremental power demand along with the reduction rate in environmental pollution can be solved by considering RES as alternative energy sources [2]. Meanwhile, RES can be considered as non-pollutant or lower exhaust emission to supply electricity to consumers [3]. Therefore the integration of the RES such as photovoltaic (PV) [4] and wind turbine [5, 6] in the power system has changed the research direction to mitigate environmental issues of conventional power dispatch [7, 8].

Power dispatch plays an essential role in the planning process and modern control power systems [9]. Thus, optimal power system operation is important in electrical networks for operating the designed flow with minimal energy cost to satisfy the system constraints [10]. Moreover, optimal power dispatch (OPD) is considered as vital planning in the power system to schedule the generator output power around a minimum cost and emission level [11]. Economic dispatch (ED) is a fundamental term used to minimize the total cost of power system operation problem related to system objectives.

Typically, pollution produced by power generation plants been neglected based on the minimum fuel cost criteria [12]. Furthermore, environmental consideration tools have to minimize the harmful emission producing from the thermal power generation [13]. The implementation of various pollution control acts, power generation companies have to worry about the cost of power generation as well as the effect of the pollutants that result from the power generating units. [14]. This term leads to get an optimal power generation for CEED by considering different system constraints in the power system [15]. In order to dispatch the overall load demand in generating units, the optimization of the CEED problem is to be necessity [16]. Consequently, solving the EED problem with RE is also essential.

This paper introduces a review of optimization methods and problem formulation for CEED considering thermal power generation, PV and wind energy source. Section 2 represents the revised view of problem formulation in accordance with individual ED and emission dispatch as well as CEED. Section 3 introduces the branches of optimization methods related to prospective system algorithm. Section 4 the detail view of integrated RES system is presented. In Section 5 the review of this paper is concluded.

2. Overview of power dispatch formulation

The formulation of Power dispatch can be classified as follow:

2.1. Economic dispatch

ED is an important optimization problem for generation cost minimization. The aim of ED is to minimize the total cost of generation whereas the operating constraints are satisfied. The fuel cost function is used to obtain optimal power output [17]. The function of overall fuel cost is expressed as follows:

$$F(P_G)=\sum_{i=1}^{N_g}C_i(P_{Gi}) \quad (1)$$

$$C_i(P_{Gi})=a_iP_{Gi}^2+b_iP_{Gi}+c_i \quad (2)$$

where $F(P_G)$ is the total cost of production, $C_i(P_{Gi})$ is the unit function of fuel cost, a_i, b_i and c_i are the unit coefficients of fuel cost and P_{Gi} is the unit output i of the real power.

Considering valve-point loading of a generator, the ED problem can be formulated as:

$$C_i(P_i)=\left(a_i+\mathrm{b}_iP_{Gi}+cP^2{}_{Gi}\right)+\left|e_i\sin\left(f_i\left(P_{Gimin}-P_{Gi}\right)\right)\right| \quad (3)$$

where a_i, b_i, c_i, e_i, and f_i are the coefficients of the ith generating unit.

2.2. Emission dispatch

By considering environmental conditions, it should minimize the harmful gases of pollutants resulted from fossil fuels burning in the thermal plant. Therefore, the emission function aims to reduce the pollutions from the powered generation [13]. The total emission can be expressed as follows:

$$E(P_G)=\sum_{i=1}^{N_g}(\alpha_i\left(P_{Gi}^2\right)+\beta_iP_{Gi}+\gamma_i)+\varepsilon_i exp\left(\lambda_iP_{Gi}\right) \quad (4)$$

where α_i, β_i, γ_i, ε_i and λ_i are the coefficients of the characteristic for ith generator emission.

2.3. Combined economic and emission dispatch

ED and emission dispatch can be combined as a multi-objective CEED to find optimal generation where reduce the emission level and fuel cost simultaneously [18]. The CEED problem can be represented as follows:

$$\min F=\sum_{i=1}^{n}\left(C_i(P_i)+h_i\times E_i(P_i)\right) \quad (5)$$

where F is an objective function that must be reduced, $C_i(P_i)$ is the fuel cost and $E_i(P_i)$ is the emission of the ith generating unit and h_i is the penalty factor.

h_i is used to convert the multi-objective function into the single objective function, where h_i can define as:

$$h_i = \frac{C_i(P_G)}{E(P_G)} \tag{6}$$

3. Optimization methods for optimal power dispatch

Optimization is a procedure to find a superior solution. Optimized power dispatch is the process of optimization to define the generating unit's schedule to meet system constraints and to supply the demand power [19].

The optimization methods of power dispatch are classified into three categories, which are hybrid, non-conventional, classical methods [19]. The conventional approach for solving ED problems is known as a classical method while non-conventional methods are used to handle the practical and non-convex ED problems. The third category is hybrid methods that combine two or more classical and/or non-conventional methods in order to improve the performance of individual methods. Figure 1 illustrates the three types of optimization methods for optimal power dispatch.

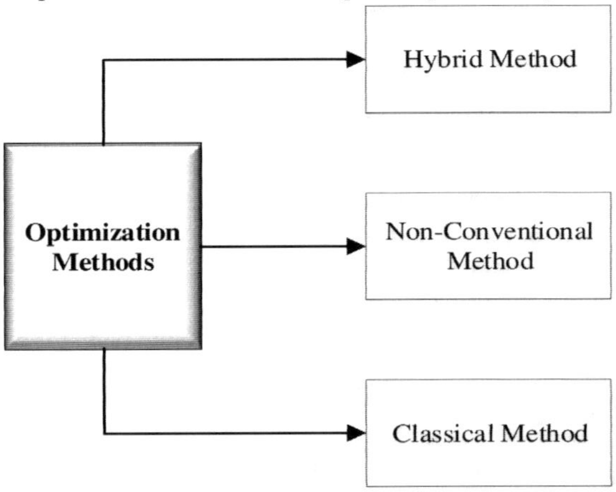

Figure 1. Types of optimization methods for optimal power dispatch.

There are a lot of classical methods have been used in order to solve ED problem such methods include newton's method [20] besides the quadratic programming technique [21] have solved ED problem with different objective functions included cost, loss, and reductions. The proposed algorithms have been achieved best optimal solution and less computational time. Interior point method [22] has considered transmission line flow limits and generation ramp rate in ED. In addition, the system operation is satisfied simultaneously. Lambda iteration method [23] was implemented on 15 and 140 units test system to prove its efficiency. The classical methods are adaptive, flexible when analysing the problem and easy to understand. On the other hand, this type has two main drawbacks: firstly, this method is not appropriate to solve nonconvex or nonsmooth problems; secondly, it generates only one solution in a single run [24, 25].

However, non-conventional methods can deal with the complicated optimization problem and developed to solve the CEED problem. Numerous methods of this type like Ant colony optimization (ACO) [10], Bat algorithm (BA) [24], Genetic Algorithm (GA) [26] and Particle swarm optimization (PSO) [27] have been solved this problem. The performance of these algorithms is better than classical optimization methods in many aspects, for instance, robust and easy to adjust according to the problem.

Table 1. Advantages and disadvantages of several optimization techniques

Ref	Optimization method	Advantages	Disadvantages
[10]	ACO	• Can find the targets under any environment • Suitable for implementation under parallel numerical computation • The reliability of the method	• The hypothetical analysis is difficult to understand. • More convergence time
[28]	PSO	• Excellent convergence characteristic. • Superior in terms of fuel cost as well as computation time. • Computational efficiency • Easy implementation	• Weak capacity of local search • The iterative process has a low rate of convergence
[29]	GA	• Flexible and powerful • Less computational time • Gives near-global optimum solution	• Response time of stable optimization cannot assure • Find a global optimum is not guaranteed
[30] [31]	Linear programming	• Adaptive and flexible when analyzing the problem. • Easy to understand.	• It is not able to solve a problem that has more than two variables. • It cannot solve the non-linear functions.
[32] [33]	Simulated Annealing	• Simple coding for the composite problems • Easy to deal with a cost function	• Need other methods in order to get an optimal solution
[34]	FBHPSO-DE algorithm	• Powerful and stable for solving the ED problem • The superiority to deal with the large-scale systems	• Its size is complex • A long period of time to execution
[35]	ACO–ABC–HS algorithm	• Effectively capable to mitigate the problem • The superiority in finding out optimum results	• Complicated structure • Need a long time for the implement

Furthermore, the hybrid method is combined with two or more algorithms in order to mitigate their weaknesses and use their strengths hence it provides better performance for solving optimization problems [18]. The proposed hybrid algorithms such as GA-PS-sequential quadratic programming [36], NM-FAPSO [36] and differential evolution algorithm-PSO [37] showed a highly efficient technique to solve the EED problem. However, the drawback of this type of algorithm sometimes is a long computational method because it uses two or more algorithms. Table 1 shows the advantages and disadvantages of several optimization techniques.

4. Review of power dispatch with renewable energy

Power dispatch is a crucial part of the power system as the responsibility is to plan optimal and effective scheduling of all generators to meet the required power demand. The power dispatcher decides which unit is the most economical to increase the power output. RE has been implemented in the power dispatch as an alternatives way to reduce pollution and saving the cost. A decades ago, the researchers have include RES into power dispatch system instead of focusing mainly on thermal [38].

The RESs have an important feature compared to conventional generation. The availability of RES installation is easy as compared to conventional power plants with the lowest transmission losses.

However, the wide continues dependence on PV and wind power face many challenges. Firstly, the PV and wind power plants are not much reliable to dispatch the power as per regulated power generation demand. Secondly, both still have an expensive investment in many areas of the world [9].

There are several related studies on EED that consider thermal generations and RE. These studies have been done to solve the CEED problem with RE. The author [39] suggested PV generation on CEED of a power system. CEED problem investigated for conventional thermal and PV plants using PSO Algorithm. The test system of this research includes the thirteen PV plants with six thermal units for testing and evaluation under the constraints of generator limits and power balance.

The same method is also used by [40] to analysis solar sharing scheduling for economic and environmental dispatch for inter-area power system dispatch utilization. In this research, the CEED problem was minimized by equality and inequality constraints and to fulfill the load demand by scheduling and sharing of PV and thermal system. Moreover, [41] optimized PV generation with transmission losses for economic and environmental dispatch using PSO algorithm to reduce emission and cost using the constraints of equality and inequality, emission constraints, and transmission losses to maximize solar availability. This research involves two tests system of the thermal generator (6 and 10 thermal) and thirteen PV units. It results to enhance in regards to the lesser overall cost and limited emission for CEED problem with the best perfect solution and lower computational time.

This research [42] proposed a quadratic programming method for hybrid microgrid including energy storage. Equality and inequality constraints used for optimizing dynamic ED. This problem implemented on the IEEE 14 bus of five electric sources that are energy storage, microturbine generator, PV, diesel generator, and the primary grid.

Furthermore, the researcher [43] used a different algorithm to optimize emission and ED for thermal and PV generation. This optimization problem solved by Euclidean affine flower pollination (eFPA) algorithm and binary (FPA). The test system has been applied to IEEE 57 and IEEE 30 bus system by using five thermal and twenty generators. This research [44] proposed new multi-objective for emission and ED using a different method which is Pareto concavity elimination transformation (PaCcET) to find the most economical operating condition and the best environmentally friendly solutions considering the constraints of equality and inequality. The test system of this research was applied to three distributed energy resources and one PV generation. Thus, this method can find the optimal set of solution solutions not only with less computation time but also with a better diverse result.

Also, [45] used a non-dominated sorting genetic algorithm II (NSGA-II) for electric vehicles taking part in the ED of thermal and wind power systems. The constraints of unit capacity, power balance, line transmission power, node voltage and thermal unit used for solving a multi-objective problem. In addition, [46] proposed hybrid energy storage for ED of off-grid PV generation system. It was used power balance constraints as well as state of charge and hydrogen level constraints to provide stable and continuous PV power. The author [7] suggested a modified harmony search (MHS) algorithm for CEED of microgrid incorporating RE sources. Three types of generators were used as a test system that is the solar unit, wind unit and combined power and heat generator under the constraints such as generation limits and power balance. MHS algorithm gave better cost reduction for various scenarios and showed its efficiency and superiority to solve the problem. Moreover, this paper [2] proposed the whale optimization algorithm (WOA) to solve CEED of RE integrated microgrid using power balance and generation constraints. The proposed approach was applied on one wind farm as well as one PV system. The results of WOA algorithm showed better quality for reducing CEED problem for all test cases and compared with different optimization techniques.

ED algorithm has been proposed by [47] to achieve optimum economic operation of the microgrid with RE sources. This microgrid consists of the wind energy system, PV system and three conventional generators. Three constraints such as power balance, ramp rate limits and generation limits have been used to ensure the optimal output for PV and wind generator. The obtained results of the proposed algorithm can guarantee that the generator's output adjustment of microgrid can be determined in the best value.

Algorithm of constrained multi-objective population extremal optimization (CMOPEO) presented by [48]. It has been proposed to find an optimal solution and improve the performance of EED incorporating RE resources by minimizing the objectives of the environment emission and total cost. However, inequality and equality constraints have been considered to solve the problem of EED. Three cases of modified IEEE 30 bus and six generators system with RE generations were used as test systems. CMOPEO algorithm can be observed as a competitive method and it was able to deal well with EED problem.

Table 2 shows the recent papers summarization on CEED considering RE sources such as PV and wind. It describes the optimization methods, objectives, constraints that utilized to solve the problem of power dispatch as well as the test system for PV and wind power.

Table 2. Summarization on CEED considering PV and wind power

No	Ref	Methods	Objectives			Test system			Constraints
			Cost	Emission	CEED	Thermal Unit	RE Sources PV	RE Sources W	
1	[40]		√	√	√	√	√		A&B
2	[41]	PSO Algorithm	√	√	√	√	√		A&B
3	[46]		√				√		A&B
4	[42]	Quadratic programming	√				√		A&B
5	[43]	eFPA-BFPA algorithm	√	√	√	√	√		A&B
6	[44]	PaCcET Allgorithm	√	√		√	√		A&B
7	[45]	NSGA- II Algorithm	√	√	√	√		√	A,B,N,L
8	[7]	MHS Algorithm			√	√	√	√	A&B
9	[2]	WOA Algorithm	√	√	√	√	√		A&B
10	[47]	ED Algorithm	√				√	√	A,B,R
11	[48]	CMOPEO Method			√	√	√	√	A,B,S

Note: W=wind; PV=photovoltaic; A=Equality constraint; B=Inequality constraint; and R=ramp rate limits; S=security constraint; N=Node voltage constraints and L=line transmission power.

5. Conclusion

This paper presented a review of optimization methods of the latest research on CEED of thermal generation and RE such as PV and wind. In power dispatch, it is necessary to consider both cost and emission level in operation and planning to ensure the power generation. Besides, the integration of RE in power dispatch becomes a more complex problem. Optimization methods were classified into three types such as hybrid, non-conventional and classical methods. Then the advantages and disadvantages of several optimization techniques were summarized. Moreover, the optimization methods, constraints and type of RE sources for the previous studies were reviewed. It presented various optimization methods applied to different test systems to minimize emission and cost for power dispatch with PV and wind. According to previous researches, it is clear that hybrid methods and non-conventional methods have high efficiency and more suitable methods for solving the CEED problem. This review study hoped can help and achieve more optimal power dispatch considering PV and wind energy.

Acknowledgments
The authors would like to thank the Ministry of Education Malaysia for supporting this research under the Fundamental Research Grant Scheme Vot No. K107 and partially sponsored by Universiti Tun Hussein Onn Malaysia.

References
[1] Velasquez MA, Barreiro-Gomez J, Quijano N, Cadena AI, Shahidehpour M. Distributed model predictive control for economic dispatch of power systems with high penetration of renewable energy resources. *International Journal of Electrical Power & Energy Systems*. 2019;**113**:607-17.
[2] Dey B, Roy SK, Bhattacharyya B. Solving multi-objective economic emission dispatch of a renewable integrated microgrid using latest bio-inspired algorithms. *Engineering Science and Technology, an International Journal*. 2019;**22**(1):55-66.
[3] Eladl AA, ElDesouky AA. Optimal economic dispatch for multi heat-electric energy source power system. *International Journal of Electrical Power & Energy Systems*. 2019;**110**:21-35.
[4] Nottrott A, Kleissl J, Washom B. Energy dispatch schedule optimization and cost benefit analysis for grid-connected, photovoltaic-battery storage systems. *Renewable Energy*. 2013;**55**:230-40.
[5] Yao D, Choi S, Tseng K, editors. Design of short-term dispatch strategy to maximize income of a wind power-energy storage generating station. 2011 IEEE PES Innovative Smart Grid Technologies; 2011: IEEE.
[6] Jin J, Zhou D, Zhou P, Miao Z. Environmental/economic power dispatch with wind power. *Renewable energy*. 2014;**71**:234-42.
[7] Elattar EE. Modified harmony search algorithm for combined economic emission dispatch of microgrid incorporating renewable sources. *Energy*. 2018;**159**:496-507.
[8] Elattar EE. Environmental economic dispatch with heat optimization in the presence of renewable energy based on modified shuffle frog leaping algorithm. *Energy*. 2019;**171**:256-69.
[9] Gandhi O, Rodríguez-Gallegos CD, Srinivasan D, editors. Review of optimization of power dispatch in renewable energy system. 2016 IEEE Innovative Smart Grid Technologies-Asia (ISGT-Asia); 2016: IEEE.
[10] Pothiya S, Ngamroo I, Kongprawechnon W. Ant colony optimisation for economic dispatch problem with non-smooth cost functions. *International Journal of Electrical Power & Energy Systems*. 2010;**32**(5):478-87.
[11] Benasla L, Belmadani A, Rahli M. Spiral optimization algorithm for solving combined economic and emission dispatch. *International Journal of Electrical Power & Energy Systems*. 2014;**62**:163-74.
[12] Mihaljan JM. The exact solution of the Rossby adjustment problem. *Tellus*. 1963;**15**(2):150-4.
[13] Hemamalini S, Simon SP. Economic/emission load dispatch using artificial bee colony algorithm. *iPi*. 2009;**1**:2.
[14] Vaishnavi VJ, Reddy AS. Multi-Objective Economic Emission Dispatch using Backtracking Search Optimization Algorithm. *International Journal of Emerging Science and Engineering (IJESE)*. 2014;**3**(2):39-46.
[15] Chaturvedi KT, Pandit M, Srivastava L, editors. On-line solution to combined economic and emission dispatch problem. 2006 IEEE International Conference on Industrial Technology; 2006: IEEE.
[16] Sita H, Reddy PU. A Comprehensive Study On Combined Economic and Emission Dispatch Optimization Problem. *IOSR Journal of Electrical and Electronics Engineering*. 2016:01-9.
[17] Younes M, Khodja F, Kherfane RL. Multi-objective economic emission dispatch solution using hybrid FFA (firefly algorithm) and considering wind power penetration. *Energy*. 2014;**67**:595-606.
[18] Kiran DV, Kumar NMG, Shashidhara SM. A Comprehensive Review on Optimization Strategies

for Combined Economic Emission Dispatch Problem. *Asian Journal of Electrical Sciences*. 2018;**7**(1):68-74.

[19] Moosavian SM, Modiri-Delshad M, Selvaraj J, Abd Rahim N. Solution techniques for optimal power dispatch problems-A review. 2016.

[20] Ramanathan R. Fast economic dispatch based on the penalty factors from Newton's method. *IEEE transactions on power apparatus and systems*. 1985(7):1624-9.

[21] Gherbi F, Lakdja F, editors. Environmentally constrained economic dispatch via quadratic programming. 2011 International Conference on Communications, Computing and Control Applications (CCCA); 2011: IEEE.

[22] Irisarri G, Kimball L, Clements K, Bagchi A, Davis P. Economic dispatch with network and ramping constraints via interior point methods. *IEEE Transactions on Power Systems*. 1998;**13**(1):236-42.

[23] Zhan J, Wu Q, Guo C, Zhou X. Fast $\lambda $-Iteration Method for Economic Dispatch With Prohibited Operating Zones. *IEEE Transactions on power systems*. 2013;**29**(2):990-1.

[24] Liang H, Liu Y, Li F, Shen Y. A multiobjective hybrid bat algorithm for combined economic/emission dispatch. *International Journal of Electrical Power & Energy Systems*. 2018;**101**:103-15.

[25] Muthuswamy R, Krishnan M, Subramanian K, Subramanian B. Environmental and economic power dispatch of thermal generators using modified NSGA-II algorithm. *International Transactions on Electrical Energy Systems*. 2015;**25**(8):1552-69.

[26] Abido M. A novel multiobjective evolutionary algorithm for environmental/economic power dispatch. *Electric power systems research*. 2003;**65**(1):71-81.

[27] Zou D, Li S, Li Z, Kong X. A new global particle swarm optimization for the economic emission dispatch with or without transmission losses. *Energy Convers Manage*. 2017;**139**:45-70.

[28] Abbas G, Gu J, Farooq U, Raza A, Asad MU, El-Hawary ME. Solution of an economic dispatch problem through particle swarm optimization: A detailed survey–Part II. *IEEE access*. 2017;**5**:24426-45.

[29] Afandi A, Fadlika I, Sendari S, Miyauchi H, Fujita G, Tutkun N, editors. Comparison of ABC and genetic algorithms on economic power dispatch. 2018 10th International Conference on Information Technology and Electrical Engineering (ICITEE); 2018: IEEE.

[30] Farag A, Al-Baiyat S, Cheng T. Economic load dispatch multiobjective optimization procedures using linear programming techniques. *IEEE Transactions on Power systems*. 1995;**10**(2):731-8.

[31] Mohan M, Kuppusamy K, Khan MA. Optimal short-term hydrothermal scheduling using decomposition approach and linear programming method. *International Journal of Electrical Power & Energy Systems*. 1992;**14**(1):39-44.

[32] Basu M. A simulated annealing-based goal-attainment method for economic emission load dispatch of fixed head hydrothermal power systems. *International Journal of Electrical Power & Energy Systems*. 2005;**27**(2):147-53.

[33] Schlünz E, Van Vuuren J. An investigation into the effectiveness of simulated annealing as a solution approach for the generator maintenance scheduling problem. *International Journal of Electrical Power & Energy Systems*. 2013;**53**:166-74.

[34] Naderi E, Azizivahed A, Narimani H, Fathi M, Narimani MR. A comprehensive study of practical economic dispatch problems by a new hybrid evolutionary algorithm. *Applied Soft Computing*. 2017;**61**:1186-206.

[35] Sen T, Mathur HD. A new approach to solve Economic Dispatch problem using a Hybrid ACO–ABC–HS optimization algorithm. *International Journal of Electrical Power & Energy Systems*. 2016;**78**:735-44.

[36] Niknam T. A new fuzzy adaptive hybrid particle swarm optimization algorithm for non-linear, non-smooth and non-convex economic dispatch problem. *Applied Energy*. 2010;**87**(1):327-39.

[37] Sayah S, Hamouda A. A hybrid differential evolution algorithm based on particle swarm

optimization for nonconvex economic dispatch problems. *Applied soft computing*. 2013;**13**(4):1608-19.

[38] Joshi PM, Verma H. An improved TLBO based economic dispatch of power generation through distributed energy resources considering environmental constraints. *Sustainable Energy, Grids and Networks*. 2019;**18**:100207.

[39] Khan NA, Awan AB, Mahmood A, Member I, Razzaq S, Zafar A, et al. Combined emission economic dispatch of power system including solar photo voltaic generation. *Energy Convers Manage*. 2015;**92**:82-91.

[40] Gupta A, Anand H, editors. Analysis of scheduling of solar sharing for economic/environmental dispatch using PSO. 2015 Annual IEEE India Conference (INDICON); 2015: IEEE.

[41] Wadhawan A, Verma P, Grover S, Anand H, editors. Economic environmental dispatch with PV generation including transmission losses using PSO. 2016 IEEE 7th Power India International Conference (PIICON); 2016: IEEE.

[42] Wibowo RS, Firmansyah KR, Aryani NK, Soeprijanto A, editors. Dynamic economic dispatch of hybrid microgrid with energy storage using quadratic programming. 2016 IEEE Region 10 Conference (TENCON); 2016: IEEE.

[43] Shilaja C, Ravi K. Optimization of emission/economic dispatch using euclidean affine flower pollination algorithm (eFPA) and binary FPA (BFPA) in solar photo voltaic generation. *Renewable Energy*. 2017;**107**:550-66.

[44] Sarfi V, Livani H, Yliniemi L, editors. A new multi-objective economic-emission dispatch in microgrids. 2017 IEEE Power & Energy Society General Meeting; 2017: IEEE.

[45] Xiaoyan Y, Chunlin G, Xuan X, Dequan H, Zhou M, editors. Research on large scale electric vehicles participating in the economic dispatch of wind and thermal power system. 2017 China International Electrical and Energy Conference (CIEEC); 2017: IEEE.

[46] Ran L, Zhengyu L, Zhen C, editors. Economic dispatch of off-grid photovoltaic generation system with hybrid energy storage. 2018 2nd IEEE Conference on Energy Internet and Energy System Integration (EI2); 2018: IEEE.

[47] Santillán-Lemus FD, Minor-Popocatl H, Aguilar-Mejía O, Tapia-Olvera R. Optimal economic dispatch in microgrids with renewable energy sources. *Energies*. 2019;**12**(1):181.

[48] Chen M-R, Zeng G-Q, Lu K-D. Constrained multi-objective population extremal optimization based economic-emission dispatch incorporating renewable energy resources. *Renewable Energy*. 2019;**143**:277-94.

IOP Conf. Series: Materials Science and Engineering 1127 (2021) 012036 doi:10.1088/1757-899X/1127/1/012036

PV Inverter with Decoupled Active and Reactive Power Control to Mitigate Grid Faults

M. Talha *, S. R. S. Raihan*, N.A. Rahim* †

* University of Malaya,Power Energy Dedicated Advanced Center(UMPEDAC) Kualalumpur, Malaysia

† Renewable Energy Research Group, King Abdulaziz University, Jeddah 21589, Saudi Arabia

E-mail: engineer_talha@hotmail.com

Abstract. Low power distributed Photovoltaic (PV) systems would be a dominant power generator in future grids. This PV penetration significantly influence the grid stability, especially in an event of grid faults. Traditional PV inverters disconnect themselves from grid on detecting a low voltage at their point of common coupling (PCC). A temporary low voltage grid fault can lead to outage, if PV inverters are not equipped with low voltage ride through (LVRT) feature. During a low voltage grid fault PV inverter can assist the grid recovery process by not only staying connected to grid but also injecting reactive power into the grid. Many grid code standards have issued guidelines to inject reactive power during a low voltage fault. Development of low voltage ride through inverters require decoupled power flow control. In addition to maximum power point operation and standardized current injection to the grid, modern PV inverters should be able to deal with LVRT and loss of grid (LoG) ride through features, as demanded by the regulating grid codes. Reduction in grid outages can be achieved, if the PV inverters stay connected during LVRT, LoG and short circuit faults. Most of previous studies on LVRT control of PV inverters have not short circuit faults and loss of grid faults at PCC. This work proposed a PV inverter controller capable of controlling complex power into the grid. A decoupled current regulator with feed forward compensation is modelled. A short circuit grid fault is also tested with the developed PV inverter. It is found that the PV inverter ride through the low voltage and short circuit faults. The system is simulated in MATLAB(Simulink), the designed controller can provide decoupled active and reactive power to the grid during the fault events.

1. Introduction

Conventionally a transient sag and swell of grid voltage is dealt centrally by a grid operator, usually at a power plant installation site, where reactive power (VAR) compensation is provided by a static compensation techniques (STATCOM) [1]. A large-scale passive capacitor/inductor arrangement can provide or absorb reactive power to stabilize the grid voltages where a single large power generator is supplying power to the grid. However, for the DG rich grid these solutions increase cost along with a demerit of slow dynamic response. They are not suitable to be installed at every low power DG. [2]. Recent raises concerns about grid reliability and stability[3]. As transient grid stability can no longer be controlled centrally by a grid operator, therefore reactive power support to the grid must be shared among all the DGs on the grid [4]. A transient fluctuation (Loss of Grid voltage) or Low voltage (Shortening of Grid Voltage), can result in long outages and overall grid failure. However, with a proper absorption or extraction of relevant reactive power during a transient event tends to prevent catastrophic grid failure[5]. In order to assist grid fault recovery, the right amount of active/reactive power should be injected by a grid tied inverter during the fault event[6]. Transient power flow can be controlled if the power references and the feedback power signals are readily available to the inverter controller [7]. Normally the inverter is working at MPPT and would insert the maximum rated current to the grid at unity power factor. An abrupt voltage dip or short circuit at the point of common coupling (PCC) would result in high current and low voltage [8]. Detecting this abnormal grid voltage, a traditional inverter would trigger an overcurrent protection and trigger a self-turnoff. If all the PV inverters on the grid behave in this way a power outage would happen. As the dominant power generating sources would have gone offline. Several minutes would be required to bring the grid online again.

On the other hand, a modern fault ride through enabled inverter would limit the output current while staying connected to the grid [9]. In another scenario a voltage fluctuation can cause temporary loss of grid for few cycles, the inverter will still sense zero volt at its PCC [10]. If the inverter is equipped with anti-islanding feature, eventually the inverter would turn off. A fault ride through inverter should discern a transient fluctuation and or islanding effectively. As modern grid evolved with majority of renewable distributed generators feeding the grid. It is necessary for inverters to ride through the transient fault event and stay connected for at least (150ms to 1.5s), as recommended by IEEE 1547 standard for connecting DGs to the grid [11]. An LVRT requirement defined by German grid codes is depicted in Figure 1 [12]. In addition to ride through feature, if an inverter provides reactive current to the grid during the transient event the grid may recover from the fault [11], thus preventing an overall grid outage and increases grid reliability. A decoupled active and reactive current reference generation and an analogous decoupled active and reactive controller would be needed to provide the grid fault ride through feature in PV inverters[1, 9]. A single-phase PV inverter is proposed in this work to provide fault ride through during the grid faults. Power decoupling and independent control of 'direct' and 'quadrature' components are the key features of the proposed inverter.

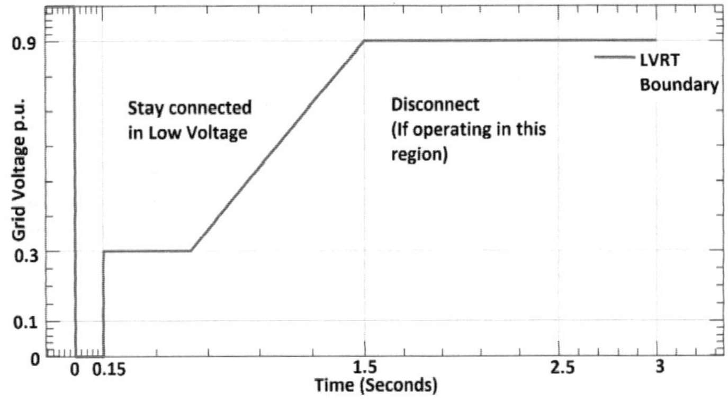

Figure 1. LVRT Grid Code Requirement

2. System Model

The block diagram in Figure 2 is used for system modelling. The system consists of PV string connected to transformerless inverter via DC-link capacitor. At the grid side H-bridge based MOSFET inverter is connected to an output L filter[13] and to the grid. An 'LCL' filter can also be utilized as it would reduce the filter size compared to 'L' filter. Voltage source grid tied inverter (VSI) can be represented by differential equation as:

$$v_{inv}(t) = Ri_{inv} + \frac{Ld}{dt}i_{inv} + v_g \quad (1)$$

Where $v_{inv} = u.\,Vdc$, is the inverter output voltage and 'u' is the modulation index, 'ig' is the grid current supplied to the grid by the inverter and 'vg'is the grid voltage. Equation (6) can be written in Laplace domain as:

$$\frac{i_{inv}(s)}{v_{inv}(s)-v_g(s)} = \frac{1}{R+sL} \quad (2)$$

Where L=L1+L2, and 'R' is the series resistance. Equation (2) suggests that it is a single pole 1st order system.

3. Power Decoupling

It is mandatory for next generation inverters to provide decoupled active or reactive power to the grid during normal and low voltage mode [14]. PQ power theory for three phase electric systems suggests that the electric signals (Phase a, Phase b, Phase c) can be transformed into equivalent two dimensional orthogonal 'αβ' coordinate system with Clarke transform.

$$\begin{bmatrix} v_\alpha \\ v_\beta \end{bmatrix} = \sqrt{\frac{2}{3}} \begin{bmatrix} 1 & -0.5 & -0.5 \\ 0 & \frac{\sqrt{3}}{2} & \frac{-\sqrt{3}}{2} \end{bmatrix} \begin{bmatrix} V_a \\ V_b \\ V_c \end{bmatrix} \quad (3)$$

Given current in 'αβ' coordinate system, the instantaneous power can be written as:

$$\begin{bmatrix} P_i \\ Q_i \end{bmatrix} = \begin{bmatrix} V_\alpha & V_\beta \\ -V_\beta & V_\alpha \end{bmatrix} \begin{bmatrix} i_\alpha \\ i_\beta \end{bmatrix} \quad (4)$$

Where 'P' and 'Q_i' is the instantaneous active and reactive power component respectively. In order to decouple the single phase active and reactive power we need orthogonal signal pair. Single phase system does not have orthogonal 'β' component, so the 'β' component is generated fictitiously. There are many methods to generate 'β' component. Amongst them the most commonly used method is 'T/4' delay method[15]. In this method an 'α' component is delayed by quarter of cycle to generate the 'β' component. Instantaneous and decoupled power components are calculated from Equation (5) and (6) [16]:

$$P = \frac{V_\alpha I_\alpha + V_\beta I_\beta}{2} \quad (5)$$

$$Q = \frac{V_\alpha I_\alpha - V_\beta I_\beta}{2} \quad (6)$$

Equations (5) and (6) suggests that once an orthogonal pair of grid voltage and current is obtained, it is possible to generate independent current references to control active and reactive power feed to the grid.

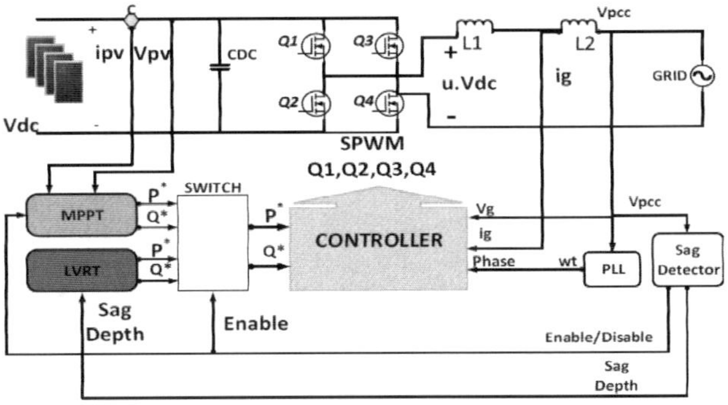

Figure 2. Transformerless PV Inverter

An overall PV system is depicted in Figure 2, the grid conditions are sensed by the controller and orthogonal signals are extracted by applying a 'T/4' delay-based algorithm on both current and voltage signals. Injected active and reactive power values are calculated according to Equation (4). The difference between injected and reference values are fed to the feed forward PI controller[17].

IOP Conf. Series: Materials Science and Engineering 1127 (2021) 012036 doi:10.1088/1757-899X/1127/1/012036

4. Controller Design

Three cascaded control loops are proposed to control PV inverter, an outer most voltage control loop, an intermediate power control loop and the innermost current control loop. PI controller is the simplest and the reliable choice with good tracking. However, variable AC quantities are needed to be converted into equivalent DC signals for acceptable steady state and transient performance [18]. Variable ac quantities (alpha, beta) in fixed reference frame can be converted to fixed quantities (direct, quadrature) in synchronously rotating reference frame as shown in Equation (7). In this way we can control of direct 'D' and quadrature 'Q' component of current independently [1]. DQ components can be obtained by Park transformation on orthogonal signals (αβ).

$$\begin{bmatrix} X_d \\ X_q \end{bmatrix} = \begin{bmatrix} \cos\theta & \sin\theta \\ -\sin\theta & \cos\theta \end{bmatrix} \begin{bmatrix} x_\alpha \\ x_\beta \end{bmatrix} \quad (7)$$

Where 'X' and 'X_q'are the DC equivalent signals of variable ac signals 'x_α'and 'x_β' respectively. The obtained dc signals can be regulated with the conventional PI controller, with much improved steady state error. For the case of voltage source inverter, we need a reference voltage as a modulating signal for PWM generator. In order to decouple this reference signal, the direct and quadrature current components are feed forwarded as shown in Figure 3 [19]. The voltage loop controls the DC-link voltage, the input of this loop is the difference between the reference DC-link voltage and its sensed value represented in Equation (8). This loop generates the active power reference as shown in Equation (10) , which is an input of an intermediate power control loop[20]. Another input of this loop is reactive power reference, which is given by the user according to relevant grid codes. Kp and Ki values can be tuned with the SISO tool in MATLAB.

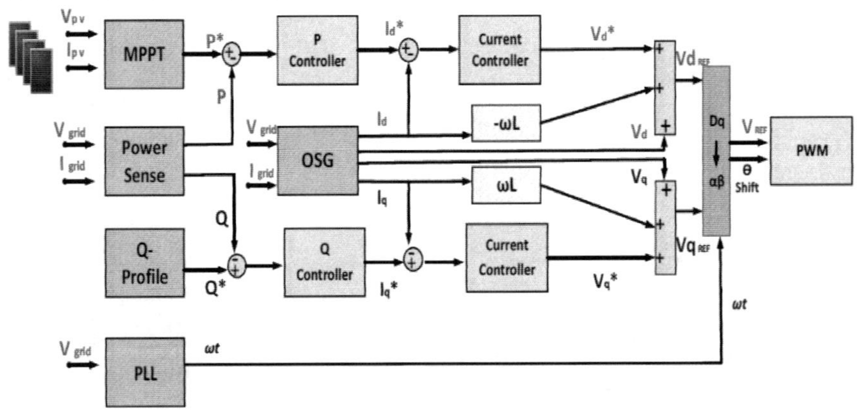

Figure 3. Controller Structure

$$e_v = e_{dc}^{\text{ref}} - V_{dc} \quad (8)$$

$$i_{ref} = k_p \cdot e_v + \frac{k_i}{s} \int e_v dt \quad (9)$$

$$P_{ref} = i_{ref} \cdot |v_g| \quad (10)$$

An intermediate power controller provides the decoupled current references in the form of direct and quadrature components. Now we need to sense the real time values of 'DQ' components, so that the error between the reference and the sensed quantities can be fed to the PI controller.

The 'Hall' sensors sense inverter current and grid voltage, these signals are delayed by quarter of a cycle to generate the 'β' component. Now'αβ' pair would be converted to synchronous 'dq' reference frame by using Equation 7, since phase angle 'θ' is already obtained by a PLL unit [16].

$$\begin{bmatrix} i_{d_{ref}} \\ i_{q_{ref}} \end{bmatrix} = \frac{1}{v_{gd}^2 + v_{gq}^2} \begin{bmatrix} v_{gd} & v_{gq} \\ v_{gq} & -v_{gd} \end{bmatrix} \begin{bmatrix} P_{ref} \\ Q_{ref} \end{bmatrix} \quad (11)$$

$$i_{ref} = i_{d_{ref}} \sin\omega t - i_{q_{ref}} \cos\omega t \quad (12)$$

The calculated 'i_{dref}'and 'i_{qre}' controls the active and reactive power respectively[7]. The error signal between calculated current and the sensed current is regulated by a PI controller.

$$G(S) = K_p + \frac{K_i}{S} \tag{13}$$

PI controller is chosen because of its superior performance and ease of implementation. All four controllers in Figure 8 used PI controller with mathematical expression in Equation (13) where Kp is proportional gain and Ki is integral gain. A current reference is generated from above explained outer loop, and it would be fed to the inner loop. As a common practice, design of multi-loop nested control is carried from the inner most loop towards the outer loops. Therefore, the current loop must be addressed in the first place. The inner most loop senses the grid current and orthogonal signal generator (OSG) generates the 'dq' current components[21, 22]. An error between observed and reference signals is fed to the current controllers as shown in Figure 3. The output of current controllers is fed to PWM modulator to generate the gating pulses for an inverter's full bridge structure as shown in Figure 3. It can be seen that there are four controller and two decoupling blocks.

5. Simulation

A simulation of transformerless inverter with power decoupling is implemented in MATLAB Simulink as shown in Figure 6. SISO tool is used to tune the controller parameters. Table 1 shows the simulated PV inverter.

Table 1. Model Specifications

Description	*Symbol*	*Value*
Inverter side filter inductor	L_1	3.5mH
Filter capacitor	C	2μF
Grid side filter inductor	L_2	50μH
Nominal grid voltage	V_g	240 V (rms)
Inverter dc link	V_{dc}	400 Vdc
Nominal grid frequency	f_g	50Hz
Nominal switching frequency	f_{sw}	3.8kHz
Nominal sampling frequency	f_s	400kHz
Nominal Power	P	1000W

The inverter is started at 0W output power, at 0.2s the power reference is set to 100% active power and 0% reactive power. The inverter's PLL synchronize its output to the grid in 2 cycles. During these 2 cycles a transient reactive power flows to the grid. The inverter feeds active power to the grid as shown in Figure 4. On the grid voltage scale the grid current (Red) is superimposed and scaled by a factor of '20' in order to view the phase difference.

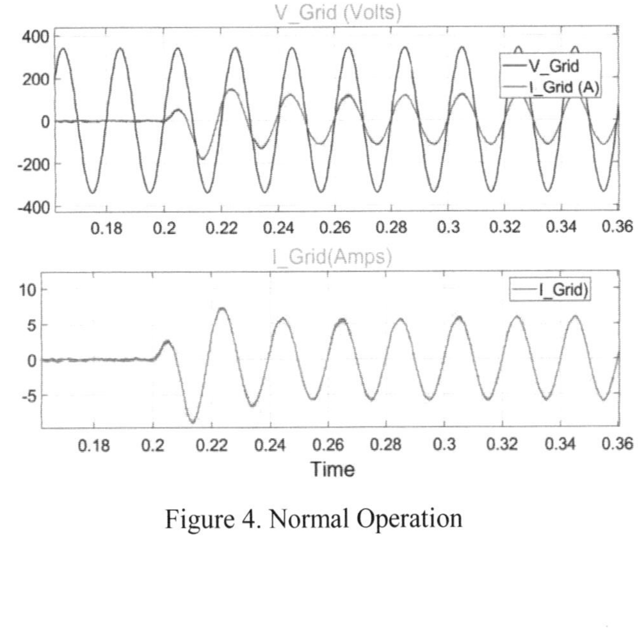

Figure 4. Normal Operation

IOP Conf. Series: Materials Science and Engineering 1127 (2021) 012036 doi:10.1088/1757-899X/1127/1/012036

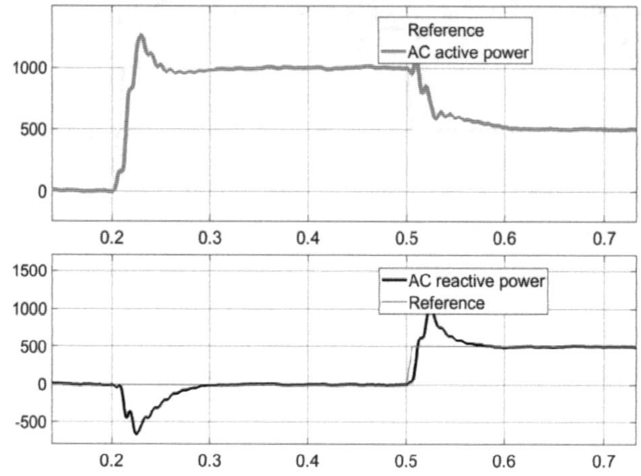

Figure 5. Active & Reactive Power Tracking

As shown in Figure 5, at 0.5s the power references are set as 50% active power and 50% as reactive power. Current is leading the voltage during a low voltage event as shown in Figure 7, thus providing the positive VARS. Figure 8 shows that the inverter follows both active (500 W) and reactive power (500VARs) references. At 1s active power has been set to 0% and reactive power to 100% (1000VARs), 90 degrees phase shift can be observed between injected grid voltage and current. At 2 Sec both active and reactive power set points are set to 0%. An overall result of the simulation setup in shown in Figure 8. The results show that by decoupling the active and reactive power and controlling it via direct and quadrature component approach provides good tracking performance.

This control approach can be utilized in modern PV systems which should be designed to provide reactive power to the grid in an event of fault. A low voltage grid fault of 50% is simulated to verify the ride through and current regulation of the control scheme. On detecting a sag, the active power is reduced and reactive power is supplied to the grid, thus low voltage ride through capability is implemented.

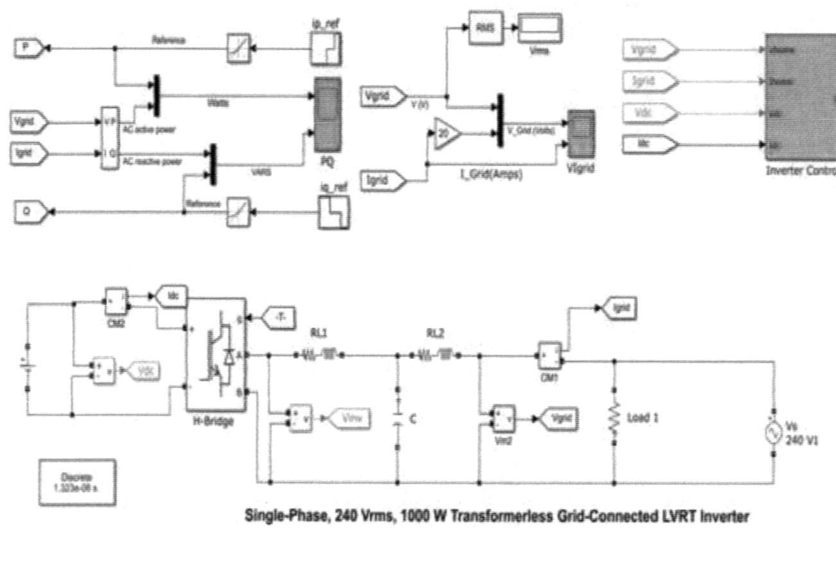

Figure 6. SIMULINK Model

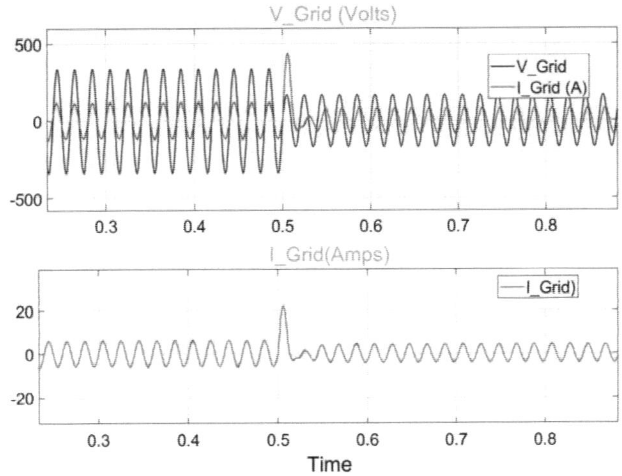

Figure 7. Low Voltage Ride Through Mode

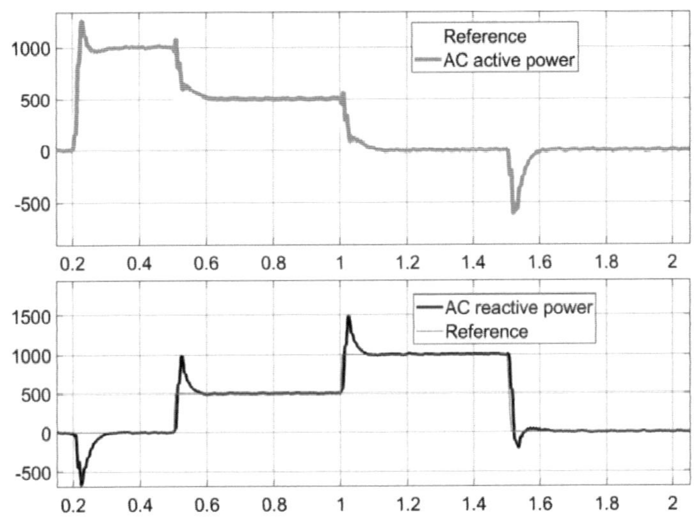

Figure 8. Complex Power Flow into Grid

A high current spike is observed on first cycle of sag occurrence. However, the current controller curtails the current and reduce the active power and increase the reactive power as shown in Figure 7. In another experiment a short circuit at the gird is also simulated. Here the controller still limits the current but initial high current spike could be dangerous as it would induce very high rate of change of current (di/dt), as shown in Figure 9. The MOSFET switches must be capable to deal with this high in rush current. Modern PV inverters should stay connected to the grid during voltage sags (shortening of grid voltage) and fluctuations (loss of grid voltage). Further research would be carried out to make PV inverter more resilient to these grid faults.

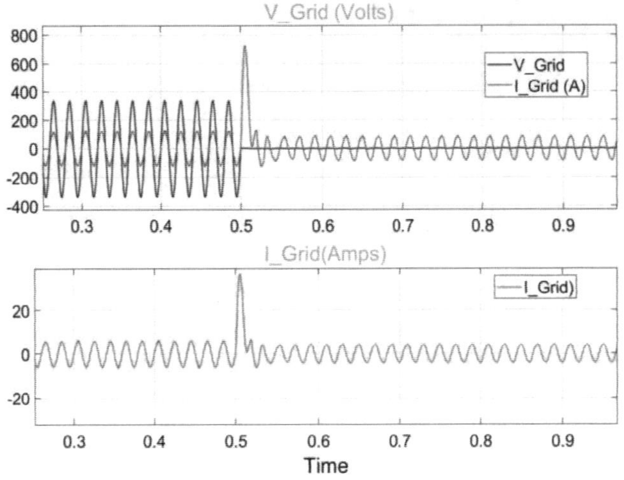

Figure 9. Zero Voltage Ride Through Mode

During both of these scenarios the active power injection to the grid would be zero, however, limited amount of current should still be injected into the grid so that the grid may be assisted in recovering from fault. The performance of current controller, sag detector, synchronization unit (PLL) must be deeply analysed to develop the grid faults resilient PV system, these systems would be explored in the future research.

6. Conclusion

In this work, a 1kW single-phase grid connected PV inverter with LVRT capability is designed and analysed. The main objective is to control active and reactive power independently in an event of grid fault. A grid connected inverter is modelled and a single-phase PQ control strategy is adopted to decouple the injected power into direct and quadrature (DQ) components. Once the DQ current components are available the active and reactive power is controlled by using PI controller with feed forward control. Result shows decoupled control of active and reactive current can be effective in providing reactive power to the grid during transient fault event. 50% and 100% grid voltage sag has been tested on the designed controller, which successfully limits the current and fulfils the low voltage ride through criteria.

Acknowledgements

The authors would like to acknowledge the support of the Ministry of Higher Education Malaysia and University of Malaya through FRGS (FP052-2017A).

References

[1] Yang Y, Blaabjerg F, Wang H. Low-voltage ride-through of single-phase transformerless photovoltaic inverters. *IEEE Trans Ind Appl*. 2013;**50**(3):1942-52.

[2] Al-Shetwi AQ, Sujod MZ. Grid-connected photovoltaic power plants: A review of the recent integration requirements in modern grid codes. *International Journal of Energy Research*. 2018;**42**(5):1849-65.

[3] Wang Y, Ren B, Zhong Q-C. Robust power flow control of grid-connected inverters. *IEEE Trans Indust Electr*. 2016;**63**(11):6887-97.

[4] Al-Shetwi AQ, Sujod MZ, Blaabjerg F, Yang Y. Fault ride-through control of grid-connected photovoltaic power plants: A review. *Solar Energy*. 2019;**180**:340-50.

[5] Yang Y, Enjeti P, Blaabjerg F, Wang H. Wide-scale adoption of photovoltaic energy: Grid code modifications are explored in the distribution grid. *IEEE Industry Applications Magazine*. 2015;**21**(5):21-31.

[6] Neumann T, Erlich I, editors. Modelling and control of photovoltaic inverter systems with respect to German grid code requirements. 2012 IEEE Power and Energy Society General Meeting; 2012: IEEE.

[7] Chatterjee A, Mohanty KB. Current control strategies for single phase grid integrated inverters for photovoltaic applications-a review. *Renewable and Sustainable Energy Reviews*. 2018;**92**:554-69.

[8] Zeb K, Uddin W, Khan MA, Ali Z, Ali MU, Christofides N, et al. A comprehensive review on inverter topologies and control strategies for grid connected photovoltaic system. *Renewable and Sustainable Energy Reviews*. 2018;**94**:1120-41.

[9] Bighash EZ, Sadeghzadeh SM, Ebrahimzadeh E, Blaabjerg F. Improving performance of LVRT capability in single-phase grid-tied PV inverters by a model-predictive controller. *International Journal of Electrical Power & Energy Systems*. 2018;**98**:176-88.

[10] Dietmannsberger M, Grumm F, Schulz D. Simultaneous implementation of LVRT capability and anti-islanding detection in three-phase inverters connected to low-voltage grids. *IEEE Transactions on Energy Conversion*. 2017;**32**(2):505-15.

[11] Photovoltaics DG, Storage E. IEEE Application Guide for IEEE Std 1547™, IEEE Standard for Interconnecting Distributed Resources with Electric Power Systems. 2009.

[12] Frias JAV, Alba JM-C, Nunez JV, Hernandez ER, editors. PV inverter requirement updated for LVRT capabilities under recent grid codes legislation. PCIM Europe 2014; International Exhibition and Conference for Power Electronics, Intelligent Motion, Renewable Energy and Energy Management; 2014: VDE.

[13] Jalili K, Bernet S. Design of $ LCL $ filters of active-front-end two-level voltage-source converters. *IEEE Trans Indust Electr*. 2009;**56**(5):1674-89.

[14] Sochor P, Tan NML, Akagi H. Low-voltage-ride-through control of a modular multilevel single-delta bridge-cell (sdbc) inverter for utility-scale photovoltaic systems. *IEEE Trans Ind Appl*. 2018;**54**(5):4739-51.

[15] Golestan S, Guerrero JM, Vidal A, Yepes AG, Doval-Gandoy J, Freijedo FD. Small-signal modeling, stability analysis and design optimization of single-phase delay-based PLLs. *IEEE Trans Pow Electr*. 2015;**31**(5):3517-27.

[16] Yang Y, Blaabjerg F, Zou Z. Benchmarking of grid fault modes in single-phase grid-connected photovoltaic systems. *IEEE Trans Ind Appl*. 2013;**49**(5):2167-76.

[17] Han Y, Fang X, Yang P, Wang C, Xu L, Guerrero JM. Stability analysis of digital-controlled single-phase inverter with synchronous reference frame voltage control. *IEEE Trans Pow Electr*. 2017;**33**(7):6333-50.

[18] Bahrani B, Rufer A, Kenzelmann S, Lopes LA. Vector control of single-phase voltage-source converters based on fictive-axis emulation. *IEEE Trans Ind Appl*. 2010;**47**(2):831-40.

[19] Monfared M, Golestan S, Guerrero JM. Analysis, design, and experimental verification of a synchronous reference frame voltage control for single-phase inverters. *IEEE Trans Indust Electr*. 2013;**61**(1):258-69.

[20] Ruan X, Wang X, Pan D, Yang D, Li W, Bao C. Control techniques for LCL-type grid-connected inverters: Springer; 2018.

[21] Bisht R, Subramaniam S, Bhattarai R, Kamalasadan S, editors. Active and reactive power control of single phase inverter with seamless transfer between grid-connected and islanded mode. 2018 IEEE Power and Energy Conference at Illinois (PECI); 2018: IEEE.

[22] Athari H, Niroomand M, Ataei M. Review and classification of control systems in grid-tied inverters. *Renewable and Sustainable Energy Reviews*. 2017;**72**:1167-76.

Investigate the Effect of Artificial Neural Network Parameters to Improve Fault Distance and Impedance Accuracy

Mohd Syukri Ali *, Ab Halim Abu Bakar*, Nasrudin Abd Rahim* †

*Higher Institution Centre of Excellence (HICoE), UM Power Energy Dedicated Advanced Centre (UMPEDAC), Level 4, Wisma R&D, University of Malaya, Jalan Pantai Baharu, 59990 Kuala Lumpur, Malaysia.,

†Renewable Energy Research Group, King Abdulaziz University, Jeddah 21589, Saudi Arabia

E-mail: syukriali@um.edu.my

Abstract. This paper investigates the effect of artificial neural network (ANN) parameters against the ANN accuracy on cable fault location. The investigation is conducted through the fault impedance and distance estimations during the occurrence of high impedance fault (HIF) in the distribution system. The measured three-phase voltage and current signals are utilized and fed into the ANN to estimate the fault impedance and distance. The accuracy of the estimated fault impedance and distance is evaluated with respect to the variation of ANN parameters. Based on the analysis, it shows that more accurate results can be obtained by utilizing the optimal value of ANN parameters.

1. Introduction

Fault is an unbalanced condition or any undesirable circumstances that will create an excessive current that leads to thermal and mechanical damage to the plant that carries it. Besides that, it can cause an extremely high temperature in arcs, which indirectly vaporize any known substances that lead to equipment destruction and fire. The system can become unstable and breakdown due to the fluctuation of the system voltage outside of their acceptable range. It can also cause the three-phase system to become unbalanced thus causing improper operation of the three-phase equipment. Most importantly, the fault will impede the power flow which causes undelivered supply to the intended load.

The occurrence of a fault is inevitable as it can be caused by the natural disaster such as storms, lightning strikes, a fallen tree trunk, animal encroachment and heat cycling. The cable degradation as well can cause the fault as the aging components and broken insulation will expose the cable. It can also be caused by human errors in the forms of accidents or other activities that carried out near to the power system equipment [1].

Fault can be classified into low impedance fault (LIF) and high impedance fault (HIF). Between these two, HIF seldom occurs as compared to LIF. It is reported that around 5% to 10% of all the fault events in the distribution system are caused by HIF. However, the actual percentage could be higher since only HIF events ending in bolted faults are recorded [2]. Even though the occurrence is seldom, but the consequences are more severe than LIF because the fault current produce due to HIF is low and unable to be identified by the conventional overcurrent protection relay [3, 4]. Besides, it is difficult to be detected because of the similarity of the operating current magnitude under normal and faulty condition. As a result, the fault will be left untreated thus increasing the risk of damage proliferation, outage time and hazard to the equipment and public.

Therefore, it is utmost important to detect and locate the fault as accurate and as fast as possible. Unfortunately detecting and locating the HIF is not an easy task especially for the underground cable due to non-visibility [5]. There are various techniques that have been proposed to detect and locate the HIF. For the HIF detection, it can be categorized into the time domain and frequency domain. In time domain HIF detection, the variation in the magnitude of the voltage and current signals over time is observed [6, 7]. Whereas in frequency domain HIF detection, the characteristic of the voltage and current signals within a frequency range is analyzed [8-10].

To estimate the HIF location, various techniques comprising of the impedance-based method, high frequency traveling wave, analytical formulation, fundamental component-based and knowledge-based methods have been introduced to tackle the HIF localization problem [1, 11, 12]. It is important to properly select the technique and it must be compatible with the distribution network topology. For example, the high frequency traveling wave technique is not suitable for the complex distribution system with lateral branches because it will cause many refractions and reflections from discontinuity points and at the lateral junctions [13]. Recently, knowledge-based method comprising of artificial neural network (ANN) [14, 15], fuzzy logic system (FLS) [16, 17], support vector regression (SVR) [18-21], core vector regression (CVR) [22] and adaptive neuro-fuzzy inference system (ANFIS) [23, 24] is preferable by the researchers to estimate the fault distance.

It was observed that ANN has been utilized to estimate the fault distance and it gives a promising result. However, the relationship between ANN parameters and the accuracy is not yet investigated. As such, in this paper the effect of changing the ANN parameters value comprising of learning rate (*lr*), momentum constants (*mc*) and the number of neurons in a hidden layer (*p*) is conducted. The performance will be evaluated based on the fitness of the estimated fault impedance and distance during the occurrence of the HIF event in the distribution system. In this investigation, a single measurement of three-phase voltage and current waveforms measured at the main substation is utilized and fed into the ANN.

2. Proposed Method

In this paper, ANN is utilized to estimate the fault impedance and distance during the occurrence of the HIF event. The performance of ANN is further enhanced by investigating the effect of varying the value of ANN parameters comprising of learning rates (*lr*), momentum constant (*mc*) and the number of neurons in the hidden layer (*p*). To investigate the effect, the value of each ANN parameters is changed one by one while the other two parameters are fixed. The performance is evaluated based on the fitness value of the estimated fault impedance and distance values.

2.1. Test System

Figure 1 shows a simplified 132/11kV distribution network developed in PSCAD/EMTDC software consisting of 33 nodes. The frequency of the system is 50Hz and the sampling frequency is 4kHz (80samples per one full cycle). Three-phase voltage and current waveforms are measured at the measurement node. In this investigation, different fault impedance values are applied at each node. There are a total of 1089 samples consisting of 11 fault impedance values (from 50Ω to 150Ω in steps of 10Ω), 33 nodes and 3 different types of fault (phase-A-to ground fault, phase-B-to ground fault and phase-C-to ground fault).

IOP Conf. Series: Materials Science and Engineering 1127 (2021) 012037 doi:10.1088/1757-899X/1127/1/012037

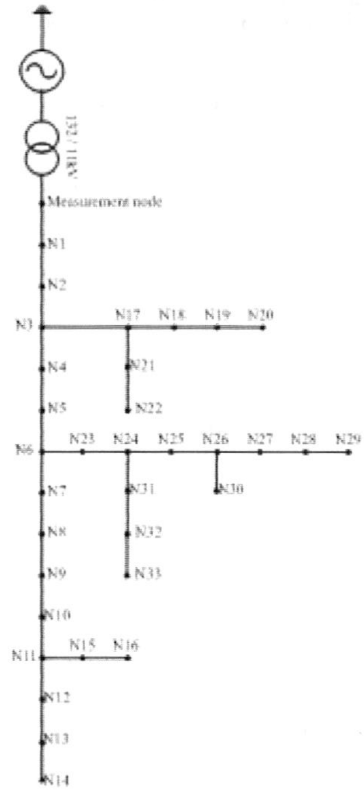

Figure 1. Distribution network consists of 33 buses

3. Investigate the Effect of ANN Parameters

In this study, the effect of changing the LMB parameters is investigated to observe the performance of ANN to estimate the fault impedance and distance due to the occurrence of HIF event. There are 3 different scenarios related to 3 different LMB parameters. For each scenario, one LMB parameter is varied while the other 2 are fixed. For example, in the first scenario, the value of *lr* is varied while the value of *mc* and *p* are fixed. After obtaining the best value of *lr*, then the value of *mc* is varied and the value of *lr* (best value) and *p* are fixed for the second scenario. In the third scenario by using the best values of *lr* and *mc*, the performance of ANN is further investigated when the value of *p* is varied.

In each scenario, the ANN training process is iterated for 100 times and only the lowest average error and maximum error of fault impedance and distance is selected. Then, the best value of LMB parameter is determined based on the fitness for each scenario. The lowest value of fitness is selected as it represents the highest accuracy of the estimated fault impedance and distance. The normalized average error, normalized maximum error and the fitness are calculated as follows:

$$\text{error}=|\text{estimated-actual}| \quad (1)$$

$$average\ error = \frac{\sum_{i=1}^{n} error_i}{n} \quad (2)$$

$$norm\_aver_k = \frac{average\ error_k}{average\ error_{max}} \quad (3)$$

$$maximum\ error = \max(error_n). \quad (4)$$

$$norm\_max_k = \frac{maximum\ error_k}{maximum\ error_{max}} \quad (5)$$

$$fitness = \frac{norm\_aver_{imp} + norm\_aver_{dist} + norm\_max_{imp} + norm\_max_{dist}}{4}$$
(6)

where n represents the total number of samples and k represent the parameters.

In the first scenario, the value of *lr* is varied while the value of *mc* and *p* are fixed. The value of *lr* is varied from 0.1 to 1 in increment by 0.1, while the value of *mc* and *p* are fixed to 0.5 and 6 respectively. The performance of ANN is evaluated based on the fitness as shown in Figure 2.

The figure indicates the result for fitness, normalized average impedance, normalized maximum impedance, normalized average distance and normalized maximum distance obtained when the value of *lr* is varied. The figure shows that the ANN performance varies with respect to the values of *lr*. It can also be observed that not all the results are the best result for that particular value of *lr*. For example, when *lr*=0.4, it gives the best result for the normalized maximum impedance, but the rest is not the best results. Therefore, it is important to choose a suitable value of *lr* in which it gives a better performance to all the results. As such, the best value of *lr* = 0.7 is selected because it gives the lowest value of fitness (0.736)

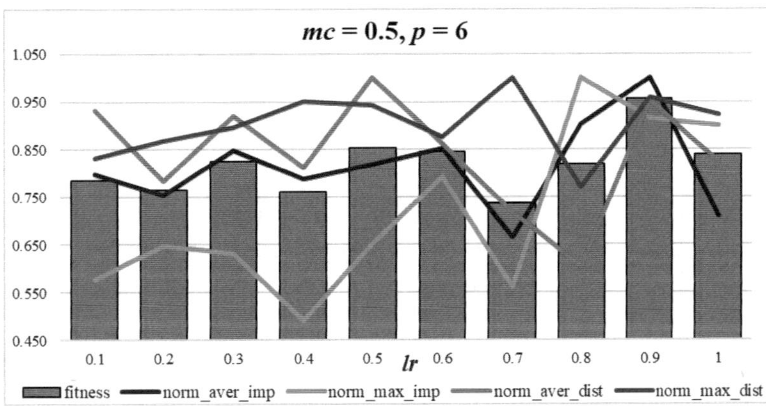

Figure 2. The performance of ANN with the variation of lr values.

Subsequently, the second scenario is conducted in which the value of *lr* is set to 0.7 based on the results obtained in the first scenario. The value of *p* is fixed to 6. Whereas the value of mc is varied from 0.1 to 1 in increment by 0.1. Figure 3 shows the results for the variation of mc values. Based on the results, it can be observed that the lowest fitness value (0.723) is obtained when *mc*=0.9.

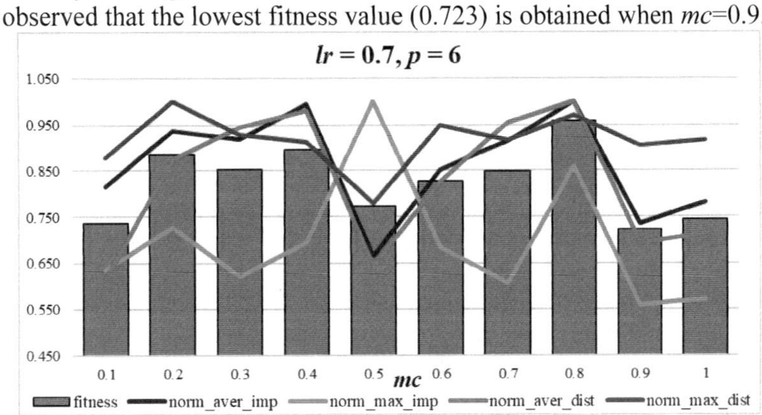

Figure 3. The performance of ANN with the variation of mc values.

For the third scenario as shown in Figure 4, the value of *p* is varied in between 3 to 12 (half to double of input data). It is observed that the performance of ANN is improved as the number of *p* is increasing. However, it should be noted that more training time is required if a greater number of *p* is utilized. Besides that, the overfitting problem can arise if too many of *p* is used. In this investigation, it can be observed that the best performance can be obtained if the value of *p*=11 is utilized with the fitness value of 0.078.

It can be concluded that these ANN parameters affect indirectly the performance of ANN. The smallest value of fitness (0.078) is obtained when the value of *lr*=0.7, *mc*=0.9 and *p*=11 is utilized. It should be noted that in this investigation the value of the parameter is set up to 1 decimal point and the analysis is conducted by changing only one parameter at the time. It is expected that more accurate results can be obtained if the value is set up to 3 decimal points for *lr* and *mc* and more than one parameter are varied concurrently. Unfortunately, it is difficult to determine the optimal values of these ANN parameters. Therefore, an optimization technique is suggested to be implemented to provide these optimal values

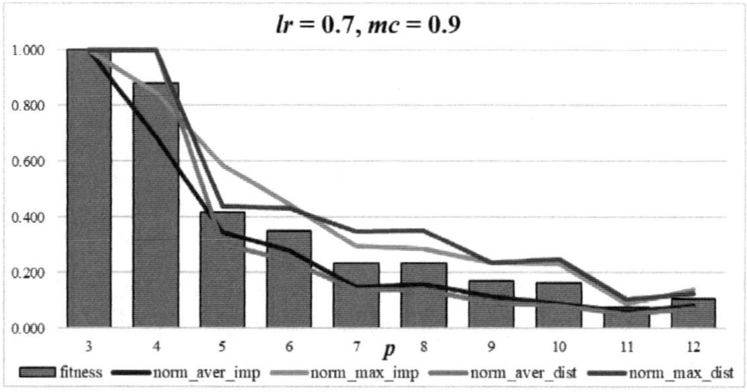

Figure 4. The performance of ANN with the variation of p values.

4. Conclusion

In this paper, an insight into the significant of ANN parameters is investigated to evaluate the ANN performance. It is observed that higher accuracy can be achieved if the optimal value of ANN parameters is utilized. To justify this hypothesis, a simple distribution network is modeled in PSCAD. Then, the high impedance fault event is simulated at each node with different fault impedance values. A single measurement of measured three-phase voltage and current signals are fed into the ANN to estimate the fault impedance and distance values. The accuracy of the ANN is measured in term of fitness. It is observed that fitness improving if the optimal value of ANN parameters is utilized. In this analysis, it is observed that the best fitness value (0.078) is obtained when the value of *lr*, *mc* and *p* are set to 0.7, 0.9 and 11 respectively.

Acknowledgements

This work was supported by the University of Malaya, Kuala Lumpur under Fundamental Research Grant Scheme (FRGS Grant No: FP093-2018A). The authors also thank the technical and financial assistance of UM Power Energy Dedicated Advanced Centre (UMPEDAC) and the Higher Institution Centre of Excellence (HICoE) Program Research Grant, UMPEDAC - 2018 (MOHE HICOE - UMPEDAC), Ministry of Education Malaysia, RU007-2018 and RU012-2019, University of Malaya.

References

[1] Filomena AD, Resener M, Salim RH, Bretas AS. Fault location for underground distribution feeders: An extended impedance-based formulation with capacitive current compensation. *International Journal of Electrical Power & Energy Systems*. 2009;**31**(9):489-96.

[2] Ghaderi A, Mohammadpour HA, Ginn HL, Shin Y. High-Impedance Fault Detection in the Distribution Network Using the Time-Frequency-Based Algorithm. *IEEE Transactions on Power Delivery*. 2015;**30**(3):1260-8.
[3] Ali MS, Abu Bakar AH, Mokhlis H, Arof H, Azil Illias H. High-impedance fault location using matching technique and wavelet transform for underground cable distribution network. *IEEJ Transactions on Electrical and Electronic Engineering*. 2014;**9**(2):176-82.
[4] Macedo JR, Resende JW, Bissochi CA, Carvalho D, Castro FC. Proposition of an interharmonic-based methodology for high-impedance fault detection in distribution systems. *IET Generation, Transmission & Distribution*. 2015;**9**(16):2593-601.
[5] Ghaderi A, Ginn HL, Mohammadpour HA. High impedance fault detection: A review. *Electric Power Systems Research*. 2017;**143**:376-88.
[6] Sharaf AM, Abu-Azab SI, editors. A smart relaying scheme for high impedance faults in distribution and utilization networks. 2000 Canadian Conference on Electrical and Computer Engineering. Conference Proceedings. Navigating to a New Era (Cat. No.00TH8492); 2000 7-10 May 2000.
[7] Ching-Lien H, Hui-Yung C, Ming-Tong C. Algorithm comparison for high impedance fault detection based on staged fault test. *IEEE Transactions on Power Delivery*. 1988;**3**(4):1427-35.
[8] Zanjani MGM, Kargar HK, Zanjani MGM, editors. High impedance fault detection of distribution network by phasor measurement units. 2012 Proceedings of 17th Conference on Electrical Power Distribution; 2012 2-3 May 2012.
[9] Yong S, Rovnyak SM. Decision tree-based methodology for high impedance fault detection. *IEEE Transactions on Power Delivery*. 2004;**19**(2):533-6.
[10] Aucoin BM, Russell BD. Distribution High Impedance Fault Detection Utilizing High Frequency Current Components. *IEEE Power Engineering Review*. 1982;**PER-2**(6):46-7.
[11] Mora-Flòrez J, Meléndez J, Carrillo-Caicedo G. Comparison of impedance based fault location methods for power distribution systems. *Electric Power Systems Research*. 2008;**78**(4):657-66.
[12] Gazzana DS, Ferreira GD, Bretas AS, Bettiol AL, Carniato A, Passos LFN, et al. An integrated technique for fault location and section identification in distribution systems. *Electric Power Systems Research*. 2014;**115**:65-73.
[13] Bakar AHA, Ali MS, Tan C, Mokhlis H, Arof H, Illias HA. High impedance fault location in 11 kV underground distribution systems using wavelet transforms. *International Journal of Electrical Power & Energy Systems*. 2014;**55**:723-30.
[14] Moshtagh J, Aggarwal RK, editors. A New Approach to Ungrounded Fault Location in a Three-Phase Underground Distribution System using Combined Neural Networks & Wavelet Analysis. Electrical and Computer Engineering, 2006. CCECE '06. Canadian Conference on; 2006 May 2006.
[15] Baqui I, Zamora I, Mazón J, Buigues G. High impedance fault detection methodology using wavelet transform and artificial neural networks. *Electric Power Systems Research*. 2011;**81**(7):1325-33.
[16] Rafinia A, Moshtagh J. A new approach to fault location in three-phase underground distribution system using combination of wavelet analysis with ANN and FLS. *International Journal of Electrical Power & Energy Systems*. 2014;**55**:261-74.
[17] Chunju F, Li KK, Chan WL, Weiyong Y, Zhaoning Z. Application of wavelet fuzzy neural network in locating single line to ground fault (SLG) in distribution lines. *International Journal of Electrical Power & Energy Systems*. 2007;**29**(6):497-503.
[18] Ye L, You D, Yin X, Wang K, Wu J. An improved fault-location method for distribution system using wavelets and support vector regression. *International Journal of Electrical Power & Energy Systems*. 2014;**55**:467-72.

IOP Conf. Series: Materials Science and Engineering 1127 (2021) 012037 doi:10.1088/1757-899X/1127/1/012037

[19] Thukaram D, Khincha HP, Vijaynarasimha HP. Artificial neural network and support vector Machine approach for locating faults in radial distribution systems. *IEEE Transactions on Power Delivery*. 2005;**20**(2):710-21.

[20] Shafiullah M, Ijaz M, Abido MA, Al-Hamouz Z, editors. Optimized support vector machine & wavelet transform for distribution grid fault location. 2017 11th IEEE International Conference on Compatibility, Power Electronics and Power Engineering (CPE-POWERENG); 2017 4-6 April 2017.

[21] Moloi K, Jordaan JA, Hamam Y, editors. High impedance fault detection technique based on Discrete Wavelet Transform and support vector machine in power distribution networks. 2017 IEEE AFRICON; 2017 18-20 Sept. 2017.

[22] Khorramdel B, Marzooghi H, Samet H, Pourahmadi-Nakhli M, Raoofat M. Fault locating in large distribution systems by empirical mode decomposition and core vector regression. *International Journal of Electrical Power & Energy Systems*. 2014;**58**:215-25.

[23] Aziz MSA, Hassan MAM, El-Zahab EA. An Artificial Intelligence Based Approach for High Impedance Faults Analysis in Distribution Networks. *International Journal of System Dynamics Aplications*. 2012;**1**(2):44-59.

[24] Barakat S, Eteiba MB, Wahba WI. Fault location in underground cables using ANFIS nets and discrete wavelet transform. *Journal of Electrical Systems and Information Technology*. 2014;**1**(3):198-211.

Parametric Evaluation of 11kV Zinc Oxide Surge Arrester using Finite Element Analysis Model

Nurul A.A. Latiff[1,*], Sameh Z.A. Dabbak[2], Hazlee A. Illias[2] and Ab H.A. Bakar[1]

[1]Higher Institution Centre of Excellence (HICoE), UM Power Energy Dedicated Advanced Centre (UMPEDAC), Level 4 Wisma R&D University of Malaya, Jalan Pantai Baharu, 59990 Kuala Lumpur, Malaysia.

[2]Department of Electrical Engineering, Faculty of Engineering, University of Malaya, Kuala Lumpur 50603, Malaysia

E-mail: ainlatiff@um.edu.my

Abstract. A surge arrester on transmission line is used to deflect lightning surge to ground, avoiding power systems from destruction. During normal condition, there is a leakage current flow through surge arresters where it is one of the main causes of surge arrester humiliation. Thus, monitoring the surge arrester condition is very significant. In this work, the effect of several parameters in a 11kV zinc oxide surge arrester design based on leakage current distribution has been studied. A comparison is made between measurements and modelling in order to validate the model that has been developed using finite element analysis. This work found that the benefit of using FEA-based model where the design parameters that affect the leakage current can be identified. Through this study, a well knowledge of leakage current can be achieved and also help in zinc oxide surge arrester design.

1. Introduction

Electrical power system continuously has an obstacle with transmission lines which is lightning surge problem. This problem can be reduced by guarding the overhead lines using counter-poise wires, grounding rods and protective devices [1]. Many kinds of protective devices can be used on overhead transmission. Expulsion gaps and protector tubes are used on the line itself while surge arresters are used at line terminations, junction of lines and substations.

The most efficient protectives devices to defend transmission lines from lightning is surge arrester where it be able to limit the peak voltages to an acceptable value that will avoid the electrical apparatus from damage. There are two types of well-known surge arrester which produces since 1965 which are silicon carbide and zinc oxide surge arrester (ZnOSA) [2]. The housing of surge arrester also can be intended in many forms and materials which provide a contribution towards the defense of the environment. Nowadays, silicone rubber is used as housing and shed material due to its excellent chemical and physical properties.

To inhibit the continuous flow of leakage currents, the surge arrester design required spark gaps but this gap is not essential for ZnOSA. This is because ZnOSA have low leakage current elements. Nevertheless, during normal condition, leakage current still flows through ZnOSA and affected by some conditions [3].

The increasing leakage current may direct to humiliation of ZnOSA and it can cause failures in electrical power system [4]. This failures will effect in high maintenance cost and time consuming where is some cases, the electrical equipment is compulsory to be changed [5]. Therefore, it is very substantial and important to avoid this breakdown [6]. This can be accomplished by examining the source of increasing leakage current [7]. In addition, the ZnOSA design also can be enhanced [8].

Modern enhancements correlated to the ZnO blocks have brought in impressive design improvements of ZnOSA. On the other hand, the design still can be improve on reducing the leakage current across the ZnOSA. Enhancement using finite element analysis (FEA) permits design engineers researchers to assemble a ZnOSA model based on real physical dimensions. Hence, the developed FEA model can be used to assess various design parameters affecting leakage current.

In this work, laboratory experiment and FEA simulation were performed to measure the leakage current across the 11kV ZnOSA. A three-dimensional (3D) surge arrester model was developed using FEA method to study the effect of surge arrester parameters on behaviour of leakage current. At first, the leakage current measurement results were compared to simulation results to validate the design. Then, the developed model was used to find the design parameters that affect the value of leakage current. The main advantage of parametric study is the most suitable affected parameter can be obtained within shorter time. Furthermore, the conducted study using the FEA model could be beneficial among surge arrester designers in designing suitable protection device. Therefore, performance of transmission lines can be improved mainly in reducing the leakage current on surge arresters.

2. Methodology

This segment explains the methodology used in this study. The leakage current measurement setup and design of the surge arrester using finite element analysis (FEA) software and its simulation are shown. The ZnOSA used in this study is 11kV with silicone housing produced by ABB as shown in Figure 1 which consists of two ZnOs blocks, a glassfibre between insulator and ZnO blocks, four silicone sheds and aluminum caps at both ends. The height of the ZnOSA is 227mm with 344mm creepage distance and maximum continuous operating voltage (MCOV) of $10kV_{rms}$.

Figure 1. 11kV zinc oxide surge arrester

2.1. Modelling of zinc oxide surge arrester

Finite element analysis (FEA) is a method for numerical solution of field problems and used in this study as a replacement of equivalent circuit model. By using FEA, the exact parameters of ZnOSA can be created. The ZnOSA was modelled in three-dimensional (3D) and physic used in this under 'AC/DC Module' is 'electric current'. It is used to generate the current density on the ZnOSA model.

The ZnOSA consist of of two zinc oxide blocks with diameter and height of 47mm respectively, a glassfibre of 10mm and the entire ZnOSA model geometry was surrounded by a layer of air with permittivity ε_r equals to 1 to allow the electric potential and leakage current distribution along the surface of surge arrester to be calculated. The complexity of the ZnOSA model was diminished by eliminating the small scale elements such as drilled holes and disc string since these elements do not influence the simulation outcomes. The 3D model is shown in Figure 2 and the electrical conductivity σ, relative permittivity ε_r were assigned as in Table 1.

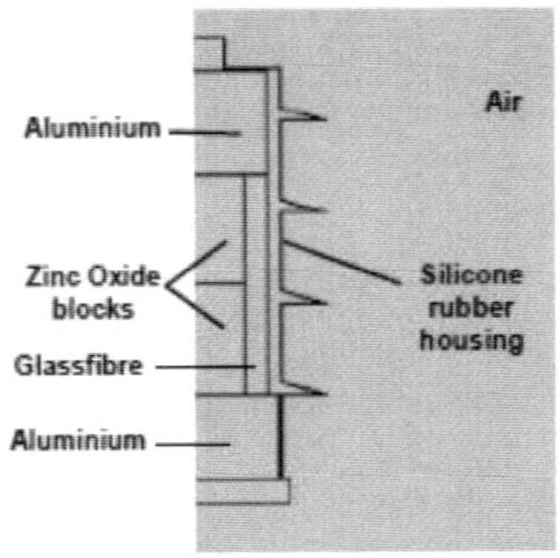

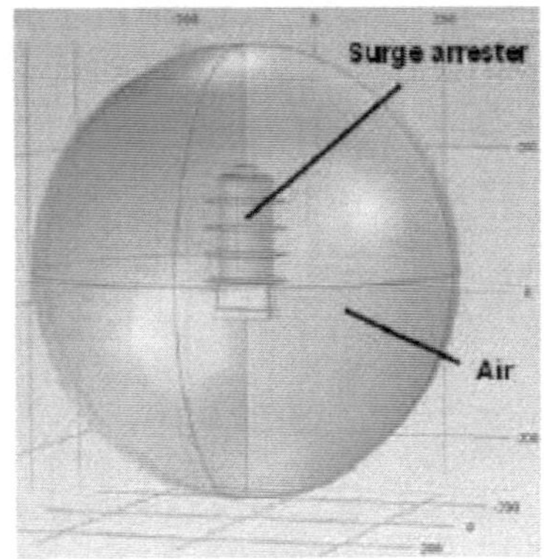

Figure 2. ZnOSA geometry (a) complete structure and (b) 3D model

The boundary conditions of the ZnO model were set with relevant interface settings and the model was meshed with solution of domain, boundary, edge and point of the model by using some elementary shape functions, which include tetrahedral, triangular, edge and vertex elements as in Figure 3 and Table 2. The simulation processes were then repeated for different parameters of surge arresters.

Table 1. Material properties

Material	Relative Permittivity, ε_r	Electrical Conductivity, σ (Sm^{-1})
Air	1	0
Silicone	11.7	$1x10^{-12}$
ZnO blocks	2250	From VI characteristics
Glassfibre	4.2	$1x10^{-14}$
Aluminum	1	$3.77x10^{7}$

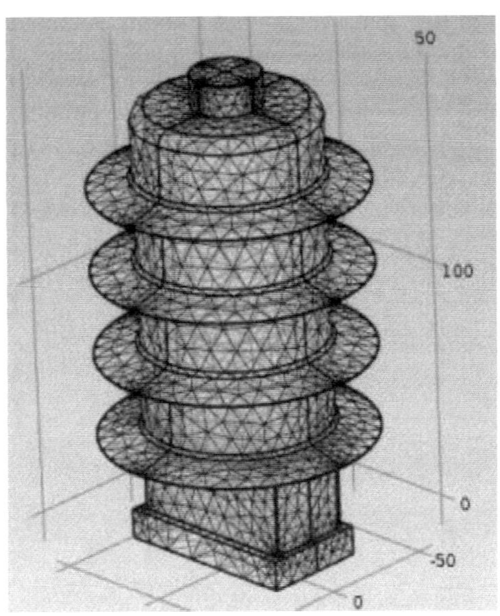

Figure 3. Meshing elements

Table 2. Mesh elements statistic

Geometric entity level	Type of element	Number of element
Domain	Tetrahedral	106194
Boundary	Triangular	14943
Edge	Edge	2792
Point	Vertex	245

2.2. Leakage current measurement setup

This subdivision emphases on leakage current measurement setup as shown in Figure 4 used in the laboratory. It consists of a sinusoidal voltage source, a step-up transformer of 0.22/100kV, a protective resistor of 2.4MW to prevent short circuit, a measuring capacitor of 1nF to observe the applied voltage, a measuring circuit, a shunt resistor of 20kW to measure the leakage currents (measure the voltage drop from the current built across the resistance) and a digital storage oscilloscope.

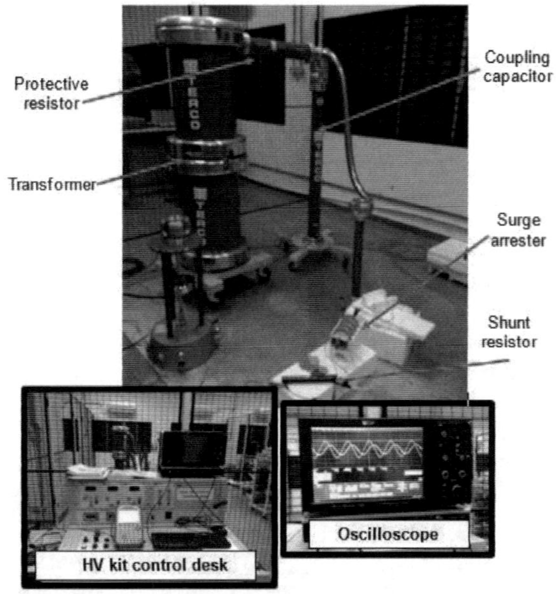

Figure 4. Leakage current measurement setup

3. Results

The feasibility of the ZnOSA FEA model has been successfully validated by having a good agreement between measurement and simulation as shown in Figure 5. Parametric assessments were performed on the effect of different design modifications related of the ZnOSA. The parameters that have been changed are ZnO width, glassfibre, thickness of silicone rubber and dielectric constant of the insulation layer.

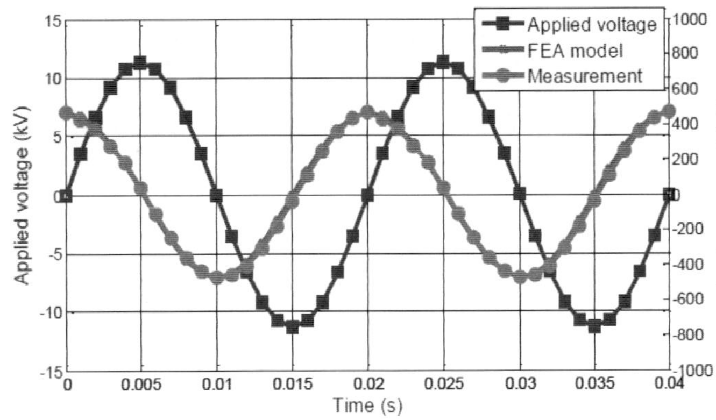

Figure 5. Comparison on ZnOSA leakage current of FEA model and measurement

3.1. *Effect of ZnO width*

The width of ZnO block was altered to study the influence on leakage current where initial value of ZnO width is 47mm and was reformed to 35mm, 40mm, 50mm and 55mm. Referring to Figure 6, the leakage current is decreases when the width of ZnO block increases which is due to larger ZnO radius yields larger cross sectional area. Its contains of ZnO grains that perform as electronic switch

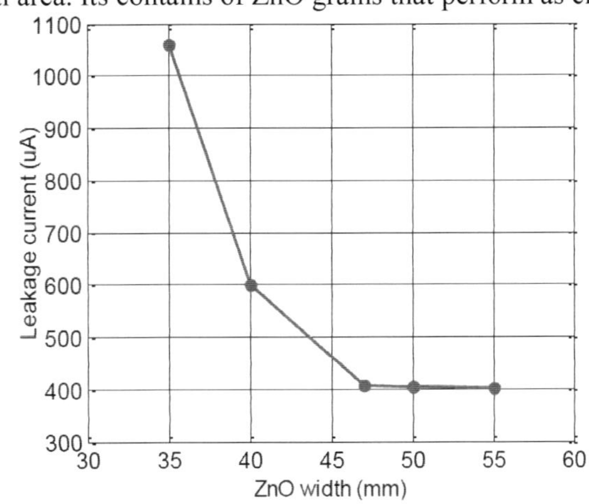

Figure 6. Leakage current for different ZnO width

3.2. *Effect of glassfibre width*

The current stress can be controlled by adding axial position of glassfibre outside the ZnO blocks. It also lessen possible sagging towards the wall of the silicone housing. Therefore, the glassfibre width was assessed with 5mm, 7mm, 10mm (initial size), 12mm and 15mm. There is a big decline on the leakage current with increasing width of the glassfibre as in Figure 7. This is because there is more size for ZnO heat to disperse into the glassfibre width and thus decreases the leakage current value.

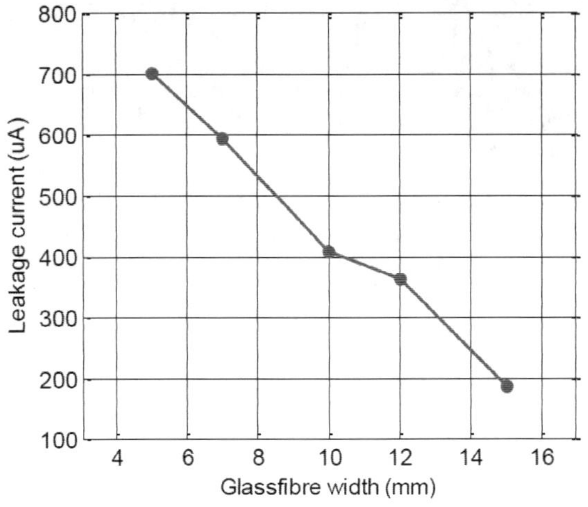

Figure 7. Leakage current for different glassfibre width

3.3. Effect of silicone housing thickness

Different thickness of silicone housing was simulated where initial thickness is 3mm and the thickness were changed to 1mm, 2mm, 5mm and 7mm. In Figure 8, it shows that there is a small difference in leakage currents. Yet, a small decrement is able to increase the lifetime of the ZnOSA and can decreases the electric field intensity of the ZnOSA housing [9].

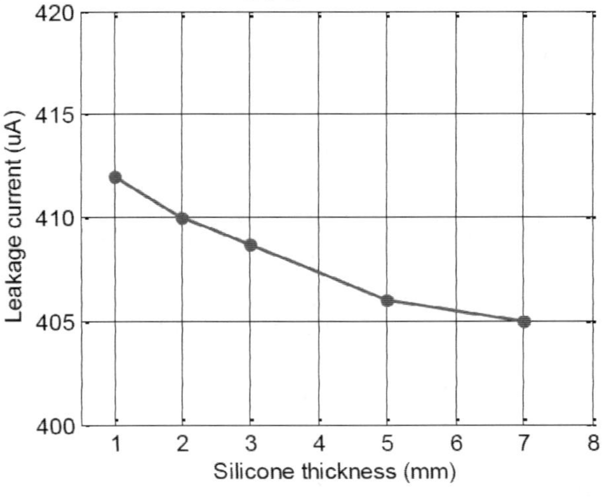

Figure 8. Leakage current for different silicone thickness

3.4. Effect of the insulation layer relative permittivity

In this work, the influence of relative permittivity of silicone rubber and glassfibre on leakage current was examined since it is important in controlling the withstand capability of ZnO blocks due to surface flashover. Figure 9 displays the results for different relative permittivity value of glassfibre and silicone rubber. The initial relative permittivity value for glassfibre and silicone rubber were 4.2 and 11.7 respectively. The results obtained show that leakage current is influenced by the housing materials where when increasing the relative permittivity of insulation layers, the leakage current decreases.

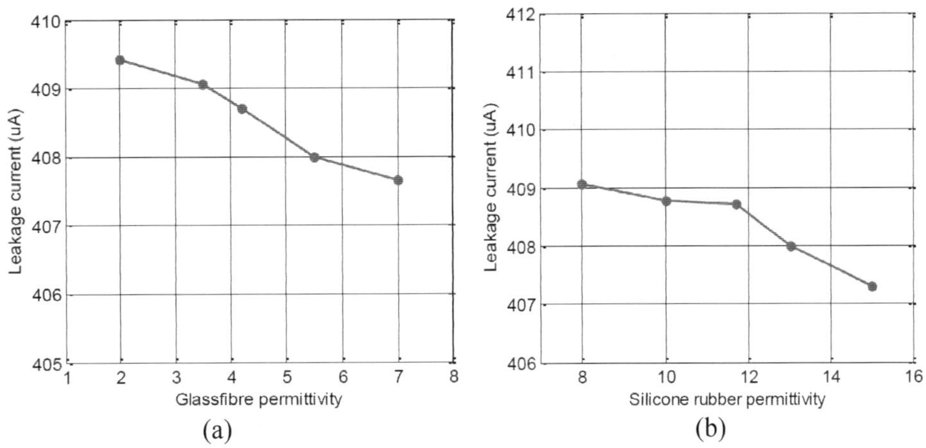

Figure 9. Leakage current for different relative permittivity of (a) glassfibre and (b) silicone rubber

4. Conclusions

In this work, an attempt is made to examine the condition of ZnOSA by means of leakage current measurement. The ZnOSA model is developed by using finite element analysis with actual dimensions. The FEA model showed an efficient accuracy compared to measurement. Hence, parametric study has been done to study the effect of leakage current under different size of design parameters. The results show that the most influenced parameter on the leakage current is zinc oxide. Therefore, the future study will related to dimension and grains structure of zinc oxide blocks. From this parametric assesment, it can assist engineers to consider the design of ZnOSA that affect the leakage current value.

Acknowledgements

The authors thank the technical and financial assistance of UM Power Energy Dedicated Advanced Centre (UMPEDAC) and the Higher Institution Centre of Excellence (HICoE) Program Research Grant, UMPEDAC - 2018 (MOHE HICOE - UMPEDAC), Ministry of Education Malaysia, RU007-2018 and RU012-2019, University of Malaya.

References

[1] Naidu M, Kamaraju V. High Voltage Engineering. McGraw-Hill; 1996.
[2] Woodworth J. History of Arresters on Power Systems 1965-Present. ArresterWorks; 2011.
[3] Hoang TT, Cho M-Y, Alam MN, Vu QT. A novel differential particle swarm optimization for parameter selection of support vector machines for monitoring metal-oxide surge arrester conditions. *Swarm and Evolutionary Computation*. 2018;**38**:120-6.
[4] Metwally IA, Eladawy M, Feilat E. Online condition monitoring of surge arresters based on third-harmonic analysis of leakage current. *IEEE Transactions on Dielectrics and Electrical Insulation*. 2017;**24**(4):2274-81.
[5] Said A. Analysis of 500 kV OHTL polluted insulator string behavior during lightning strokes. *International Journal of Electrical Power & Energy Systems*. 2018;**95**:405-16.
[6] Lu J, Xie P, Fang Z, Hu J. Electro-thermal modeling of metal-oxide arrester under power frequency applied voltages. *Energies*. 2018;**11**(6):1610.
[7] Latiff NA, Illias HA, Bakar AH, Dabbak SZ. Measurement and modelling of leakage current behaviour in ZnO surge arresters under various applied voltage amplitudes and pollution conditions. *Energies*. 2018;**11**(4):875.
[8] Latiff NAA, Illias HA, Abd Halim S, Dabbak SZ. Design technique for leakage current reduction in surge arrester using gravitational search algorithm and imperialist competitive algorithm. *COMPEL-The international journal for computation and mathematics in electrical and electronic engineering*. 2018.

[9] Illias HA, Abd Halim S, Mokhlis H. Determination of surge arrester discharge energy using finite element analysis method. *IET Science, Measurement & Technology*. 2015;**9**(6):693-701.

IOP Conf. Series: Materials Science and Engineering 1127 (2021) 012039 doi:10.1088/1757-899X/1127/1/012039

Feasibility Analysis of Solar Power Generation System for Office Building in Academic Institution

Muhammad Firdaus Mohd Zublie [1,2,4], Md. Hasanuzzaman [1,2], Nasrudin Abd. Rahim [1,3]

[1]Higher Institution Centre of Excellence (HICoE), UM Power Energy Dedicated Advanced Center (UMPEDAC), Level 4, Wisma R&D, University of Malaya, Jalan Pantai Baharu, 59990 Kuala Lumpur, Malaysia

[2]Institute for Advanced Studies, Universiti of Malaya, 50603 Kuala Lumpur, Malaysia

[3]Renewable Energy Research Group, King Abdulaziz University, Jeddah 21589, Saudi Arabia

[4]Politeknik Sultan Azlan Shah, 35950, Behrang, Perak, Malaysia

E-mail: muhammad_firdaus@psas.edu.my

Abstract. This paper study the potential of energy-saving through lighting located at the office building in an academic institution known as Politeknik Sultan Azlan Shah (PSAS). Exercise through energy audit has been conducted at 3 main academic building to identify energy usage of lighting. 3 academic building involved is Mechanical Engineering Department (MED), Civil Engineering Department (CED), and the Electrical Engineering Department (EED). This study analysis 5 types of lighting to identify which types of lighting use the highest amount of energy in academic buildings. The 5 types of lighting involved in this study are Fluorescent T8 36W, Fluorescent T8 18W, Compact Fluorescent (CFL) 24W, Metal halide (MH) 400W and Ttube 18W. Energy savings are calculated based on the installation of hybrid solar in 3 related buildings. From an energy audit, the energy-saving at-least 5% from previous data has been estimated through the installation of a solar photovoltaic (PV) system. Through feasibility analysis, it is found that energy-saving potential obtained in the year 2019 for MED, CED and EED are about 4299 kWh, 2,319 kWh and 3,269 kWh respectively. Besides, the energy bill saving obtained is about RM 1,569, RM 846 and RM 3,269 respectively for MED, CED, and EED.

1. Introduction

Global warming is a real challenge and accepted as a global issue. An active action to handle this challenge must be taken. One of the methods to achieve the target of reducing the carbon emissions by buildings in Malaysia with encouraging renewable energy source use, which also known as a clean energy source. Using this source in the power system can reduce greenhouse gas emissions, which is identified as a major contributing factor to global warming [1]. In Malaysia, the lighting is identified as one of the major energy consumers in the buildings [2] where is for an academic building, an energy consumption approximately 19% [3]. However, in the academic building in PSAS, there are 5 types of lighting with different power rated and specification. A study conducted by Alajmi [4] on higher institution learning building in Kuwait using energy audit strategy to gather information about energy usage and calculate the potential energy saving that can be achieved. Azucena et al [5] conducted the study at university of Mexico and only focus their research on energy consumption per equipment such as lighting, research equipment, cooling, and water heating only. Another study by Mohammad Ahsan et al [6] focuses their energy audit on lamps, motors and other equipment. Based on a literature review [4-7] using an energy audit is the best strategy to collect data about high energy-consuming equipment in the educational building. In this study, an energy audit has been conducted on lighting only and this setting is acceptable because a lot of literature review mention about the lighting is one of their research target. Analyse and identify energy- consuming of each lighting will show which types of lighting that need to be focus for purpose of energy-saving. Energy audit findings of facilitate the use of energy and the potential of energy savings for lighting is calculated [7] for PSAS. In effort to reduce the energy consumption of in the academic building, the implementation of solar power generation system is currently the most widely installed in the building sector [8]. The research by Clara et al [9] has been present a case study involving an office building installed an integrated PV solar system in Norway. While, Ruben et al [10] mentioned most universities in Indonesia applied solar energy with off-grid system. The author presents the feasibility analysis of photovoltaic power installation using HOMER software. However, relying only on solar energy cannot meet electricity demand effectively because the supplied it is inefficient and variable. Solar power system depends on weather and requires backed up by traditional fossil fuel energy [11]. Installing a hybrid solar system is one of the practical solutions for stand- alone or off-grid power generation systems. This paper will analyses follow the previous work by [9] but using a different methodology to analyse the data. Based on the energy audit data, the energy forecast is done but not using the HOMER [10]. The aim of this study to analyse the potential of energy savings through lighting located at the office building. The 5 types categories of lighting involved in this study such as Fluorescent T8 36W, Fluorescent T8 18W, Compact Fluorescent (CFL) 24W, Metal halide (MH) 400W and Ttube 18W. Through the energy audit data of lighting, the feasibility analysis has been calculated by installing a rooftop solar power generation system in Politeknik Sultan Azlan Shah. The total energy consumption of lighting in 3 main academic buildings in PSAS is about 21% of whole energy consumption at 3 departments involved. Focusing on the lighting for energy- saving purposes through feasibility analysis has been confirmed gives a significant impact on the PSAS.

2. Methodology

2.1. Baseline Data

The baseline data is referred to data of energy consumption is measured using power meter was installed in each academic building PSAS. This reference value will compare with the calculated energy consumption to minimize the error percentage between measured and calculated data. Baseline data is referring to measured data for energy consumption (kWh) in 2019. Based on baseline data, the prediction of energy consumption and energy saving in PSAS for future years has been estimated. The data collected is referring to average energy consumption data (kWh) for 12 months in 2018. All the data has been extracting via the online Energy Monitoring System (EMS) owned by PSAS. Table 1 shows the baseline data for energy consumption in the year 2018 is imported from EMS involving 3 main academic buildings.

Table 1. Baseline data for year 2018 import from EMS

Building	Energy Consumption (kWh)/Year
Mechanical Engineering Department **(MED)**	327,324
Civil Engineering Department **(CED)**	173,468

Electrical Engineering Department **(EED)**	442,349
Total Energy Consumption **(MED + CED + EED)**	943,141

2.2. Data collection through energy audit

The gather of data follows the methodology implement by [6] where energy audit has been conducted in 3 main academic department at PSAS. 3 academic building involved is Mechanical Engineering Department (MED), Civil Engineering Department (CED), and the Electrical Engineering Department (EED). Most researchers [12-14] implement the energy audit methodology to identify which types of equipment consuming the highest energy. In this research, types of lighting use the highest amount of energy in academic buildings has been identified. The walk-through energy audit collect the information about types of lighting, quantity, power rated and operating hours. The lighting is divided into 5 types of categories such as Fluorescent T8 36W, Fluorescent T8 18W, Compact Fluorescent (CFL) 24W, Metal halide (MH) 400W and Ttube 18W. Table 2, Table 3 and Table 4 show the types of lighting, a total of quantity, rating power and operating hours based on academic calendar PSAS 2018 and normal routine activity at MED, CED and EED.

Table 2. Types of lighting, total of quantity, rating power and operating hours for normal routine activity at MED.

No	**Types of Lighting**	**Quantity**	**Power rated**	**Operating Hours**
			(kW)	**(h/year)**
1	T8 36W	1209	0.036	708
2	T8 18W	204	0.018	708
3	CFL 24W	23	0.024	708
4	MH 400W	155	0.4	708
5	Ttube 18W	650	0.018	708
TOTAL		**2241**	**0.496**	**708**

Table 3. Types of lighting, total of quantity, rating power and operating hours for normal routine activity at CED.

No	**Types of Lighting**	**Quantity**	**Power rated**	**Operating Hours**
			(kW)	**(h/year)**
1	T8 36W	404	0.036	708
2	T8 18W	52	0.018	708
3	CFL 24W	35	0.024	828
4	MH 400W	75	0.4	828
5	Ttube 18W	775	0.018	708
TOTAL		**1341**	**0.496**	**756**

Table 4. Types of lighting, total of quantity, rating power and operating hours for normal routine activity at EED.

No	**Types of Lighting**	**Quantity**	**Power rated**	**Operating Hours**
			(kW)	**(h/year)**
1	T8 36W	1260	0.036	828
2	T8 18W	135	0.018	708
3	CFL 24W	13	0.024	828
4	MH 400W	42	0.4	828
5	Ttube 18W	800	0.018	828

TOTAL	2250	0.496	804

2.3. *Equations*
The estimation of energy consumption by the lighting in the PSAS is calculated using as the following equation (1) [6, 7].

$$EC_L = P_r \times O_h \times L_f \qquad (1)$$

where EC_L is the annual energy consumption of lighting (kWh), P_r is power rated (kW), O_h is annual operating hours, L_f is the loading factor. The loading factor for lighting considered in ideal condition and set as 1 [7]. The estimated operating hours for lighting has been calculated by considering the total operating hour per day in normal routine is 8 hour. The number of days the lighting in ON condition for 1 year is calculated based on 32 weeks of lectures and plus others 4 weeks for the final examination. The estimation of annual energy saving for lighting in the PSAS is calculated using as the following equation (2)

$$AES_L = P_r \times Havg \times R_p \qquad (2)$$

where AES_L is the energy-saving estimation of per year (kWh), Pr is power rated of lighting, Hy is representing an operating hour operation by year, L is the loading factor of lighting and Rp is a percentage of lighting replacement. Energy-saving for each type of lighting in this research has been calculated based on the percentage of lamps quantities is replaced. Annual bill savings in annually using new lighting are calculated using the following equation in (3): [6]

$$ABS = AES_L \times E_c \qquad (3)$$

Where ABS is annual bill savings (RM), AES_L is energy saving per year (kWh) and E_c is the energy cost (RM/kWh). Tariff rate for PSAS refer to tariff C1 for medium voltage general commercial tariff.

3. Results and Discussion
Analysis of energy audit data for lighting has been conducted in 3 main academic building. The number of lamps or quantity, power rated, operating hours, energy consumption and energy bill are also recorded and calculated. The energy consumption and energy bill for all types of lighting in 3 academic buildings have been calculated by year by using Equation (1). The calculation estimation of energy bills using the TNB tariff rates category C1. The tariff under category medium voltage general commercial using one rate for all kWh per month at a charge rate of RM0.365 [15].

3.1. *Energy consumption and energy bill at MED*
By using the data from Tables 2, 3 and 4 and calculated using equation (1) then the details of total energy consumption and energy bills for 5 types of lighting are shown in Tables 5, 6 and 7 respectively for MED, CED, EED.

Table 5. Energy consumption and energy bill for lighting at MED

No	Types of Lighting	Energy consumption/year (kWh)	Energy bill/year (RM)
1	T8 36W	30,815	11,247
2	T8 18W	2,600	949
3	CFL 24W	391	143
4	MH 400W	43,896	16,022
5	Ttube 18W	8,284	3,024
	TOTAL	**85,985**	**31,385**

Table 6. Energy consumption and energy bill for lighting at CED

No	Types of Lighting	Energy consumption/year (kWh)	Energy bill/year (RM)

1	T8 36W	10,297	3,758
2	T8 18W	663	242
3	CFL 24W	696	254
4	MH 400W	24,840	9,067
5	Ttube 18W	9,877	3,605
	TOTAL	**46,372**	**16,926**

Table 7. Energy consumption and energy bill for lighting at EED

No	**Types of Lighting**	**Energy consumption/year (kWh)**	**Energy bill/year (RM)**
1	T8 36W	37,558	13,709
2	T8 18W	1,720	628
3	CFL 24W	258	94
4	MH 400W	13,910	5,077
5	Ttube 18W	11,923	4,352
	TOTAL	**65,370**	**23,860**

3.2. *Percentage of apportionment for the lighting in each department*

The percentage of apportionment for the lighting is shown in Table 8 where the energy consumption for the whole building in PSAS obtains from measured data using a power meter and stated in (A) column. Total energy consumption for lighting obtain from an estimation calculation through energy audit data and is represented in (B) column. Based on the result shows in Table 8, the lighting using about 21% of whole energy consumption at 3 departments involved.

Table 8. The percentage of apportionment for the lighting in 3 main department

Building	**Total (kWh)/year for whole building (A)**	**Total (kWh)/year lighting only (B)**	**Percentage B/A (%)**
MED	327,324	85,985	26
CED	173,468	46,372	27
EED	442,349	65,370	15
(MED + CED + EED)	**943,141**	**197,727**	**21**

3.3. *Feasibility analysis of solar power generation system*

The feasibility analysis of the solar power generation using a rooftop solar photovoltaic (PV) system was conducted and the energy-saving potential in MED, CED and EED has been calculated. By assuming the power generated by solar PV for self-consumption [16] where electricity generated is entirely for own use and not exported to the grid [17], energy-saving (kWh) and bill saving (RM) per year for lighting have been calculated. The feasibility analysis has been conducted using solar PV monocrystalline module 250W. By assuming the average of the peak sun hour for Malaysia is 4 hours means one solar panel is generating 1 kWh per day and about 30 kWh per month. According to Table 5, the annual energy consumption for lighting in MED is 85,985 kWh per year and the energy bill is RM 31,385 per year. Means energy demand for lighting in MED is approximate 60 kW per day and the energy consumption is about 239 kWh per day.

Referring to Table 6, the annual energy consumption for lighting in CED is 46,372 kWh per year and the energy bill is RM 16,926 per year. The energy demand for lighting in CED is approximate 32 kW per day and the energy consumption is about 129 kWh per day. The energy demand for CED is less almost half compared with MED. In EED, the annual energy consumption for lighting is 65,370 kWh per year and the energy bill is RM 23,860 per year as shown in Table 7. The energy demand for lighting in EED about 45 kW per day and the energy consumption is about 182 kWh per day. The information about the value of energy demand will help in determining the number of solar PV panel need to be installed to meet the requirement of energy goal. By using the data from Tables 2, 3 and 4 and calculated using equation (2) and (3) then the total energy saving and energy bills are shown in Tables 9, 10 and 11 respectively for MED, CED, EED. The forecast for energy-saving and bill saving has been calculated based on an increment of the percentage due to of the energy supplied by solar PV. That percentage of increment for energy supplied by solar PV (5%, 10%, and 15%) based on Malaysian Polytechnic POLYGreen Blueprint 2015 [18]. This document stated all the polytechnic institutions need to reduce their utility costs minimum by 5% per year. The percentage represents the savings will gained for each year from 2019 to 2021.

Figure 1 and 2 shows the graph of a forecast for energy-saving and bill saving per year respectively at MED, CED, and EED if install solar photovoltaic for self-consumption. The results in Figures 1 and 2 show that increment of the total energy supplied by solar will decrease energy consumption supplied by TNB and proportionally increase the amount of bill saving. The solar power generation system also involves PWM solar charge controller 12/24V and the energy produced by solar PV will be stored in the battery sealed lead acid 12V 330Ah. The output of direct current from the battery will convert to the alternating current through 3000W inverter 12V 330 Ah. The output of the inverter will be used to operating the lighting system.

Table 9. Forecast for energy and bill saving per year at MED if install solar photovoltaic for self-consumption

Energy supplied by solar PV	Energy saving per year (kWh)	Bill saving per year (RM)
5%	4,299	1,569
10%	8,599	3,138
15%	12,898	4,708

Table 10. Forecast for energy and bill saving per year at CED if install solar photovoltaic for self-consumption

Energy supplied by solar PV	Energy saving per year (kWh)	Bill saving per year (RM)
5%	2,319	846
10%	4,637	1,693
15%	6,956	2,539

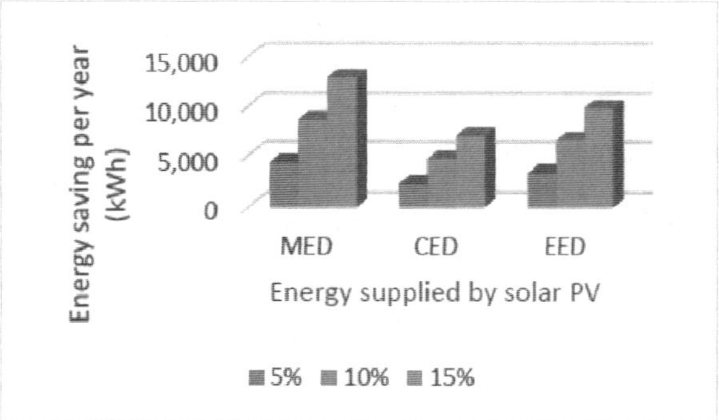

Figure 1. Forecast for energy saving per year at MED, CED, and EED if install solar photovoltaic for self-consumption

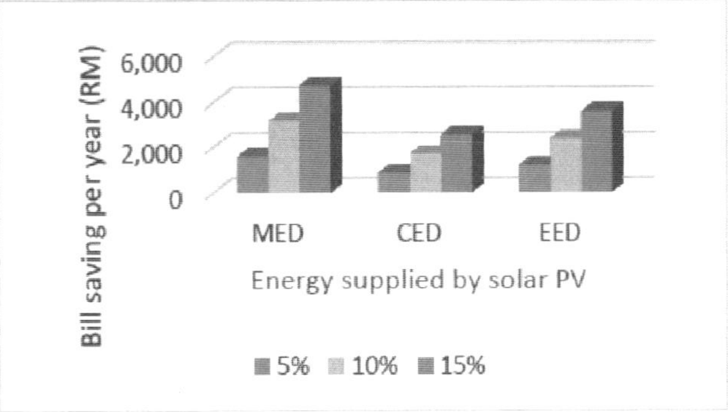

Figure 2. Forecast for bill saving per year at MED, CED, and EED if install solar photovoltaic for self-consumption

Table 11. Forecast for energy and bill saving per year at EED if install solar photovoltaic for self-consumption

Energy supplied by solar PV	Energy saving per year (kWh)	Bill saving per year (RM)
5%	3,269	1,193
10%	6,537	2,386
15%	9,806	3,579

3.4. Carbon Emission Reduction

The installation of rooftop solar PV power generation produces carbon emission reduction. The estimated value on the potential of carbon emission CO2 which has been eliminated due to installing the solar PV power generation system is shown in Table 12, 13 and 14 for MED, CED, and EED respectively. The energy produced by solar PV is equal to energy-saving obtained by departments. To calculate the CO2 reduction, the energy-saving per year multiplying with an emission factor of 0.693 kg of CO2 per kWh [19].

Table 12. CO2 reduction per year for MED

Energy saving per year (kWh)	Carbon emission reduction (kg of CO2)
4,299	2,979
8,599	5,959
12,898	8,938

Table 13. CO2 reduction per year for CED

Energy saving per year (kWh)	Carbon emission reduction (kg of CO2)
2,319	1,607
4,637	3,214
6,956	4,820

Table 14. CO2 reduction per year for EED

Energy saving per year (kWh)	Carbon emission reduction (kg of CO2)
3,269	2,265
6,537	4,530
9,806	6,795

4. Conclusion

Based on the research, the total energy consumption of lighting in 3 main academic buildings in PSAS is 197,727 kWh. It is about 21% of whole energy consumption at 3 departments involved. The estimation of energy consumption per year is about 85,985 kWh, 46,372 kWh, and 65,370 kWh while energy bill per year is about RM 31,385, RM 16,926, and RM 23,860. The energy consumption and energy bill above refer respectively for MED, CED, and EED. Through feasibility analysis, it is found that energy-saving potential obtained in the year 2019 for MED, CED and EED are about 4299 kWh, 2,319 kWh, and 3,269 kWh respectively. Besides, the energy bill saving obtained is about RM 1,569, RM 846 and RM 3,269 respectively for MED, CED, and EED.

Acknowledgements

The authors would like to thank the financial and technical assistance provided by AUN-SEED/Net Japan ASEAN Collaborative Education Programme (JACEP)

References

[1] Shaaban M, Tan W-S, Abdullah MP. A hybrid renewable energy system for a longhouse. *ELEKTRIKA-Journal of Electrical Engineering*. 2017;**16**(3):23-9.
[2] Kamaruzzaman SN, Zulkifli N, editors. A review of the lighting performance in buildings through energy efficiency. 2nd International Conference on Research in Science, Engineering and Technology (ICRSET'2014); 2014.
[3] Ali AN, Tee. BT, editors. Lighting simulation for energy analysis in laboratory building. Proceedings of Mechanical Engineering Research Day 2017; 2017; Melaka, Malaysia.
[4] Alajmi A. Energy audit of an educational building in a hot summer climate. *Energy and Buildings*. 2012;**47**:122-30.
[5] Escobedo A, Briceño S, Juárez H, Castillo D, Imaz M, Sheinbaum C. Energy consumption and GHG emission scenarios of a university campus in Mexico. *Energy for sustainable development*. 2014;**18**:49-57.
[6] Habib MA, Hasanuzzaman M, Hosenuzzaman M, Salman A, Mehadi MR. Energy consumption, energy saving and emission reduction of a garment industrial building in Bangladesh. *Energy*. 2016;**112**:91-100.
[7] Roslizar A, Alghoul M, Bakhtyar B, Asim N, Sopian K. Annual energy usage reduction and cost savings of a school: End-use energy analysis. *The Scientific World Journal*. 2014;**2014**.
[8] Tsalikis G, Martinopoulos G. Solar energy systems potential for nearly net zero energy residential buildings. *Solar Energy*. 2015;**115**:743-56.
[9] Good CS, Lobaccaro G, Hårklau S. Optimization of solar energy potential for buildings in urban areas–a Norwegian case study. *Energy procedia*. 2014;**58**:166-71.
[10] Kristiawan RB, Widiastuti I, Suharno S, editors. Technical and economical feasibility analysis of photovoltaic power installation on a university campus in indonesia. MATEC Web of Conferences; 2018: EDP Sciences.
[11] Lofthouse J, Simmons RT, Yonk RM. Reliability of renewable energy: Solar. *Institute of Political Economy, UtahState University*. 2015.
[12] Kumar A, Ranjan S, Singh MBK, Kumari P, Ramesh L. Electrical energy audit in residential house. *Procedia Technology*. 2015;**21**:625-30.
[13] Saidur R, Rahim N, Masjuki HH, Mekhilef S, Ping H, Jamaluddin M. End-use energy analysis in the Malaysian industrial sector. *Energy*. 2009;**34**(2):153-8.
[14] Paucar M, Amancha P, Viera E, San Antonio T, Salazar D. Implementation of a methodology to perform an energy audit with academic purpose. *International Journal of Applied Engineering Research*. 2017;**12**(24):14908-13.
[15] Tenaga Nasional Berhad. Pricing & Tariff Tenaga Nasional Berhad [last accessed 1 Sept. 2018. Available from: https://www.tnb.com.my/commercial-industrial/pricing-tariffs1.]
[16] Rodrigues S, Torabikalaki R, Faria F, Cafôfo N, Chen X, Ivaki AR, et al. Economic feasibility analysis of small scale PV systems in different countries. *Solar Energy*. 2016;**131**:81-95.

[17] Commission. E. Guidelines on the connection of Solar Photovoltaic Installation for self-consumption. 2017.
[18] MOHE. Malaysian Polytechnic Polygreen Blueprint 2015. Ministry of Higher Education; 2015.
[19] Abdul-Azeez I. Measuring and monitoring carbon emission to promote low-carbon development in Johor bahru. *Massachusetts Institute of Technology*. 2016.

IOP Conf. Series: Materials Science and Engineering 1127 (2021) 012040 doi:10.1088/1757-899X/1127/1/012040

Design of a small scale fluidized-bed incinerator for MSW with ability to utilize HHO as auxiliary fuel

Osama Majeed Butt, Muhammad Shakeel Ahmad, Hang Seng Che, Nasrudin Abd Rahim

UM Power Energy Dedicated Advanced Center (UMPEDAC), University of Malaya, Kuala Lumpur, Malaysia

E-mail: osamabutt26@ymail.com

Abstract. Waste incineration has become a mature technology and widely accepted due to its environment friendliness, easy operation and ability to reduce more than 95% mass fraction. There are some inherent limitations associated with incinerators such as high fuel consumption and cite feasibility. In this study, we propose a novel design concept of small scale fluidized bed incinerator for household use, with the ability to consume brown's gas (HHO) as primary fuel for incinerate waste. In principle, the HHO gas generated through photovoltaic (PV) integrated water electrolysis system would be feed from the bottom to provide heat energy to the waste. Theoretical design of fluidized bed type incinerator has been presented with water electrolyzer system. It was calculated that 150 lph of hydrogen is required for this proposed incinerator system which can handle 5 kg of waste.

1. Introduction

Municipal Solid Waste is considering to be trash, garbage which includes paper, plastic, food waste, wood, nappies, ash etc. Growth in urbanization and technologies tends to increase in the generation of the waste. Different techniques like landfill, incineration, biological conversion are being used from decades to treat this waste. All of these treatment methods are consider to be a source of sustainable energy with a promise of 'waste to energy'. According to International Energy Agency, one ton of MSW should have 5 to 20 MJ/Kg of calorific value for an effective energy production [1].

Earlier, open dumping landfill or incineration was used to treat waste which was a threat to the environment without considering the potential of energy and environmental impacts. Rapid increase the population and industrial development has led to the new alternative techniques from conventional for waste management with an approach of waste to energy. There are three basic types of waste to energy techniques [2]: thermal conversion, biochemical conversion and landfill. The most common and effective waste to energy technique is incineration. This technique does not only help in managing the increasing volume of waste but can also has potential to have significant role in recovering energy that can be used as an alternative for the traditional fuels [3]. As 80% need of the global primary energy is met by fossil fuel [4], incineration technology has a promise in offsetting fossil fuel usage by increasing renewable share while addressing the waste treatment at the same time [5]. In this method, waste is treated in a furnace at high temperature which reduces waste mass by 70% and volume by 90% [6]. Composition of the waste with its proximate characteristics [7] and heat content (shown in figure 1) determine the feasibility and requirement of the auxiliary fuel to incinerate the waste [8, 9]. Amount of energy release is also dependent on calorific value of the waste.

2. Potential of Incineration as Waste Management Technology

Advantages and disadvantages of waste management strategies are assessed by life-cycle inventory analysis and their impact. Incineration technology has a promise to reduce both waste volume and demand of space for landfill in an efficient manner. This technology has gain interest as waste generation is increasing day by day with human activities and normally burn out near its source of generation [10]. Therefore, it is advantageous as incineration facility can be installed near the centre of gravity of waste generation which in result reduce the transportation cost of waste for treatment. Use of resultant ash from MSW incinerator for suitability of green environment not only reduce the cost but also reduces the requirement of landfill capacity. Although incineration technology for municipal solid waste is advantageous when lack of suitable landfill sites or long transportation is required, but results in high cost. There are certain requirements which are required for incineration. Also it has been seen that composition of waste is not suitable for auto-combustion that is generated in developing countries as from figure 1, which required auxiliary fuel for incineration [9]. Also the incineration plant is complex which requires skilled staff. Flue gases released from incinerator can be hazardous for environment if not handle properly. Residue of the incinerator should be dispose of properly in controlled and well-operated landfills otherwise it could be harmful for ground and underground water [11].

This study is based on design of small scale incinerator which will be capable of handling and treating municipal solid waste at domestic level. It will not only help in managing waste at its point of generation but also help in producing renewable energy which can be used for several heating purposes. It was calculated that 150 litters per hour of the hydrogen is required as auxiliary fuel for incinerating 5 Kg of waste. This proposed system is considering to be more efficient as hydrogen has more calorific value than other fossil fuels and will reduce green-house gas emission.

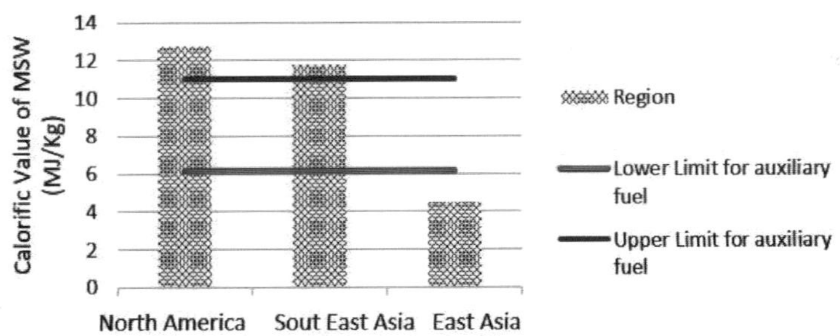

Figure 1. Heat content in waste of three different regions of world. Requirement of auxiliary fuel based on this heat content [9]

3. Experimental

For last many years, advancement in fluidized bed technology change the status from laboratory and pilot plants to commercial. There are several reasons which made fluidized bed incineration method more advantageous over conventional systems. These advantages include an effective burning of all types of fuel, from low calorific value to high calorific value, under stable conditions. Even various slurry sewage sludge can be handled in fluidized-beds [12].

An incineration technique is proposed in this study based on fluidized bed technology. As fluidized bed technology is best for the waste having high moisture contents [13]. Therefore, a drying technique is additionally added for this purpose which is done by circulating or bubbling hot air [14]. This technique is able to reduce 80% of the moisture content to less than 20% which can be than incinerated easily [14].

3.1. *Theoretical Design*

A rectangular vessel with rounded edges is designed for the incineration purpose which is equipped with a plate at the bottom of the vessel for a proper distribution of air for circulation. This plate holds the inert material to make a surface known as 'bed'. Commonly, sand is used for making this bed. A chamber is placed at the bottom of the bed for a smooth and uniform distribution of the air throughout the sand bed. An auxiliary burner is attached to the bed which keeps the temperature up to 800°C. Depending upon the velocity of the air flow it works on two modes: bubbling and circulating. As there is no moving part is included in the design so this technology can be used for continuous and semi-continuous operations. Due to the compact furnace and large heat content of the fluidized material, starting and stopping becomes very easy. A conceptual design is shown in figure 2.

For this system solid waste is to be feed after being shredded to suitable size. This shredded material is placed above the bed by using a feeding hoper as shown in figure 2. A second burner is placed above the material (as shown in figure 2) to start the incineration process. This burner is also known an as secondary burner. By increasing the velocity of the air that is passed through sand bed, it becomes fluidized. On increasing the air velocity more, the sand bed starts to expand and allow the excess air to pass through it in bubbles. This bubbling boils the mix of particles and air thoroughly by quickly forming a thermal equilibrium between the particles of waste and air.

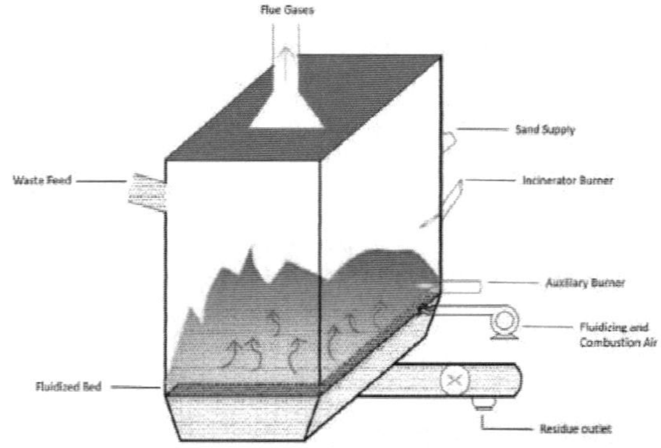

Figure 2. Proposed design of fluidized-bed incinerator for treating municipal solid waste

3.2. *Prospective green technology*

Aim of this system was to design a green technology which not only address the waste treatment issue but also become a significant green technology. As continuous supply of fuel is required for fluidized bed incinerator to work properly. Conventionally, fossil fuels (oil, gas, coal) is being used in this regard. In this system this fossil fuel is being replaced by hydrogen. Moreover, this hydrogen is further generated by water electrolysis process in which electrolyzer is being energized by mean PV panels. Due to this hydrogen generation this system is considering to be a standalone, which not only solve the dependency problem of incineration on fossil fuel but also generate fuel (hydrogen) itself which it required for the system. As hydrogen is considered to be green fuel so it is seen that greenhouse emission would be low from the conventional technique. Also the calorific value of hydrogen is more than the fossil fuel [15] as shown in table 1, so this proposed system as shown in figure 3 is considering to be more efficient. In this way this whole system is consider to be a step towards green sustainable technology.

Table 1. Calorific value of different types of fuel

Sr.	Type of fuel	Calorific Value (kJ/kg)
1.	Kerosene	43124
2.	Diesel	42600
3.	Natural Gas	38000 – 50000
4.	Charcoal	33000
5.	Hydrogen	150000

As shown in figure 3, an incinerator was designed for treating the waste by batch method. A maximum of 5 kg of shredded waste can be treated in one batch for an hour. For burning and heating of the incinerator, one electrolyzer was used which was connected to auxiliary and incinerator burners. Amount of hydrogen required for this system was also calculated in order to design electrolyzer.

Initial Hydrogen required for moisture 55% content [16]:

Moisture content by weight in 5kg waste	=	2.75 kg
Energy to evaporate 1 kg of water	=	600 kJ
Energy to evaporate 2.75 kg of water	=	1650 kJ
Hydrogen required for 2.75 kg water to evaporate	=	1.65/1200
	=	0.013 kg

1kg of Hydrogen	=	11.126 m3
0.013 kg of Hydrogen	=	0.0152 m3
Hydrogen in litters	=	150 litters

Designed electrolyzer have a capacity of 150 lph of hydrogen generation. To energize these electrolyzer one PV panels of 300 watts was used. As for Malaysia, it was considering 5.1 hr/day of sunlight. So for a stable system, a battery was introduced of 12Volt, 65 Ah. A control unit was use to charge, discharge the battery and responsible for running electrolyzer in efficient way.

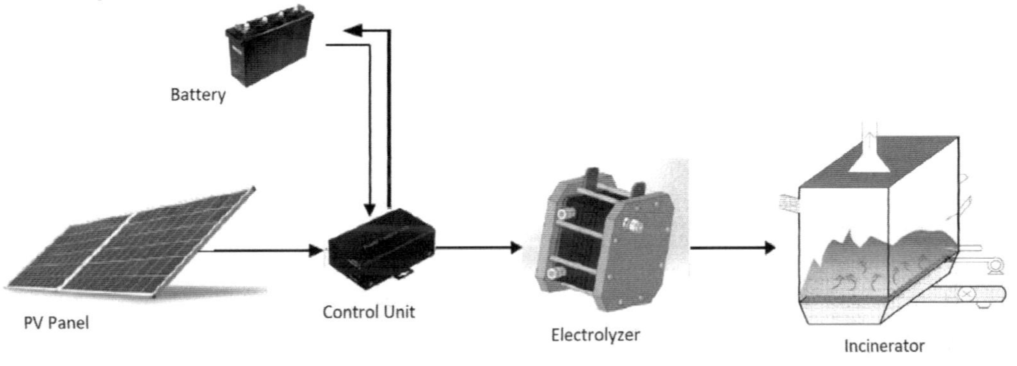

Figure 3. Proposed incineration system coupled with hydrogen electrolyzer

4. Results

A significant advantage was seen of the proposed system over conventional incineration was its ability to reduce undesirable gas emission which includes oxides of sulfur, nitrogen, carbon monoxide and other unburnt hydrocarbons. Formation of nitrogen oxides limits due to relatively low uniform temperature (800°C – 1000°C). Oxygen mixed with the hydrogen fuel provides a favourable environment for fluidized bed for complete combustion which minimize the production of carbon monoxide and emission of unburnt hydrocarbon. Moreover, incineration process has a potential to recover heat energy during the process. Following advantages were observed for this proposed system:

- Eliminate dependency on fossil fuel
- Have a potential of stand-a-alone system
- Effectively fluidizing the sand bed due to high calorific value of hydrogen
- Low green-house emission

5. Conclusion

A variety of waste can be treated by using this proposed system which includes municipal solid waste, agriculture waste, paper, organic waste etc. This system is theoretically favourable for the waste with high moisture contents and low calorific value. As these type of waste need a constant heating environment which hydrogen has a potential due to its high calorific value. Hydrogen showed promising results as an alternative for the fossil fuel which in future can be used for different applications. Integration of the incinerator with PV also paved path for incineration technology towards sustainability.

Acknowledgements
The authors thank the technical and financial assistance of UM Power Energy Dedicated Advanced Centre (UMPEDAC) and the Higher Institution Centre of Excellence (HICoE) Program Research Grant, UMPEDAC - 2018.

References

[1] Melikoglu M. Vision 2023: Assessing the feasibility of electricity and biogas production from municipal solid waste in Turkey. *Renewable and Sustainable Energy Reviews*. 2013;**19**:52-63.

[2] Kalyani KA, Pandey KK. Waste to energy status in India: A short review. *Renewable and sustainable energy reviews*. 2014;**31**:113-20.

[3] Cucchiella F, D'Adamo I, Gastaldi M. Sustainable waste management: Waste to energy plant as an alternative to landfill. *Energy Convers Manage*. 2017;**131**:18-31.

[4] OECD I. Energy and Air Pollution: World Energy Outlook Special Report 2016. 2016.

[5] Tester JW, Drake EM, Driscoll MJ, Golay MW, Peters WA. Sustainable energy: choosing among options: MIT press; 2012.

[6] Kumar S. Technology options for municipal solid waste-to-energy project. *TERI Information Monitor on Environmental Science*. 2000;**5**(1):1-11.

[7] Lu J-W, Zhang S, Hai J, Lei M. Status and perspectives of municipal solid waste incineration in China: A comparison with developed regions. *Waste Manage (Oxford)*. 2017;**69**:170-86.

[8] Tsui T-H, Wong JW. A critical review: emerging bioeconomy and waste-to-energy technologies for sustainable municipal solid waste management. *Waste Disposal & Sustainable Energy*. 2019;**1**(3):151-67.

[9] Sebastian RM, Kumar D, Alappat BJ. A technique to quantify incinerability of municipal solid waste. *Resources, Conservation and Recycling*. 2019;**140**:286-96.

[10] Wendt JO, editor Combustion science for incineration technology. Symposium (International) on Combustion; 1994: Elsevier.

[11] Rand T, Haukohl J, Marxen U. Municipal solid waste incineration: a decision maker's guide. The World Bank; 2000.

[12] Mininni G, Braguglia C, Marani D. Partitioning of Cr, Cu, Pb and Zn in sewage sludge incineration by rotary kiln and fluidized bed furnaces. *Water Sci Technol*. 2000;**41**(8):61-8.

[13] Baeyens J, Van Puyvelde F. Fluidized bed incineration of sewage sludge: a strategy for the design of the incinerator and the future for incinerator ash utilization. *J Hazard Mater*. 1994;**37**(1):179-90.

[14] Li S, Li Y, Lu Q, Zhu J, Yao Y, Bao S. Integrated drying and incineration of wet sewage sludge in combined bubbling and circulating fluidized bed units. *Waste Manage (Oxford)*. 2014;**34**(12):2561-6.

[15] Soares C. Gas turbines: a handbook of air, land and sea applications: Elsevier; 2011.

[16] Kathirvale S, Yunus MNM, Sopian K, Samsuddin AH. Energy potential from municipal solid waste in Malaysia. *Renewable energy*. 2004;**29**(4):559-67.

Photovoltaic performances and lifetime analysis of TiO2/rGO DSSCs sensitized with Roselle and N719 dyes

A Omar*, M S Fakir*, K S Hamdan*, N H Rased*, N A Rahim* †

*Higher Institution Centre of Excellence (HICoE), UM Power Energy Dedicated Advanced Centre (UMPEDAC), Level 4, Wisma R&D, University of Malaya, Jalan Pantai Baharu, 59990 Kuala Lumpur, Malaysia,

†Renewable Energy Research Group, King Abdulaziz University, Jeddah 21589, Saudi Arabia

E-mail: azimahomar@um.edu.my

Abstract. Two titanium dioxide/reduced graphene oxide (TiO_2/rGO) based dye-sensitized solar cells (DSSCs) were fabricated and separately sensitized with natural dye extracted from Roselle flowers and ruthenium-based dye of N719. The current-voltage (I-V) performances were measured using Visiontec Solar I-V tester at standard illumination of AM1.5 and irradiance level of 100 mW/cm^2. The lifetime of the TiO_2/rGO based DSSCs were furtherly investigated by measuring the I-V performances of the cells on the 5th day and 15th day. The highest power conversion efficiency (PCE) of 1.233 % was measured for the photoanode sensitized with N719 dye on the 15th day. The PCE of Roselle dye dropped from 0.992 % on the 5th day to 0.897 % on the 15th day. Dye degradation and low electron injection have influenced the unstable photovoltaic performances of Roselle dye even with the presence of electron conducting path provided by the rGO particles.

1. Introduction

Nowadays, the primary sources of fossil fuels generate greenhouse gases which mostly carbon dioxide. Further implication of this includes environmental pollution and global temperature warming [1]. On the other hand, dye-sensitized solar cells (DSSCs) are the ideal type of solar power source to generate electrical power. The fabrication of them involve environmentally friendly process, simple fabrication work, low cost and relatively high efficiency [2, 3]. The main component of DSSCs is the dye that is capable to produce electrons. However, the usages of ruthenium complex dyes are not suitable for environmentally friendly concept of DSSC. Although the ruthenium dyes help in delivering high power conversion efficiencies of the DSSCs, they are also reported as expensive compound, hazardous and highly volatile [1, 4]. Moreover, the demand to use natural dyes which are non-toxic and cost-effective dyes need to be considered. The sources are cheaper, easy to get in abundance and workable in the third-generation solar cells. On the other hand, many reports have been published on using the natural dyes extracted from leaves, fruits, flowers and roots in the fabrication work of DSSCs [2]. For example, Al-Alwani et al. (2018) extracted *Alternanthera dentata* leaves and *Musa acuminata* bracts as sensitizers and obtained power conversion efficiency (PCE) of 0.15 % and 0.31 %, respectively [5]. Li et al. (2013), employed red cabbage as a natural dye and achieved PCE of 2.908 % for 24 h immersion time [6].

Although the PCE generated by the natural dyes is less than the DSSCs sensitized with metal-based complex dyes, the usage of it is promising and still have a room to be improved. For example, the photoanode part can be modified by adding a graphene-based material such as reduced graphene oxide (rGO) to improve the electron transport in DSSCs [7]. Ghann et al. (2019) reported that graphene materials added into TiO_2 paste provides a porous photoanode film and therefore increase the DSSC's efficiency due to higher dye loading and increased photon absorption level [8]. Moreover, Ramamoorthy et al. (2017) claimed that rGO attached to the TiO_2 surface layer provides an electron transport channel which could prevent from recombination inside the cell [9].

In the present work, two TiO_2/rGO based DSSCs were fabricated and separately sensitized with natural dye of Roselle flowers and synthetic dye of N719. Their photovoltaic performances were analyzed as per their photovoltaic parameters, current-voltage (I-V) and power-voltage (P-V) curves. Thus, the photovoltaic properties of short-circuit current density, J_{sc}, open-circuit voltage, V_{oc}, fill-factor, *FF*, power conversion efficiency, *PCE* (or η) and maximum power, P_{max} of the DSSCs were furtherly measured on 5th day and 15th day to analyze the lifetime and stability of the cells.

2. Experimental Work

2.1. Materials

Titanium (IV) oxide nanopowder, Ethanol, acetone, poly(ethylene)glycol-block (PEG), Triton X-100 and rGO were purchased from Sigma-Aldrich Corporation. Fluorine-doped tin oxide (FTO) coated glasses (resistivity ~25Ω/sq) were purchased from Ossila (Sheffield, United Kingdom). N719 ruthenium powder, liquid electrolyte of MPN100 and platinum (Pt) paste were purchased from Solaronix SA (Aubonne, Switzerland). Roselle flowers were obtained from Malacca, Malaysia.

2.2. Preparation of Roselle Natural Dye

Roselle petals were washed with water and dried. 6 g of cleaned petals were crushed using mortar and pestle. The paste was mixed in the solution of 20 ml deionized (DI) water and 6 ml ethanol. The mixture was stirred on hotplate at 40 °C for 24 h. The red color solution was filtered through Whatman No. 43 paper to obtain an anthocyanin source. Then, the dye solution was wrapped with an aluminium foil to prevent from light and kept inside the refrigerator. Figure 1 shows the process to extract the Roselle dye.

Figure 1. Natural dye extraction from Roselle flowers.

2.3. *Preparation of N719 Ruthenium Dye*
17.83 mg of Ruthenizer powder was dissolved in 50 ml absolute ethanol to get 0.3 mM ethanolic solution of N719 dye. The mixture was dissolved in the glass bottle and wrapped with aluminium foil to prevent exposure from any light sources. The solution was then ultrasonicated for 30 min to fully dissolve the traces of Ru-powder. The solution was kept for 24 h in the dark cabinet before use.

2.4. *Preparation of TiO_2/rGO Photoanodes*
6 g of TiO_2 nanopowder was dissolved in deionized (DI) water followed by acetone and PEG. The solution was vigorously stirred on the hot plate at 60 °C for 24 h. Then, 0.1 ml of Triton X-100 were added to the mixture to improve the TiO_2 films adhesion on the glass substrate. The sonicated and dried of 2 mg rGO were added in the prepared TiO_2 paste. The paste was vigorously stirred for 24 h to obtain a gel-like solution. The obtained TiO_2/rGO paste were deposited on the conductive side of the cleaned FTO substrates by a doctor blade technique. The photoanodes area of 0.35 cm^2 were dried at room temperature for 5 min before annealed at 400 °C for 30 min in oven. After the photoanodes were cooled down to room temperature, they were separately sensitized in the Roselle and N719 dyes for 24 h. Longer immersion time is to ensure the dyes were fully adsorbed throughout the TiO_2/rGO mesoporous layer. After 24 h, any excess dye on the photoanodes were cleaned with ethanol.

2.5. *Fabrication of TiO_2/rGO based DSSC*
For construction of complete DSSCs, the counter electrodes were platinized with semi-transparent platinum (Pt) paste onto the cleaned FTO substrates. The substrates were then annealed at 400 °C for 1 h in the oven. The dye-covered TiO_2/rGO photoanode and Pt counter electrode were assembled into a sandwich structure cell with a parafilm spacer attached in between them. Two small paper clamps were used to clamp the cell. After injection of the liquid electrolyte at the small hole of the spacer, the cell was sealed with the parafilm spacer.

2.6. *I-V Measurement of TiO_2/rGO based DSSC*
The current-voltage (I-V) parameters of the TiO_2/rGO based DSSCs were measured by IV Tester System (VS-6820, IVT Solar) at AM1.5 condition and under 100 mW/cm^2 solar light generated from Xenon arc lamp (XES Series Class AAA solar simulator, XES-1600SE-200S). After the I-V measurements, the DSSC's power conversion efficiency were calculated by the following equations [10]:

$$FF = \frac{V_m J_m}{V_{oc} J_{sc}} \quad (1)$$

$$\eta = \frac{V_{oc} \cdot J_{sc} \cdot FF}{P_{in}} \times 100\% \quad (2)$$

where P_{in} is the radiation power incident on the cell, J_{sc} is short-circuit current density at zero voltage, V_{oc} is the open-circuit voltage at zero current density, J_m is the maximum current density, V_m is the maximum voltage and *FF* is the fill factor.

In order to study the lifetime of the cells, the fabricated TiO_2/rGO-DSSCs based on Roselle and N719 dyes were kept in a dark container and at a room temperature. The I-V performances were measured on the 5th day and 15th day. Thus, the photovoltaic parameters were recorded for further analysis and comparison.

3. Results and Discussion

3.1. *Photovoltaic Performances on the 1st Day*
Table 1 shows the photovoltaic performances of TiO_2/rGO photoanode sensitized with Roselle and N719 dyes. The performances on the 1st day were also illustrated by J-V and P-V curves as shown in

Figure 2. It is observed that TiO_2/rGO based DSSC of Roselle dye achieved PCE of 0.779 % higher than the photoanode sensitized with N719 dye (PCE = 0.338 %).

Moreover, the percentage difference in V_{oc} values between the two dyes are not significant which only 17.24 %. The other photovoltaic parameters that influence this increment include J_{sc} and FF. Higher J_{sc} is due to good electron transfer mechanism inside the cell. Wongcharee et al. (2007) reported that higher interaction between TiO_2 molecules and anthocyanin leads to a better electron mechanism due to shorter distance between dye skeleton of cyanidin and delphinidin and the juncture connected to the TiO_2 surface [11]. This factor is also supported by an excellent electrical conductivity of rGO that increase the amount of electron transfer from the anthocyanin molecules to the conduction band of TiO_2 [9, 12, 13]. Figure 3 shows chemical structures of cyanidin and delphinidin in the Roselle dye [11].

Table 1. Photovoltaic parameters of TiO_2/rGO based DSSC on the 1st day.

Dye	J_{sc} (mA/cm^2)	V_{oc} (V)	FF (%)	PCE (%)	P_{max} (mW/cm^2)
Roselle flower	1.226	0.794	80.0	0.779	0.779
N719	0.951	0.668	53.16	0.338	0.338

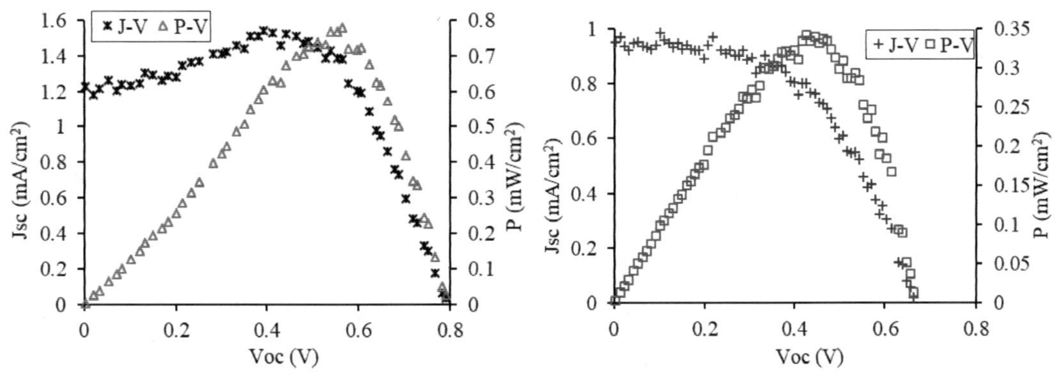

(a) (b)

Figure 2. J-V and P-V curves of TiO_2/rGO photoanodes sensitized with (a) Roselle and (b) N719 dyes on the 1st day.

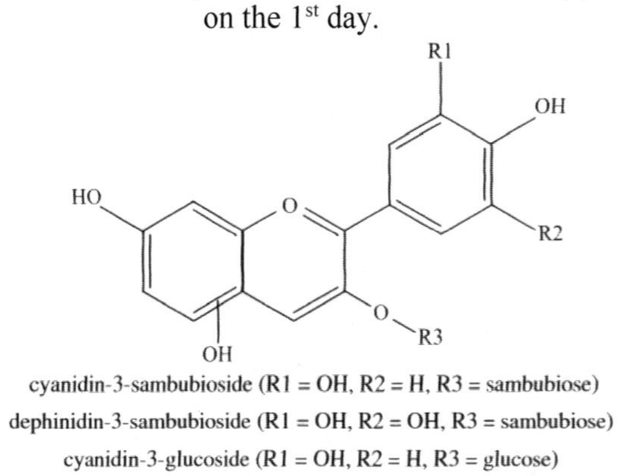

Figure 3. Chemical structures of cyanidin and delphinidin in the Roselle dye [11].

3.2. *Photovoltaic Performance on the 5th day*

In order to study the lifetime of the fabricated DSSCs, the I-V performance were measured again on the 5th day. Table 2 and Figure 4 show the photovoltaic performances and J-V and P-V curves of the TiO_2/rGO based DSSCs, respectively. It was observed that the efficiency of the photoanodes immersed in the Roselle and N719 dyes are slightly higher compared to the results obtained on the first day of testing these solar cells. The percentage increase of PCE values were measured around 27.34 % and 60.95 % for Roselle and N719 dyes, respectively.

Moreover, the maximum power, P_{max} generated by the Roselle dye is almost 1 mW/cm^2 compared to only 0.5 mW/cm^2 for the N719 dye sample. On the other hand, a small percentage difference of 24.33 % on the J_{sc} values for Roselle and N719 dyes were measured. Bell et al. (2011) reported that the incorporation of rGO in the TiO_2 paste extend the electron lifetime by a factor of four and generate the photogenerated current up to ten times [14]. These factors resulting in the increase of electron transfer rate and reduces electron recombination in the DSSC [12].

Table 2. Photovoltaic parameters of TiO_2/rGO based DSSC on the 5th day.

Dye	J_{sc} (mA/cm^2)	V_{oc} (V)	FF (%)	PCE (%)	P_{max} (mW/cm^2)
Roselle flower	1.383	0.837	85.71	0.992	0.992
N719	1.083	1.009	49.80	0.544	0.544

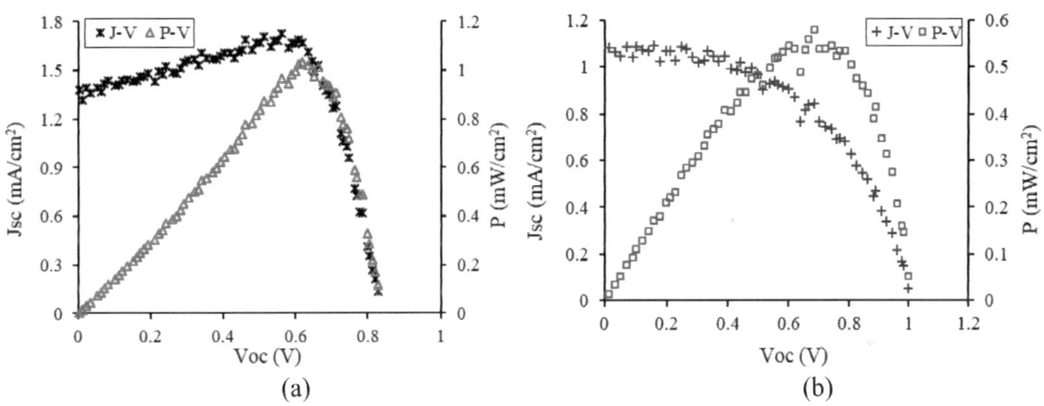

Figure 4. J-V and P-V curves of TiO_2/rGO photoanodes sensitized with (a) Roselle and (b) N719 dyes on the 5th day.

3.3. *Photovoltaic Performance on the 15th day*

The effect of incorporating of rGO in the TiO_2 paste were further measured on the 15th day of their photovoltaic performance. The photovoltaic parameters and J-V and P-V curves are presented in Table 3 and Figure 5, respectively.

Table 3. Photovoltaic parameters of TiO_2/rGO based DSSC on the 15th day.

Dye	J_{sc} (mA/cm^2)	V_{oc} (V)	FF (%)	PCE (%)	P_{max} (mW/cm^2)
Roselle flower	1.14	0.87	90.46	0.897	0.897
N719	1.283	1.25	76.90	1.233	1.233

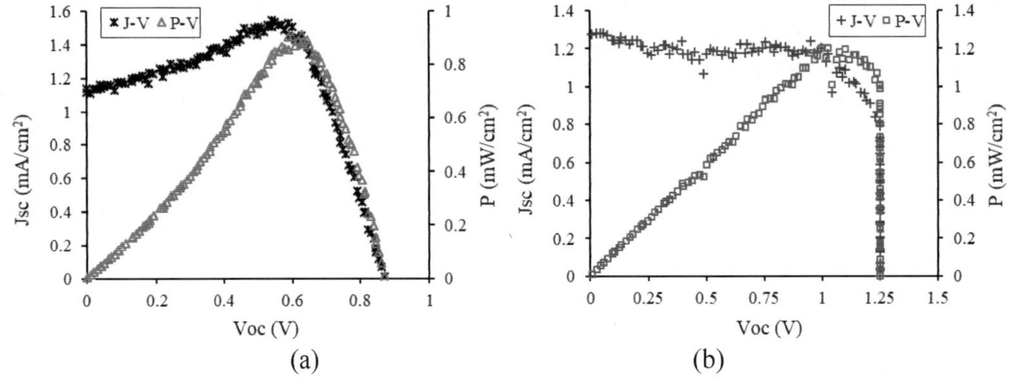

Figure 5. J-V and P-V curves of TiO_2/rGO photoanodes sensitized with (a) Roselle and (b) N719 dyes on the 15th day.

As time goes on, it was found that the J_{sc} value of Roselle dye is slightly decreases to 1.14 mA/cm^2 compared to the value measured on the 5th day (J_{sc} = 1.383 mA/cm^2). The decreased in J_{sc} value is due to high charge recombination event in the electrolyte at the TiO_2/dye interface [12]. Moreover, although the *FF* value increases to 90 %, the PCE value decreases to 0.897 % after 14 days.

The PCE percentage difference between the Roselle and N719 dyes were calculated around 31.55 % which is lesser than 58.33 % measured on the 5th day. Furthermore, the N719 dye sample shows tremendous PCE performance where the values increase from 0.544 % on the 5th day to 1.233 % on the 15th day. The percentage difference was calculated as large as 77.55 % between the values measured on the 15th and 5th day.

3.4. Photovoltaic Performances and Lifetime Comparison Between Roselle vs N719 Dyes

From the obtained photovoltaic performances in the previous sections, the results can be categorized for each of the photovoltaic parameters of J_{sc}, V_{oc}, *FF* and *PCE*. Figure 6 shows the lifetime comparison between Roselle and N719 dyes for each of the parameters. It is observed that the values of J_{sc}, V_{oc}, and *PCE* for N719 are gradually increase even after 2 weeks or 336 hours. In contrast, the *FF* value drops to 49.80 % on the 5th day before it increases back to 76.90 % on the 15th day. Higher fill factor signifies that the internal resistance inside the TiO_2/rGO photoanode sensitized with N719 is low. Ding et al. (2015) reported that good fill factor is referred to as having low total series resistance of the solar cell [15].

On the other hand, the photovoltaic parameters of TiO_2/rGO sensitized with Roselle dye were found as unstable. For example, the J_{sc} and efficiency values were found increase on the 5th day but decrease on the 15th day. In general, the J_{sc} parameter represents electron density in the conduction band (CB) of TiO_2 where it leads to higher electron population in the CB of FTO [9]. The increased of J_{sc} could also be due to the increased of dye molecules that still retained in the cells even after so many days. Both phenomena indicate large amount of electron injection to the CB of TiO_2 that directly influences the overall power conversion efficiency [9]. Cheng et al. (2013) reported that increment of J_{sc} is due to the enhancement of dye absorption amount in the photoanodes and high surface area of the photoanodes [16]. Nevertheless, the decreased of J_{sc} values on the 15th day are due to low amount of electrons flow through the external circuit of the DSSC device. Another possibility includes electron recombination event by the tri-iodide ions at the TiO_2/dye interface where it degrades the photovoltaic performance of the cell [17].

Moreover, the V_{oc} parameter for both dyes increase as the lifetime of the cells increased from day one to day fifteen although it is quite low for Roselle dye sample. V_{oc} is an important parameter where it represents a difference between the conduction band (CB) of TiO_2 and the redox potential of electrolyte [9]. Furthermore, the presence of rGO also facilitates the photogenerated electrons in the CB of TiO_2 to flow into the external circuit of DSSC where it provides as an electron conducting path that suppresses the recombination but increases the J_{sc} values.

From this analysis, Roselle dye seems not stable yet in their lifetime due to dye degradation and low electron injection after 336 hours. Iqbal et al. (2019) reported that natural dyes demonstrate lower power conversion efficiencies than the ruthenium-based dyes due to the dye degradation [18]. Moreover, Tekerek et al. (2011) mentioned that weak binding energy of natural dye with the TiO_2 film and low charge-transfer absorption in the visible light range also degrade their photovoltaic performance [19].

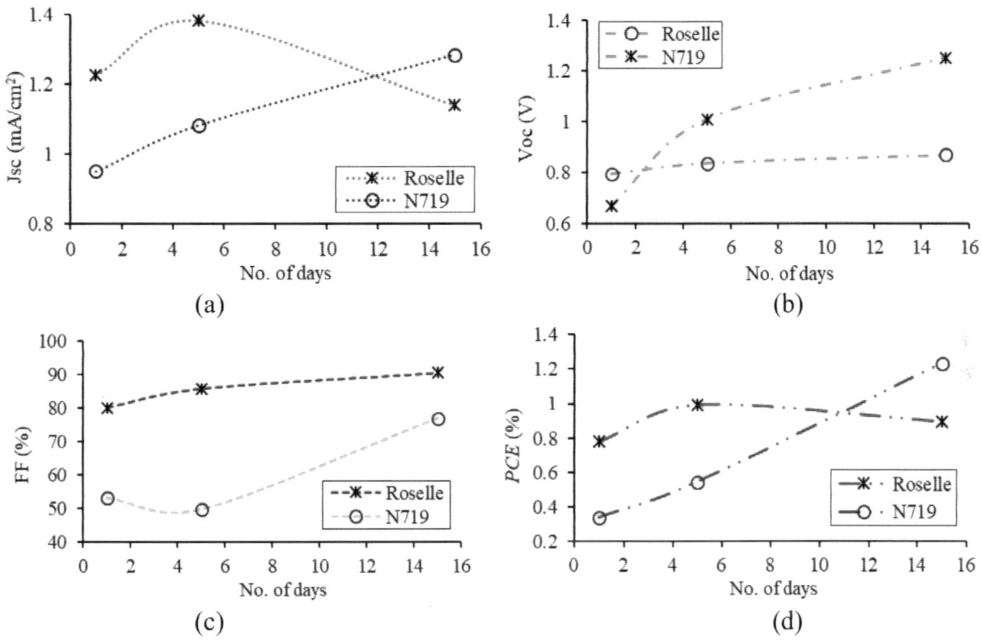

Figure 6. Each photovoltaic parameter of (a) J_{sc} (b) V_{oc} (c) FF and (d) η versus number of days between Roselle and N719 dyes.

4. Conclusion

The anthocyanin extracted from *Hibiscus sabdariffa L.* (Roselle flowers) planted in Malacca, Malaysia was used as the natural dye for TiO_2/rGO based DSSCs. The photovoltaic performances and lifetime analysis of the cells were compared with the TiO_2/rGO photoanodes sensitized with synthetic N719 dye. The highest PCE values was achieved by the DSSC with N719 dye at 1.233 % on the 15th day. The PCE difference between the Roselle and N719 dyes were measured around 31.55 %. Each photovoltaic parameter of J_{sc}, V_{oc}, FF and PCE were evaluated to analyze the cell's lifetime. After 2 weeks, the TiO_2/rGO DSSC sensitized with N719 dye showed improvement in the photovoltaic performances compared to the Roselle dye sample. The existence of rGO has retained the amount of dye adsorption and the photogenerated electrons flow in the TiO_2/rGO based DSSC. This research work requires further investigation on the structures, morphologies, optical, chemical and electrochemical impedance properties of TiO_2/rGO photoanodes.

Acknowledgements

The authors thank the technical and financial assistance of UM Power Energy Dedicated Advanced Centre (UMPEDAC) and the Higher Institution Centre of Excellence (HICoE) Program Research Grant, UMPEDAC - 2018 (MOHE HICoE - UMPEDAC), Ministry of Education Malaysia, RU007-2018 and RU012-2019, University of Malaya.

References

[1] Hemmatzadeh R, Mohammadi A. Improving optical absorptivity of natural dyes for fabrication

of efficient dye-sensitized solar cells. *Journal of theoretical and applied physics*. 2013;**7**(1):1-7.

[2] Hosseinnezhad M, Rouhani S, Gharanjig K. Extraction and application of natural pigments for fabrication of green dye-sensitized solar cells. *Opto-electronics review*. 2018;**26**(2):165-71.

[3] Al-Ghamdi AA, Gupta R, Kahol P, Wageh S, Al-Turki Y, El Shirbeeny W, et al. Improved solar efficiency by introducing graphene oxide in purple cabbage dye sensitized TiO2 based solar cell. *Solid State Commun*. 2014;**183**:56-9.

[4] Phinjaturus K, Maiaugree W, Suriharn B, Pimanpaeng S, Amornkitbamrung V, Swatsitang E. Dye-sensitized solar cells based on purple corn sensitizers. *Appl Surf Sci*. 2016;**380**:101-7.

[5] Al-Alwani MA, Ludin NA, Mohamad AB, Kadhum AAH, Mukhlus A. Application of dyes extracted from Alternanthera dentata leaves and Musa acuminata bracts as natural sensitizers for dye-sensitized solar cells. *Spectrochimica acta part A: molecular and biomolecular spectroscopy*. 2018;**192**:487-98.

[6] Li Y, Ku S-H, Chen S-M, Ali MA, AlHemaid FMA. Photoelectrochemistry for Red Cabbage Extract as Natural Dye to Develop a Dye-Sensitized Solar Cells. *International Journal of Electrochemical Science* 2013;**8**(1):1237 - 45.

[7] Casaluci S, Gemmi M, Pellegrini V, Di Carlo A, Bonaccorso F. Graphene-based large area dye-sensitized solar cell modules. *Nanoscale*. 2016;**8**(9):5368-78.

[8] Ghann WE, Kang H, Uddin J, Chowdhury FA, Khondaker SI, Moniruzzaman M, et al. Synthesis and characterization of reduced graphene oxide and their application in dye-sensitized solar cells. *ChemEngineering*. 2019;**3**(1):7.

[9] Ramamoorthy R, Karthika K, Dayana AM, Maheswari G, Eswaramoorthi V, Pavithra N, et al. Reduced graphene oxide embedded titanium dioxide nanocomposite as novel photoanode material in natural dye-sensitized solar cells. *Journal of Materials Science: Materials in Electronics*. 2017;**28**(18):13678-89.

[10] Ammar AM, Mohamed HS, Yousef MM, Abdel-Hafez GM, Hassanien AS, Khalil AS. Dye-sensitized solar cells (DSSCs) based on extracted natural dyes. *Journal of Nanomaterials*. 2019;**2019**.

[11] Wongcharee K, Meeyoo V, Chavadej S. Dye-sensitized solar cell using natural dyes extracted from rosella and blue pea flowers. *Sol Energy Mater Sol Cells*. 2007;**91**(7):566-71.

[12] Jiao L, Fanourakis S, editors. Effects of thermally reduced graphene oxide in the photoanode on the properties of dye sensitized solar cells. IOP Conference Series: Earth and Environmental Science; 2018: IOP Publishing.

[13] San Esteban ACM, Enriquez EP. Graphene–anthocyanin mixture as photosensitizer for dye-sensitized solar cell. *Solar Energy*. 2013;**98**:392-9.

[14] Bell NJ, Ng YH, Du A, Coster H, Smith SC, Amal R. Understanding the enhancement in photoelectrochemical properties of photocatalytically prepared TiO2-reduced graphene oxide composite. *The Journal of Physical Chemistry C*. 2011;**115**(13):6004-9.

[15] Ding H, Zhang S, Chen J-T, Hu X-P, Du Z-F, Qiu Y-X, et al. Reduction of graphene oxide at room temperature with vitamin C for RGO–TiO2 photoanodes in dye-sensitized solar cell. *Thin Solid Films*. 2015;**584**:29-36.

[16] Cheng G, Akhtar MS, Yang O-B, Stadler FJ. Novel preparation of anatase TiO2@ reduced graphene oxide hybrids for high-performance dye-sensitized solar cells. *ACS applied materials & interfaces*. 2013;**5**(14):6635-42.

[17] Liu Q, Wang J. Dye-sensitized solar cells based on surficial TiO2 modification. *Solar Energy*. 2019;**184**:454-65.

[18] Iqbal MZ, Ali SR, Khan S. Progress in dye sensitized solar cell by incorporating natural photosensitizers. *Solar Energy*. 2019;**181**:490-509.

[19] Tekerek S, Kudret A, Alver Ü. Dye-sensitized solar cells fabricated with black raspberry, black carrot and rosella juice. *Indian Journal of Physics*. 2011;**85**(10):1469-76.

P(HEMA-co-EA) as host polymer for flexible dye-sensitized solar cell (DSSC) electrolyte

N H Rased*, S R S Raihan*, N A Rahim*

*Higher Institution Centre of Excellence (HICoE) UM Power Energy Dedicated Advanced Centre (UMPEDAC), Level 4 of, Wisma R&D University Malaya (UM), Federal Territory of Kuala Lumpur, 59100 Kuala Lumpur.

E-mail: nhidayahrased@um.edu.my

Abstract. An electrolyte is the main component of a dye-sensitized solar cell (DSSC) which influences the efficiency of the DSSC. In order to prevent leakage problem and adhesive problem, new solid-state polymer electrolyte proposed rather than using a liquid electrolyte. For this research, new copolymers using 2-hydroxyethyl methacrylate (HEMA) and Ethyl Acrylate (EA) monomers are randomly copolymerized via UV-cure polymerization method with different ratio of each monomer (10,30,50,70,90) towards development as a solid-state electrolyte in flexible DSSC. In this study, the best ratio of the copolymer HEMA-co-EA will be the host polymer for electrolyte in DSSC. The most promising characteristics as a host in polymer electrolytes are due to its smooth cross-sectional surface and lowest glass transition temperature. Therefore, a best ratio of p(HEMA-co-EA) will be stirred with sodium iodide (NaI), Tetrahydro folic acid (THF) and iodine crystal for 24 hours to form a homogenous solution of an electrolyte. This p(HEMA-co-EA) will incorporate with different weight ratios of sodium iodide (NaI). This research reported that after 1500 s only three ratios of new host polymer p(HEMA-co-EA) were successfully copolymerized completely. All three ratio will be characterized by physical appearance, FTIR, DSC and XRD. By naked eyes, only 3 ratios of monomer (50HEMA:50EA, 70HEMA:30EA and 90HEMA:10EA) gave brittle structure which proved the copolymerization process completely success. For FTIR, those three ratios show the breakdown of a double bond at the HEMA monomer structure proved the complete copolymerization process. DSC shows that all three ratio shows the glass transition temperature (Tg) and only 50HEMA:50EA gave melting temperature (Tm) at 192.49°C and XRD confirmed the phase structure and crystallinity of three ratios. All these characterizations show that HEMA and EA monomers can be successfully random copolymerize with a three suitable ratio of each monomer and been used as an electrolyte for flexible DSSC.

1. The first section in your paper

As we know, Silicon is non-renewable energy source and will be finished one day. Due to this reason, the fabrication of Silicon solar cell/panel become more expensive nowadays and alternating way to replace silicone sources been studies by many researchers [1]. For the reasons, Professor M. Grätzel in 1991 inverted low manufacturing cost and high efficiency device to capture sunlight and converts the light to energy [2]. The device is known as dye-sensitized solar cell (DSSC) [2]. DSSC is belong to third generation of solar cells and become attracted due to their simple device fabrication processes in ambient conditions.

A typical DSSC consists of photoelectrode and counter electrode at the back contact of the photovoltaic cell as the catalyst surface deposited onto the thin layer of TCO substrate. There are a few types of transparent conductive oxide glass substrate that are commonly used in DSSC such as indium-doped tin oxide (ITO), fluorine doped tin oxide (FTO), or aluminium-doped zinc oxide (AZO) [3]. So much attention has recently been focused on the fabrication of flexible DSSC, where plastic is used as the substrate ([4],[5],[1]). So far, liquid electrolyte for DSSC electrolyte gave highest 12.3% energy conversion efficiency [6]. However, liquid electrolyte had a few problems such as dye degradation, leakage problem and creates electrode corrosion which limits the commercialization potential of DSSC as well as flexible DSSC. By concerning these problems, new quasi- solid-state polymer electrolyte from acrylic-based are developed to give better adhesive properties and increased the conversion efficiency due to the unique retention and absorption properties hydrophobic hydrogel materials [7]. In order to synthesis quasi-solid electrolyte for DSSC, the modification of host polymer is need. UV-cure preparation method is one of very fast curing process for acrylic-based polymer that lead to lower-energy consumption and high productivity [7]. For this research, the promising ratio of p(HEMA-co-EA) host polymer been discussed.

Although acrylic-based monomer had been used mainly in biomedical application[8] such in vivo use[9] and biomedical industry such as intraocular lenses, corneal prosthesis and contact lenses in ophthalmology [10] etc., these acrylic-based monomers also gained attention for solar cell application [11- 15] from other researchers. Example of acrylic-based monomer that had potential in solar cell application especially as host polymer of DSSC's electrolyte are hydroxyethyl methacrylate (HEMA) [16-20], poly(methyl methacrylate (PMMA) [13, 20-25], hydroxyethyl acrylate(HEA) [26], Ethyl acrylate (EA) [25], butyl acrylate (BA) [13], Glycidyl methacrylate(GMA) [27], poly-acrylic acid (PAA) [13, 28] poly(glycidyl acrylate) (PGA) [29], Methyl acrylate (PMA) [30, 31] and many more. Acrylic monomers had proposed in this research since its stability against UV radiation. Acrylate, methacrylate, acrylonitrile and many more are examples of acrylic monomers types. Among all these types of acrylic monomers, methacrylate monomer is a versatile monomer group. Methacrylate monomers mostly used to produce copolymers because of its very reactive double bond and easily to polymerize with other type of acrylic monomers [32].To the best of our knowledge, there is no report on poly(2-hydroxyethyl methacrylate-co-ethyl acrylate p(HEMA-co-EA) since the morphology image (applied for biomedical application) reported by Diego, Olmedilla [33] shows porosity structure which expected give larger surface area and increase the adhesive properties for electron transfer if been use as electrolyte for flexible DSSC.

2. Methodology

2.1. Materials

2-hydroxyethyl methacrylate (HEMA) (brand Aldrich, clear liquid form) density 1073 g/mL at 25 degree-lit. Ethyl Acrylate (EA) density 0.918g/cm3 at 25 degree liquid colourless. 2, 2-dimethoxy-2-phenylacetophenone (DMPA) (99%) with density 1.1 ± 0.1 g/cm were purchased from Aldrich. Tetrahydrofuran (THF) was obtained by J.T. Baker. All materials used without purification.

2.2. Preparation of host polymer p(HEMA-co-EA) with different ratio of each monomer

(x) wt.% of HEMA and (100-x) wt.% EA monomer were mixed in five different ratios (where x+ 10,30,50,70 and 90) in beaker. Then, photo-initiator (0.016g DMPA) added into the 1g copolymer (HEMA-co-EA) [32, 34, 35]. The difference of copolymerization ratio of the two monomers cannot be too small because there is no significant effect to the polymer chain. The mixture was stirred at 40°C with 80 rpm until the solution became clear for a few seconds. The clear solution poured onto Teflon mould and exposed to UV light with the presence of nitrogen gas about 1500 s. Figure 1 shows the chemical structure for copolymer p(HEMA-co- EA).

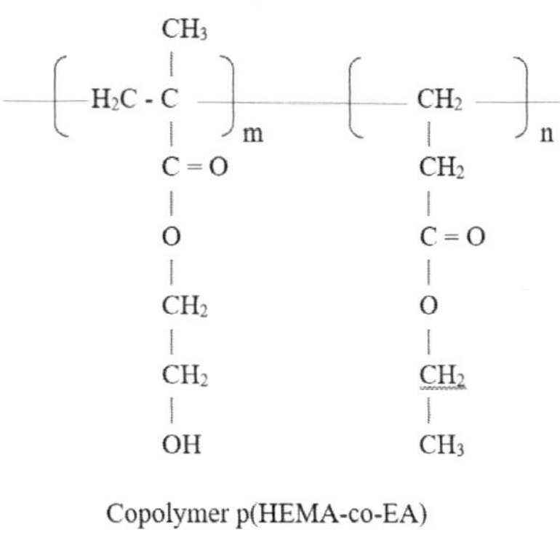

Figure 1. Chemical structure of p(HEMA-co-EA)

2.3. Characterization of host polymer p(HEMA-co-EA)

In order to identify the most potential copolymer ratio to be used as host polymer electrolytes, Fourier-transform infrared spectroscopy (FTIR), Differential Scanning Calorimetry (DSC) and X-ray Diffraction (XRD) were carried out.

2.3.1. Fourier-transform infrared spectroscopy (FTIR) Perkin Elmer Frontier FTIR spectrometer with attenuated Total Reflectance (ATR) accessory was used to record FTIR spectra in 4000-450 cm^{-1} range for the sample. ATR suitable used for solid sample.

2.3.2. Thermal analysis. Thermal measurement in this research using differential scanning calorimetry (DSC Q20, TA Instrument) at gas flow rate 30ml/min from 10°C to 230°C. DSC measures the heat flow, temperature and phase transitions in materials as plotted graph of time and temperature. Besides, DSC used to characterize thermal transitions of materials and provide useful informations about melting point, glass transition temperature, crystallinity, heat capacity and thermal stability [36].

2.3.3. X-ray Diffraction (XRD) XRD is a versatile, non-destructive technique for identification of various crystalline forms (phase) of compound samples. Modern computer-controlled diffractometer system uses automatic routines to measure, record and interpret the unique diffractogram produced by individual constituents in even highly complex mixture. XRD spectra (brand Malvern Panalytical, model Empyrean) data was recorded at wavelength 5-100° (2Θ).

3. Results and discussions

3.1. Physical

As shown in Figure 2, the colour of the different ratio changed after exposed under UV-light in 1500s. The 90HEMA:10EA sample shows the faster brittle structure with the most yellow in colour. However, the 10HEMA:90EA and 30HEMA:70EA samples were in gel state, white in colour and became agglomerated even after more than 1500 s. This is due to the incomplete copolymerization process even after 1500 s. Differ with 50HEMA:50EA, the copolymer became stronger and can be broke out after 1500 s and seen in white colour. By increasing the HEMA ratio, the copolymer become more darker yellow and more brittle after 1500 s. The physical characterization was done on the copolymer since the copolymer containing EA more 50 wt.% are poor in mechanical properties become gel-like (Figure 2, (a) and (b)) below, hence they are not suitable use for preparation of solid polymer electrolyte.

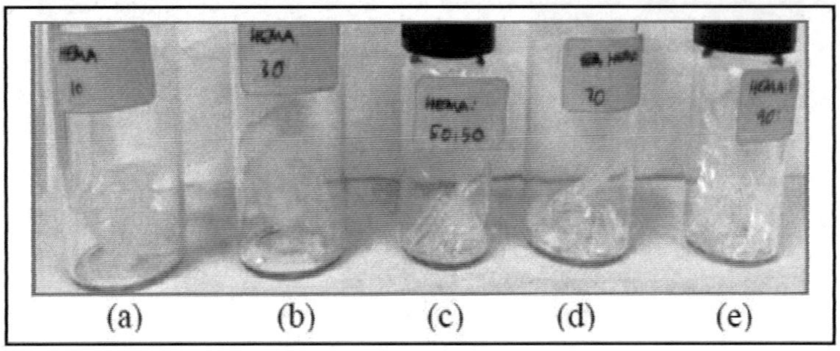

Figure 2. Photo of (a)10HEMA:90EA, (b)30HEMA:70EA, (c)50HEMA:50EA, (d)70HEMA:30EA, (e)90HEMA:10EA

3.2. FTIR

Figure 3 shows full FTIR spectra of HEMA, EA monomers, and HEMA-co-EA copolymers with different ratios.

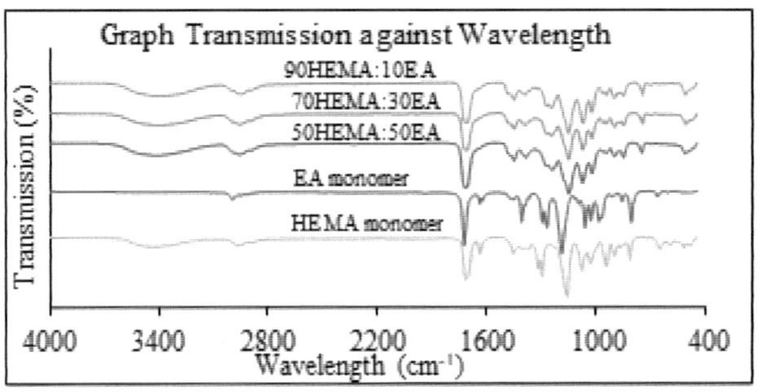

Figure 3. Full FTIR spectra for all samples.

As shown in Figure 4, the intensity peak at wave number 3400 cm^{-1} become decrease as the ratio of HEMA increase. The FTIR spectra of HEMA monomer shows strong broad peak absorption at wavelength ~3400 cm^{-1} but no absorption peak shown for EA monomer. This is due to the present of OH stretching in HEMA structure. For pure HEMA, the broad peak (-OH) present at 3422 cm^{-1} and shifted to lower wavenumber after copolymer with EA monomer. Then, it shows that OH (alcohols phenols) group remain the properties of HEMA even after copolymer p(HEMA-co-EA) formed. C- H stretching vibration of ethelene ($O\text{-}C_2H_5$) and methylene ($C\text{-}CH_3$) group show peaks at 2800 to 3000 cm^{-1} peaks region. Similar observation reported by Venkatesan, Liu [38].

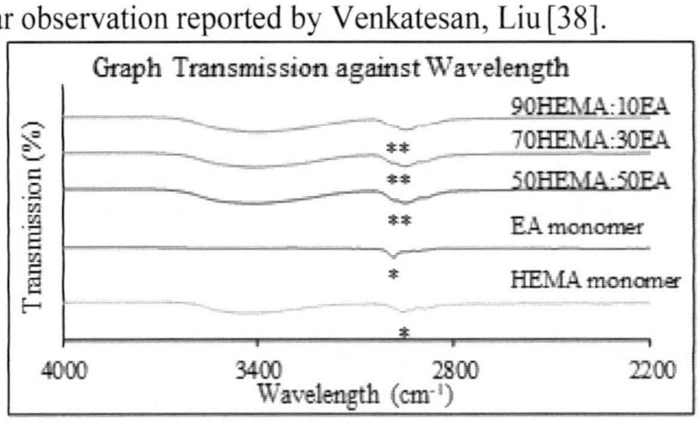

Figure 4. FTIR spectra for wavelength 4000 to 2200 cm^{-1}.

The double bond for HEMA and EA monomer presented in Figure 5 can be seen at wavelength ~1638 cm^{-1}. After copolymerized both monomers with different ratio, the peak at 1638 cm^{-1} disappeared for all 3 sample (50HEMA:50EA), (70HEMA:30EA) and (90HEMA:10EA) which that free- radical reaction under UV light was well performed since the C=C stretching vibration peak with the intensity of absorption weak to medium around 1,600~1,680 cm^{-1} of all monomers demolished after successfully copolymerized [13, 32, 37]. Carbonyl group (C=O) and ether group (C-O-C) gave strong absorption peaks at 1720 cm^{-1} and 1158 cm^{-1} respectively in Figure 5 and Figure 6 respectively. The peaks are shifted to the lower wavenumbers due to the interaction between monomer HEMA and EA [32].

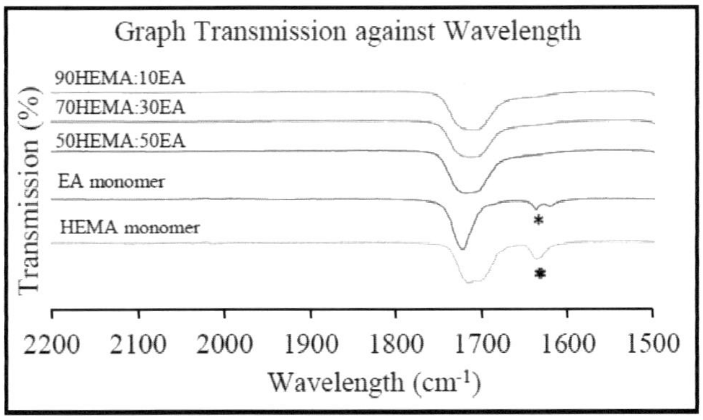

Figure 5. FTIR spectra for wavelength 2200 to 1500 cm^{-1}.

Peaks located at 1298 cm^{-1}, 1640 cm^{-1} and 1320 cm^{-1} based to vinyl group =CH deformation/wagging [39]. Medium C-H bend (alkane) at peak ~1370 to 1453 cm^{-1}. Then the strong absorption peak at 1162,1186,1141,1141 and 1150 cm^{-1} for HEMA, EA, 50HEMA:50EA and both 70HEMA:30EA, 90HEMA:10EA respectively indicates C-O stretching. Significantly, as shown in Figure 6, there is medium peak at 750 cm^{-1} for three different ratio of copolymer p(HEMA-co- EA) but none peak presented for EA and HEMA monomer. This is due to the present of C-H bending group after the dissappears of C=C (double) because of the copolymerization process between two monomers. Peak spectra show the absorption peak at 664 cm^{-1} for EA monomer and 653 cm^{-1} for HEMA monomer. Since the ratio of monomer HEMA is more than ratio monomer EA in p(HEMA-co-EA), the FTIR spectra more likely follow the FTIR of the HEMA. Moreover, C-H bond present due to the break bond at $C-CH_2$ group showed by demolished of 939cm^{-1} peaks proves the successful copolymerization process [32].

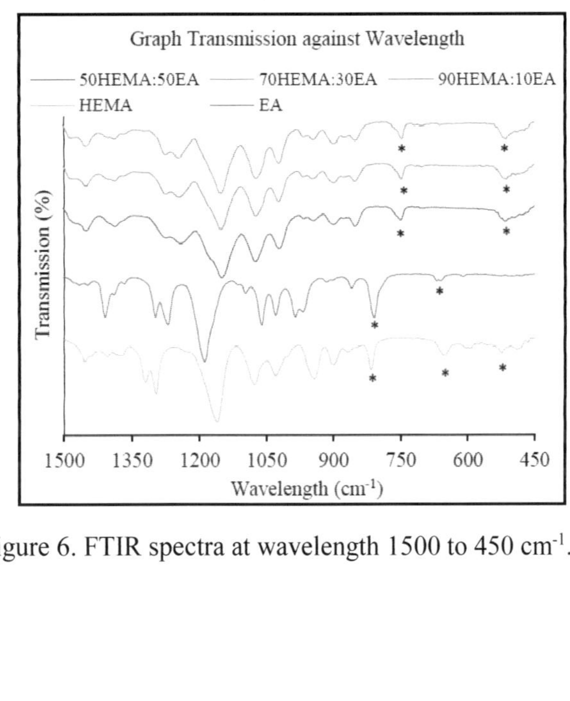

Figure 6. FTIR spectra at wavelength 1500 to 450 cm^{-1}.

3.3. Thermal analysis (DSC)

DSC curves of copolymer HEMA-co-EA with different ratio were shown in Figure 7. Tg is defined as the temperature at which the polymer experiences the transition from rubbery to rigid states [40]. The free volume created by continuous segmental motion of polymer chain will increase the ionic conductivity of polymer. The amorphous properties help the migration of ion and facilitates the movement of ions [41]. Among three sample, 50HEMA:50EA shows the lowest Tg value which is 45.8 °C. While Tg value for 70HEMA:30EA is 53.9 °C and 90HEMA:10EA gives 57.8 °C value of Tg. The value of Tg were calculated as midpoint at gradient as marked (•) in Figure 7. Nature of amorphous polymer gave lower Tg value. Low Tg caused the polymer chains to produce faster segmental motion and bond rotations. So that increased the ionic mobility in polymer electrolyte. A more disordered arrangement of polymer matrix gave more amorphous region of polymer matrix. The disorder will influence the flexibility of polymer backbone then increases the mobility of charge carriers [37]. With the increasing HEMA monomer ration in p(HEMA-co-EA), the Tg value of copolymer become increased. The nature of amorphous polymer will be confirmed by XRD. Higher melting point and higher heat of fusion indicates higher crystallinity. The melting point only exist for 50HEMA:50EA at temperature 192.49 °C. This shows that 50HEMA:50EA have semi-crystalline structure that effected when the presence of heat flow since it has lowest Tg value and the only shows the melting temperature point marked as X in Figure 7.

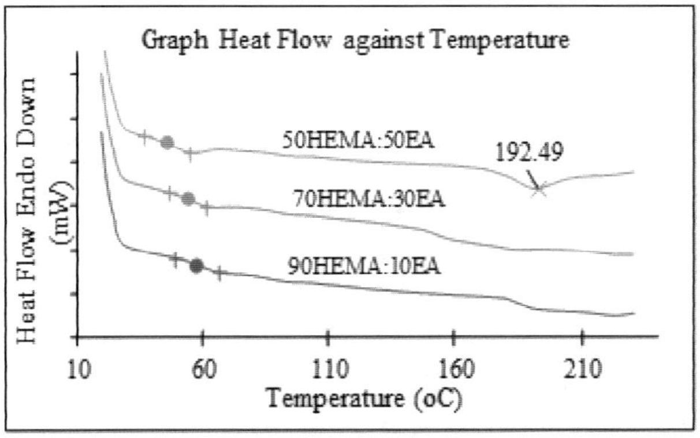

Figure 7. DSC curves for p(HEMA-co-EA) by different ratio of each monomer.

3.4. XRD

The XRD patterns of p(HEMA-co-EA) were recorded in the angle range of 2θ (5° < 2θ < 100°) and shown in Figure 8. For amorphous polymer, the XRD peaks will be broad and polymer with crystalline region will have sharp XRD peaks [42]. No peak observed at 50°–80° part of the spectra for any of the samples. Only two significant broad peaks (around 2Θ: 7° and around 18-20°) were observed in the all three samples which is a strong indication that the samples are amorphous. Shanti, Bella [13] reported that the synthesized copolymer is a amorphous type since the broad hump shown at short range order. If the crystallites of the copolymer are very small, the peaks of the XRD pattern will broaden, resulting larger FWHM. However, all samples gave small FWHM which is less than 1° (2Θ). Results of 90HEMA:10EA and 50HEMA:50EA showed semi-crystalline morphology since both ratio shows two sharp peaks with small FWHM as listed in Table 1.

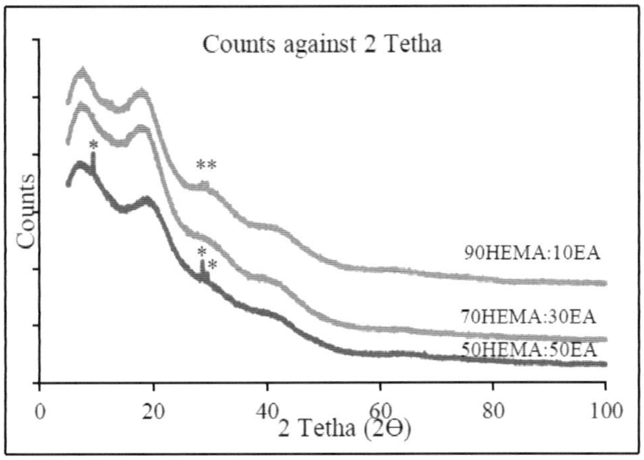

Figure 8. XRD spectra

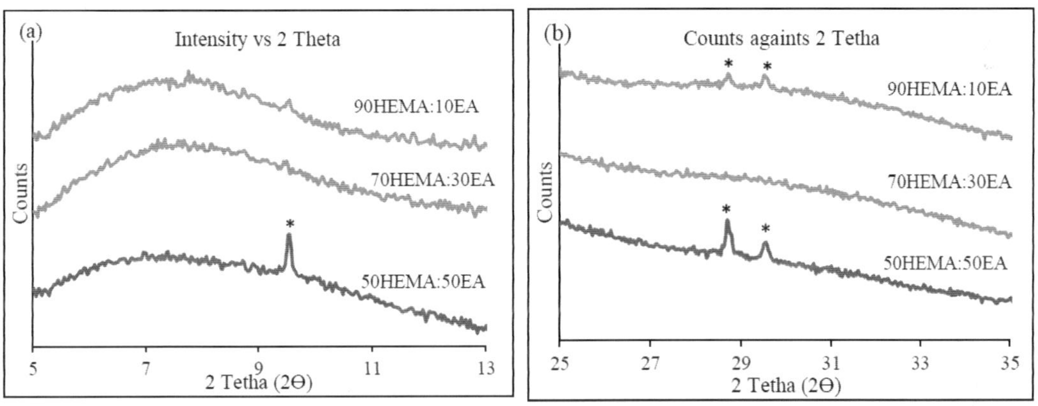

Figure 9. XRD spectra for two specific range 2Θ (a) Range 5° to 13° (b) Range 25° to 35°.

Using Debye-Scherrer equation (2), the average crystalline size of p(HEMA-co-EA) can be calculated:

$$D= (k \lambda)/ (\beta \cos \theta) \quad (1)$$

where λ is the X-ray wavelength (1.5418 Å), K is the Scherrer constant with value from 0.9 to 1 (shape factor), β is the width of the XRD peak at half height and θ is the Bragg angle. Using equation (2) and considering the dominant peaks, the average crystalline size was calculated and tabulated in Table 1. FWHM data was recorded from XRD (brand Malvern Panalytical, model Empyrean) report. Two small peaks were observed at 28.7° (2Θ) and 29.5° (2Θ) which suggest that both 50HEMA:50EA and 90HEMA:10EA start to form crystalline structure [38]. The Tg values point up the weakening of the intra and intermolecular forces between chains as a result of copolymerization process [43]. For this research, the Tg value increase as the ration of HEMA increased in p(HEMA-co-EA). This explained the inter and intermolecular chain of the copolymer that synthesized. The crystalline size of p(HEMA-co-EA) at 42° peak position was not been calculate since the change in the peak intensity was insignificant. So, the 50HEMA:50EA and 90HEMA:10EA can be grouped in semi crystalline structure with the presence of broad peak and small intense peak.

Table 1. Calculated crystalline size for significant small FWHM peaks.

Sample	Peak position (2Θ)	FWHM (2Θ)	Lamda, λ (nm)	Crystalline size, D (nm)
50HEMA: 50EA	9.5400	0.0768	0.15418	108.46
	28.7055	0.1023	0.15418	83.75
	29.5252	0.1279	0.15418	67.12
90HEMA: 10EA	9.5336	0.1535	0.15418	54.26
	28.7139	0.1535	0.15418	55.82
	29.5181	0.1023	0.15418	83.96

4. Conclusion

HEMA and EA monomers were randomly copolymerized via UV-cure method with five different ratios. However, only three ratios of monomers were successful completely copolymerize randomly under UV-cure method with the present of photo-initiator for 1500 s which are 90HEMA:10EA, 70HEMA:30EA and 50HEMA:50EA. These three ratios were continued for other characterizations method. FTIR spectra shows the disappearance of main peak indicate the break of C=C (double bond) in each monomer confirmed the successful copolymerization of both monomer by UV-cure method. The observation by XRD indicated that 90HEMA:10EA and 50HEMA:50EA were the semi- crystalline copolymer synthesized in this study and DSC confirmed the 50HEMA:50EA has crystalline structure due to the presented melting temperature value. Thus, from this research, both HEMA and EA monomers successfully can be copolymerized by easy UV-cure method and future to be used as host polymer in flexible DSSC's electrolyte rather than been used in biomedical application.

Acknowledgements

The authors would like to thank HICoE Ministry of Education and SATU Joint Research Grant (ST008-2019) for financial support.

References

[1] Iftikhar, H., et al., Progress on Electrolytes Development in Dye-Sensitized Solar Cells. Materials, 2019. 12(12): p. 1998.
[2] O'Regan, B. and M. Grätzel, A low-cost, high-efficiency solar cell based on dye-sensitized colloidal TiO2 films. Nature, 1991. 353: p. 737.
[3] Su'ait, M.S., M.Y.A. Rahman, and A. Ahmad, Review on polymer electrolyte in dye-sensitized solar cells (DSSCs). Solar Energy, 2015. 115: p. 452-470.
[4] Weerasinghe, H.C., et al., Fabrication of efficient solar cells on plastic substrates using binder-free ball milled titania slurries. Journal of Photochemistry and Photobiology a-Chemistry, 2009. 206(1): p. 64-70.
[5] Bittner, F., T. Oekermann, and M. Wark, Scale-Up of the Electrodeposition of ZnO/Eosin Y Hybrid Thin Films for the Fabrication of Flexible Dye-Sensitized Solar Cell Modules. Materials, 2018. 11(2): p. 232.
[6] Yella, A., et al., Porphyrin-sensitized solar cells with cobalt (II/III)-based redox electrolyte exceed 12 percent efficiency (vol 334, pg 629, 2011). Science, 2011. 334(6060): p. 1203-1203.
[7] Asif, A. and W.F. Shi, UV curable waterborne polyurethane acrylate dispersions based on hyperbranched aliphatic polyester: effect of molecular structure on physical and thermal properties. Polymers for Advanced Technologies, 2004. 15(11): p. 669-675.
[8] Montheard, J.P., M. Chatzopoulos, and D. Chappard, 2- Hydroxyethyl Methacrylate (Hema) - Chemical- Properties and Applications in Biomedical Fields. Journal of Macromolecular Science-Reviews in Macromolecular Chemistry and Physics, 1992. C32(1): p. 1-34.
[9] Kemal, E., K.O. Adesanya, and S. Deb, Phosphate based 2-hydroxyethyl methacrylate hydrogels for biomedical applications. Journal of Materials Chemistry, 2011. 21(7): p. 2237-2245.
[10] Chirila T, H.D., Biomaterials and Regenerative Medicine in Ophthalmology. 2nd ed ed. 2016,

UK: Elsevier.
[11] Tang, Z.Y., et al., Preparation of poly(acrylic acid)/gelatin/polyaniline gel-electrolyte and its application in quasi-solid-state dye-sensitized solar cells. Journal of Power Sources, 2012. 203: p. 282-287.
[12] Li, P.J., et al., The application of P(MMA-co- MAA)/PEG polyblend gel electrolyte in quasi-solid state dye-sensitized solar cell at higher temperature. Electrochimica Acta, 2007. 53(2): p. 903-908.
[13] Shanti, R., et al., Poly(methyl methacrylate-co-butyl acrylate-co-acrylicacid): Physico-chemical characterization and targeted dye sensitized solar cell application. Materials & Design, 2016. 108: p. 560-569.
[14] Su'ait, M.S., A. Ahmad, and M.Y.A. Rahman, Ionic conductivity studies of 49% poly(methyl methacrylate)- grafted natural rubber-based solid polymer electrolytes. Ionics, 2009. 15(4): p. 497-500.
[15] Imperiyka, M., et al., Preparation and Characterization of Polymer Electrolyte of Glycidyl Methacrylate-Methyl Methacrylate-LiClO4 Plasticized with Ethylene Carbonate. International Journal of Polymer Science, 2014.
[16] Xia, K., et al., High efficiency quasi-solid state dye- sensitized solar cells based on a novel mixed-plasticizer modified polymer electrolyte. Electrochimica Acta, 2015. 153: p. 28-32.
[17] Li, Q.H., et al., Quasi-solid-state dye-sensitized solar cells from hydrophobic poly(hydroxyethyl methacrylate/glycerin)/polyaniline gel electrolyte. Materials Chemistry and Physics, 2014. 144(3): p. 287-292.
[18] Dai, Y.H., et al., In Situ Quaternized Gel Electrolyte For Quasi-Solid-State Dye-Sensitized Solar Cell. Advanced Materials, Pts 1-3, 2012. 415-417: p. 1586-1589.
[19] Winther-Jensen, O., et al., In situ Photopolymerization of a Gel Ionic Liquid Electrolyte in the Presence of Iodine and Its Use in Dye Sensitized Solar Cells. Macromolecular Rapid Communications, 2010. 31(5): p.479-483.
[20] [20] Dai, Y.H., et al., Influence of Polymer Gel Electrolyte on the Performance of Dye-Sensitized Solar Cells Analyzed by Electrochemical Impedance Spectroscopy. Acta Physico-Chimica Sinica, 2012. 28(11): p. 2669-2675.
[21] Xu, T.T., et al., Environmental effects on the ionic conductivity of poly(methyl methacrylate) (PMMA)- based quasi-solid-state electrolyte. Ionics, 2018. 24(9):p. 2621-2629.
[22] Ganesan, S., et al., Low-cost tetra ethylene glycol derivatives in polymer blend electrolytes for dye- sensitized solar cells with high photovoltaic conversion efficiencies. Materials Science and Engineering B- Advanced Functional Solid-State Materials, 2018. 229:p. 37-43.
[23] Sundararajan, V., et al., Exploring the effect of novel N - butyl-6-methylquinolinium bis(trifluoromethylsulfonyl)imide ionic liquid addition topoly(methyl methacrylate-co-methacrylic) acid electrolyte system as employed in gel-state dye sensitized solar cells. Electrochimica Acta, 2017. 240: p. 361-370.
[24] Mohan, K., et al., A highly stable and efficient quasi solid state dye sensitized solar cell based on Polymethyl methacrylate (PMMA)/Carbon black (CB) polymer gel electrolyte with improved open circuit voltage. Electrochimica Acta, 2017. 247: p. 216-228.
[25] Seidalilir, Z., et al., High-Performance Gel-Type Dye- Sensitized Solar Cells Using Poly (methyl methacrylate- co-ethylacrylate)-Based Polymer Gel Electrolyte with Superior Enduring Stability. Journal of the Electrochemical Society, 2015. 162(14): p. H922-H928.
[26] Li, Q.H., et al., Enhanced photovoltaic performances of quasi-solid-state dye-sensitized solar cells using a novel conducting gel electrolyte. Journal of Power Sources, 2014. 248: p. 923-930.
[27] Bella, F., M. Imperiyka, and A. Ahmad, Photochemically produced quasi-linear copolymers for stable and efficient electrolytes in dye-sensitized solar cells. Journal of Photochemistry and Photobiology a- Chemistry, 2014. 289: p. 73-80.
[28] Yuan, S.S., et al., Multifunctional graphene incorporated conducting gel electrolytes in enhancing

photovoltaic performances of quasi-solid-state dye-sensitized solar cells. Journal of Power Sources, 2014. 260: p. 225-232.

[29] Rokesh, K., A. Sambandam, and K. Jothivnkatachalam, Polymer Electrolytes in Dye Sensitized Solar Cells. Vol. 4. 2015. 262-271.

[30] Azmar, A., M.D. Rozana, and T. Winie, Characterization of PMA-TPAI and PVAc-TPAI solid polymer electrolytes and application in dye-sensitized solar cell. Journal of Applied Polymer Science, 2018. 135(47).

[31] Winie, T., A. Azmar, and M.D. Rozana, Ionic liquid effect for efficiency improvement in poly(methyl acrylate)/poly(vinyl acetate)-based dye-sensitized solar cells. High Performance Polymers, 2018. 30(8): p. 937-948.

[32] Dzulkurnain, N.A., et al., Characterization of Random Methacrylate Copolymers Synthesized Using Free- Radical Bulk Polymerization Method. International Journal of Electrochemical Science, 2015. 10(1): p. 84-92.

[33] Diego, R.B., et al., Acrylic scaffolds with interconnected spherical pores and controlled hydrophilicity for tissue engineering. Journal of Materials Science: Materials in Medicine, 2005. 16(8): p. 693-698.

[34] Imperiyka, M., et al., A UV-prepared linear polymer electrolyte membrane for dye-sensitized solar cells. Physica B-Condensed Matter, 2014. 450: p. 151-154.

[35] Imperiyka, M., et al., Photo-polymerization of methacrylate based polymer electrolyte for dye-sensitized solar cell. Journal of Polymer Engineering, 2014. 34(8): p. 695-702.

[36] Rayung, M., et al., Characteristics of ionically conducting jatropha oil-based polyurethane acrylate gel electrolyte doped with potassium iodide. Materials Chemistry and Physics, 2019. 222: p. 110-117.

[37] Chang, S., et al., Synthesis of a novel alkaline- developable photosensitive copolymer based on MMA, MAA, SM, and 2-HEMA-grafted GMA copolymer for an innovative photo-imageable dry-peelable temporary protective plastisol. Journal of Polymer Research, 2013. 20(4).

[38] Venkatesan, S., et al., Highly efficient quasi-solid-state dye-sensitized solar cells using polyethylene oxide (PEO) and poly(methyl methacrylate) (PMMA)-based printable electrolytes. Journal of Materials Chemistry A, 2018. 6(21): p. 10085-10094.

[39] Qin, H. and M.J. Panzer, Chemically Cross-Linked Poly(2-hydroxyethyl methacrylate)-Supported Deep Eutectic Solvent Gel Electrolytes for Eco-Friendly Supercapacitors. Chemelectrochem, 2017. 4(10): p. 2556-2562.

[40] Callister, W.D., Jr., Materials science and engineering: an introduction. 4th ed. ed. 1997, New York: John Wiley & Sons.

[41] [41] Johan, M.R., et al., Effects of Al2O3 nanofiller and EC plasticizer on the ionic conductivity enhancement of solid PEO–LiCF3SO3 solid polymer electrolyte. Solid State Ionics, 2011. 196(1): p. 41-47.

[42] Yang, J.M., et al., Two step modification of poly(vinyl alcohol) by UV radiation with 2-hydroxy ethyl methacrylate and sol-gel process for the application of polymer electrolyte membrane. Journal of Membrane Science, 2009. 341(1-2): p. 186-194.

[43] Fares, M. and A. Othman, Smart pH-sensitive Alternating Copolymers of (Methylacrylamide-Hydroxyethylmethacrylate); Kinetic and Physical Properties. Journal of Macromolecular Science®, 2010. Part A: Pure and Applied Chemistry: p. 61-70.

AUTHOR INDEX